OIL SPILL PREVENTION AND REMOVAL HANDBOOK

OIL SPILL PREVENTION AND REMOVAL HANDBOOK

Marshall Sittig

NOYES DATA CORPORATION

Park Ridge, New Jersey London, England

1974

Library of Congress Catalog Card Number: 74-82351
ISBN: 0-8155-0543-4
Printed in the United States

Published in the United States of America by
Noyes Data Corporation
Noyes Building, Park Ridge, New Jersey 07656

FOREWORD

By the very nature of the subject matter covered in this book, its categorization spills over into three fields. Since it deals with cleaning up a dirtied marine environment, it naturally falls into Pollution Technology Review and because it deals with our major energy source among the fossil fuels it most certainly can be considered an Energy Technology Review. It also quite literally spills over into Ocean Technology since the mishaps discussed occur both on ships at sea and at offshore drilling sites.

The book is based mostly on authoritative government reports and U.S. patents. We are fortunate in the United States to be receiving direct help not only from the numerous surveys but also from active research and development programs which are being supported by the Federal government. The book attempts to organize and clarify the many ways and means made available in this literature for the removal of the contaminating oil from both the ocean's surfaces and its beaches.

Here are condensed vital data from government sources of information that are scattered and difficult to pull together. Important processes are interpreted and explained by examples from 320 U.S. patents and 5 foreign patents. One should have to go no further than this condensed information to establish a sound background for action against oil spills.

Advanced composition and production methods developed by Noyes Data are employed to bring new durably bound books to the reader in a minimum of time. Special techniques are used to close the gap between "manuscript" and "completed book." Industrial technology is progressing so rapidly that time-honored, conventional typesetting, binding and shipping methods are no longer suitable. Delays in the conventional book publishing cycle have been bypassed to provide the user with an effective and convenient means of reviewing up-to-date information in depth.

The Table of Contents is organized in such a way as to serve as a subject index and provides easy access to the information contained in this book. Special attention is called to the list of companies producing materials and equipment useful in cleaning up oil spills and contaminated beaches.

15 Reasons Why the U.S. Patent Office Literature Is Important to You –

1. The U.S. patent literature is the largest and most comprehensive collection of technical information in the world. There is more practical commercial process information assembled here than is available from any other source.

2. The technical information obtained from the patent literature is extremely comprehensive; sufficient information must be included to avoid rejection for "insufficient disclosure."

3. The patent literature is a prime source of basic commercially utilizable information. This information is overlooked by those who rely primarily on the periodical journal literature.

4. An important feature of the patent literature is that it can serve to avoid duplication of research and development.

5. Patents, unlike periodical literature, are bound by definition to contain new information, data and ideas.

6. It can serve as a source of new ideas in a different but related field, and may be outside the patent protection offered the original invention.

7. Since claims are narrowly defined, much valuable information is included that may be outside the legal protection afforded by the claims.

8. Patents discuss the difficulties associated with previous research, development or production techniques, and offer a specific method of overcoming problems. This gives clues to current process information that has not been published in periodicals or books.

9. Can aid in process design by providing a selection of alternate techniques. A powerful research and engineering tool.

10. Obtain licenses – many U.S. chemical patents have not been developed commercially.

11. Patents provide an excellent starting point for the next investigator.

12. Frequently, innovations derived from research are first disclosed in the patent literature, prior to coverage in the periodical literature.

13. Patents offer a most valuable method of keeping abreast of latest technologies, serving an individual's own "current awareness" program.

14. Copies of U.S. patents are easily obtained from the U.S. Patent Office at 50¢ a copy.

15. It is a creative source of ideas for those with imagination.

CONTENTS AND SUBJECT INDEX

INTRODUCTION

Oil spills, which have been in the public eye in a big way since the Torrey Canyon accident in 1967 and the Santa Barbara Channel incident of 1969 threaten to be much bigger news. In early 1974, the U.S. Government was considering offering oil leases on the outer continental shelf (OCS) off the Atlantic Coast as part of a ten-fold increase in offshore land open to oil exploration. The expansion from 1 million acres of OCS oil leases to 10 million acres in areas ranging from the U.S. East Coast to the Gulf of Alaska means that much more potential for oil spills from wells and tankers exists.

When this increase in offshore drilling and transport from offshore locations to shore terminals is coupled with the increases in sizes of supertankers and potential disastrous spills from supertanker accidents, one has a problem of monumental proportions. As applied to water quality considerations, the pollutant "oil" can cover a number of types of materials: Petroleum (crude oil), petroleum products (gasoline, diesel fuel, lubricating oils, fuel oils, etc.) and, to a lesser degree, edible materials such as vegetable oils and animal fats. These diverse substances are distinguished by the following characteristics: Mostly organic chemical in composition, mixtures of molecules, mostly insoluble in water, generally lighter than water and liquid or fluid semisolid. Any type of material fulfilling all of these characteristics will certainly be characterized as oil pollution when encountered in the aquatic environment. However, without question, the problem of oil pollution is predominantly petroleum-oriented.

Effective oil spill prevention is the best method for reducing the problem of oil spills. Thorough training programs, properly maintained equipment, adequate alarm systems, strict adherence to industry and governmental codes and regulations, all make essential contributions to the prevention of spills. If despite these preventive measures a spill does occur, it follows that the less oil spilled, the easier the cleanup job and the better for all. In most cases, the action necessary to limit the spill is obvious: close the valve that has been accidentally opened; cease pumping through the ruptured oil line; or repair the leak. If petroleum is escaping from a shoreside facility, use sandbags or throw up a temporary dike to prevent drainage into the water. In the case of a grounded tanker or barge, transfer the oil to another vessel. The proper corrective action is most often obvious, but if not taken promptly, what should be a minor spill may escalate into a major headache.

On May 26, 1967, the President of the United States, Lyndon B. Johnson, directed the Secretary of the Interior and the Secretary of Transportation to examine how the resources of the Nation could best be mobilized against the pollution of water by spills of oil and other hazardous substances. Referring to the Torrey Canyon and Cape Cod incidents earlier that year, the President considered it "imperative that we take prompt action to prevent similar catastrophes in the future and to insure that the Nation is fully equipped to minimize the threat from such accidents to health, safety, and our natural resources." He asked for a thorough assessment of existing technical and legal resources, and for recommendations toward an effective national and international program. A report was prepared by the Secretary of the Interior and the Secretary of Transportation entitled *Oil Pollution – A Report to the President* which marked a significant point in the attempt to prevent and control oil spills.

The report dealt primarily with water pollution by oil, but where appropriate addressed itself to other hazardous substances as well. It reflected a conviction that the problem of oil pollution must be faced and solved as part of the current general effort to improve the quality of life. Just as the world cannot afford to accept the slow poisoning of its air, or the fouling of its cities and countryside, it cannot abandon to pollution the treasure of its waters and shorelines. Oil in its many forms, oil in vast quantities, is one of the necessities of modern industrial society. Under control, serving its intended purpose, oil is efficient, versatile, productive. Out of control, it can be one of the most devastating substances in the environment. Spilled into water it spreads havoc for miles around.

The destructive characteristics of oil out of control, and the inadequacy of current measures for dealing with it, have never been better illustrated than when the Torrey Canyon, with 119,000 tons of crude oil in her tanks, ran aground and broke up off the Southern coast of England in March 1967. The desperate efforts of the British and French to cope with the tragedy captured the attention and sympathy of people all over the world. Oil spills, large and small, as well as the careless or accidental release of other hazardous materials into the environment, have long been of concern to pollution control authorities. Once an area has been contaminated by oil, the whole character of the environment is changed. Afloat, even a relatively small quantity of oil goes where the water goes. By its nature, oil on water is a seeker. Once it has encountered something solid to cling to, whether it be a beach, a rock, a piling, the feathers of a duck or gull, or a bather's hair, it does not readily let go.

Cleaning up an oil-contaminated area is time-consuming, difficult, costly. To the costs of the cleanup must be added the costs of the oil invasion itself, the destruction of fish and other wildlife, damage to property, contamination of public water supplies, and any number of other material and esthetic losses. Depending on the quantities and kinds of oil involved, these losses may extend for months or years, sometimes for decades, with correspondingly heavy costs of restoring the area to its prior condition. The risks of contamination by oil and other hazardous substances are as numerous and varied as the uses made of the many materials involved and the means of transporting them. These risks involve terminals, waterside chemical and other industrial plants, loading docks, refineries, tankers, freighters, barges, pipelines, tank cars, trucks, filling stations, everywhere that oil is used, stored, or moved. All are subject to mechanical failures compounded by human carelessness and mistakes. There are countless opportunities for oil to get out of control.

The world today is not fully prepared to deal effectively with spills of oil or other hazardous materials, large or small, and much less with a Torrey Canyon type disaster. Because sizable spills are not uncommon, and major spills are an ever-present danger, effective steps must be taken to reduce this vulnerability. All preventive measures cost money. The cost of preventive measures which might be incorporated in ships or industrial establishments or operating equipment must be weighed agains the costs, both tangible and intangible, that arise from disastrous spills. Such cost evaluations may be expected to guide the development of ships and industrial facilities, with effects on their future size and operating characteristics. On the whole, economics and good sense commend an effort to prevent pollution rather than accept the costs of its occurring. For this reason, preventive action is stressed.

In spite of preventive efforts there will still be spills, either due to human carelessness or to calamities beyond human control. For those reasons, attention must always be paid to improved cleanup measures. Present cleanup procedures leave much to be desired and are often too expensive, too tardy, too ineffective, and too destructive of the marine and land environment. There is room for a great deal of improvement in present techniques, and there is a need for the development of better ones. As suggested by the New England Interstate Water Pollution Control Commission, in the event of any oil spill, the responsible vessel shall be held until responsibility for such a spill is determined, the oil adequately removed and the vessel operator has provided evidence of financial responsibility for such environmental damage as may have occurred.

The oil pollution problem has significant international aspects. First, accidental or deliberate spills which threaten one country's coasts may occur outside the territorial waters of that country. Despite this fact, each country must be able to act quickly against a threat that develops in international waters so that it may take whatever immediate preventive or remedial steps are necessary. Secondly, vessels which discharge oil may be outside the registry of an affected coastal nation and thus not be within the direct and simple application of the nation's laws. For this reason, attention has been given for some years to international cooperation in control of oil pollution. As an example, the Subcommittee on Oil Pollution of the Intergovernmental Maritime Consultative Organization (IMCO) has done continuing work in this field. Through this Subcommittee, IMCO is presently examining a series of proposals which might be adopted internationally to minimize the threat of future spills. Further efforts to reduce oil pollution should include seeking expanded use of international regulation because any other solution would be incomplete. For these reasons, there is need for rapid international action, as well as urgent steps to be taken domestically.

An important factor in any oil spill is the potential fire hazard. Light petroleum products, such as gasoline, benzene, and naphtha, are the most flammable. These lighter petroleum products spread rapidly on water and, because of their high volatility, evaporate quickly. In open water, where no fire hazard is involved, wind and water action are sufficient to disperse the products naturally. Near a tanker, pier, terminal, or other location where the fire danger is serious, spills should be confined and fire-preventive foam spread on the surface of the oil. When the fire hazard no longer exists, the foam and gasoline mixture should be pumped into a suitable container and disposed of in the most appropriate manner. Heavier oil products present a less serious fire hazard, since their higher ignition point makes them difficult to set ablaze. The volatile fractions of crude evaporate and dissolve quickly, and the remaining crude is difficult to ignite. Attempts to burn off crude oil on the open sea, therefore, are usually unsuccessful.

In the course of the preparation of this volume, major dependence has been placed on the report entitled *Control of Oil and Other Hazardous Materials,* a publication of the training program of the Office of Water Programs of the Environmental Protection Agency, published as report PB 213,880 by National Technical Information Services, Springfield, Virginia (December 1971).

HISTORICAL ASPECTS

Petroleum (rock oil) has been known to man for at least six millenia. Earliest known usage was in the vicinity of the oil seeps in the Black and Caspian seas, where oil was employed for some of the same uses it is put to now, viz., cooking, heating, lubrication, road-building, etc. These seepages were regarded as obnoxious by the local populace, suggesting that the concept of aquatic oil pollution is not new. Natural seepages are found throughout the world and are believed to constitute an environmental insult. When compared to the volume of oil contamination resulting from current technological exploitation of petroleum, the estimated quantity of oil introduced into the aquatic environment as a result of natural seepages is relatively minor, however, as outlined in Report LBL-I, Vol 2, *Instrumentation for Environmental Monitoring: Water,* Berkeley, University of California (February 1, 1973).

Early recorded comment on inland oil pollution characterized the Caspian Sea in 1754 as greatly spoiled due to leakage of oil carried in wooden-hulled ships. Similar problems existed with oil barges on the Volga until stone ballast was added to increase the draft and hence the hydrostatic pressure outside the hull. Contamination of water resources by oily matter, primarily petroleum and its products, is a natural consequence of the exponential growth of the petroleum industry. For example, in 1860, one year after the first successful domestic producing well was established, world petroleum production ran around 0.5 million barrels. A century later it had grown to 7 billion barrels. It has more than doubled in the decade since.

One of the early major domestic oil-spill incidents involved the tanker Santa Rita in San Francisco Bay in 1907. A sizeable spillage of fuel oil cargo was ignited by sparks from a dock locomotive, and the resulting fire endangered nearby shipping. The steady climb in oil pollution levels in the decades to follow appeared not to crystallize public concern until the tanker sinkings along the coast during World War II resulted in beach pollution too serious to ignore. As noted by D.E. Kash et al in *Energy Under the Oceans,* Norman, Oklahoma, The University of Oklahoma Press (1973), the trend is to some combination of imports and drilling on the outer continental shelf (OCS) which will in any event give rise to the vastly increased possibility of oil spills over what would be the case for primarily domestic production.

In 1970, imports provided 12% of U.S. energy supply. The possible range of imports in 1985 is extreme. In the lowest demand-highest domestic production situation, imports would represent only 3 to 4% of supply. In the highest demand-lowest domestic production situation, imports would supply approximately 40% of domestic energy. The most likely situation is that imports will supply between 20 and 28% of U.S. total energy. This is between 9 and 14 million barrels of oil per day and between 3.2 and 3.9 trillion cubic feet of gas per year.

Although state-controlled offshore lands hold promise for increased production, their percentage of 1985 domestic production is much less significant than the OCS. Development in these areas may be difficult also because of local opposition. The OCS is likely to contribute the major portion of the oil and gas produced offshore in 1985. Thus considerable emphasis has been placed in this volume on prevention and control of oil spills from offshore production.

Continued development of OCS oil and gas will thus take place within the context of continuing demands for environmental quality. OCS oil and gas operations appear to be identified in the minds of many citizens with the environmental concerns generated in the late sixties and early seventies. Union's blowout at Santa Barbara is repeatedly mentioned as a major catalyst for the environmental movement and therefore as a major turning point for public policy. (Although referred to as "Union's blowout," Union Oil Company had three partners: Gulf, Mobil, and Texaco.) In California, for example, the moratorium on leasing state offshore lands imposed after Santa Barbara is still in effect; and on November 7, 1972, California voters passed Proposition 20, the Coastal Zone Conservation Act. This Act established a statewide Coastal Zone Conservation Commission and six regional Commissions to oversee a state plan for the preservation, protection, restoration, and enhancement of the environment and ecology of the coastal zone.

However, plans for OCS operations all along the Atlantic Coast from Maine to Florida and in the Gulf of Alaska are proceeding as described in a comprehensive report by the Council on Environmental Quality entitled *OCS Oil and Gas — An Environmental Appraisal,* Washington, D.C. (April 18, 1974).

THE MAGNITUDE OF THE PROBLEM

Petroleum in its many forms has been on the move in the United States since 1859 when the first commercially successful oil well was developed in Pennsylvania. First transported by wagon and log raft, oil is now en route from the oil field to refinery to consumer by pipe, water, rail and highway. During 1970 more than 4.5 billion barrels of petroleum moved as crude oil from the production field through the refineries and the as refined products to the consumer. Ocean tankers, barges, pipelines, railroad tank cars and tank trucks are the vital elements of the complex and diversified transportation system required to move this volume of oil. Estimated amounts of oil lost as well as the sources of this form of pollution are shown in Table 1. It is interesting to note that the loss of waste or used oil from vehicles (crankcase oil) may be the largest single source of oil pollution, larger, in fact, than the total volume from all sources lost directly to the oceans.

TABLE 1: ESTIMATES OF OIL INTRODUCED INTO WORLD'S WATERS AND POTENTIAL LOSSES TO WATERS, 1969

	Metric Tons per Year	Percent of Total
Tankers (normal operations)		
Using control measures (80%)	30,000	
Not using control measures (20%)	500,000	
Total	530,000	10.7
Other ships (bilges, etc.)	500,000	10.1
Offshore production (normal operations)	100,000	2.0
Accidental spills		
Ships	100,000	2.0
Nonships	100,000	2.0
Refineries and petrochemical	300,000	6.0
Subtotal	1,630,000	
Potential losses to water from industrial and automotive (not fuel)		
Highway vehicle spent oils	1,800,000	36.6
Industrial plus all other vehicles	1,500,000	30.6
Subtotal	3,300,000	
Total	4,930,000	

Note: Oil from pleasure craft and natural seeps not included.

Source: Report PB 213,880

However, more recent figures from a National Academy of Science Panel on *Inputs, Fates and Effects of Petroleum in the Marine Environment* differ in some respects as shown in Table 2. This data was reported in Chemical and Engineering News for May 20, 1974, page 7.

TABLE 2: NAS ESTIMATES OF INPUTS AND FATES OF PETROLEUM IN THE MARINE ENVIRONMENT

	Metric Tons per Year
Tankers, oil terminals and other transportation — related sources	2,100,000
River and urban runoff	1,900,000
Atmospheric fallout	600,000
Natural seeps	600,000
Industrial wastes	300,000
Municipal wastes	300,000
Coastal refineries	200,000
Offshore production	100,000
Total	6,100,000

Source: National Academy of Sciences

In 1970 the Federal government estimated that there were more than 7,500 oil spills annually into the country's waters, arising from such diverse sources as:

Vessels (spills, oily ballast/bilge discharge, sportcraft)
Pipelines
Land- and water-based storage tanks
Refining and other manufacturing operations
Jettisoning of fuel tanks by aircraft
On and offshore loading and unloading terminals
On and offshore drilling and production operations
Miscellany

Current estimates of annual oil spills run as much as 50% higher according to Report LBL-I, Vol. 2, *Instrumentation for Environmental Monitoring — Water,* Berkeley, University of California (February 1973). Apart from the qualifying statistics, sources of aquatic oil pollution have been listed additionally as: Civic and salvage dumps, garages, unburned fuel from ships' funnels, gasoline service stations, municipal sewage systems, and unburned fuel from vehicles, locomotives. The citation of oil-spill occurrences along with quantitative estimates must be accompanied by the understanding that the data are not necessarily either precise or complete. The instances and amounts may well increase as techniques for reporting this information improve. Over a short period they may serve to illustrate the dimensions of the oil pollution problem. This is shown in Table 3.

TABLE 3: REPORTED OIL SPILLS IN U.S. WATERS OVER 100 BARRELS*

	1968	1969
Vessels	347	532
Shore Facilities	295	331
Unidentified	72	144
Total	714	1,007

*(i.e., 4,200 Gallons)

Source: Report LBL-I, Vol. 2

Movement of one barrel of petroleum from the oil field to the consumer may require 10 to 15 transfers between as many as six different transport modes. Each mode is subject to accidents, and at the transfer points spill frequency is extremely high. Approximately one barrel of product is lost for each one million transported. About 7,500 oil spills are now occurring annually with an estimated total loss of 500,000 barrels. Most of the spilled oil is discharged to water. In addition to spills, the potential for discharges from normal vessel operation is in excess of 700,000 barrels of oil annually.

A United States fleet of 387 tankers and 2,900 barges presently operates in the Nation's waterways. In worldwide oil traffic, United States vessels make up only 5% of tanker traffic. United States and foreign flag tankers were responsible for approximately 80 to 90% of the oil spilled in 1969. Today's world tank ship fleet amounts to about 4,000 ships with a total dead weight tonnage of 170 million long tons. Only about 400 ships are American registered. These tankers range in tonnage from 2,000 tons to 312,000 tons and Japan, the world's largest shipbuilder, is building a 470,000 ton tanker.

Supertankers are a recent phenomena. The first 100,000 ton tanker went into operation in late 1959. Today there are over 200 tankers of 100,000 tons or greater in size and about 240 under construction. Only 22% of these tankers are owned by oil companies.

Another 18% are owned by the governments of the world and the remainder are privately owned. These privately owned tankers are contracted for by the oil companies. There are two basic trades; the crude oil trade and the oil products trade. The supertankers are used almost exclusively in the crude oil trade. Shifting of ships from one trade to another are infrequent, however, when they occur the tanks must be thoroughly cleaned especially for transfer from the crude to the products trade. Tank trucks and railroad tank cars together account for less than 1% of total product lost. Tank trucks are the last leg of the transportation system, delivering refined products to the retail consumer. Approximately 158,000 tank trucks are on the road. In areas not served by pipelines or navigable waterways, over 81,000 railroad tank cars take over the transport of crude and refined petroleum products.

It has been reported that human error accounted for 88% of the total number of spill incidents. The means of rectifying this situation is by better training and education, improved engineering, and when all else fails enforcement by a responsible State or Federal agency. Another important factor in the cause of spills by vessels is the stopping ability of the tankers under crash stop conditions, vessel in full reverse. It has been reported that the most important factor in connection with collision and stranding, the two most dreaded casualties, is the crash stop ability. Unfortunately, the ability of tankers to come to a crash stop has decreased as their size has increased. For the 400,000 tonner, the straight-line stopping distance for a crash stop would be four to five miles and would take approximately 30 minutes. During this period of backing full, the ship's master is unable to steer her or regulate the speed. If the engines are not put full astern but on stop it takes up to one hour for the Universe Ireland to come to a stop. Figure 1 shows the relation between vessel size and the time required for a crash stop.

FIGURE 1: RELATION OF TANKER SIZES TO TIME REQUIRED FOR AN EMERGENCY STOP

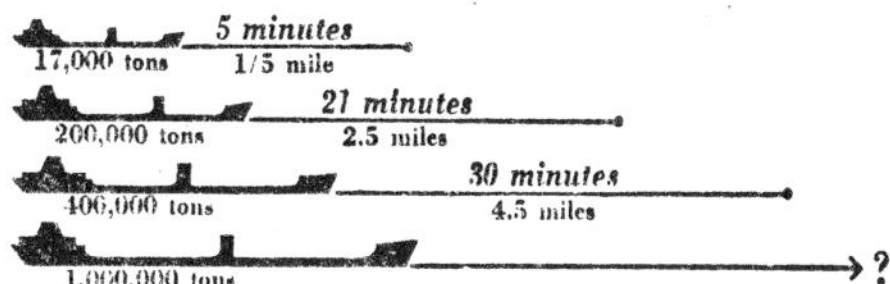

Source: Report PB 213,880

Dillingham Corporation under contract to API, conducted a statistical study to develop an understanding of the basic characteristics of major oil spills, defined as a spill of 2,000 barrels (84,000 gallons) or more of a heavy (or persistent) oil which will not naturally evaporate or disperse rapidly in the environment, and thus define the nature and scope of the problem. This report, entitled *Systems Study of Oil Spill Cleanup Procedures,* La Jolla, California, Dillingham Corporation (February 1970) is available from the American Petroleum Institute, (Washington D.C.) in 2 volumes. Based on an analysis of 38 past spills, which occured during the period 1956 to 1969, they reported that: (A) Source — 75% were associated with vessels, principally tankers. (B) Composition — 90% involved crude or residual oils. (C) Volume — 70% of the spills were greater than 5,000 barrels with a median spill volume of 25,000 barrels. (D) Distance offshore — 80% occurred within 10 miles of shore and the oil would reach shore within one day. (E) Duration — 75% of the spill incidents lasted more than five days with a median duration of 17 days. (F) Extent — 80% contaminated less than 20 miles of coastline with a median extent of four miles of coast. (G) Coastline — 85% occurred off shoreline considered to be recreational. (H) Distance from port — 75% occurred within 25 miles of the nearest port.

Item (D) above is probably one of the most significant factors revealed by this study. As shown in Figure 2, 50% of the offshore spills occurred less than one mile from shore. Since oil appears to drift at approximately 3% of the wind velocity, and with an assumed average wind of 15 knots, the oil slick would drift at approximately 0.5 knots. Thus, with 50% of the spills less than one mile offshore, an onshore wind could move oil onto shore in two hours. The question that now arises is how does one mobilize shoreline protection equipment during this short length of time?

FIGURE 2: DISTANCE FROM SHORE AND RESPONSE TIME AVAILABLE FOR SHORELINE PROTECTION—DATA FROM 25 INCIDENTS

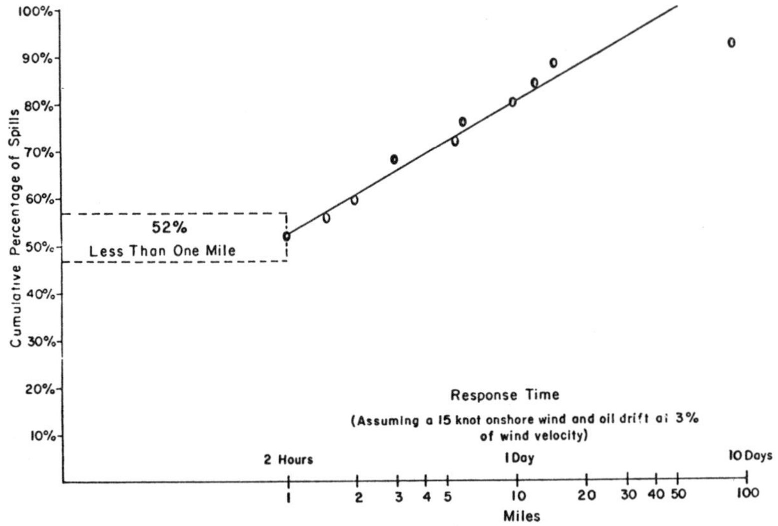

Source: Dillingham Corporation Report to A.P.I. (1970)

For reference purposes, major spill incidents along with significant characteristics, are shown in Table 4.

TABLE 4: MAJOR OIL SPILL INCIDENTS

Name	Date	Cause of spill	Material	Volume (barrels)
ALGOL, tanker	02-09-69	Grounding	#6 F. O.	4,000
ANDRON, tanker	05-05-68	Sinking	Crude	117,000
ANNE MILDRED BROVIG, tanker	02-20-66	Collision	Crude	125,000
ARGEA PRIMA, tanker	07-17-62	Grounding	Crude	28,000
ARROW, tanker	02-04-70	Grounding	Residual	36,000
BENEDICTE, tanker	05-31-69	Collision	Crude	14,000
Bridgeport, Conn., terminal	6-15-70	Pumping	#2 F. O.	20,000
Chester Creek, pipeline	08-08-69	Break	#2 F. O.	3,500
CHRYSSI P. GOULANDRIS, tanker	01-13-67	Unknown	Crude	2,600
Dutch Coast Spill	02-16-69	Unknown	Residual	1,000
ESSO ESSEN, tanker	04-29-68	Grounding	Crude	30,000
ESSO HAMBURG, tanker	01-29-70			10,000
FLORIDA, barge	09-16-69	Grounding	#2 F. O.	4,000
GENERAL COLOCOTRONIS, tanker	03-07-68	Grounding	Crude	30,000

(continued)

TABLE 4: (continued)

Name	Date	Cause of spill	Material	Volume (barrels)
HAMILTON TRADER, tanker	04-30-69	Collision	Residual	5,000
HESS HUSTLER, tank barge	11-12-68	Grounding	#6 F. O.	40
Humboldt Bay, refinery	12- -68	Hose failure	Diesel	1,400
KENAI PENINSULA, tanker	11-05-68	Collision	Crude	1,000
KEO, tanker	11-05-69	Hull failure	#4 F. O.	210,000
Louisiana, Chevron platform	04-10-70	Fire	Crude	60,000
Louisiana, Shell platform	12-1-70	Fire	Crude	Unknown
MARTITA, tanker	09-20-62	Collision	Bunker C	4,300
Moron, refinery	03-29-68	Pumping	Crude	16,000
New Castle, power station	1963-65	Leak	Residual	40
OCEAN EAGLE, tanker	03-03-68	Grounding	Crude	83,400
R. C. STONER, tanker	09-06-67	Grounding	Mixed	143,300
Refinery Loading Site	1962	Hose failure	Crude	2,000
ROBERT L. POLLING, tank barge	05-10-69	Collision	#2 F. O.	4,700
Santa Barbara, platform	01-28-69	Natural faults	Crude	100,000
Schuylkill River, Berks Assoc.	11-13-70	Lagoon failure	Waste Crankcase	70,000
Sewaren, N. J., storage tank	10-31-69	Tank failure	Crude	200,000
Ship Shoal, drill rig	03-16-69	Storm shifting	Crude	2,400
Staten Island, N.Y., 2 Esso barges	05-22-70	Collision	#6 F. O.	10,000
USS SHANGRI-LA (CVA-38)	1965	Pumping	NSFO	200
TAMPICO, tanker	03- -57	Grounding	Diesel	60,000
TIM, tank barge	02-18-68	Sinking	#6 F. O.	7,000
TORREY CANYON, tanker	03-18-67	Grounding	Crude	700,000
Waikiki Beach	04-21-68	Unknown	Bunker C	
Waterford Beach	01-18-69	Unknown	#6 F. O.	
WITWATER, tanker	12-13-68	Hull failure	Mixed	15,000
WORLD GLORY, tanker	06-13-68	Hull failure	Crude	322,000

Source: Report PB 213,880

Oil Spill data are not released to the public on a regular basis, however, the Environmental Protection Agency (EPA) reported the details concerning major oil spills within U.S. jurisdiction for the period of January 1, 1971 to June 30, 1971 in Table 5.

TABLE 5: MAJOR SPILLS WITHIN U.S. JURISDICTION FOR SIX MONTHS IN 1971

SOURCE	DATE	CAUSE	MATERIAL	VOLUME BBLS.*
Transmission Line	Jan 6	Rupture	Crude	4,048
Pipeline	Jan 8	Rupture	Crude	10,000
Navy barge	Jan 14	Sinking	Diesel	1,619
Oregon Standard	Jan 20	Collision	Bunker	20,000
Esso Gettysburg	Jan 23	Grounding	#2 Fuel Oil	8,600
Storage Tank	Jan 31	Tank Rupture	Crude	63,000
Idaho Standard	Feb 1	Grounding	Gas & Diesel	519
Barge	Feb 5	Grounding	Fuel Oil	976
Barge	Feb 16	Puncture	Gasoline	1,800
Storm Sewer	Feb 28	Unknown	Jet Fuel	3,571
Tug and Barge	Mar 1	Collision	Gasoline	1,786
Pipeline	Mar 5	Rupture	Crude	1,300
Pipeline	Mar 9	Unknown	Crude	3,700
Tank	Mar 23	Overflow	#6 Fuel Oil	1,428
Pipe	Mar 28	Leaking valve	Crude	500
Tank	Mar 30	Rupture	Gasoline	7,143
Pipe	Apr 5	Rupture	Bunker	1,650
Pipe	Apr 6	Rupture	Gasoline	1,429
Barge	Apr 26	Overflow	Diesel	5,550
Barge	May 11	Grounding	Kerosene	1,100
Tank	Jun 4	Puncture	JP4	1,071
Barge	Jun 12	Grounding	Crude	600
Dock	Jun 24	Unknown	Unknown	1,553
Pipe	Jun 25	Rupture	Crude	1,000

*1 BBL (barrel) = 42 U.S. gallons

Source: Report LBL-I, Vol. 2

THE FATE OF SPILLED OIL

The physical behavior of oil in the aquatic environment is incompletely understood. The thickness of an unconstrained oil slick depends on the characteristics of the oil and the temperature of the water. It may vary from a thickness of only a few thousandths of an inch in temperate waters to as much as one-quarter of an inch in arctic waters. Persistence, or residence time, is the time that oil is detectable in the water, sediments, or biota. However, criteria and techniques for determining or estimating residence time vary considerably among investigators, and reported persistence can depend as much on the sensitivity of detection methods as upon how long the oil in fact remains. Visual observation, the least sensitive, is employed most frequently. Although some studies are based on chemical analyses and bioassays, lack of uniform observation and detection methods confuses the question of oil persistence. Although visual observations can provide useful data, until methods are standardized, these gross data should be interpreted as underestimating oil persistence.

Petroleum in sea water is altered chemically by evaporation, dissolution, microbial action, chemical oxidation, and photochemical reactions, often collectively called weathering. How fast oil degrades is markedly influenced by light, temperature, nutrients and inorganic substances, winds, tides, currents, and waves. They all affect the microbial degradation, evaporation, dissolution, dispersal, and sedimentation processes. Degradation rates, appear to vary with the composition of the oil. The more toxic fractions are generally less susceptible to microbial degradation. The heavy residuals that do not degrade may be deposited in sediments or they may float as tar lumps or tar balls.

The tar-asphalt residue of the weathering of oil is a product of a complicated multiprocess phenomenon. The main processes in roughly the order of occurrence after a spill are spreading, evaporation, dissolution and emulsification, auto-oxidation, microbiological degradation, sinking and resurfacing after which the process repeats itself. While these processes are occurring, the slick may also be moving. Estimates of oil persistence are quite varied, even within a given habitat. Data are not standardized in format or type. Few studies are analytical, and few provide information on the hydrocarbons present in the sediments or local biota. Oil and its breakdown products may remain in sediments indeterminately. Or they may be churned up by turbulence to recontaminate a recovering area.

The persistence problem may be somewhat different in Alaska than in the Atlantic, for example. The generally lower marine and coastal temperatures of Alaska will slow microbial action, not only because bacterial metabolism is slower but also because oil is more viscous at lower temperatures, which in turn causes thicker films and clumping and thus impedes bacterial attack. In addition, limited winter daylight reduces photochemical oxidation. Any oil in sediments of the Alaskan Gulf is expected to remain longer than in most Atlantic waters. Some weathering in Alaskan waters has been reported. It is aided by turbulence. Some examples of the fate of oil from recent spills are quoted by the Council on Environmental Quality as follows:

San Francisco. Tanker collision, January 18, 1971. An estimated 20,000 barrels of Bunker C (residual oil) was released. The oil entered two types of shallow habitats, rocky shores and mussel reefs, within several hours. An August 1971 survey showed reef mussels contaminated with oil, implying that oil persisted in the mussel reef zone for at least 6 months. The investigators estimated that all signs of the oil would disappear from the rocky shores within 2 years of the spill. Because the survey employed only visual observation, 2 years is considered an underestimate.

Chedabucto Bay, N.S., Canada. February 4, 1971: A spill released approximately 108,000 barrels of No. 6 (residual) fuel oil, contaminating two intertidal habitats, sandy beach and rocky shore. A study conducted 26 months later showed that the mud bank had lost little of the originally observed oil. The salt marshes and estuarine lagoons retained 50% of the original oil content. The data indicate that residence time of oil could be at least 3 years in the salt marsh and mud sediments.

West Falmouth, Massachusetts. September 16, 1969. The barge Florida ran aground and ruptured its hull off West Falmouth and released an estimated 4,500 barrels of No. 2 fuel oil. Two years later, oil was reported in sediments and bottom organisms. On the basis of gas-liquid chromatography, 30% of the oil in the sediments in April 1971 was considered aromatic. Oil found in the marshes 5 feet below the surface was predicted to persist in the sediments for at least 3 more years (5 years after the spill).

Wreck Cove, Washington. January 6, 1972: The unmanned troopship General M.C. Meigs broke loose while under tow and went aground, releasing approximately 3,000 barrels of Navy special (residual) fuel oil. A storm broke the oil into globules 5 to 30 centimeters in diameter when observed on the beach. Oil was trapped for several months in the upper tidal pools of the rocky ledges.

Santa Barbara. January 28, 1969: A blowout lasted several weeks and released an estimated 33,000 barrels of oil. Samples taken in the sediments on March 31, May 1, and June 13, 1970, showed no evidence of a reduced oil content over this period. From 1972 to 1973 the sandy beaches were reported recovered from oil contamination, but weathered oil on the cobbles in the upper intertidal zone in February 1973 may be linked to the spill. These data indicate that residence time of oil could be at least 3 years in the salt marsh and mud sediments.

Southwest England. March 18, 1967: Torrey Canyon — About 700,000 to 860,000 barrels of Kuwait (heavy) crude oil was spilled when the Torrey Canyon broke up at sea. Large amounts of emulsifiers were used in cleanup operations. In a few areas emulsifiers were not immediately used and oil was dispersed. The study indicated an estimated oil persistence of at least several months in both the rocky and sandy shores.

Casco Bay, Maine. November 25, 1963: Some 20,000 to 25,000 barrels of Iranian Agha-Jari (heavy) crude oil was spilled in Casco Bay. Color photographs taken in 1970 and 1972 were used to ascertain the presence of oil residue on rocks in Simmond's lobster pond. Chemical anaylses of both sediments and soft shell clams on July 20, 1972, evidenced contamination 10 years after the spill.

Physical Movement

Oil can move great distances in water. As it moves, it changes chemically. Oil is transported by the ocean currents, surface winds, and surface waves. It may ultimately drift out to sea or come ashore. It may come into contact with various marine organisms. Freshly spilled oil is considerably different than weathered oil, i.e., oil that has been in the water for some time and has given up many of its volatile and soluble components. Nevertheless, weathered oil may damage birds and marine organisms and may remain in the sediments.

In the absence of current or debris, an oil slick will move in the direction of the wind at a rate about 3 to 4% of the wind velocity. In the absence of wind or debris, an oil slick will move in the same direction with the same speed as the water current. The actual movement will be due to some combination of wind and current. Figure 3 shows the expected 4-hour shape of a 50,000 barrel spill typical of a tanker-type accident after 4 hours as described in *Control of Oil Spills* by J.G. Herbich, Texas A & M University (March 1972).

In general, surface crudes spread to very thin films and drift with the wind with a bias to the right of the wind in the northern hemisphere. According to the American Petroleum Institute at a concentration of 25 gallons per square mile, oil is barely visible; at twice that level (3.0×10^{-6} inch thickness) it appears as a silvery sheen. By doubling the concentration once again to 100 gallons per square mile, the first evidence of color appears. The wavelength of the color is indicative of the film thickness, and may be used to estimate the volume of oil present. After discharge on the sea, oil forms into large patches surrounded by a film of, presumably, the more mobile volatile components. Sometimes evaporation, solution, or microbial degradation removes the film within hours, leaving slowly hardening pebble-sized lumps (tarballs) which resist further decomposition. The deleterious water-in-oil emulsion (dubbed chocolate mousse by the French) which washed ashore from the Torrey Canyon was tentatively attributed to the British use of detergent dispersing agents but now appears to be inherent with this and certain other crude oils.

FIGURE 3: EXPECTED SHAPE OF 50,000 BARREL SPILL AFTER 4 HOURS

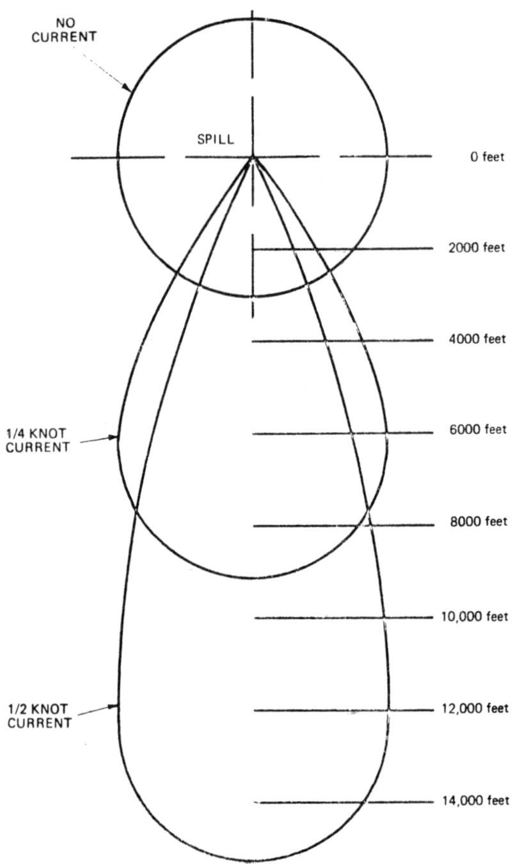

Source: Texas A & M University

Natural Degradation

The major processes involved in the degradation of oil are as follows, as outlined in EPA Report PB 213,880.

(A) Spreading: This process, the first to occur, thins the slick out to a few millimeters or less and is dependent on several parameters, among them, viscosity of the oil, surface tension of the oil and water, and time.

(B) Evaporation: Evaporation is the process by which the low molecular weight compounds

of relatively low boiling point are volatilized into the atmosphere. The rate of this process is also governed by many parameters, among them are viscosity of the oil, type of oil, and weather conditions, such as wind and sea state. The major loss due to evaporation occurs during the first few days.

(C) Dissolution: Dissolution is the process by which the low molecular weight compounds and polar compounds are lost by the oil to the large volume of water under and around it. The rate of this process is also governed by many parameters including the type of oil, viscosity of the oil, the amount of oxidation the oil has undergone before, during and after the spill and the weather conditions such as wind, and sea state. Although this process starts immediately, it is a long term one and continues throughout the duration of the total weathering process since the oxidation and microbiological degradation processes constantly produce polar compounds which are finally dissolved in the water.

(D) Emulsification: Emulsification is the process by which one liquid is dispersed into another immiscible liquid in droplets of optically measurable size. In the case of oil, the emulsion can be either an oil-in-water emulsion or a water-in-oil emulsion.

(E) Auto-Oxidation: Auto-oxidation is the light catalyzed reaction by which hydrocarbons react with atmospheric oxygen to form ketones, aldehydes, alcohols and carboxylic acids which are all polar compounds and, therefore, can either dissolve in the water or act as emulsifying agents or detergents.

(F) Microbiological degradation: Microbiological degradation is a multifaceted process. Certain bacteria, actinomycetes, filamentous fungi, and yeasts utilize hydrocarbons and chemically oxidized hydrocarbons as food sources.

Aerobic microbial oxidation — Most of the microorganisms which oxidize hydrocarbons require oxygen in either the free or dissolved form. When the oxidation of the oil occurs at the air-water interface there is usually sufficient oxygen to allow the maximum biological degradation to occur. However, areas of activity beneath the surface in the water column or in bottom muds are severely limited by the supply of oxygen.

Anaerobic microbial oxidation — A few organisms are known which oxidize hydrocarbons when little or no dissolved or free oxygen is present. These utilize nitrate or sulfate as their oxygen source. *Psuedomonas aeruginosa,* for example, utilizes n-octane or n-hexadecane while reducing nitrate to nitrite. Many sulfate-reducing bacteria are known.

(G) Sinking: Evaporation, dissolution and oxidation of lighter hydrocarbons may cause the oil to increase its density. When this happens to a sufficient degree the oil will sink to the bottom where anaerobic microbial oxidation will be the main process of degradation.

(H) Resurfacing: If the density of the oil mass is reduced to a sufficient degree by anaerobic oxidation, the oil will float again and the processes above will again occur until the oil has either completely disappeared or has reached some land mass.

One mechanism for removal of oil at the surface of water is bacterial oxidation. Horn and his colleagues at Woods Hole Oceanographic Insitution found oxygen consumption of a floating oil lump at 10°C to be about the equivalent of oxidation of 7×10^{-6} g/hr/cm^2 of petroleum as noted in *Patterns and Perspectives in Environmental Science,* Washington, D.C., National Science Board (1972). Since a sphere has a surface:volume relationship of 3/r, this observation tells us that floating oil with a density of 1.0 if divided into spherical particles of radius 21×10^{-6} cm, will be completely consumed in one hour at 10°C. One can reasonably expect this value to increase to 42×10^{-6} cm at about 18°C and to double again at about 26°C. By the same arithmetic, a film of oil 7×10^{-6} cm thick will be consumed in an hour if the bacteria thrive only on one surface at 10°C, but in half this time if they can attack both surfaces at once.

It may be more illuminating to consider these rates in terms of years (8,765 hours). At

10°C, a layer of oil attacked on only one surface will be consumed at the rate of 0.6 mm per year. This figure may be compared with Emery's (K.O. Emery, Woods Hole Oceanographic Institution) 880 tons per 78,000 square kilometers in a year, which is approximately 1.1×10^{-5} mm/year, or his 135,000 tons per year of petroleum-like substances produced by phytoplankton, which is 1.7×10^{-3} mm.

These rate computations allow one to draw several conclusions. One is that the practice of adding emulsifiers to floating oil to facilitate its dispersal into small units will also facilitate its natural oxidation as long as the emulsifiers are not bactericidal. Another is that keeping the oil at the sea surface, where ambient temperatures are highest, will minimize the time required for its natural oxidation. And another is that oil will be more persistent in polar latitudes than in temperate or tropical latitudes. Still another is that both natural accumulations of petroleum components in marine sediments and production of similar compounds by phytoplankton take place at rates much below the natural ability of the systems at the sea surface to oxidize floating oil residues.

Inasmuch as bacteria form an important food source for the ciliary and mucus feeders in the marine plankton, and since observation shows that bacterial growth is enhanced in the presence of the combination of solid surface and source of fixed carbon offered by floating oil lumps, it seems inconsistent to refer to their presence as chronic pollution. This was concluded in the report of the National Science Board referred to above.

Containment

Containing an oil spill, that is, preventing its spread over the surface of the water, has a two-fold aim. First, holding the oil near the ship, pier, or terminal from which it was spilled localizes the problem and minimizes pollution. Second, containment makes it easier to remove the oil from the water. If the oil is not contained and removed, wind or current may carry it a great distance and compound the problem. If, for example, the oil should be carried to beaches, recreation areas, or fishing grounds, considerable damage to resources and property may result. Cleanup would also require more specialized equipment and materials. Oil is contained by means of a barrier, usually one of three general types: Floating booms, underwater bubble barriers, and chemical barriers.

Booms contain an oil slick by encircling, sweeping or directing it to a collection point or by a combination of these efforts. The boom's floatation system may be a part of or separate from the containment section. Typically a fence or skirt extending above and below the water surface forms the containment section. The effectiveness of a boom is limited by waves, winds, and currents. Most booms fail because as the oil accumulates, currents carry it under the boom. Further, in turbulent seas when wave heights exceed 10 feet, a true oil slick does not form. Rather, oil droplets are dispersed in the water column and neither containment nor cleanup is possible.

To date the best boom performance that has been reported (Exxon) is containment in 6 to 8 foot seas with 20 mile winds and 1.25 mile currents. The U.S. Coast Guard has developed a mechanical containment and recovery system for the open ocean with a capability of 50 barrels per minute. Work continues to obtain improved performance, but because of turbulence, the effectiveness of booms and recovery systems is nearing its upper limit. Only in relatively calm waters are booms generally effective. In open seas, oil spill containment is severely limited.

Also, the possibility of occurrence of pollution from leakage of such lighter than water contaminants in unpredictable fashion in any part of the world makes it desirable that the contaminant containment device be of such nature that it can be readily flown to any site where it may be required and promptly assembled and stabilized in position. The containment device should be a complete packaged unit that can be air freighted from strategic locations to the point of need, and therefore the weight and bulk limitations of the package must be within the ranges which can be readily handled by air freight or helicopters available for transporting such devices.

Planned Removal

A number of factors influence the choice of oil spill removal techniques, including type and amount of petroleum products spilled, water currents, weather, equipment available, and such special considerations as the presence of marine life, closeness to drinking water intakes, and importance of the area as an estuary, fishing ground, wildlife habitat, or bathing beach. Once a spill has been contained, clean-up measures can be instituted. The two major techniques used today are mechanical gathering and sorbent recovery. Mechanical recovery may be used without restriction in carrying out the contingency plan, while sorbents may be used providing that these materials do not in themselves or in combination with the oil increase the pollution hazard. Other methods include dispersing agents, sinking agents, biological agents, and burning agents, which may be used under limited circumstances and with specific approval of the appropriate federal and local agencies. This requirement has tended to inhibit their use.

Mechanical cleanup devices are also generally limited to calm waters. They operate at low recovery rates, generally 1 to 5 barrels per minute, and thus are of limited effectiveness for large spills. Straw, manufactured fibers, and absorbent clays may be spread on a slick, mixed with the oil, and collected. Straw is considered the most cost-effective sorbent because it holds five times its weight in oil and costs $25 to $50 per ton. However there are serious logistical problems in spreading and collecting the sorbents as well as disposing of the oil-contaminated materials.

Use of dispersants in U.S. coastal waters is strictly limited by the National Contingency Plan because of their potentially toxic effects on marine organisms. Dispersants are, however, used widely in the United Kingdom except in designated critical environmental areas. Use of burning and sinking agents is also limited by the Plan. These materials have been limited because they do more harm to the marine biota than the oil. Both Government and industry research and development have resulted in improved oil spill containment and cleanup devices. The floating boom is the primary containment device to date. Cleanup is effected mainly by mechanical gathering and sorbent recovery. Two other cleanup techniques involve use of dispersing agents and combustion. Of course, natural physical and biological processes always come into play, especially if the spill is far from shore.

LEGISLATIVE AND REGULATORY BACKGROUND

The control of oil pollution is the subject of a number of laws. The importance Congress attaches to the prevention and abatement of oil pollution is shown in the Federal Water Pollution Control Act of 1956 (33 U.S.C. 466 *et seq.*) as amended by the Water Quality Improvement Act of 1970. More space is devoted to oil pollution (Section 11) than to any other subject. The amended act specifically deals with the problem of oil discharges and provides for national effluent standards for oil. Further amendments pertaining to oil pollution are contained in the Federal Water Pollution Control Act Amendments of 1972 and in the Ports and Waterways Safety Act of 1972.

There are also OCS orders under the Outer Continental Shelf Land Act which are under the jurisdiction of the U.S. Geologic Survey as noted in *Outer Continental Shelf Resource Development Safety,* Report PB 215,629, Springfield, Va., Nat. Tech. Information Service (Dec. 1972). The laws of water pollution have also been reviewed in *"U.S. Coast Guard Oil Pollution Investigation and Control School Investigator's Manual,"* Report AD 758,510, Springfield, Va., Nat. Tech. Information Service (Nov. 1972).

The principle U.S. federal surveillance activities for oil and other visible pollutants are conducted by the U.S. Coast Guard and U.S. Army Corps of Engineers. These agencies have responsibilities in the enforcement of statutes dealing with several phases of pollution control and have substantial numbers of persons engaged in water-related activities. The Environmental Protection Agency, through Technical Support Laboratories located within ten regions, supplies laboratory services necessary to establish the nature and source of oil spills. Cooperation with other federal and state agencies is an essential feature. At the direction of PL 91-224 the Council on Environmental Quality is responsible for preparing the National Oil and Hazardous Substances Pollution Contingency Plan to coordinate governmental response to oil spills.

Section 402 of the Federal Water Pollution Control Act (FWPCA) Amendment of 1972 establishes a permit system for discharges into the navigable waters of the U.S. The term "navigable waters" is defined by the Amendments as "waters of the United States, including the territorial sea." Section 403, "Ocean Discharge Criteria" extends the Section 402 permit system to the contiguous zone and the oceans.

The Environmental Protection Agency (EPA) is required by Section 403 to set standards under which a permit can be issued. Before issuing such a permit, the Administrator of EPA is required, within 180 days after enactment of the Amendments, to establish guidelines on the effect of disposal of pollutants on human health and welfare, on marine life, and on recreational and economic values, together with guidelines for determining the persistence of the pollutant and other possible locations for its disposal. It is likely that a substantial amount of new information will be required before such sweeping guidelines can can be established. Thus it is particularly significant that Section 403 also provides that, "in any event where insufficient information exists on any proposed discharge to make a reasonable judgment on any of the guidelines established pursuant to this subsection no permit shall be issued under section 402 of this Act."

Applicability of the general performance standards established by the Amendments may be limited to three miles because of the Amendments' definition of "navigable waters." Owing to this restricted definition, performance standards provided for in the Amendments to alleviate polluting discharges into navigable waters may not be applicable seaward beyond the three-mile limit. (At least one OCS operator interprets the Amendments as being applicable on the OCS and has submitted descriptions of its platform discharges to EPA.) Specifically, requirements for the use of "best practicable control technology" by 1977 and "best available technology" by 1983, as well as the national goal of eliminating all polluting discharges by 1985, may not be applicable to the waters above the OCS. Together, these limitations may lead the Administrator of EPA to utilize the "no permit shall be issued" provision of Section 403(c)(2), in which case discharges into OCS waters would continue to be controlled exclusively by OCS orders.

The oil and hazardous substance liability provisions of Section 311 of the Amendments are basically the same as the prior law on this subject. However, there is one important modification. Liability has been added for the removal of any hazardous material discharged into the navigable waters and the contiguous zone, the latter being defined as "the entire zone established or to be established by the United States under Article 24 of the Convention on the Territorial Sea and the Contiguous Zone." (The Convention established a maximum limit of 12 miles for contiguous zones.) However, since Section 311 defines an "offshore facility" as one that is located within the "navigable waters" of the United States, the provisions for the control of pollution from hazardous substances are inapplicable to offshore facilities located seaward of the three-mile limit. And further, notwithstanding its broad definition of "discharge," the provisions of Section 311 are intended to apply to the discharge of oil from any offshore facility if the discharge is not in harmful quantities and is pursuant to, and not in violation of, a permit issued for the facility under Section 402 of the Amendments.

Although the Amendments have only recently been enacted and have not been tested, it would be desirable to incorporate at least two changes: (1) extend the jurisdictional scope of the Amendments to include the contiguous zone in order to make oil and hazardous substances provisions clearly applicable to those offshore facilities located seaward of the territorial sea; and (2) extend the 1977 and 1983 performance standards and the 1985 national goal to eliminate all types of polluting discharges into waters of the United States by redefining "navigable waters" to include at least the contiguous zone, according to D.E. Kash et al in *"Energy Under The Oceans,"* Norman, Oklahoma, The University of Oklahoma Press (1973).

EFFECTS OF OIL SPILLS ON THE ENVIRONMENT

Damages caused by oil pollution are both significant and diverse. Such pollution can destroy or limit marine life, ruin wildlife habitat, kill birds, limit or destroy the recreational value of beach areas, contaminate water supplies, and create fire hazards. The suffering and mortality caused to birds by oil pollution, the coating of public and private property with layers of oil, is extremely conspicuous and attracts great public attention and sympathy. We hear conflicting statements from competent and respected scientists regarding the biological effects of these incidents. Why this divergence of opinion? Primarily, because data upon which to base a sound opinion is either incomplete, superficial or both.

It is difficult to evaluate the effects of oil since it is not a single substance but a complicated and variable mixture of literally thousands of chemical compounds. This fact is clearly demonstrated in Table 6 which shows that the toxicity of crude oils alone can cause mortality in the organism tested from as high as 89% to a low of 1%. These constituents of oil share many common properties; however, they also differ considerably in many properties which influence their effects on the environment. Among these are the following.

Toxicity: Many low boiling aromatic hydrocarbons are lethal poisons to almost any organism while some higher boiling paraffin hydrocarbons are essentially nontoxic to most forms of life.

Solubility: Benzene derivatives may be soluble in water at concentrations around 100 ppm; naphthalenes at 30 ppm, while higher molecular weight hydrocarbons may be essentially insoluble in water. Solubility, of course, will significantly influence the toxicity of a component of oil.

Biodegradability: This varies widely according to such molecular features as hydrocarbon chain length and degree of branching. The rate of degradation, of course, will influence the persistence of environmental effects.

Volatility, Density, and Surface Activity: These will determine whether oil components or an oil mixture will tend to evaporate, sink, or easily disperse into the water column. It has been reported that two-thirds of Nigerian and two-fifths of Venezuelean crude oil will evaporate after a few days.

Carcinogenicity: Some components of crude, refined, and waste oils are known to have cancer-inducing properties.

TABLE 6: CHEMICAL AND PHYSICAL PROPERTIES OF CRUDE OILS 1-10 WITH FIGURES OF RELATIVE TOXICITY AT DIFFERENT TEMPERATURES

	1	2	3	4	5	6	7	8	9	10
Specific gravity at 16°C/16°C	0.797	0.843	0.794	0.876	0.851	0.869	0.854	0.89	0.851	0.971
Sulfur content, wt %	0.05	0.20	<0.1	0.96	1.7	2.5	1.33	0.96	0.75	2.59
Asphaltenes content, wt %	<0.05	<0.05	0	2.8	0.5	1.4	0.7	0.17	0.7	5.8
Pour point, °C	--	--	--	--	-26	-32	-21	-1	7	15
Wax content, wt %	27	9	13.5	7	6	5.5	7	6.5	11	3
Viscosity at 21°C, cs	--	5.1	--	19.07	8.2	17.0	8.6	28.9	12.4	3,495
Viscosity at 38°C, cs	2.34	3.0	1.65	8.32	5.5	9.6	5.6	13.4	6.45	739
Total distillate to 149°C, wt %	26.7	20.1	36.2	14.6	17.3	15.3	17.9	10.7	26.3	2.8
Total distillate to 232°C, wt %	42.7	37.2	53.2	26.8	31.5	27.6	32.3	21.8	39.9	7.6
Total distillate to 343°C, wt %	69.7	67.5	75.6	48.3	51.7	44.6	52.5	43.5	60.5	22.7
Total distillate to 371°C, wt %	75.8	72.5	79.8	53.6	56.4	48.7	57.4	48.4	65.1	27.4
SG at 16°C/16°C of C_5, 149°C cut	0.744	0.749	0.731	0.737	0.705	0.703	0.718	0.742	0.717	0.736
SG at 16°C/16°C of C_5, 371°C residue	0.842	0.924	0.889	0.961	0.957	0.975	0.958	0.961	0.933	1.013
Aromatic content of C_5, 149°C cut, wt %	13.5	10	17	17	8	7	7.5	18.5	13	10
% Mortality in *L. littoralis** at 3°C	45	47	24	55	82	71	76	79	7	66
% Mortality in *L. littoralis** at 16°C	89	74	83	63	56	72	48	32	21	1
% Mortality in *L. littoralis** at 26°C	66	47	60	38	56	45	62	53	15	2

*Intertidal Gastropods

Source: Report PB 213, 880

Oil pollution, whether it be due to the spill or discharge of a crude oil or a refined product, may damage the marine environment many different ways, among which are:

Direct kill of organisms through coating and asphyxiation.

Direct kill through contact poisoning of organisms.

Direct kill through exposure to the water-soluble toxic components of oil at some distance in space and time from the accident.

Destruction of the food sources of higher species.

Destruction of the generally more sensitive juvenile forms of organisms.

Incorporation of sublethal amounts of oil and oil products into organisms resulting in reduced resistance to infection and other stresses (the principal cause of death in birds surviving the immediate exposure to oil).

Destruction of food values through the incorporation of oil and oil products, into fisheries resources.

Incorporation of carcinogens into the marine food chain and human food sources.

Low level effects that may interrupt any of the numerous events necessary for the propagation of marine species and for the survival of those species which stand higher in the marine food web.

Because of their low density, relative to seawater, crude oil and distillates should float; however, both the experiences of the Torrey Canyon and of the West Falmouth oil spill have shown oil on the sea floor. Oil in inshore and offshore sediments is not readily biodegraded; it can move with the sediments and can contaminate unpolluted areas long after an accident.

EFFECTS ON MAN

The effects of aquatic oil pollution on man himself are quite minimal for the simple reason that he finds the prospect of oily water so aesthetically displeasing that he avoids personal contact with it. For further information see the National Technical Advisory Committee on Water Quality Criteria Report of 1968, Wash., D.C., U.S. Govt. Printing Office (1968), under the Subcommittee for Recreation and Aesthetics. Contamination of domestic water supplies in particular must be avoided. The Subcommittee for Public Water Supplies made the following statement:

> "It is very important that water for public water supply be free of oil and grease. The difficulty of obtaining representative samples of these materials from water makes it virtually impossible to express criteria in numerical units. Since even very small quantities of oil and grease may cause troublesome taste and odor problems, the Subcommittee desires that none of this material be present in public water supplies."

Aesthetic considerations predominate over toxicological considerations, as the following excerpt from J.E. McKee & H.W. Wolf, *Water Quality Criteria,* Sacramento, Calif., State Water Resources Control Board (1963) serves to illustrate:

> "An exhaustive search of literature to determine the toxic components of refinery wastewaters toward humans revealed a paucity of toxicological data on the ingestion of such substances by humans or by test animals. From the approximate order of magnitude based on data from the fields of occupational health and industrial hygiene, it was concluded that any tolerable health concentrations for oily substances far exceed the limits of taste and odor. It appears, therefore, that hazards to human health will not arise from drinking oil-polluted waters, for they will become esthetically objectionable at concentrations far below the chronic toxicity level."

Removing oil contaminants from domestic water supplies may be costly, as McKee & Wolf go on to point out.

> "Floating or emulsified oil in raw water supplies will complicate the coagulation, flocculation, and sedimentation processes at a treatment plant. Oil-coated floc may not settle properly. If free or emulsified oil reaches sand filters or ion exchange beds it will coat the grains, decrease the effectiveness of filtration, and interfere with backwashing. Taste and odor-producing compounds will require the greater use of activated carbon or heavy chlorination. Again, however, the taste and odor factor will control the threshold or limiting concentration of oily material acceptable in a domestic water supply."

EFFECTS ON AGRICULTURE

The Subcommittee on Water Quality for Agricultural Uses made no specific recommendation in regard to contamination by oily substances. Some data are available on agricultural effects due to oil Limited application of crude petroleum has been reported to improve rather than interfere with crops. Farm animals have been reported to develop a fondness for crude oil and suffer adverse effects, including death, as a result. An effluent standard of 30 ppm of oil emulsified in water was set by the Ohio Department of Health for a creek, due to its use by grazing cattle.

EFFECTS ON INDUSTRY

The Subcommittee for Water Quality Requirements for Industrial Water Supplies noted: "Steam generation and cooling are unique water uses in that they are required in almost every industry." Both functions are adversely affected by contamination with oil: "In steam production, the presence of oil in boiler feedwater may cause foaming, priming, overheating of tubes resulting in blistering or failure, and poor transmission of heat from the metal to the water." As a result, the American Boiler Manufacturers Association in its standard guarantee on steam purity specifies that "the total quantity of oil or grease, or substance which is extractable either by sulfuric ether or by chloroform, shall not exceed 7 ppm in the boiler water when the sample being tested is acidified to one percent hydrochloric acid, or 7 ppm in the feedwater when the sample being tested is first concentrated at low temperature and pressure to the same ppm total solids as the boiler water." Additional examples of adverse industrial effects of oily water are reported by McKee & Wolf in the reference cited earlier in this section.

EFFECTS ON WILDLIFE

Examples of wildlife flora and fauna which have benefited from the release of oily substances into their environment are few indeed. They appear to be limited to bacteria and fungi which feed on components of the oily matter, or to kelp which flourished apparently as a result of oil-implicated extermination of natural predators. Otherwise, evidence is abundant that contamination of the aquatic environment with oily material is an adverse event, particularly for fauna. Numerous graphic examples are described in lay terms by Marx in his chapter "Oils vs Marine Life" in *Oilspill*, San Francisco, Calif., Sierra Club (1971). A more technical review is presented by Revelle et al. McKee and Wolf summarize:

> "Oil films may interfere with gas exchange, coat bodies of birds and fish, impart a taste to fish flesh, exert a direct toxic action on some organisms as a result of water-soluble components, and interfere with fish-food organisms and the natural food cycle. Oil from surface films becomes adsorbed on clay particles, settles to the bottom, and there remains a source of pollution, for it may be stirred up to float again or may leach toxic principles."

Most of the available toxicity data are reported as the median tolerance limit (TL_m), the concentration that kills 50% of the test organisms within a specified time span, usually in 96 hours or less. In many cases, the differences are great between TL_m concentrations and concentrations that are low enough to permit reproduction and growth. Aromatic hydrocarbons are generally implicated as the toxic components present in petroleum and its products.

Surface oil films interfere with the respiration of aquatic insects (a fact used for control of the mosquito). Oil is also a wellknown hazard to waterfowl by destroying the buoyancy and insulation of their feathers. The mortality rate of waterfowl which encounter oil slicks approaches 100%. The effects of marine oil contamination on flora and fauna have been even more striking.

Waterfowl

Marine birds, especially diving birds, appear to be the most vulnerable of the living resources to the effects of oil spillage. Harm to the birds from contact with oil is reported to be the result of a breaking down of the natural insulating oils and waxes shielding the birds from water and loss of body heat, as well as due to plumage damage and ingestion of oil or an oil dispersant mixture. In addition, birds may be harmed indirectly through contamination of nesting grounds or through interruption of their food chain by destruction of marine life on which the birds feed.

The possible effects of the spillage on the bird population will vary with the season. For example, young birds during the late nesting season and flightless adults during the molting season may be particularly vulnerable along the shore. Conversely, various groups of migratory birds may avoid exposure because of their absence at the time of the spill. Nonmigratory birds will be the hardest hit with the possibility of eliminating an entire colony.

Bird kills in the thousands may result from a specific spill incident. Efforts to cleanse or rehabilitate contaminated birds have generally been unsuccessful with less than 20% of the treated birds surviving. Estimates of the ability of bird groups to repopulate differ greatly; however, there is a general consensus that the numbers of many marine bird types are vastly reduced from 30 years ago.

Shellfish

Shellfish including mollusks such as clams, oysters and scallops along with crabs, lobsters, and shrimp appear to be the segment of marine life most directly affected by oil spillage in the coastal zone. Most of these types will survive contamination by heavy oil alone, however the flavor of the flesh will be tainted. Lighter petroleum fractions, such as diesel or gasoline, appear to be more fatal, and some species such as clams may experience significant mortalities. Fortunately, in most spill incidents the effects on shellfish appear to be fairly temporary, and even in those situations where high mortalities were observed at the time of the incident, recovery appears to have taken place within a period of six months to two years.

Fish

Finfish generally appear to be unaffected by the presence of spilled oil as their mobility permits them to avoid areas with high oil or chemical concentrations. Danger to fish is probably limited to possible harm to eggs, larvae, or juveniles which seasonally may be found concentrated in the upper water layers or in shallow areas near shore.

Marine Mammals

Relatively few observations of any direct effect of oil spills on larger marine mammals such as whales, seals, and sea lions have been made. These animals appear to be able to sense and avoid oil on the surface of the water, and various accounts of oil-covered animals do not appear to be substantiated. Possible effects to young or disabled mammals are viewed as being comparable to normally occurring mortalities, and spilled oil is generally considered to result in minimal harm to marine mammals.

The Marine Food Chain

The effects of oil spillage on the marine food chain or food web (which consists of plants, bacteria and small marine organisms) is not well understood because of the wide fluctuations and cycles that occur naturally and are totally independent of the effects of oil. Lower marine plants appear to be fairly tolerant to contamination by oil and where destruction has taken place, have repopulated rapidly although in proportions varying from original numbers. Some forms of algae and diatoms appear to be stimulated in growth by a certain amount of oil. Various bacterial organisms will also feed on available oil and multiply, thus providing energy to the protozoan level of the food chain.

EFFECTS ON MARSHES

Cowell reports that the short-term effect of oil on a marsh is that the oil adheres firmly to the plants and hardly any is washed off by the tide, except where there are puddles of oil. Leaves may remain green under the oil film for a few days, but eventually they become yellow and die. Plants recover by producing new shoots, a few of which can usually be seen within 8 weeks of pollution, unless large quantities of oil have soaked into the plant bases and soil. Seedlings and annuals rarely recover.

In the long term, recovery from oil spillages has been observed many times. The cases cover different salt marsh communities, different types, volumes and degrees of weathering of oil, and pollution at different times of year. Vegetative recovery from experimental spraying at different times of year has been observed. The evidence indicates that marshes recover well from a single oil spillage, or from successive oil spillages provided these are separated by long time intervals.

PREVENTION AND CONTROL OF SPILLS FROM OFFSHORE PRODUCTION

More than 100 companies are now active in offshore petroleum, operating off the coasts of 75 countries. The value of the work in progress is about $3 billion and at the end of 1970, the total investment in the offshore oil industry amounted to about $20 billion.

Daily offshore U.S. production is in the range of 1.30 million barrels of oil, representing 14.5% of the domestic total. Proved reserves of offshore oil stand at 85 billion barrels, or about 20% of the estimated world inventory of 425 billion barrels.

In 1969 approximately 1,160 wells were drilled in the offshore U.S. areas. At the present time there are 2,564 offshore platforms, structures and facilities in the Gulf of Mexico, off the California coast and in the waters of Alaska. Of this number, 764 are in navigable (3 miles from shore) waters. Even though water is not discharged from all these platforms, structures and facilities, they do constitute a potential source of pollution due to blowouts and other accidents.

Blowouts continue to be the most challenging accidents in OCS petroleum operations. Response to a blowout depends on whether the well is being drilled or is producing, and whether the escaping oil or gas is burning. If possible, the well is capped. When damage to the wellhead prevents capping or when serious environmental pollution will result if blowing oil is not consumed, the escaping gas and oil are allowed to burn at the surface and relief wells are drilled to the producing zone or zones. Mud, water, and/or cement are pumped down the relief well to close the formation and stop the blowout.

The drawbacks to this technique are that it is costly in both dollars and time. For an offshore well, relief drilling can often require up to four months or more to complete. (An Alaskan offshore well blew for 14 months in 1962-63 when weather conditions impeded efforts to regain control.) In addition, production lost during drilling can result in costs comparable to those of the relief wells themselves.

From 1953 through 1972, when nearly all the wells were drilled in the U.S. OCS, 43 major accidents occurred. Nineteen were associated with drilling, 15 with production and 4 with pipelines. Over the 19 years, there has been an average rate of 0.005 (0.5%) drilling and production accidents per successful OCS well drilled. During the same period, 8 blowouts were recorded in state waters.

The frequency of OCS accidents generally increased as activity increased until 1968, when the accident frequency peaked. It has been decreasing since then. The 1969 Santa Barbara blowout, in releasing from 18,500 to 780,000 barrels of oil, raised serious questions on the adequacy of OCS technology. Since Santa Barbara, three major production platform accidents have occurred in the Gulf of Mexico. In the Shell accident (1970), estimates of oil lost range from 53,000 to 130,000 barrels. The Chevron accident (1970) resulted in loss of 30,500 barrels. Finally, the Amoco accident (1971) resulted in loss of 400 to 500 barrels.

The diminishing number of drilling accidents since 1968 reflects improvements in both technology and practice. The frequency of production accidents has not decreased so markedly, perhaps because oil offshore production facilities and pipelines do not, in all instances, meet the specifications now called for in new facilities and pipelines. Table 7 shows the location, case, date and extent of major oil spills from offshore production facilities in the period 1964-72.

TABLE 7: MAJOR OIL SPILLS FROM OFFSHORE PRODUCTION FACILITIES, 1964-1972*

	Cause	Date	Amount reported (barrels)
Offshore platforms			
Union "A," Santa Barbara	Blowout	January 28, 1969	77,400
Shell ST 26 "B," La.	Fire	December 1, 1970	52,400
Chevron MP 41 "C," La.	Fire	March 10, 1970	30,950
MP gathering net and storage, La.	Storm	August 17, 1969	12,200
Signal SS 149 "B," La.	Hurricane	October 3, 1964	5,000
Platform, 15 miles offshore	–	July 20, 1972	4,000
Continental EI 208 "A," La.	Collision	April 8, 1964	2,600
Mobil SS 72, La.	Storm	March 16, 1969	2,500
Tenneco SS 198 "A," La.	Hurricane	October 3, 1964	1,600
Offshore pipelines			
West Delta, La.	Anchor dragging	October 15, 1967	157,000
Persian Gulf	Break	April 20, 1970	95,000
Coastal channel, La.	Hit by tug prop	October 18, 1970	25,000
Chevron MP 299, La.	Unknown	February 11, 1969	7,400
Gulf ST 131, La.	Anchor dragging	March 12, 1968	6,000
Coastal channel, La.	Equipment failure	December 12, 1972	3,800
Coastal waters, La.	Leak	March 17, 1971	3,700
Coastal channel, Tex.	Leak	November 30, 1971	1,000
Coastal channel, La.	Leak	September 28, 1971	1,000

*Over 1,000 barrels.

Source: MIT Department of Ocean Engineering, 1974, *Analysis of Oil Spill Statistics.*

By 1980 exploratory drilling and possibly production will be extended to coastal shores and slopes of about 120 countries. The estimated value during that year will be about $14 billion and the cumulative value of offshore operations will be more than $50 billion.

Current practice in the offshore oil industry is to group platforms, structures and facilities, located in particular fields, for economy in operation. Thus, for example, one main platform may serve as the supporting facility for ten or more structures. This main platform then provides living quarters, office space, treatment facilities and control centers for the other structures in the group. At an average bare-bones cost of $8 million, real estate on these main platforms runs about $40 per square foot with an average surface area of 200,000 square feet. These surface areas and cost figures play an important role in the selection equipment for wastewater treatment, for example; the problem of offshore wastewater treatment will be discussed in a section which follows.

Several routine OCS operations result in chronic discharges of oil and other materials to the water. Unlike that for accidental spills, their probability is 1.0, they have a 100% chance of occurring. Some scientists believe that over the life of a field these intentional releases may damage the environment as much as the large accidental oil spill.

Securing platforms with pilings or anchors, anchoring vessels, and burying pipelines offshore disturbs bottom sediments and increases turbidity. In most drilling operations, cleaned drilling mud and drill cuttings are discharged overboard. Drill cuttings are shattered and pulverized sediment and native rock. Drilling mud may consist of such substances as bentonite clay, caustic soda, organic polymer, proprietary defoamer, and ferrochrome lignosulfonate. During the course of drilling an average 15,000 foot well, approximately 110 tons of commercial mud components and 950 tons of drill cuttings are discharged overboard. In its environmental impact statement on the proposed OCS lease sale in the northeast Gulf of Mexico, the Bureau of Land Management estimated that a maximum of 1 million tons of drill cuttings and 123,000 tons of mud would be discharged in the area as a result of drilling 1,120 wells to an average depth of 15,000 feet.

During operations, waters from the geological formations are often produced. The waters may be fresh or may contain mineral salts such as iron, calcium, magnesium, sodium, and chloride. Their discharge increases the mineral content and lowers dissolved oxygen levels in the area of operations. The waters often contain small amounts of oil.

Once a well is completed and connected to production facilities, production may begin. If oil, gas, and other materials are produced, they must be separated. The oil is separated, metered, and pumped to shore by pipeline, to offshore storage tanks for eventual transfer to a tanker, or directly to a tanker. The gas is separated; if it contains water, it is dehydrated by contacting it with glycol; and then it is pressurized, metered, and pumped to shore by pipeline. Where there is no gas pipeline or OCS gas production is not economical under prevailing market conditions, the gas is pressurized and reinjected into the reservoir. The flow diagram for a typical production facility is presented in Figure 4.

FIGURE 4: A TYPICAL PRODUCTION FACILITY WITH SAFETY EQUIPMENT

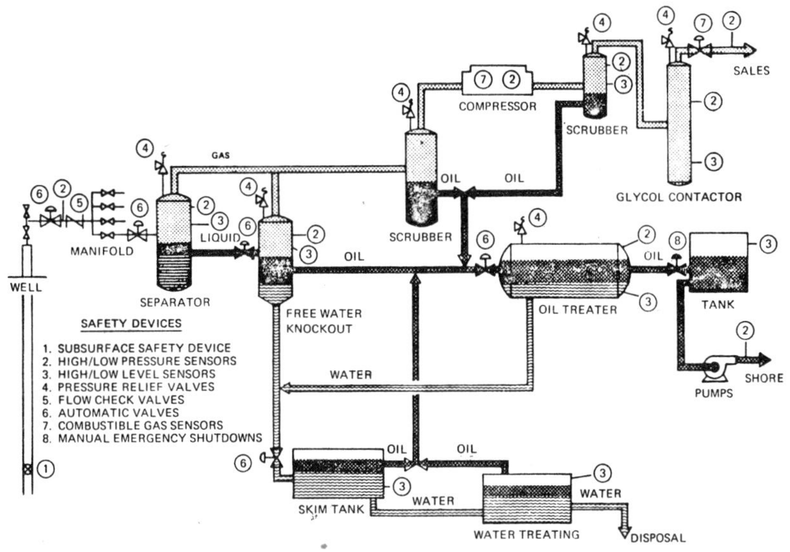
SALES
COMPRESSOR
SCRUBBER
GAS
OIL
GLYCOL CONTACTOR
MANIFOLD
LIQUID
SCRUBBER
OIL TREATER
TANK
WELL
SEPARATOR
FREE WATER KNOCKOUT
SAFETY DEVICES
1. SUBSURFACE SAFETY DEVICE
2. HIGH/LOW PRESSURE SENSORS
3. HIGH/LOW LEVEL SENSORS
4. PRESSURE RELIEF VALVES
5. FLOW CHECK VALVES
6. AUTOMATIC VALVES
7. COMBUSTIBLE GAS SENSORS
8. MANUAL EMERGENCY SHUTDOWNS
WATER
SHORE
PUMPS
SKIM TANK
WATER TREATING
DISPOSAL

Source: C.C. Taylor, Exxon Co.

When water is produced with the oil, separation is required. Consistent with OCS Order No. 8, separated water may be discharged into the ocean. The maximum allowable oil content is 100 parts per million; the average allowable oil content is 50 parts per million or less. Sand produced with the oil may be discarded into the ocean after the oil had been removed, as required in OCS Order No. 7.

Because of possible explosions and fire, storms, and earthquakes, many devices are installed to warn of impending or existing dangers and to control or stop the flow of gas and oil if trouble is sensed. Some of the safety devices with which fixed platform production facilities are equipped are pressure, level, and combustible gas sensors; manual, automatic, and pressure relief valves; and fire detection and fighting equipment. In addition, each well is equipped with a subsurface safety valve which can shut the well down in case of surface equipment failure. Required safety and pollution control equipment and procedures are described in OCS Order No. 8.

A set of check lists which identify system points of spill vulnerability in drilling, production and gathering/distribution systems has been presented in EPA Report EPA-R2-73-280a, Wash., D.C., U.S. EPA (August 1973). Application of these guidelines to a specific facility can improve spill prevention.

MUD CLEANING

Texaco Process

A process developed by *L.P. Teague; U.S. Patent 3,693,733; September 26, 1972; assigned to Texaco Inc.* involves treating well drilling cuttings that normally result from the boring of an oil or gas well in an offshore body of water. The treating process includes the sequential separation and washing of the drilling cuttings to free them of water contaminating components. The washing is achieved through a detergent circulatory system in which particulated cuttings are removed from the mud. The cuttings are then washed to remove possible water contaminating components therefrom. Thereafter, the drilling cuttings are rinsed and returned to the body of water, free of both water polluting elements and the detergent washing agent. Figure 5 shows the essential features of this process.

Normally, the well bore cuttings treated according to the disclosed process, are carried from a well bore **4** during the drilling operation so long as the drilling mud flows. Thus, as drill string **6** is rotatably driven to urge the drill bit **7** downward, liquefied mud is forced under pressure through the drill string **6** to exit at the lower drill bit **7**. The mud thereby lubricates the downhole operation, and in passing upwardly through the annulus **5** between drill string **6** and the well bore wall, carries with it various forms of drilling cuttings as heretofore mentioned.

Further in regard to the drilling mud, as is generally known, the composition of the mud is usually compounded to the particular drilling situation and condition. More specifically, the weight and the chemical makeup of the mud are initially determined and subsequently altered as needed and as the drilling progresses.

While not presently shown in great detail, the mud flow is urged under pressure from well bore **5**, upwardly to the drilling deck of the offshore platform, and discharged as an effluent stream by way of line **9**, into a tank **8** that is ancillary to shaker **10**. From the tank the mud mixture overflows onto the perforated face **11** of shaker **10**.

Shaker **10** as shown, comprises a vibratory or stationary type separator having a tilted screen working surface **11** upon which the mud mixture overflows from the tank. The mesh of the screen utilized on shaker face **11** is variable, being contingent on the characteristics of the substratum being drilled and the type of drilling cuttings being carried by the mud flow.

FIGURE 5: SCHEME FOR CONTROL OF POLLUTION FROM DRILL CUTTINGS AT AN OFFSHORE DRILLING SITE

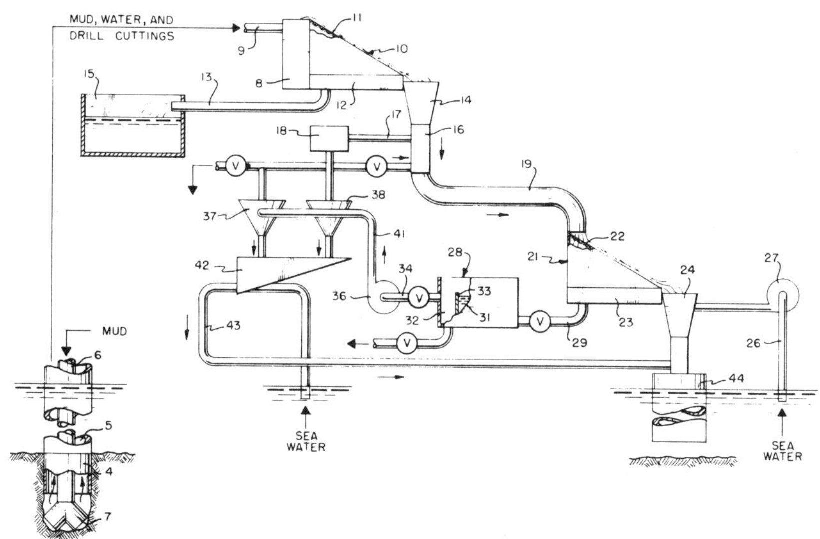

Source: L.P. Teague; U.S. Patent 3,693,733; September 26, 1972

In the shaker, the liquefied mud vehicle will by and large traverse the screen openings and is received in an underpositioned pan **12**. The latter is communicated by a conduit **13**, with a mud storage tank **15**. The remainder of the mixture deposited on shaker face **11**, and which does not pass through the screen, will comprise essentially an aggregate of solids being of sufficient size to remain at the screen surface. Solid matter, through the screen's vibratory action or through gravity flow, advances along the screen face to be discharged at the lower side thereof.

A collector **14** communicated with shaker face **11** receives the stream of drilling cuttings which in essence comprises a conglomerate of solid matter as well as some liquid. This flowable mass further embodies the previously mentioned nonwater-soluble, oily base constituents which normally cling to the cuttings. A wash chamber **16** is communicated with the collector discharge outlet to receive a stream of unprocessed drilling cuttings. Wash and spray chamber **16** includes a compartment adapted to receive the downwardly passing drilling cuttings, with means in the compartment to retain the cuttings sufficiently long to be brought into contact with the liquid detergent.

The wash chamber, in the presently disclosed arrangement, a spray chamber, further includes a spray nozzle system **17** disposed thereabout and appropriately arranged to deliver detergent streams against the cuttings. The spray nozzle system is communicated with a pressurized liquid detergent source represented by reservoir **18**.

Toward cleaning or scouring the cuttings of oily matter, the cuttings as an alternative can be immersed in a bath rather than being sprayed. The apparatus used in this latter step will be adapted in accordance with the consistency and the volume flow of cuttings as well as with other features of the drilling process.

Within spray chamber **16**, detergent is brought into contact with the cuttings under sufficient pressure and/or turbulence to remove substantially all of the extraneous matter clinging thereto. An elongated conduit **19** directs a stream of liquid detergent and drilling cuttings from spray chamber **16** whereby to physically expose the cuttings to the cleaning and separating action.

For this use, and toward achieving the necessary scouring and cleaning function, the detergent liquid can include any of a number of commercial solutions as for example, a biodegradable phosphate-free detergent.

Conduit **19** is connected at the discharge end thereof to a second separator **21**. The latter, as in the instance of shaker **10**, is a vibratory unit having a screen-type face **22** across which the detergent and cuttings flow is directed. The mesh size or openings of the screen face **22** are usually smaller than the mesh of screen face **11**, and of a sufficient size to pass the liquid detergent therethrough and into shaker reservoir **23**. The remaining cuttings stream, substantially free of detergent and other liquid, falls from screen top **22** and into a discharge chamber **24**.

Chamber **24** includes a receptacle to receive and retain the flow of cleaned cuttings for further cleaning. Receptacle **24** in the instant arrangement is communicated with sea water drawn from the immediate area by conduit **26** and pump **27**, or from an alternate source of water. After further cleaning by contact with sea water, the cleaned cuttings are discharged into a downcomer **44**. The member comprises in its simplest form an elongated tubular conduit that extends downwardly beneath the water's surface terminating short of the sea floor. Cuttings deposited at the upper end thereby are directed toward the floor where they tend to settle without the concern of prompting a water polluting situation on the surface.

Solution, including detergent separated from the drilling cuttings within separator **21**, is received in reservoir **23**. The latter includes an outlet communicated with the inlet of skimmer tank **28** by a valved connecting line **29**. Skimmer tank **28** embodies a first compartment **31** into which the detergent is fed and into which additional detergent can be added if such addition is required for reconstituting the material.

A second compartment **32** is communicated with the first compartment across a transverse panel **33**. Compartment **32** is provided with an outlet to receive detergent in valved line **34**, which in turn is communicated with the suction of detergent pump **36**. The discharge of the pump **36** is communicated with one or more hydrocyclone units **37** and **38** or similar fluid separating units, by way of line **41**. The function of the units **37** and **38** is to provide a final separation of detergent from any remaining materials in the flow stream.

Hydrocyclone units **37** and **38** function to centrifuge detergent from any remaining mud, water, and/or other fluidized or particulated components. The separated and cleaned fine solids pass upwardly into manifold **42** and are carried by line **43** into conduit **44**. The detergent material, essentially free now of solids, discharges into wash chamber **16** to again contact the incoming mud and cutting flow.

Open ended downcomer conduit **44**, as noted herein is disposed in the body of water, normally depending from the offshore drilling platform. The conduit **44** preferably is positioned with its lower open end spaced from the floor at the offshore location, or provided in the alternate with openings formed about the lower end. The conduit or caisson upper end is open to the atmosphere and disposed in alignment with the discharge opening of cuttings collector **24**. Thus, in the course of the process, substantially clean cuttings are fed into the caisson upper end. The clean cuttings thus enter the water and flow downwardly by gravity through the caisson, to be deposited at the ocean floor.

Use of the disclosed method for treating drilling cuttings results in a cleaner operation as well as a more economical one. The method serves to maintain a nonpolluting condition at the offshore production or drilling site and also permits maximum recovery of both drilling mud and washing detergent for the subsequent reuse of both items.

LEAKAGE CONTROL AT WELL HEAD

Blair Process

A process developed by *T.O. Blair; U.S. Patent 3,658,181; April 25, 1972* is one in which a plurality of perforate cones are secured at longitudinally spaced intervals to an elongated cable. The lowermost cone is arranged over an underwater source of oil leakage and the upper end of the cable terminates at an oil collecting chamber adjacent the surface of the water. Leaking oil thus is reduced to small bubbles or streams by passage upward through the perforate cones which also direct the oil inward toward the cable, forming a column of oil which is collected at the collecting chamber. Figure 6 shows a suitable form of apparatus for the conduct of the process.

FIGURE 6: PERFORATE CONE APPARATUS FOR COLLECTION OF UNDERWATER OIL LEAKS

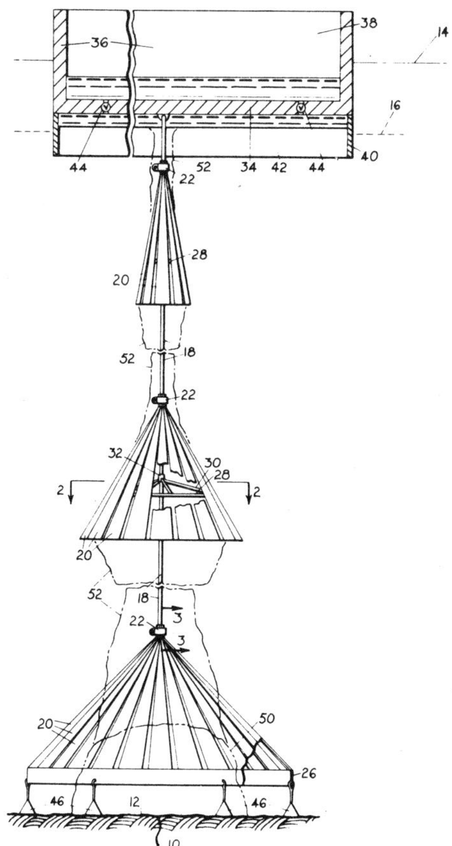

Source: T.O. Blair; U.S. Patent 3,658,181; April 25, 1972

The apparatus is shown in position to collect oil leaking from a fracture **10** in the ocean floor **12** a distance below the surface **14**. The surface of the ocean is turbulent to a variable depth, indicated by the broken line **16**, which depth is dependent upon weather conditions. Below that depth the ocean water is relatively quiescent.

The apparatus includes an elongated core member **18**, illustrated in the drawings as an elongated flexible metal cable. Depending upon circumstances of use, the core member may be provided in the form of a rod, pipe or other suitable member. In any event the core member is adapted to extend downward in the water substantially to the depth of an oil leak. Thus, in the embodiment illustrated, the cable is adapted to extend substantially to the ocean floor.

A plurality of perforate cone members are secured to the core member at longitudinally spaced intervals. In the embodiment illustrated in Figure 6, each cone member is formed of a plurality of circumferentially spaced vanes **20** which are secured at their upper ends to the core member, as by means of the split clamp member **22** which functions, upon tightening of the clamping bolt **24**, to draw the upper ends of the vanes securely to the core member. The vanes diverge downwardly from the core member to provide a cone which flares downward to its maximum peripheral dimension.

In this regard it is to be understood that the cone member may be circular in transverse dimension, or it may be square, rectangular or any other desired configuration. Means is provided for securing the vanes in spaced-apart relation to provide the cone with a perforate structure. In the lowermost cone illustrated this means is provided by the peripheral skirt **26** to which the bottom ends of the vanes are secured in spaced-apart relation.

In the cones above the lowermost cone, such means is provided by the peripheral ring **28** located intermediate the ends of the vanes and to which the vanes are secured in spaced-apart relation. Radiating inward from the ring are a plurality of circumferentially spaced spokes **30** which are secured at their inner ends to a central hollow sleeve **32** slidably receiving the cable core member therethrough. The arrangement of spokes thus serves to maintain the cone centered about the core member. The cone members may be made of wire rods, screening, expanded metal, perforate sheet metal, or other forms of material, as desired.

The upper end of the core member is surrounded by means for collecting the leakage oil. In the embodiment illustrated the oil collecting means is combined with oil storage means in the form of a barge having a bottom **34** and a peripheral wall **36** defining an oil storage chamber **38**. A peripheral skirt **40** extends downward from the bottom **34** to define with the latter an oil collecting chamber **42**. The skirt extends downward from the bottom of the barge a distance sufficient to extend below the turbulence level **16** of the ocean, so as to insure against the loss of collected oil in the chamber **42**. The upper end of the core member is secured to the bottom wall of the barge substantially at the center of the latter.

Means is provided for transferring collected oil from the chamber **42** to the storage chamber **38**. In the embodiment illustrated such means is provided by one or more shut-off valves **44** each in a passageway extending through the bottom **34** of the barge. The valves normally are maintained closed. However, when a desired quantity of oil has been collected in the chamber **42** the valves may be opened to allow the oil to transfer upward into the storage chamber **38** of the barge. When the barge has been filled with oil it may be replaced with an empty one, or the storage oil may be transferred to an oil tanker, by conventional pumping procedures. The operation of the apparatus illustrated is as follows.

The barge is moved into position so that the lowermost cone member overlies the fracture in the ocean floor. A plurality of weights **46** may be attached to the lowermost cone member to assure maintenance of the position of the latter properly over the fracture. As the oil leaks from the fracture it generally forms a large bubble **50**, as illustrated. As the bubble progresses upward through the slits or other forms of perforations in the lowermost cone member, it is broken up into small bubbles or streams. Further, since the oil tends to progress upward and inward along the vanes toward the axis of the core member, the body of oil passing through the cone member is drawn inward, decreasing in diameter. This column **52** of oil thus progresses upward about the core member. Since the column of oil may tend to spread upward to increasing diameter, for example because of slight current flow of water in the quiescent depths of the ocean, the next adjacent cone member is provided with a maximum peripheral dimension sufficient to contain the expanded column of oil and to reconcentrate it to smaller diameter as the column passes through the perforate cone member. This procedure is repeated successively as the oil column progresses upward toward the surface of the ocean, whereupon the oil is trapped within the collecting chamber **42**.

Cerebro-Dynamics Process

A process developed by *A.M. Crucet; U.S. Patent 3,653,215; April 4, 1972; assigned to Cerebro-Dynamics* provides an expansible oil collector and method for isolating oil escaping from an underwater source. The collector is in the form of a buoyant ring with an anchor ring suspended below the buoyant ring by cables. A thin, flexible wall or shield interconnects the anchor ring with the buoyant ring. When the collector is positioned over an underwater source of oil leakage, the anchor ring is lowered by means of the cables on the buoyant ring until the anchor ring rests on the underwater surface and encircles the source of leakage. The fluid collects at the surface of the water in the interior of the buoyant ring. Figure 7 illustrates one form of such an apparatus.

FIGURE 7: EXPANSIBLE COLLECTOR FOR UNDERWATER OIL LEAKS

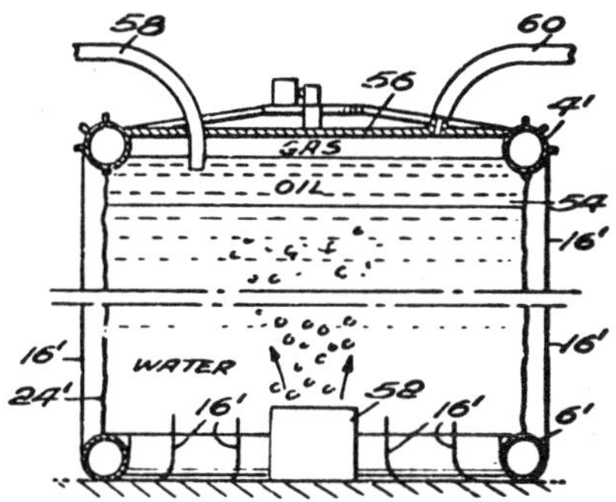

Source: A.M. Crucet; U.S. Patent 3,653,215; April 4, 1972

The structure of the device includes a float ring **4'** and an anchor ring **6'** which is supported by cables **16'**. The shield **24'** encloses the source of oil leakage **52**. Gas is also being evolved at the source **52**. The oil accumulates in a layer **54** on the water and above the oil layer, the gas is trapped by a canopy **56** which is secured around its perimeter to the float ring **4'**, the oil is pumped out of the device through a pipe **58** and a pipe **60** is connected with the canopy **56** for drawing gas from the space under the canopy. With this arrangement, both liquid and gas may be readily isolated and removed without contaminating adjacent areas.

The device of this process can be readily transported from one location to another by floating the device on the surface of the water. Furthermore, the device can be placed in service quickly over a source of oil leakage. The shield **24** isolates the oil and gas flow from underwater currents which otherwise might divert the flow of oil away from the float ring **4**. This device is good for areas where substantial cross-currents are encountered.

Esso Production Research Co. Processes

A process developed by *L.D. Woody, Jr.; U.S. Patent 3,653,214; April 4, 1972; assigned to Esso Production Research Company* provides a barrier apparatus for containing oil accumulation on a water surface. An elongated buoyant member having a generally triangular cross-section with such triangle preferably having slightly rounded identical sides is arranged in the water such that a line from the vertex of the triangle perpendicular to the base thereof substantially coincides with the water level. The length of such line is several wave lengths in magnitude in order to act as a dampener to wave amplitude. The length of the base of such triangle is sufficient (in cooperation with the length of the line perpendicular to the base) to inhibit or prevent oil from flowing over the barrier member and inhibit or prevent oil from becoming trapped beneath the barrier member under normal heave thereof.

Figure 8 shows an offshore platform **8** used for oil and/or gas well drilling or production operations surrounded by a lightweight floatable barrier member **9**. Buoys or floats **10**, suitably anchored, may be used to maintain the barrier member in proper position about platform **8**.

FIGURE 8: BARRIER APPARATUS FOR SURFACE OIL COLLECTION AT OFFSHORE DRILLING SITE

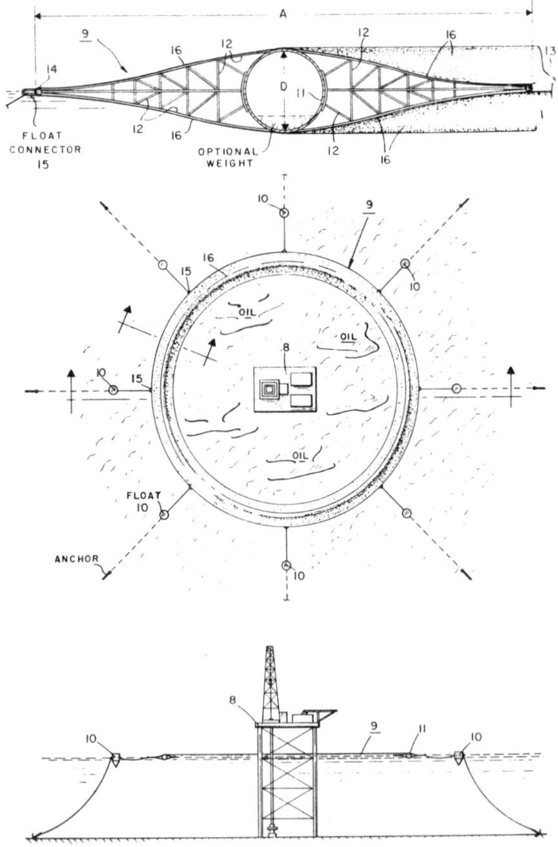

Source: L.D. Woody, Jr.; U.S. Patent 3,653,214; April 4, 1972

The structure of one embodiment of barrier member **9** is shown more clearly in the sectional detail. A central tubular member **11** is attached to light weight framing ribs **12** which extend horizontally on each side of tubular member **11** to support tension straps or ring members **13** and **14**. Loops **15** are fixed to ring member **14** at spaced-apart intervals for connecting anchor and float lines to the barrier member. The entire structure, including the ring members, ribs, and tubular member, is covered with a leakproof material **16**.

As seen in the sectional detail the barrier member cross-section comprises two generally isosceles triangles having a common base and slightly rounded or oval sides arranged in the water such that a line from vertex to vertex of the triangles perpendicular to the base thereof substantially coincides with the water level. One-half of such line, that is the distance from ring member **14** or ring member **13** to the center of tubular member **11** or one-half the distance **A** is several wave lengths in magnitude in order to dampen wave amplitude. For example, if the wave length were 6 feet the length of the distance one-half **A** might appropriately be 36 feet. The diameter **D** of tubular member **11**, that is, the height of the barrier member at the center thereof or the length of the base of the triangles, is sufficiently long below the water line to inhibit or prevent oil from becoming trapped beneath barrier **9** and sufficiently long above the water line, in cooperation with the horizontally extending portion of barrier member **9**, to dampen wave amplitude. As an example, for a dimension of 36 feet for the distance one-half **A** the diameter of tubular member **11** might appropriately be 6 feet. As shown in the sectional detail, the tubular member **11** may be provided with additional weight if desired to assure proper positioning in the water.

A process developed by *T.W. Childers; U.S. Patent 3,670,814; June 20, 1972; assigned to Esso Production Research Company* is one in which oil emanating from a submerged well structure which includes a wellhead and a production manifold is recovered and oil pollution of surface waters and neighboring shorelines is prevented by confining the fugitive oil underwater above an oil-water interface under a roof having side curtains and extending over the well structure. One part of the roof is fixed above the production manifold, and another part of the roof is removably secured over the wellhead and includes structure for coacting with a remotely operated running-in device for removably installing and retrieving such roof part.

A fluid conduit opened by an oil-water interface-level detecting device drains oil below the roof to a production conduit of the well structure, for example, the production manifold, by gravity flow. The interface level detecting device opens the conduit when the interface reaches a predetermined minimum level below the inlet of the fluid conduit. When the interface climbs to a selected higher level below the inlet with drainage of oil into the conduit, the interface level detecting device closes the inlet valve, preventing sea water from entering the fluid conduit. Means responsive to the interface level shut in production from the well structure if confined oil accumulates to a predetermined maximum interface level below the inlet of the fluid conduit.

Attainment of the predetermined minimum level, as indicated by the interface level detector device, may actuate electrical circuitry provided for connection to an indicator placed at a remote accessible location, signaling the occurrence of oil leakage from the submerged well structure and the accumulation of the predetermined minimum volume of fugitive oil. The fluid conduit may have valve means to prevent backflow of oil into the confining area from the well structure production conduit. Valve means may regulate the discharge of oil into the well structure from the fluid conduit. Pumping means may be utilized to force oil from the fluid conduit into the well structure when the production line pressure in the well structure exceeds the hydrostatic pressure at the level of the well structure.

Another process developed by *T.W. Childers; U.S. Patent 3,770,052; November 6, 1973; assigned to Esso Production Research Company* is a process for installing and forming an inverted drip pan over an underwater wellhead and its associated production manifold. One part of the pan is fixed above the production manifold, and another part of the pan is removably secured over the wellhead by use of means on the wellhead roof for coacting with a remotely operated running-in device for removably installing and retrieving such roof part. Figure 9 shows the form of apparatus which may be used in this and the preceding Childers patents.

FIGURE 9: UNDERWATER COLLECTION PAN DEVICE FOR OIL ESCAPING FROM WELLHEAD

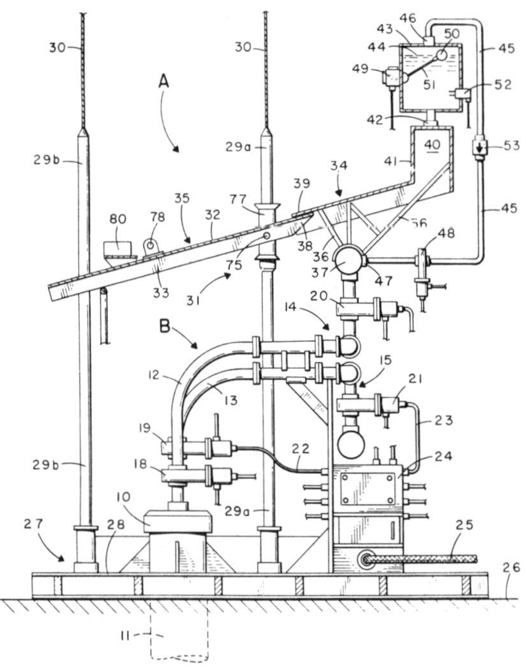

Source: T.W. Childers; U.S. Patent 3,770,052; November 6, 1973

The submerged well structure **B** includes a wellhead **10**, which surmounts a well **11** illustrated as a dual completion well; production lines **12, 13,** which produce dual completion well **11**; and production manifolds **14, 15,** into which production from production lines **12, 13,** feeds by way of jumper pipes **16, 17.** Production manifolds **14, 15** connect into pipelines which lead to storage tanks which may be submerged, located above the ocean's surface on a fixed or floating platform, or onshore. Production through production lines **12, 13** is regulated by production control valves **18, 19,** and feed through production manifolds **14, 15** is regulated by production manifold valves **20, 21.** Production control valves **18, 19** and production manifold valves **20, 21** are linked by hydraulic lines (e.g., lines **22, 23**) to an electrohydraulic production control module **24**, which transmits and receives monitoring and control signals, respectively, to and from a remote accessible site, such as a fixed or floating platform or an onshore station, by means of a multiconductor signal power cable **25**.

The submerged well structure and its associated production control equipment are supported on ocean floor **26** by means of a template structure **27**, which includes a base **28** mounting front and rear guideposts **29a** and **29b**, respectively, from which guide wires **30** are extended to a fixed or floating platform on the ocean's surface. The underwater pollution control device **A** includes a roof **31** extending over the well structure **B**. Roof **31** takes the form of an inverted pan, with a top surface **32** and side curtains **33**, and includes a portion **34** over the wellhead. The manifold cover pan **34** is affixed over production manifolds **14, 15,** suitably by structural members **36** secured to the header portion **37** of the uppermost manifold **14.** Wellhead cover pan **35** is removably secured above the wellhead by a structure hereinafter described coacting with the guideposts of the template, in order that the wellhead cover pan **35** may be removed for vertical reentry to perform work-over operations and the like.

In the embodiment illustrated, roof **31** has a sloped configuration over wellhead structure **B**, the roof ascending from the portion **35** covering the wellhead to the portion **34** covering the manifold. The forward portion **38** of the wellhead cover pan **35** is inserted under the rear portion **39** of the manifold cover pan **34** to form a continuous and upwardly inclining passage from wellhead cover pan **35** to manifold cover pan **34** by which oil fluids, of lighter specific gravity than water, and fugitive from well structure **B**, are directed to flow by gravity upward under manifold cover pan **34**. Manifold cover pan **34** includes as oil fluid collecting structure a transverse passage **40** provided by a gutter **41** formed in the forward portion of manifold cover pan **34**. Gutter **41** converges from side curtains **33** of manifold cover pan **34** to drain into a standpipe **42** which empties into a roof receiving chamber **43**. For a multiwellhead template, one manifold cover pan **34** may receive oil percolating upward from a plurality of wellhead cover pans **35** in which centrally peaked gutter **41** delivers tributary oil from any roof portion **34** or **35** to the central roof receiving chamber **43**.

Instead of being inclined upwardly in the same direction as wellhead cover pan **35** so as to direct fugitive oil fluids to a collecting structure **40** at an uppermost end remote from the near portion **39** of the manifold cover pan, the manifold cover pan may be horizontally oriented or may be inclined downwardly from the near portion **39** near the forward end **38** to the wellhead cover pan. If inclined downwardly, the transverse passage **40** would be situated at near portion **39** in an uppermost location in the roof **31**. The forward or leading end **38** of the wellhead cover pan would continue to be placed under the near edge **39** of the manifold cover pan, such that fugitive oil passing upwardly under wellhead cover pan **35** would empty into the transverse collecting structure **40**.

Oil confined under roof **31** above oil-water interface **44** is removed from its underwater confinement and discharged into well structure **B** free of sea water by means of a fluid conduit **45** connecting roof **31** and well structure **B**. Fluid conduit **45** has an inlet **46** opening into the receiving chamber **43** of roof **31** and an outlet **47** opening into a production line of well structure **B**, suitably downhole production tubing, wellhead production lines **12, 13** or, as illustrated, into header portion **37** of production manifold **14**.

Admission into inlet **46** of oil above interface **44** is controlled by a normally closed inlet valve **48** operatively associated with an oil-water interface level detecting device **49** which opens inlet valve **48** when the oil-water interface reaches a predetermined minimum level below inlet **46** and closes inlet valve **48** when interface **44** climbs to a selected higher level below inlet **47** with drainage of oil into fluid conduit **45**. Suitably, interface level detecting device **49** may include a relay for electrically actuating valve **48** on closure of the relay by downward movement of float **50** on pivot arm **51** to the predetermined minimum location within side curtains **33** of roof **31** below inlet **46**. (Other oil-water interface level detecting devices suitably include solid state electrical microswitches, capacitance probe-type oil detectors, magnetic permeability switches and the like, as known to the art, although the float actuated mechanical link is preferred for its long service life and reliability.)

When inlet **46** is opened, oil above interface **44** is driven into fluid conduit **45** by the heavier sea water if line pressure in production manifold **14** is less than the hydrostatic pressure of the water at the level of the underwater structure. If line pressure is greater than the hydrostatic pressure at that level, oil from the production manifold may backflow through fluid conduit **45**, driving interface level **44** down within receiving chamber **43** to a predetermined maximum level to which interface level detecting device **49** responds, as by closure of a second stage relay, and engages circuitry leading to production control unit **24** to shut in production from well structure **B**. Response of interface level detecting device **49** to attainment of the predetermined maximum level may also engage circuitry which may be provided for electrical connection to an indicator installed at a remote accessible location to signify accumulation under roof **31** of a predetermined maximum volume of oil above the predetermined maximum level.

Alternatively, as illustrated, a capacitance probe **52** or other level detecting means may be separately installed in the lower portion of the receiving chamber **43**, instead of being integrated with the structure of the device **49** detecting the predetermined minimum level, for connection to electrical circuitry leading to a remote indicator and/or operative connection to production control unit **24**. A check valve **53** may suitably be provided in fluid conduit **45** to prevent backflow into chamber **43** of oil from production manifold **14**. If so, when inlet **46** is opened on attainment of the predetermined minimum volume and forward flow through fluid conduit **45** is prevented by a higher pressure in the production line, well structure **B** will be shut in by continued collection of fugitive oil by roof **31** if the predetermined maximum volume interface level is attained.

As previously stated, manifold cover pan **34** is affixed over production manifolds **14, 15.** Suitably, manifold cover pan **34** may be attached to its mount prior to lowering of the well structure **B** on template **27** into place on ocean bottom **26**.

Removable wellhead cover pan **35** is installed after well structure **B** and manifold cover pan **34** are set in place on ocean bottom **26**. Wellhead cover pan **35** is pivotally hinged in its forward portion by means of a plurality of hinge pins **75** seated in mounting rings affixed to confronting sides of guide sleeves **77**. The top surface **32** of wellhead cover pan **35** mounts a crossbar **78** between upstanding ears secured to the rear of the center of wellhead cover pan **35**. A semicircular stop **80** is braced at an acute angle with top surface **32** by a brace.

Fre-Del Engineering Corporation Process

A process developed by *E.C. Greenwood; U.S. Patent 3,592,005; July 13, 1971; assigned to Fre-Del Engineering Corporation* is one in which a floating barrier is anchored in place in a generally circular shape around an offshore oil rig. The barrier consists of a plurality of rigid segments extending above and below the surface of the water and attached to each other by flexible couplings which permit movement in both horizontal and vertical planes. At each joint, a sheet of flexible material seals the joint from an interchange of oil and water from the inside of the barrier to the outside. Figure 10a shows such a barrier in place around a drilling platform.

There is shown a drilling platform **10** positioned on the floor of the ocean and extending above the surface of the water **12**. Also shown is a barrier, generally designated **13**, made up of a plurality of segments **14**. Barrier **13** may have a diameter, for example, of from 500 to 1,000 feet which would permit it to contain any oil **15** which may leak, spill or blow out from the wells below platform **10**. A plurality of anchors **16** (not shown) connected to segments **14** of barrier **13** by steel cables **17** are located at intervals which are frequent enough to hold barrier **13** in place in the presence of expected currents, winds and wave forces. Steel cables **17**, however, are loose enough or extensible enough to provide for changes in depth due to tides. If found necessary, anchors **16** may be set in concrete caissons poured on the ocean floor.

Since oil is lighter than water, any oil leaking, spilling or blowing out from an underwater well within the confines of barrier **13** will flow to the top and be contained by the barrier. For this reason barrier **13** is made of a plurality of segments **14** which are light enough in weight to float on the surface of water **12**.

FIGURE 10: OIL BARRIER FOR OFFSHORE OIL RIGS

a.

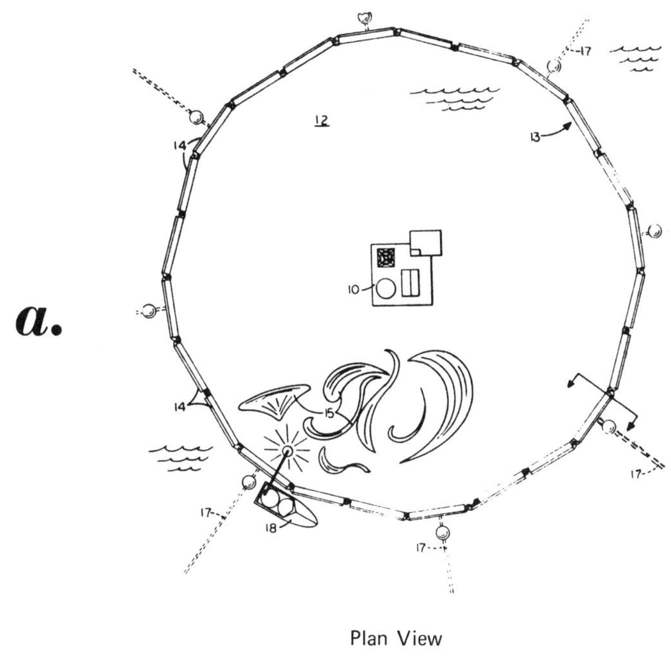

Plan View

b.

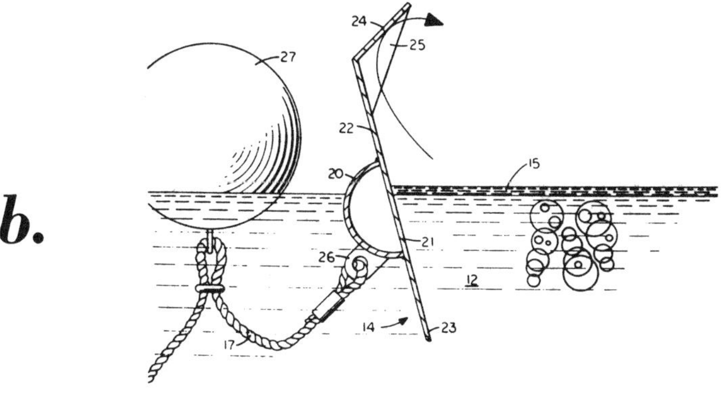

Sectional View

Source: E.C. Greenwood; U.S. Patent 3,592,005; July 13, 1971

Figure 10b is a cross-sectional view of the barrier used. As shown there, segments **14** extend both above and below the surface. According to a first embodiment, segments **14** may consist of half of a large diameter pipe **20**, the open side of which is welded adjacent one end of a flat plate **21**. The pipe, containing air, will float on the surface of the water. A first section **22** of the plate will extend above the surface of the water and a second section **23** of the plate will extend below the surface of the water. A typical height may be 2 to 3 feet above the surface of the water and 4 to 6 feet below the surface of the water. The segments may be as short as 2 or 3 feet in cases where very rough seas are usual or as long as 10 or more feet where ocean conditions are generally calm.

In addition, a splash plate **24** may be welded at the top edge of section **22** of plate **21** to prevent wave action from forcing any oil floating on the surface of the water from splashing over the top of the barrier. A plurality of spaced gussets **25** may be utilized to support the splash plate.

Figure 10b also shows steel cables **17** attached to a ring **26** on pipe **20** to hold barrier **13** anchored in place around the drilling platform. A plurality of buoys **27** attached to the cables and spaced from the pipe may be provided to the support cables so that the weight of the cables does not pull the the barrier underwater. In this manner, a loop may be provided between the cable support buoy and the pipe to provide for changes in depth of the barriers due to tides.

With the barrier in place, any oil seeping, leaking or blowing out from the well beneath the platform will be effectively contained within the confines of the barrier. From this trapped location, oil may easily be pumped from the surface onto tanker **18** where it may be simply separated from the water and carried to shore. For example, the famous Santa Barbara Channel oil blowout which occurred during January 1969, could have been contained in a 500 foot diameter circular barrier since the 500 barrels per day which were emanating from the ocean floor would have only caused an oil accumulation of approximately ½ inch per 24 hours. A siphoning ship outside of barrier **13** could have easily cleaned this up each day.

Headrick Process

A process developed by *E.E. Headrick; U.S. Patent 3,567,019; March 2, 1971* is one in which a light weight, elongated, flexible, tubular structure is provided for use in confining leaking oil and other lighter-than-water substances to a predetermined area. At an offshore location the ends of the structure are drawn together to create a closed figure surrounding the point where oil is surfacing. The structure is divided into two chambers, one of which is filled with a liquid having at least the specific gravity of the liquid in which the structure floats. The second chamber of the structure is adapted to be distended such that a barrier is created extending above and below the surface producing a surface interlock with the liquid below the floating substance to prevent the substance from slipping beneath the barrier. Figure 11 is a cross-sectional view of such a barrier.

FIGURE 11: FLOATING FLEXIBLE BARRIER FOR SURFACE CONTAINMENT OF OIL FROM UNDERSEA LEAKS

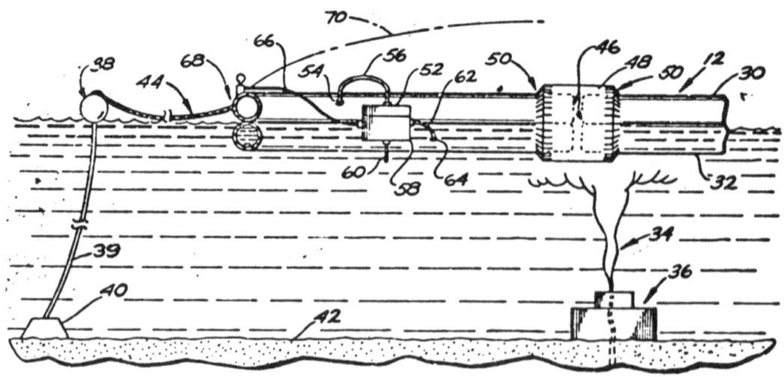

Source: E.E. Headrick; U.S. Patent 3,567,019; March 2, 1971

As shown therein, the barrier **12** comprises two tubes **30** and **32** vertically disposed relative to one another and joined together at the point of tangency. This is accomplished by a fused junction where the tubes are extruded as a unitary structure or by encircling clamps or cables to maintain the tubes in the proper relative orientation. It is a prime requirement that the tubes be maintained in a fixed relationship so that the lower tube **32** extends beneath the surface **34** of the water in which the barrier floats to create a water surface interlock while the upper tube **30** extends a predetermined distance above the surface to create a catch basin encircling an oil leak **34** from a wellhead **36**. The barrier is maintained in position by an anchor buoy **38** which is attached by a cable **39** to a weight **40** resting or secured to the ocean floor **42**. A line or cable **44** extending from the buoy is attached to the barrier.

In storage the two tubes are stacked or coiled in a flat, deflated condition. The opposite ends **46** of each tube are sealed to make each tube a closed airtight chamber. When being deployed in an offshore location, the barrier is placed in the water in the pattern of a closed geometric figure surrounding the point where the oil leak is surfacing. In relatively quiet water with slow subsurface currents, a vertical projection downward from the periphery of the closed figure normally surrounds the location of the oil leak on the ocean floor. The end portions **46** are brought together and secured in an end to end relation by means of a sleeve **48** which is adapted to be slipped over and attached to each end of the barrier at a point slightly removed from the actual ends thereof by means of laces **50** which are drawn tightly together to complete the closure of the geometric figure. Other means of securing the ends of the barrier together will also be apparent to those skilled in the art.

The upper tube is then distended with air by means of a compressor **52** which is connected to a valve **54** in the wall of tube **30** by an air hose **56** or by other buoyant means. Subsequently a pump **58** draws water through tube **60** from the body of water in which the barrier is located and pumps it through hose **62** and valve **64** into the lower tube until it is filled with water. Filling tube **60** with a liquid of the same specific gravity as the body in which it is located causes the barrier to settle in the water until the point of juncture between the two tubes is approximately level with the surface of the water. The compressor and the pump can also be arranged to function in an "on demand" mode such that if an air or water leak should develop in tube **30** or **32** respectively, the loss of pressure is immediately sensed, calling the compressor or the pump into operation to maintain a sufficient outward pressure to prevent the collapse of either of the tubes, particularly tube **30**, until repairs can be effected. An electrical connection **66** to a signalling device **68** energizes the device when the "on demand" capacity of either the compressor or the pump is called into operation.

It is also contemplated that a blower can be utilized in place of the compressor for filling and maintaining positive pressure in tube **30**. In this embodiment the blower would operate in a continuous mode and eliminate the need for an on demand feature. Where the barrier is used to surround the area of a leak from a fissure or subsurface wellhead where no drilling platform is present, it is also contemplated that a hood or canopy **70** attached to the barrier in an airtight manner (shown in phantom) can be provided to create a dome which can be used to capture and hold natural gases bubbling up from the leak until an accumulator can be positioned adjacent the barrier or protector to recover the collected gases.

The barrier need not weigh in excess of 5 or 10 pounds per linear foot and normally may weigh as little as 2 to 3 pounds per linear foot. Thus, where the barrier encircles an area having a diameter of 200 feet, the total weight of the barrier may be as low as about 1,200 pounds and certainly need not exceed approximately 6,000 pounds. In this manner the barrier can be rapidly transferred to a vessel for transportation to the location of a leak and subsequent easy placement in a surrounding relationship relative to the location of oil surfacing from the leak. Materials satisfying these requirements include vinyl and polyethylene.

Hyde Process

A process developed by *W.H. Hyde; U.S. Patent 3,681,923; August 8, 1972* involves a method and apparatus for controlling subnatant seepage of oil as from an underwater oil well or the like. The method includes the steps of collecting the seepage within an underwater receptacle located along the floor of a body of water in which the seepage occurs utilizing the floor as one of the confining walls of the receptacle. The seepage thus collected is conveyed upwardly in a confined state to the surface of the body of water and is deposited into a receiver from which the collected oil can be transported to a reservoir. The apparatus includes a receptacle open at its bottom to overlie the particular floor area of body of water at which seepage is or may be present so as to collect or confine the seepage along the floor which serves as the bottom wall of the receptacle.

A seepage conduit connected with the receptacle may be disposed circumjacent the outer conduit of an underwater well with which the apparatus is associated, and such conduit functions to carry the seepage from the receptacle to the surface of the water at which it empties into a receiver or reservoir from which the oil is pumped into a container for storage and processing as, e.g., separation of the oil from water admixed therewith. Figure 12 shows the essentials of the apparatus employed.

FIGURE 12: SEA-FLOOR COLLECTOR AND SEA SURFACE REMOVAL APPARATUS FOR COLLECTION OF OIL SEEPAGE

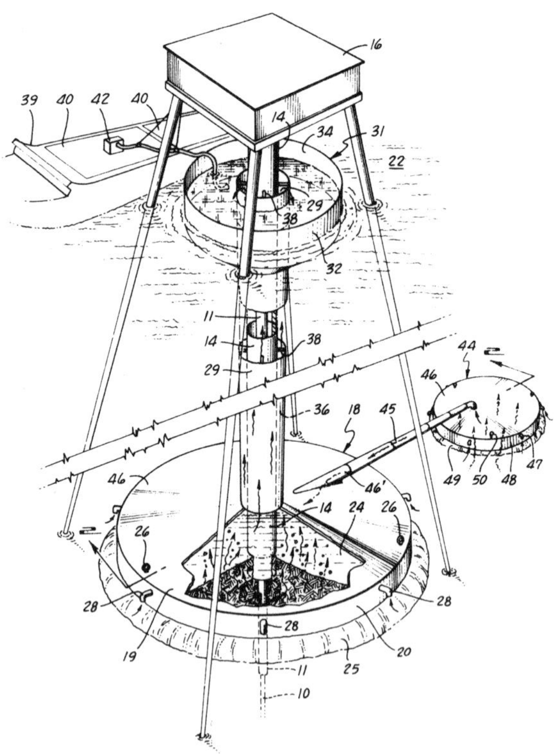

Source: W.H. Hyde; U.S. Patent 3,681,923; August 8, 1972

The oil well has the usual conductor tube or pipe **10** driven into the ground to an oil bearing strata or elevation, and it extends upwardly therefrom to a location above the surface of the water so as to carry oil from the oil-bearing depths to a reservoir or other storage or processing means located at the water surface. Circumjacent the conductor tube **10**, is a surface casing **11** that also projects into the ground below the surface of the water but not necessarily to the same depth as that of the conductor tube **10**. The space defined between the inner and outer conduits **10** and **11** is usually filled with a material such as concrete or sand and the choice may depend upon the particular elevation at which the material is to reside. In a typical oil well, the conductor tube may have a diameter of from 16 to 36 inches and, evidently, the casing **11** must have a somewhat greater diameter since it confines the tube **10** therewithin.

Surrounding the casing **11** is an outer conduit **14** that is often referred to as a snorkel tube because it extends upwardly from an embedded condition within the ocean floor (or floor of any other body of water) to a location above the body of water to provide an atmospheric environment within its interior. That is to say, the snorkel tube **14** is evacuated of water so as to be free therefrom, and irrespective of its length provides an atmospheric environment in which personnel may work both to install the necessary pumping and ancillary equipment and thereafter to maintain the same.

Apparently then, the snorkel tube must have a substantial diameter, and in most cases, the diameter thereof is within the range of about 36 to 72 inches, and the wall thickness of the tube will depend upon its diameter and the pressure forces it is to withstand and by way of example is most frequently in the range of from ¾ of an inch to 2 inches. For purposes of simplifying the figure, the usual well equipment has been omitted from the large space defined between the casing **11** and the snorkel tube, although the usual ladder with which the snorkel tube is equipped is partially shown.

The snorkel tube or any other outer conduit provided by the oil well is sometimes referred generally hereinafter as the "well conduit" or "outer well conduit", and it is a very stable structure driven into the ocean bed to a depth (generally to the point of resist) at which it is quite rigid and requires little, if any, support intermediate the ocean floor and the surface of the body of water thereabove. At its upper end, the outer well conduit usually provides the support for a platform or rigging depicted only generally and denoted with the numeral **16**. As indicated hereinbefore, respecting this process the oil well may be conventional and the brief description thereof given is in no sense intended to be definitive but only generally descriptive and is included for environmental purposes.

The seepage control apparatus includes a main receptacle **18** having a top well **19** and depending side walls **20**. The side walls are adapted to seat against and perhaps penetrate to some depth the floor underlying the body of water **22** with which the control apparatus is associated. Thus, the floor forms the bottom wall of the receptacle **18** and defines with the top and side walls **19** and **20** thereof a chamber **24**. Desirably, ingress and egress of liquid into and out of the chamber **24** from about the lower edge of the side walls **24** is inhibited, and for this purpose a base seal **25** of mud or concrete is provided along the side wall adjacent the floor.

The receptacle **18** may be as large as necessary or desired and it is intended to overlie the floor area at which seepage of oil occurs. Quite frequently, such seepage appears in the vicinity of an oil well or in close proximity thereto and, therefore, the receptacle is conveniently disposed in circumjacent relation with the oil well and particularly with the outer conduit thereof. Depending on its size and the depth at which it is to be positioned, the receptacle **18** may be constructed under water or it may be constructed in whole or in part above water and lowered into position. For this purpose, the receptacle may be provided with a plurality of connectors or lowering rings **26** which may be secured to the top wall **19** and are adapted to have cables releasably affixed thereto by means of which the receptacle can be lowered.

The receptacle need not be constructed of materials having the great strength required to withstand considerable pressure differences because the chamber **24** is to be filled with liquid so as to have within its interior the pressure existing at the elevation at which it is located. For this purpose (and to facilitate seepage flow as described subsequently) the receptacle **18** is provided with a vent means in open communication with the body of water **22**, and such vent means may take the form of one or more conduit-equipped openings **28**.

A seepage conduit **29** is connected with the receptacle **18** for removing oil (i.e., usually a mixture of oil and water) from the chamber **24** and for conveying such liquid mass to a suitable container located remote from the receptacle. In the apparatus shown, such seepage conduit **29** is circumjacent the outer well conduit **14** and connects with the receptacle **18** through a large opening in the top wall **19** thereof. The seepage conduit extends upwardly to an elevation above the level of the water **22** and communicates thereat with a container which, in the form shown, is a receiver **31** that surrounds the upper end portion of the conduit and has a side wall **32** that rises to a height above the surface of the body of water so that wave action and any other surface disturbances will not interfere with the contents of the receiver chamber **34** defined in part by the side wall **32** and in part by a bottom wall that converges inwardly and downwardly from the side wall and is fixedly secured to the conduit **29** so as to prevent ingress and egress of water into and from the chamber **34** from about the conduit **29**. As shown, the receiver **31** is partially immersed in the body of water **22** and is therefore at least in part supported by the buoyance thereof.

Since the seepage conduit **29** conveys liquid from the receptacle **18** to the receiver **31** it will be appreciated that a substantially unobstructed flow passage must be defined within the conduit. As explained heretofore, the conduit **29** is circumjacent the outer well conduit **14** and, accordingly, the seepage conduit must be somewhat larger in diameter than the well conduit so as to define a flow space or passage **36** therebetween. However, because of the relatively large diameter of these two conduits, it is not necessary that the diameters thereof differ greatly in order to establish a large-area flow space of adequate capacity. The symmetry of such space **36** is permanently established by guide or support structure **38** interposed therebetween the conduits within the flow space **36**.

Such guide structure or stabilizing means may take any convenient form and in the apparatus illustrated, it constitutes a plurality of angularly spaced spider structures or spacers located at each of a plurality of vertically spaced locations along the length of the conduit **29**. Each of the spacers **38** may be welded or otherwise fixedly secured in position. As in the case of the receptacle **18**, the seepage conduit **29** need not be constructed of exceptionally strong material since there is essentially no pressure differential as between the interior and exterior thereof. However, the presence of the spacers **38** serves to reinforce and thereby stiffen the conduit **29**. Depending upon the particular environment, the conduit **29** may be either rigid or flexible.

Oil may be transported from the receiver chamber **34** to a suitable reservoir which can take any convenient form and may be located in the body of water **22** so as to be relatively proximate the receiver **31** or it might be located on shore. As shown, the reservoir is provided by a barge **39** which may support or be subdivided into a plurality of containers **40** which are flow connected with the receiver **31** by a conduit or pipe system that usually will include a suitable pump mechanism generally indicated with the numeral **42**. The pump mechanism **42** may be mounted upon the barge or, in certain instances, may be supported on the rigging or platform **16** of the oil well.

In certain cases, a fissure through which oil seeps to the surface of the floor **21** may occur at a location not enclosed within the main receptacle **18** but may be sufficiently close thereto that the facilities of a control apparatus already installed can be utilized in controlling the seepage. In such cases, it may be advantageous to construct or lower, as the case may be, a satellite or auxiliary receptacle **44** onto the floor **21** underlying the body of of water **22** and to connect such satellite receptacle to the main receptacle **18** by a suitable flow conduit or duct **45**. In anticipation of this possibility, the main receptacle **18** may be provided with one or more openings **46'** therealong that are ordinarily capped but which can be opened and connected with such ducts **45** should this be required. Evidently, one or more satellite or auxiliary receptacles **44** may be employed in association with any main receptacle **18** and in terms of its essential structure, each may be a substantial duplicate thereof.

Thus, each receptacle **44** may have a top wall **46** and depending side walls **48** adapted to seat against the floor **21** and be sealingly related thereto by concrete or other seal means **49**. A plurality of connectors or lowering rings **50** are provided along the top wall **46** for purposes of lowering the receptacle **44** as heretofore explained with reference to the receptacle **18**. Each auxiliary receptacle **44** may be equipped with vent means **47** as heretofore explained although in some instances the vent means associated with the receptacle **18** may be relied on for pressure equalization within the receptacle maintaining a free upward flow of seepage through the conduit **29**.

The seepage control apparatus is conveniently installed or erected at the time that an oil well is constructed, but as explained hereinbefore, it can be built in structural association with an oil well already in existence and for this purpose any conventional construction techniques usually employed in the oil well industry can be followed.

Laval-Peters Process

A process developed by *C.C. Laval, Jr. and P.P. Peters; U.S. Patent 3,664,136; May 23, 1972;* utilizes a device for collecting oil leakage from formations beneath a body of water constructed of flexible, impervious sheet material for compact storage that can be readily erected for use by inflation. The device has a pair of predetermined upper and lower torus-shaped envelopes connected by a frusto-conical shroud with the lower envelope being of a larger diameter and adapted to be filled with a nonbuoyant material to distend the same and to open the shroud to its frusto-conical form in circumscribing relation to a point of oil leakage. The upper envelope is relatively smaller and is adapted to be inflated with air to make it buoyant and to circumscribe an opening having a discharge conduit extended therefrom for transferring the oil leaking from the bottom of the body of water internally of the collecting device to the surface.

FIGURE 13: COLLECTING DEVICE FOR SUBMARINE OIL LEAKAGE

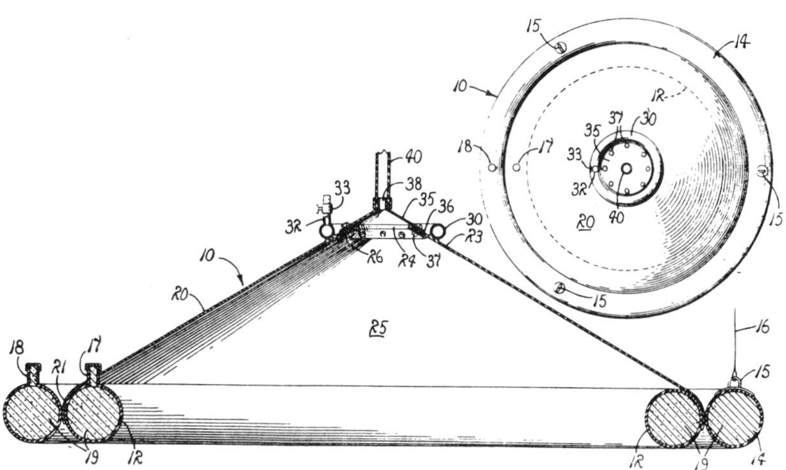

Source: C.C. Laval, Jr. and P.P. Peters; U.S. Patent 3,664,136; May 23, 1972

Figure 13 shows a plan and enlarged sectional elevation of the device used. The device is constructed substantially entirely of a flexible impervious sheet material, such as one of the polyethylene plastics, fabric impregnated with neoprene or Hypalon, rubber, silicon modified rubber and the like, so that it can be folded for compact storage and conveniently erected for use by inflation. The collecting device provides a pair of substantially similar inner and outer torus-shaped lower envelopes **12** and **14**, respectively, which are peripherally connected in concentrically nested relation by the application of a suitable adhesive therebetween or by suitable vulcanizing methods, as dictated by the particular material employed.

A plurality of hook engaging lift rings **15** are mounted on the upper peripheral portion of the outer envelope **14** for use in elevationally supporting and positioning the envelopes by a plurality of cables **16**, not shown, releasably connectable to the rings. Each of the inner and outer lower envelopes **12** and **14** has a substantially upwardly extended capped filler tube **17** and **18**, respectively, of a somewhat more rigid material or thickness than the envelopes, through which the interior of the envelopes is adapted to be filled with a nonbuoyant, flowable material **19**.

In most circumstances, a nonbuoyant, flowable material is utilized which can be pumped into and out of the envelopes. This permits ready weighted distention of the device for use and correspondingly ready collapse for removal, transport and/or storage. For this purpose, glycerine; mercury; acrolein dimer; acetic anhydride; many acids, such as acetic; heavy liquid plastics; heavy chlorine liquid compounds, such as ethylene dichloride, and trichloroethane; many esters, such as diethyl sulfate, ethylene carbonate and epoxide; glycols and triols, such as ethylene glycol; chromifax solvent; some alcohols, such as 2-mercaptoethanol; and some aldehydes, such as acetaldol can be utilized successfully. Usually the heavier the liquid the better because the desired ballasting is achieved with a smaller quantity having to be handled.

In all instances, the nonbuoyant flowable material must be compatible with the material forming the envelopes. Thus, if a heavy acid is utilized, acid resistant envelopes should be employed. Also, it is normally preferable to employ a liquid which does not react with water if the envelopes leak or which is not excessively exothermic if such leakage occurs. Where a permanent installation is desired, sand, ready mix concrete or the like is utilized. Such material serves to distend the envelopes to their annular operating condition and to provide sufficient ballast for sinking the envelopes to the bottom of a body of water in circumscribing relation to a leaking submarine oil formation. Such flowable particulate or semiamorphous weighting material also serves to conform the envelopes to any irregularities in the bottom so as to achieve a substantial seal thereagainst.

The collecting device further includes a substantially frusto-conical shroud **20** having a lower annular end **21** connected by vulcanizing or the like to the outer periphery of the inner lower envelope **12**. The shroud is not connected to the upper portion of the envelope but is wrapped outwardly thereover and is secured at a point between the two envelopes. This minimizes undue strain on the connection which results when it is made too high and the inner envelope rolls inwardly due to tide or wave action. The shroud has an opposite upper annular end **23** of a substantially smaller diameter than the lower end which circumscribes a constricted upper opening **24** from an oil collecting chamber **25** within the shroud. A substantially rigid annular ring, preferably constructed of stainless steel sheet material, is encased within a thickened portion of the upper end of the shroud to provide sufficient rigidity to maintain the circular configuration of the opening **24**.

An upper torus-shaped envelope **30**. of a substantially smaller diameter than the lower envelopes **12** and **14**, is disposed in circumscribing relation to the upper end **23** of the shroud **20** with its lower periphery secured thereto by vulcanizing or the like. A substantially vertically disposed air filler tube **32** is extended from the upper periphery of the upper envelope and provides a suitable valve, schematically indicated by the reference numeral **33**, for introducing and maintaining a volume of air within the envelope for rendering it sufficiently buoyant to maintain the shroud in a taut, frusto-conical configuration in circumscribing relation to the oil collecting chamber **25** therein when the collecting device is submerged in a body of water. A frusto-conical or funnel-shaped cap of substantially rigid sheet material such as stainless steel or the like is disposed in covering relation to the upper opening **24** from the shroud **20**. The cap provides a lower annular edge **36** which is disposed in overlapping relation to the upper end **23** of the shroud within the upper envelope **30** and is rigidly secured to the shroud by a plurality of connecting bolts **37**. The cap further includes an upwardly extended tubular connector member **38** which affords releasable connection for the lower end of an elongated oil discharge conduit **40** having an upper end, not shown, connectable to a suitable suction producing pump or the like, not shown, for creating a low pressure within the collecting chamber **25** of the shroud **20**.

Lejeune Process

A process developed by *L.M. Lejeune; U.S. Patent 3,674,150; July 4, 1972* is a process for preventing offshore oil well pollution resulting from offshore oil well blowouts or pipe line ruptures. The apparatus includes a sloping cylindrical member having a conical bottom wall open at its upper end. The device is centered over the leak so that the oil passes up into the device through the open top of the conical bottom wall. In the case of a low pressure leak the oil flows down the upper face of the conical wall and is collected in a sump within the apparatus from which it can be pumped. In the case of high pressure blowouts valve plates are provided which can be hydraulically closed to contain the oil spout which then flows into a sump in the apparatus surrounding the conical bottom wall. This sump also may be emptied by pumping. In both instances fire extinguishing gases may be pumped into the device in case ignition of the gas and oil should occur. One form of such an apparatus is shown in Figure 14.

FIGURE 14: COLLECTION DEVICE FOR COLLECTING OIL FROM SUBSURFACE LEAKS

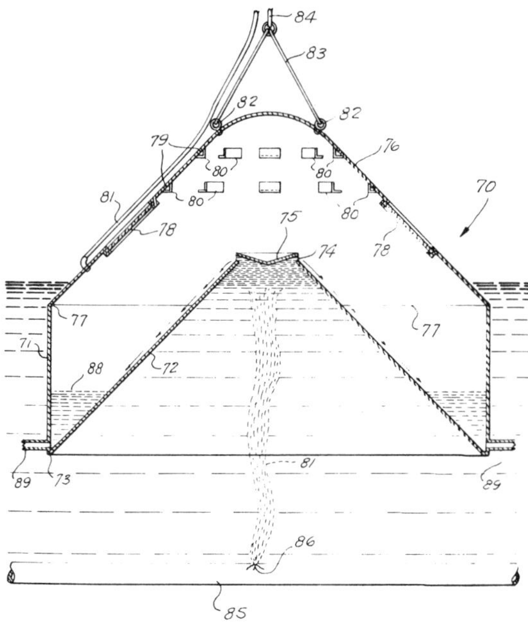

Source: L.M. Lejeune; U.S. Patent 3,674,150; July 4, 1972

The apparatus **70** includes a cylindrical wall **71** having a conical wall **72** secured to the lower edge thereof at **73** extending upwardly therein. The conical wall has a circular opening **74** at its upper end kept in spaced apart relation thereto with an inverted conical cap **75**.

A closed dome **76** is secured to the cylindrical wall **71** annularly at **77** and extends upwardly therefrom. Observation windows **78** are mounted in the dome so that the interior of the apparatus is visible. A plurality of gas vent holes **79** are formed in the dome with each having an angle baffle **80** associated therewith. The angle baffles are secured to the inner side of the dome.

A conduit **81** extends through the dome and leads to a source of fire extinguishing gas such as CO_2 or nitrogen. Lifting eyes **82** are secured ot the dome adjacent the top thereof and a lifting yoke **83** connects the eyes to a cable **84** extending from a crane or hoist. The apparatus is used in conjunction with a submarine oil pipe line **85** which has ruptured at **86** to release oil **87** which goes upwardly therefrom. The conical wall and the cylindrical wall form a sump **88** therebetween to collect the oil. Conduits **89** extend from the sump to a point of storage for the oil.

In the use and operation of the apparatus, it is lifted by a crane or hoist and lowered over the rupture in a pipe line until the oil flowing upwardly therefrom strikes the cap and is deflected through the circular opening where it flows down the inside of the conical wall and is collected in the sump. The oil may then be pumped from the sump through the conduits **89** to a point of storage. In the event that the oil becomes ignited within the apparatus fire extinguishing gases such as CO_2 and nitrogen are pumped through the conduit **81** to extinguish the flames. The specific linkages connecting the hydraulic rams **21**, **31**, and **45**, respectively, to the choke **19**, valve **30** and valve **44** are conventional and have been indicated by broken lines.

Madej Process

A process developed by *T.A. Madej; U.S. Patent 3,666,100; May 30, 1972* for collecting oil from an underwater leak includes the steps of (1) detecting the location of the leak, (2) submerging an inverted collector shell under the water to a position directly over and enclosing the leak so that oil, being lighter than water, rises from the source of the leak into the collector shell and displaces water in the collector shell to partially fill the collector shell with oil, (3) providing a conduit leading from the submerged collector shell upward through the water to a pump and from a pump to a storage receptacle, and (4) pumping oil from the collector shell through the conduit to the storage receptacle with the pump. Figure 15 shows the schematic arrangement of apparatus used in this process.

An oil leak from a fault **10** is shown, the fault extending to the ocean bottom **12** from an underground oil pocket **14** below the ocean in an offshore location. The pocket is under an oil well tower **16** which has a well pipe **18**. The first step of the method is to locate the fault or other source of an oil leak, and this is done by conventional techniques.

A collector shell **20** which is shaped something like an inverted funnel is then submerged from the surface of the ocean or other body of water to a position where it covers and encloses the upper end of the fault **10** as shown. Preferably the open bottom of the collector shell is located on or just above the ocean floor **12** directly over the mouth of the fault. The collector shell may be lowered by means of suitable rigging from a ship or boat and may be guided into position over the mouth of the fault by divers or television camera.

When the collector shell is in the proper position over the fault, a conduit **22** is connected to an open top **23** of the collector shell. The conduit **22** extends from the collector shell to the surface of the body of water, and in this embodiment the upper end of the conduit **22** is connected to a pump **24** which is mounted on the tower **16** above the surface of the water. From the pump, an extension **22'** of the conduit **22** leads to a storage receptacle which may be provided in a ship **26**. If desired, however, a storage receptacle could be provided in a fixed location, and if the well is not too far from shore, the storage receptacle could be located on shore.

FIGURE 15: COLLECTION AND REMOVAL DEVICE FOR OIL FROM UNDERWATER LEAKS

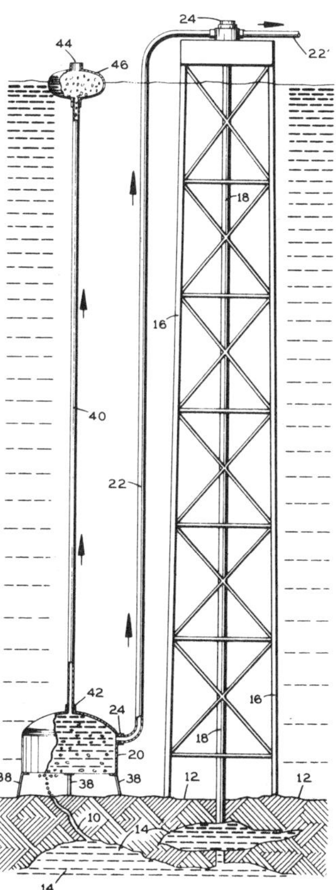

Source: T.A. Madej; U.S. Patent 3,666,100; May 30, 1972

After the collector shell is in place, oil rises from the fault directly into the collector shell and displaces some water from the collector shell to partially fill the shell. Oil or a mixture of oil and water is then pumped from the collector shell through the conduit **22** to the storage receptacle **26** by means of the pump. The pumping of oil from the collector shell **20** is continued at a rate sufficient to draw off all oil leaking from the fault. While this collecting of oil is continuing, the well may be plugged as by pumping drilling mud into it by conventional techniques, the collection of oil by means of the collector shell and pump **24** serving to prevent contamination of the water until such time as the well can be plugged in a manner to block the path or flow of oil to the fault.

The figure also shows in dashed lines how a collector shell **30** may be positioned around a leak through a hole **32** in the well piping **18**. The collector shell is again funnel-shaped and may be in two halves joined around the piping to position the collector shell directly over and partially enclosing the hole. The conduit extends from the collector shell to the pump in accordance with the previous description and it will be assumed that the same conduit **22'** leads from the pump **24** to the storage receptacle in the ship **26**. Just as in the case where oil is leaking from a source below the floor **12** of the ocean, the first step where there is a leak in the piping **18** is to locate the leak. Next the halves of the funnel or collector shell **30** are submerged and assembled in a position where the collector shell **30** partially encloses the opening **32**. Oil rises from the opening **32** into the collector shell and displaces some water from the shell. Oil is pumped from the shell through the conduit **34** to the storage receptacle in the ship **26** by means of the pump **24**. This collection of oil is continued until such time as the leak is stopped.

In the case of the shell **30**, access may be had to the hole **32** from underneath the collector shell. In the case of the collector shell **20**, access may be had to the fault **10** through a door **36** provided in the side of the collector shell **20** as shown.

Miranda Process

A process developed by *S.W. Miranda; U.S. Patent 3,643,741; February 22, 1972* is one in which oil flow from an open fissure on the ocean bottom is controlled by polymerizing in place in the fissure a resin composition which binds the oil or by placing a dome over the fissure to capture the oil. The dome may be tapped and the oil collecting in the dome recovered.

In the upper view in Figure 16, a portion of the ocean bottom **10** overlying a pool **12** of oil is illustrated. Fissures **14** have developed and oil **15** escaping from the fissure rises to the surface **16** of the water to form an oil slick. A sealant plug **18** is formed in the fissure by applying resin forming liquid to the open end **22** of the fissure **14** from a hose **24** terminating in a nozzle **26**. In the case of a two-part liquid resin forming system, hose **24** is coaxial and the liquids mix in the nozzle just before being applied to the opening **22**.

The hoses **24** are supplied from a liquid supply apparatus **28** which may be carried by a submersible craft **30**, such as a one or two man or larger submarine or by a surface ship **32**. The supply apparatus can include separate containers of the resin forming liquids and a compressor to apply a sufficient pressure to the containers to force the liquids through the hose **24** and dispense the liquids into the opening **22**. Many different types of commercially available liquid dispensing systems are suitable for use.

The lower view in the figure illustrates the application of this process to sealing oil leaks associated directly with an offshore oil well. The well **40** includes a platform **42** supported on pilings **44** which are seated in the ocean floor. A pump **46** is supported on the platform over the well casing **48**. The casing lines the hole bored into the oil pool **12**. A fissure **14** has developed adjacent the casing **48**.

In the case of a break **50** in the casing **48** causing leakage into the fissure **14**, a slant hole **52** is bored into the casing **48** at a location below the break **50**. The nozzle **26** of the hose **24** is inserted into the slant hole and the resin-forming liquid is pumped through slant hole **52** into the casing to form a plug **54** which seals off the leak. After the break **50** in the casing **48** is repaired, the plug can be drilled out and the well returned to a producing status.

If the fissure penetrates into the well pool causing oil leakage, a slant hole **58** can be drilled into the fissure **14** and liquid pumped from hose **24** into the fissure to form a plug **60** which seals the fissure.

The resin-forming liquids may be of many types but preferably are selected from the preformulated polymerizing systems which include a polyol component, preferably containing a hydroxyalkyl substituted tertiary amine, a polyol reactive resin and a mixture of nonionic surfactants. When these ingredients are combined in the presence of oil they polymerize and react with and bind up to 70% by weight of oil to form a solid mass which can function to seal or plug a fissure. The preformulated polymerizing system is adapted to compensate for properties of a wide variety of oils and is formulated such that merely by changing the the ratio of components or adding other ingredients thereto, the properties and curing time of the final solidified material is varied at will.

FIGURE 16: SCHEMATIC OF TECHNIQUE FOR SEALING UNDERWATER FISSURES USING SURFACE AND SUBMERSIBLE VESSELS

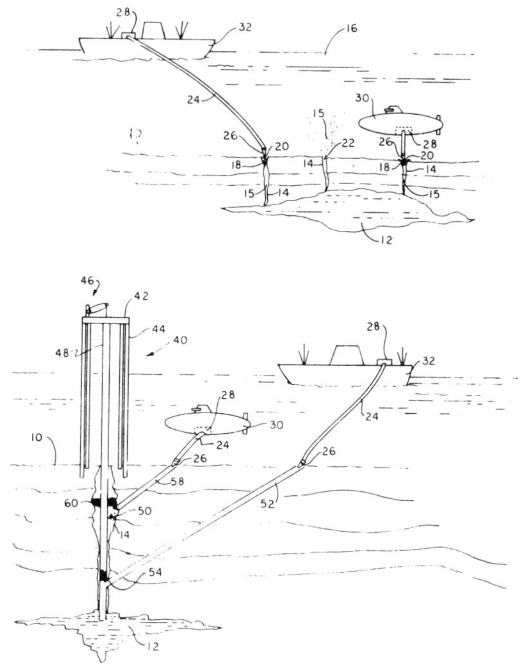

Source: S.W. Miranda; U.S. Patent 3,643,741; February 22, 1972

Missud Process

A process developed by *L. Missud; U.S. Patent 3,599,434; August 17, 1971* is one utilizing a floatable tube and skirt assembly encircling an offshore oil drilling rig for collecting oil released by leakage during the drilling operation. It is movable between an operative oil-confining position and an inoperative position in noninterfering relation to the drilling shaft. The tube is connected to the rig structure by telescopically extensible arms, there being (in one embodiment) a plurality of floatable lift canisters connected to the arms and tube, each canister having perforated walls and an inner inflatable bag, an air pump on the rig being connected to the canisters and the tube. When all the canister bags and the tube are deflated, they will sink to an inoperative position, pivotally guided inwardly to that position by the telescopic arms, the arms also serving to guide the tube into its operative floating position. A drawstring arrangement contracts the skirt into compact form against the tube, in noninterfering relation to the drilling shaft.

In the simplified diagram in Figure 17, the offshore drilling rig **10** is shown in an offshore oil well drilling position, the rig containing the conventional derrick **11**, platform **12**, columns **13** and revolvable pipe shaft **14** attached at its lowermost point to a drilling bit not shown. The oil-confining device of this process comprises the inflatable floatable tube **15** supporting the skirt **16** encircling the rig, the skirt being made of flexible or limp material and carrying at the bottom periphery a flexible weighted means. The tube is supported by drawstrings **23**. A suction pump **67** and an air pump **55** connect by pipe **54** to the canister bags mentioned above.

FIGURE 17: OFFSHORE DRILLING PLATFORM WITH OIL CONTAINMENT MEANS IN PLACE

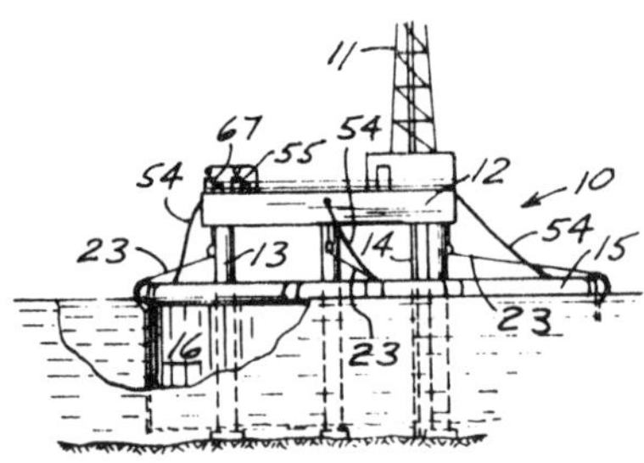

Source: L. Missud; U.S. Patent 3,599,434; August 17, 1971

Offshore Recovery System Inc. Process

A process developed by *B.H. Cunningham; U.S. Patent 3,745,773; July 17, 1973; assigned to Offshore Recovery System Inc.* is one in which a safety offshore drilling and pumping platform is provided wherein the platform is equipped with a large catchment basin which is spaced above the ocean floor and which has an opening therein so that drilling and pumping operations can be conducted within the catchment basin. In the event of an oil leak means are provided for recovering oil which rises in the catchment basin, saving the oil and preventing pollution. Figure 18 shows the essential apparatus employed. The upper view is a perspective and the lower view is a cross-section.

Referring now to the drawing by reference characters there is shown an oil well drilling platform **1** having support legs **3**, each of which has a foundation element **5** extending to the bottom of the ocean floor. Adjustable jacks **7** are employed so that the platform may be leveled on an irregular bottom.

The catchment basin itself can take the form of an inverted funnel having a conical section **9** leading to a tubular section **11**. Although the funnel has been shown as being conical it is obvious that it can be in various configurations and might have a rectangular bottom and it is not necessarily symmetrical. In other words the tubular portion **11** might be off to one side of the converging section.

The top of the tubular section **11** has an opening therein so that a drilling string or oil recovery pipe **13** can pass therethrough. Preferably a seal **15** is provided at the top of the funnel so that escaping gas can be caught. For this purpose a pipe **17** may enter the tubular portion and the structure may include one or more gas recovery vessels **19**.

Near the top of the converging portion of the catchment funnel a pipe **21** is provided which leads to a pump **23** and a separation vessel **25**. In the event of leakage, the oil will naturally collect at the top of the funnel and in the tubular portion thereof and can be pumped out through **21** into the separation vessel where two phases are formed, an upper oil phase **27** and a lower aqueous phase **29**. The oil can be recovered through pipe **31** and taken off to storage. The aqueous phase will probably retain some oil so preferably the water is released through pipe **33** into the catchment basin to prevent pollution.

It should be particularly noted that the bottom of the catchment basin **9** does not extend to the bottom of the formation so that a substantial amount of space as at **35** is left between the bottom of the ocean floor and the bottom of the catchment basin as oil is removed through the pipe **21**. The oil flows upwardly into the catchment basin by gravity alone and does not provide any positive pressure with the catchment basin which might cause oil to be forced out. Further, oil is only pumped after it reaches the top of the catchment basin which largely prevents emulsification of the oil and water.

FIGURE 18: APPARATUS FOR OFFSHORE DRILLING INSIDE CATCHMENT BASIN TO CONTAIN OIL LEAKAGE

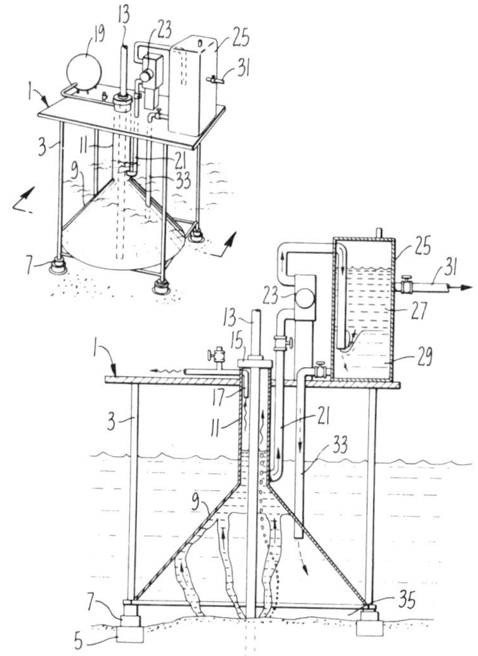

Source: B.H. Cunningham; U.S. Patent 3,745,773; July 17, 1973

The device of this process permits drilling for oil in sensitive areas where even a minor leak might cause a disaster. Since the catchment basin is always in place, even the initial leakage is caught. Since the catchment basin forms part of the drilling structure, it is always in place so that after the drilling is over and the oil is merely being pumped, it will still prevent any leakage which might occur during the life of the well.

Many variations can be made on the exact structure. The catchment basin might be rectangular, square, or in various configurations depending upon the particular application of the device. Although a single, tubular structure **11** has been shown for a single drilling rig, it is obvious that this might be enlarged so that several drills or pipes might run into the vessel. Further, a single catchment basin might have a number of upright sections as at **11** so that plurality of wells could be serviced from a single catchment basin.

Rowland Process

A design developed by *D.H. Rowland; U.S. Patent 3,576,108; April 27, 1971* is one in which elongated sections of flexible buoyant tubes are connected together in end-to-end relation for extending offshore along a coastline surrounding an oil tanker or an oil well location. The wall of the tube sections is formed in accordion-pleated fashion for readily conforming to variations in the surface of the supporting water. Anchor means maintains the assembled tubes in approximate location while other means prevents longitudinal expansion of the tubes beyond the yield point of their material. Figure 19 shows such a barrier in place around an offshore drilling rig and also shows a detail of barrier construction.

FIGURE 19: OFFSHORE WELL STRUCTURE WITH OIL CONTAINMENT BARRIER IN PLACE

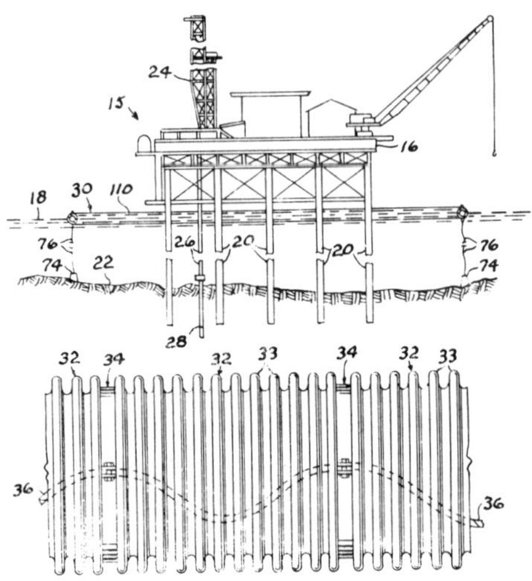

Source: D.H. Rowland; U.S. Patent 3,576,108; April 27, 1971

The reference numeral **15** indicates, as a whole, a substantially conventional offshore drilling rig comprising a horizontal platform **16** supported above the surface of the ocean or water **18** by a plurality of elongated legs or piles **20** extending downwardly through the water and into the surface of the earth **22**. The drilling rig includes a derrick **24** for handling drill pipe **26**, or the like, extending downwardly into a borehole **28**.

The numeral **30** indicates a floating dam or boom, as a whole, comprising an elongated endless fluidtight section of tubing circumferentially surrounding the drilling rig **15** in radial outwardly spaced relation. The boom **30** is preferably formed by a plurality of tubular sections **32** interconnected by clamp means **34** in end-to-end relation. The outside diameter of each section is relatively large, for example 20 inches. Longitudinally each section **32** is substantially greater than its diameter. The wall of each tubular section is relatively thin when compared to the diameter and is preferably formed in accordion roll pleated fashion, except at its extreme end portions, describing a plurality of equally spaced circumferential U-shaped folds **33** permitting a bellows-type action and increasing the flexibility of the tube in conforming to variations in the surface of the water and returning to its relaxed position.

The radial spacing between the innermost limit of the accordion-pleated wall, defining the inside diameter of each section **32**, and the outermost surface of the folds **33**, is relatively small when compared to the diameter, for example, 2 inches. In its relaxed position, the folds, or accordion-pleating assumes the position shown. The spacing between the folds **33**, in relaxed position is substantially equal to the width of each fold **33** measured longitudinally of the section.

This stretching or expanding action is limited by a flexible strand **36** extending through each tubular section and connected with the clamp means **34**. The expanding and contracting action of the pleated wall permits longitudinal bending action of the boom **30** in conforming to the surface of the water **18**.

Since the specific gravity of the rubber-like material forming the tubular section **32** is substantially equal to water, the boom **30** normally floats mostly above the water **18** and forms an efficient barrier of oil. It is desirable that the tubular sections be partially submerged so that the depending circumferential arc portion of the respective tubular section extends a sufficient distance below the surface of the water to form an effective barrier for the oil. Therefore, each tubular section **32**, after connecting one end to another section, is loaded or partially filled with ballast, such as a quantity of water. The assembled boom **30** is maintained in approximate position by anchors **74** connected in spaced-apart relation to selected ones of the clamp means **34** by flexible lines **76**.

In operation oil rising to the surface of the water **18** around the drilling rig as a result of seepage, blowout, or any other source, is trapped within the confines of the boom **30**, as indicated by the lines **110** due to the difference in the specific gravity of oil and water. The oil will rise to a level within the boom higher than the surface of the water and under such conditions the oil may then be picked up as by the use of a suction pump, or the like, not shown. In the event the boom is used as an elongated length extending in offshore relation to a shoreline, the boom simply forms a wall or dam preventing movement of oil toward the shoreline.

Siegel Process

A process developed by *G. Siegel, U.S. Patent 3,610,194; October 5, 1971* utilizes a prefabricated offshore fluid storage facility for one or more fluids that may be moved in a collapsed condition to a desired location on a body of water, and then expanded and anchored beneath the surface of the water to store one or more fluids therein, the specific gravities of which are less than that of the water in which the facility is anchored above fissures, cracks or portions of faults in a submerged land area, it will capture and retain hydrocarbon products escaping therefrom.

Texaco Process

A process developed by *P.L. Paull and F.C. Armistead; U.S. Patent 3,548,605; Dec. 22, 1970; assigned to Texaco Development Corporation* utilizes a self-contained submersible structure adapted to be lowered through a body of water to form a provisional passage for crude oil and/or gas escaping from a well or substratum fissure. The structure includes a support frame forming a fluid guiding and entrapping means, together with a collapsible conduit which communicates the source of escaping fluid with the water's surface whereby to form a confined pool.

Figure 20 shows, in the upper portion, the submersible vehicle suspended from a surface craft above a leak in the ocean floor and, in the lower view, the submerged apparatus in operating condition with the canopy extending from the underwater oil source to the water's surface.

In the upper view, the flow control apparatus **10** is shown in a partially submerged position, being supportably guided by cables **15** from a floating barge **12**, while being lowered onto a fluid escaping fissure **13** at the ocean floor. The device includes a frame formed of structural and buoyant members. A fluid gathering, funnel shaped collector **14** as well as a collapsed canopy **16** are carried on the frame. Canopy **16** is stored and retained in a collapsed condition as the frame is lowered to the ocean floor. Thereafter, and as shown in the lower view, canopy **16** is released from its storage compartment and

FIGURE 20: DIAGRAMS SHOWING PLACEMENT AND UTILIZATION OF PORTABLE OIL CONTAINMENT STRUCTURE ON SEA FLOOR

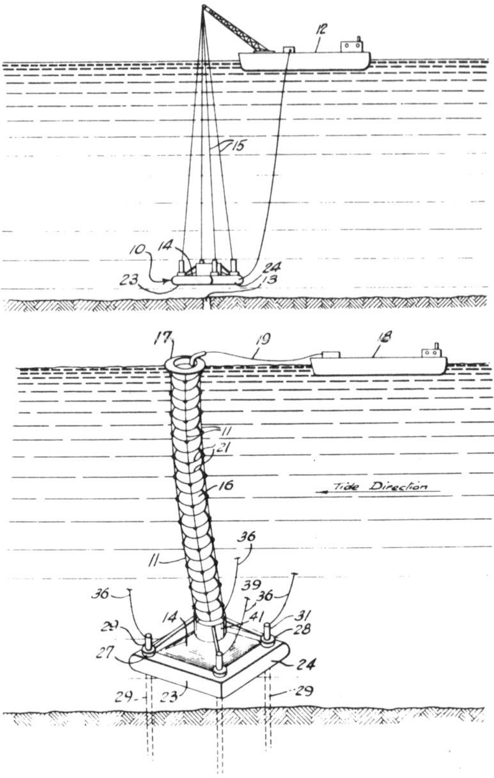

Source: P.L. Paull and F.C. Armistead; U.S. Patent 3,548,605; December 22, 1970

drawn to the ocean's surface by a float collar **17**. The canopy thus defines a closed, flexible passage communicating the fissure area with the water's surface. Normally, a pool comprising a mixture of water and crude oil is gathered at the upper end of the nonrigid passage. The pool is continuously and controllably evacuated by a floating pipe line **19** connected to a pumping system carried on adjacently anchored barge **18**.

Crude oil is then extracted from the mixture, stored in the barge, and the water returned to the ocean. To facilitate proper functioning of conduit **16**, the latter, because of its flexibility, is provided with a plurality of longitudinally spaced support rings **21** which partially rigidize and strengthen the conduit. Guide cables **11** carried on a reel type takeup **27** on the frame, interconnect the respective rigidizing rings to float collar **17** whereby the ascent and descent of the latter is remotely controlled.

The support frame includes a base formed by buoyant members such as cylindrical tubes **23** and **24** connected at their respective extremities to define a lower peripheral enclosure. Means is incorporated into the base for anchoring the unit to the substratum of a body of water. The anchoring means may take the form of any of several appropriate devices such as piling or the like.

In this arrangement, where the substratum is sufficiently fluid to permit the entrance of a pile anchor, one or more of the latter are used. To accommodate the anchors, each corner of the frame base is provided with a remotely controlled jacking mechanism **27** and **28** operably engaged with a pile anchor **29** and **31** registered vertically in the mechanism and having an open end at the frame lower side.

Pile anchor **29** is of a familiar structure and comprises in essence an elongated fluid tight cylindrical body having a lower open end and a closed weighted upper end. The upper end is communicated by an exhaust valve or fitting having an outlet adapted to receive a hose **36** or similar member extending through the water's surface. Thus, by regulating the internal atmosphere of the pile anchor from the water's surface, the latter can be drawn into a fluid-like substratum to a sufficient holding depth. For a more firm substratum, this type of anchor would of course be inapplicable in which instance driven piles could be employed.

Jacking mechanism **27** operably connected to pile anchor **29**, includes a ratchet or similar mechanical device adapted to engage longitudinally spaced grooves on the pile periphery. Thus, remote actuation of the jacking mechanism permits the pile to be adjusted up or down as needed to bring the frame into fixed, substantially tight engagement with the usually soft upper layer of the substratum.

A fluid guide, positioned within and held by the support frame, includes the funnel-like collector **14** having a relatively broad open rim disposed adjacent to the base lower end. The inwardly tapered walls of the collector define a substantially frusto-conical figure which leads to, and terminates at a relatively constricted opening **36** at the upper end.

A storage compartment **39** for collapsible canopy **16** held within the support frame, includes an upwardly extending cylindrical shell **41** having the lower edge connected to the outer surface of the collector at a peripheral joint. Annular compartment **39** defined between spaced apart walls of shell **41** and a retainer, is adapted to receive canopy **16** in a collapsed condition on the compartment's lower bulkhead. There the canopy is stored prior to being put into use for conducting a fluid flow to the ocean's surface. Canopy compartment **39** is shown positioned centrally of the support frame by a plurality of structural members and braces extending from the base inwardly to engage the outer wall of shell **41**.

Trindle Process

A process developed by *B.A. Trindle; U.S. Patent 3,592,008; July 13, 1971* utilizes a flotation confinement apparatus for disposition on a body of water which apparatus is particularly useful around offshore oil rigs and in cooperation with a shoreline to retain or to exclude oil or other undesirable matter until such matter can be removed or otherwise eliminated. The floating barrier **12** used in this process is shown in cross-section in Figure 21.

FIGURE 21: SECTION OF CONTAINMENT BARRIER SUITABLE FOR USE AROUND OFFSHORE RIG

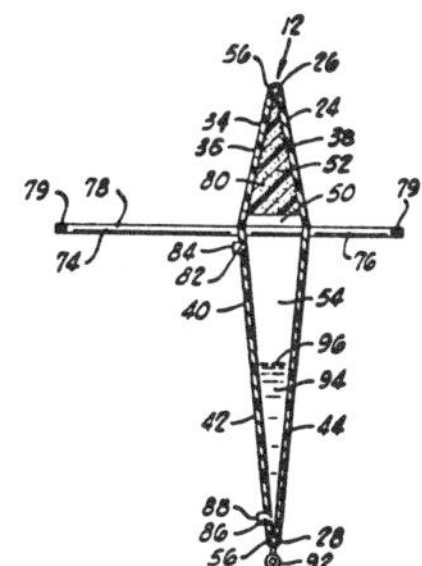

Source: B.A. Trindle; U.S. Patent 3,592,008; July 13, 1971

It basically comprises a rectangular shaped shell **24** having an elongated diamond shaped cross-section and an upper and lower edge **26** and **28**, respectively.

The shell **24**, as shown has an upper triangular shaped section **34** having sides **36** and **38** thereon, and a lower triangular shaped section **40** having sides **42** and **44** thereon. The upper and lower triangular shaped sections **34** and **40** are separated by plate **50** which, in cross-section, form a common base for the upper and lower triangular shaped sections. One end of the side **36** of the upper triangular section **34** is joined to one end of the side **38**, thereby forming the upper side **26** of the shell **24**. The ends of the sides **36** and **38**, opposite the ends which are joined, are each secured to one side of the plate **50**, thereby forming a hollow upper chamber **52** in the upper triangular section **34**. One end of the side **42** of the lower triangular section **40** is joined to one end of the side **44**, thereby forming the lower side **28** of the shell **24**. The ends of the sides **42** and **44** opposite the ends which are joined, are each secured to one side of the plate **50**, thereby forming a hollow lower chamber **54** in the lower triangular section **40**.

The end **30** of the shell is secured to one end of the sides **36** and **38** of the upper triangular section **34**, to one end of the sides **42** and **44** of the lower triangular section **40** and to one end of the plate, thereby enclosing one end of the upper and lower chambers **52** and **54**. The end **32** of the shell is secured to one end of the sides **36** and **38** of the upper triangular shaped section **34**, to one end of the sides **42** and **44** of the lower triangular shaped section **40**, and to one end of the plate, opposite the ends secured to the end **30** of the shell. thereby enclosing one end of the upper and lower chambers **52** and **54**. It is apparent from the foregoing that the plate **50** isolates the upper chamber from the lower chamber, for reasons which will be described in detail below.

Since the ends **30** and **32** of the shell, the sides **36** and **38** of the upper triangular shaped section **34**, and the sides **42** and **44** of the lower triangular shaped section **40** will be either immersed in water, or exposed to the water, the sides and ends mentioned above should be constructed of material suitable for this particular type of application. In a preferred form, the sides and ends of the shell are constructed of a laminated fiber glass material which affords desirable strength, resilience and lightweight qualities. It should be noted too, that in one form, the sides of the shell could be of a one-piece construction for purposes of economy and manufacture.

A metal frame **56**, which may be constructed of pipe, is secured to the shell and extends around the outer periphery thereof. The metal frame **56** provides the basic structural strength, both lateral and longitudinal support, to the flotation section **12**.

In a preferred form, the guide arm **74** is adjustable so that the long arm **78** and the short arm **76** may be alternately extended from the opposite sides of the flotation section **12**. However, if the guide arm is not adjustable in this manner, but rather is secured in place in the flotation section, it is necessary to provide a flotation section **12a** to be used in cooperation with the flotation section in forming the flotation confinement apparatus.

The upper hollow chamber **52**, in a preferred form, is filled with a flotation material **80**. It has been found that a polyurethane foam plastic provides an excellent flotation material to be used in this portion of the shell.

A fill port **82**, having a plug **84** removably disposed therein, is provided in the side **42** of the lower triangular section **40**, and it is disposed adjacent the connection of the side **42** with the plate **50**. A drain port **86**, having a plug **88** removably disposed therein, is provided in the lower portion of the side **42** generally adjacent the side **28** of the shell **24**. The purpose of the fill port **82** is to permit the filling of the lower chamber **54** with ballast, from the lower chamber **54**.

A plurality of eyebolts **92** are secured on the lower side **28** of the shell **24**, and these are provided to accommodate an anchor or additional weights which may be required during the operation of the flotation confinement apparatus as will be described below.

In a preferred form of using the flotation sections to form a protective periphery, the lower chamber **54** is initially filled through port **82** with a ballast **94**, such as water, to a level **96**. The level **96**, or in other words the volume of ballast **94** which is put in the lower chamber **54**, will depend on the requirements of the particular application. It is apparent, that the greater the volume of ballast **94** which is put in the lower chamber **54**, the lower the shell **24** will ride in the water, or in other words, the closer the topside **26** of the shell **24** will be to the waterline. In this manner the volume of debris combined within the protective periphery is adjustable and controllable.

A plurality of flotation sections are loaded on a barge or similar water vehicle and transported to the particular area of water where the flotation confinement apparatus is to be used. A flotation section **12** is lowered into the water in a position where a portion of the shell generally adjacent the upper side **26** thereof extends above the waterline. A flotation section **12a** is lowered into the water in the same manner, and it is positioned such that the end of one flotation section is generally adjacent the end of an adjacent flotation section **12**. The flotation sections are moved into an interlocking and adjoining relationship. In one constructed form, the flotation sections are about 20 feet in length and ten feet in overall height. These dimensions will vary with requirements.

Tuttle-Lister Process

A process developed by *R.L. Tuttle and G.T. Lister; U.S. Patent 3,630,033; Dec. 28, 1971* utilizes an apparatus for controlling oil slicks which incorporates a plurality of modular flotation tanks joined together to form a closed structure, with a large gate therein to allow the structure to be positioned around an oil slick and control curtains extending vertically downward from the structure beneath the surface of the water and around the oil slick. Figure 22 shows the overall apparatus in place as well as a detail of one pontoon and of the skirt attached to the pontoon.

Referring now in detail to Figure 22, the apparatus is comprised of a plurality of modular pontoons **10**. Each pontoon has secured thereto fastening flanges **12**. The fastening flanges each have apertures **14** therein. The pontoons can then be fastened one to another by aligning the flanges and placing a bolt through apertures **14**. The pontoons are then joined by the flanges as described to form assembly **18**, which encloses an area **20**.

Due to the fact that the apparatus is designed for use on the sea, it will be subject to constant ocean swells. To prevent undue strain on the assembly and eventual material failure and breaking apart of the assembly, ball and socket connectors can be secured to the pontoons **10**.

In order to allow the assembly, which is closed on all sides to prevent the escape of any oil, to be positioned about an oil slick, it is necessary to replace the fastening flanges on a pair of adjoining pontoons with hingeing means (not shown) **36**, at providing the control apparatus with a large gate **38**, as shown. With gate **38** in the opened position, the assembly can be positioned about a slick or desired area. Upon encircling the desired area, the gate is closed. The gate also provides an access by boat to an oil well within the enclosed area.

To contain the oil slick enclosed by the assembly, control curtains **40** are fastened to the pontoons. The control curtains extend vertically downward from the pontoons beneath the surface of the water, with weights **42** being attached to the lower portion of the curtains at **44** to maintain the control curtains in the vertical position. Each pontoon has a mounting plate **46** secured thereto, each mounting plate having a plurality of apertures **48** therein. The control curtains are then fastened to mounting plates **46** by use of bolt means **47** which extend through apertures **48** in the mounting means **46** and through apertures **52** in the upper portion of the control curtains. The control curtains are constructed out of a durable and flexible material such as rubber sheeting. Due to the fact that such sheeting typically comes in certain lengths, the control curtains must be sectioned to fit a particular assembly. To prevent any loss of oil between the sections, the sections are fastened to mounting plates **46** in overlapping relationship to one another and rigid strips **54** are

FIGURE 22: SLICK CONTAINMENT APPARATUS IN POSITION AROUND RIG AND DETAILS OF FLOAT AND OF CURTAIN BARRIER FOR OIL CONTAINMENT

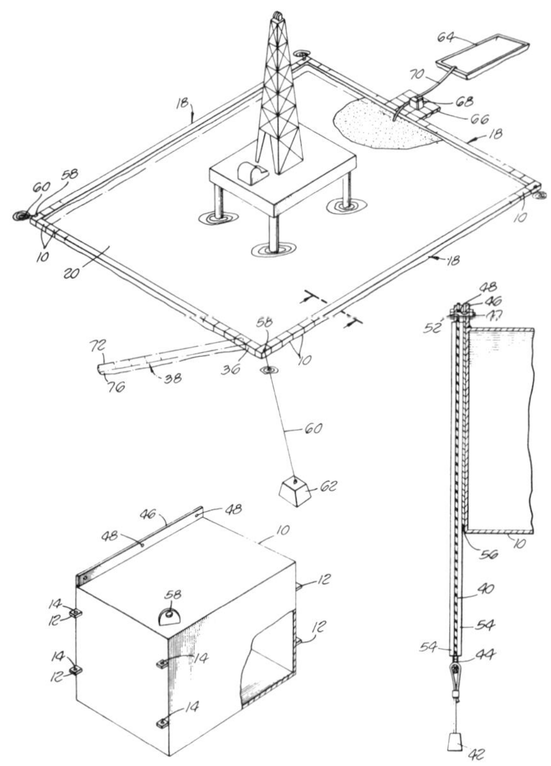

Source: R.L. Tuttle and G.T. Lister; U.S. Patent 3,630,033; December 28, 1971

positioned on either side of the overlapping sheets, creating self-sealing gaskets which prevent oil from escaping the enclosed area. Shields **56** are secured to the pontoons to prevent the control curtains from being pinched between adjacent pontoons thereby weakening the curtains.

Line attachments **58** are attached to the pontoons. A line **60** is secured to attachments **58** and can be used to move the assembly in the water. In order for the assembly to hold a given position about a designated area, anchor means **62** are attached to lines **60**, thereby securing the assembly in fixed relation to the ocean floor. In operation, the enclosed area may fill with oil and it may become necessary to pump the oil out of the area encircled by the assembly into a barge or second storage area, designated **64**. To facilitate this operation, a pumping platform **66** is provided which consists of several pontoons fastened together and located adjacent pontoons **10** in the assembly. The size of the platform is determined by the number of pontoons used.

A pumping means **68** is positioned on platform **64**, a hose **70** is then run from the enclosed area **20** to the pumping means and from the pumping means to the barge or second storage area **64**. The oil is then pumped from the assembly into the barge or second storage area.

It should be noted that the assembly can itself be used as a second storage area. When moving the assembly over the water towards the oil slick or area to be encircled, gate **38** is in the opened position and the entire assembly is towed by lead pontoons **72**. The lead pontoons **72** would be positioned at the locking point **76** of gates **38**.

Van't Hof Process

A process developed by *G. Van't Hof; U.S. Patent 3,611,728; October 12, 1971* utilizes a flexible or elastic structure for containing and storing a liquid such as oil which floats on the surface of another liquid such as water. It comprises a plurality of buoyant vertical members which are joined together in substantially liquid-tight side-by-side relationship to form a continuous wall having a first end and a second end. Means are provided for joining a portion of the wall adjacent the first end to a portion of the wall adjacent the second end to form thereby an open-ended enclosure which extends above and beneath the surface of the water.

The enclosure that is formed is substantially cylindrical in shape and has diametral dimensions adapted to meet the particular conditions of a real and volumetric requirements for a particular site. The height of the vertical members is controlled by similar considerations. For example, to provide a confining area around a small drilling platform, vertical members 12 feet in height can be joined together to form a continuous wall which, when made into an enclosure, has a diameter of about 100 feet and surrounds the drilling platform.

The enclosure can be positioned relative to the water surface so that approximately 3 to 4 feet of the vertical elements extend above the water surface and the remaining 8 to 9 feet extend beneath the water surface. The portion extending beneath the water surface and a small part of the portion extending above the water comprise the confining or storage portion, whereas the balance of the upper portion has a height sufficient to prevent escape of the contained liquid due to current or wave action.

Preferably, the vertical members of a buoyant material such as wood or plastic are joined together, as by cables, at a shore facility. The continuous wall that is formed is then towed or carried in a folded state to the particular site where the enclosure is to be made. The enclosure is formed at the site by overlapping several members at opposite ends of the wall and bolting them together. The cables are prestressed by use of hydraulic jacks or other methods to form an encompassing cylindrically shaped enclosure. The height of the enclosure above the water surface may be controlled by attaching buoyancy tanks or weights to the enclosure as required.

The structure is flexible and may be readily transported from one site to another. In place, even though it is anchored down, limited movement of the entire enclosure occurs responsive to wave action so that the effect of the forces produced by wave action upon the structure is minimized. Additionally, the use of separate elements to form the ring provides a spring action, imparting thereby a flexibility or elasticity to the enclosure which further improves its capability to withstand wave action.

Figure 23 is a schematic representation of an offshore oil drilling site having a floating platform **36** supporting a drilling rig **38**. Oil well casing **40** extends from the drilling platform to beneath the surface of the ocean floor **42**. A buoyant enclosure **44** formed in accordance with the process encompasses the floating platform.

Oil leaking from a fissure at the well floats upwardly through the water as depicted by bubbles **46**. Despite the presence of ocean currents, the enclosure can be positioned over the source of leakage so that it confines and stores the oil reaching the surface of the water, as shown at **48** within the enclosure. The stored oil can then be removed as required by pumping into barges or tankers. It will be appreciated that, in the event the oil leakage is occurring at some point remote from the drilling platform, the enclosure can be positioned at the location on the water where the oil surfaces.

FIGURE 23: ONE TYPE OF FLEXIBLE CONTINUOUS OIL CONTAINMENT WALL IN PLACE AROUND OFFSHORE PLATFORM

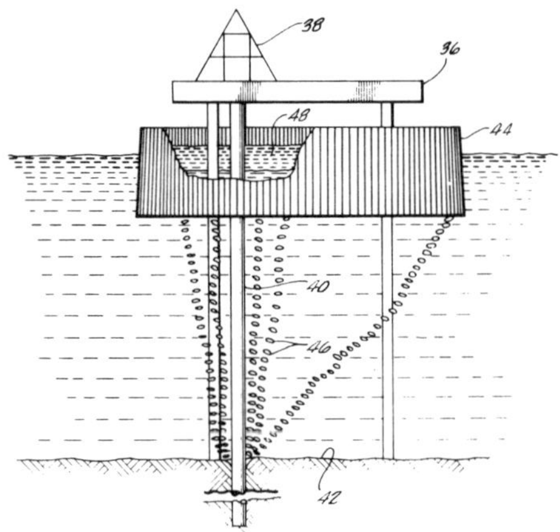

Source: G. Van't Hof; U.S. Patent 3,611,728; October 12, 1971

Verdin Process

A process developed by *S.M. Verdin; U.S. Patent 3,554,290; January 12, 1971* involves extinguishing gas and oil well fires, particularly in multiple-well, offshore installations, and preventing loss of oil to the surrounding area after fire extinction. The apparatus comprises a hood adapted to be placed over the well site, the hood having a curved interior upper wall which deflects the gushing oil into a catch basin from which it can be pumped away. Figure 24 shows the transportable well enclosure structure as well as its operation at the offshore well site.

In the illustrated embodiment the enclosure **10** is defined by two spaced-apart vertical side walls **12**, a vertical end wall **14** and an inclined top wall **16**. The size of the enclosure **10** is such that it will cover the largest well site that it is designed for. For example, a typical offshore platform might be 60 feet square and 60 feet above water level and support wellheads. The walls **12**, **14** and **16** may be of sufficient thickness to render the enclosure self-supporting, or they may be of thinner construction reinforced with internal or external bracing.

An important feature of the enclosure is that it include an interior oil-deflecting surface capable of deflecting upwardly gushing oil in a lateral direction toward a collecting basin **18**. In the illustrated embodiment the inner surface of the top wall **16** provides the deflecting surface. The top wall **16** may be visualized as having the shape of a 45° pipe bend which has been cut longitudinally along its axis. Accordingly, the lower surface of the top wall **16** is concave from its lower end to its upper end and from side-to-side. While it is necessary that the deflecting surface be generally curved from its lower end to its upper end, the illustrated shape is not critical. If desired, the deflecting surface may be defined by a separate structure suitably supported from the walls which define the enclosure.

The oil catch basin extends from one side wall **12** to the other and is disposed adjacent the rear wall **14** in a position to catch the oil which has been deflected upwardly and rearwardly by the wall **16**, as shown by the arrows **20** in Figure 24. Also schematically shown is an offshore platform **22** supported above the level of the sea **24** by legs **26** and

FIGURE 24: DESIGN OF HOOD FOR OIL CONTAINMENT AND FIRE EXTINGUISHMENT AT OFFSHORE INSTALLATION

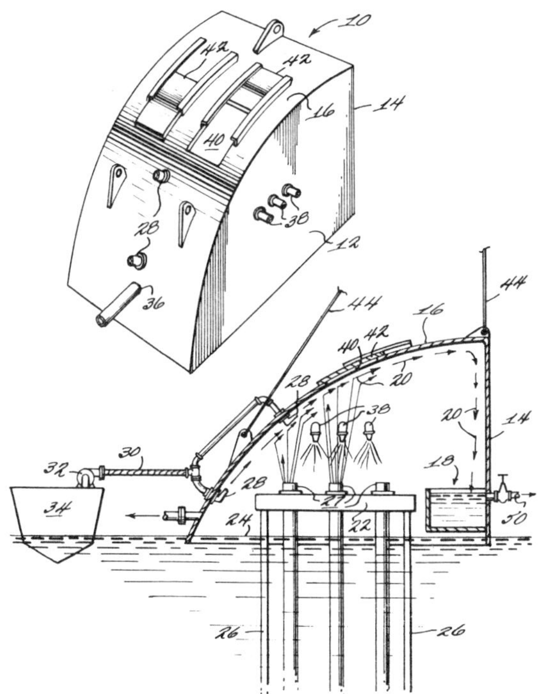

Source: S.M. Verdin; U.S. Patent 3,554,290; January 12, 1971

having a plurality of wellheads **27** projecting therefrom. When a wellhead has been ruptured or broken oil flows from the head under such pressure that it spurts or gushes to a considerable height. When the enclosure has been placed over the site, the oil will strike the lower deflecting surface of the wall **16**.

In order to aid in deflecting the oil along the deflecting surface streams of water may be injected along the surface by means of suitable jets **28**. Valved water supply pipes **30** are connectable to a pressurized source of water, such as pumps **32** carried by a service vessel **34** and adapted to draw water from the sea.

It is usually necessary to provide one or more vent pipes, such as shown at **36**, for releasing gas from the enclosure, since many offshore wells are dual gas-oil wells. Vent pipes are particularly necessary when the enclosure is lowered into contact with the sea thereby preventing escape of gas around the lower edges of the enclosure. In use the vent pipe **36** is connected to a long pipe **37** to convey the gas to a remote location where it can be burned. The enclosure may also include pipes **38** for injecting fire extinguishing liquids such as water or drilling mud on the platform. These pipes which may be mounted on the sidewalls **12** can be supplied with pressurized liquid from pumps on the service vessel.

When all the necessary equipment for capping the wellheads **27** has been assembled, the capping operation will be carried out through the walls of the enclosure, for example through hatches **40** which are fitted with sliding covers **42**. As previously explained the capping operation involves the lowering of a blowout preventer assembly over the damaged open end of a wellhead and operating the assembly to cut off the flow of oil. The assembly and other equipment are brought to the site by the service vessel or other vessel equipped with one or more cranes, and after one of the covers is slid back, the capping operations are effected through the respective hatch. The hatches should be relatively small so as to permit as little oil as possible to escape during the capping operation. The power mechanism for sliding the covers may be carried by the enclosure or by the service vessel.

Warren Petroleum Corporation Process

A process developed by *R.V. Phelps; U.S. Patent 3,657,895; April 25, 1972; assigned to Warren Petroleum Corporation* utilizes a platform for offshore oil wells having a curbing around the periphery of the deck of the platform. The deck slopes downwardly from the curbing into a central opening to drain all oil spilled on the deck into the opening. A cylindrical sleeve open at its lower end to admission of water extends downwardly, preferably to the marine floor, from the opening. The diameter of the sleeve is at least as large, and preferably in the range of 20 to 50 feet, as the opening whereby all oil or other liquids draining into the opening is confined within the sleeve. The platform can be entirely of steel, steel framework mounted on a concrete substructure, or of concrete modules assembled at the well site.

Zielinski Process

A process developed by *R.O. Zielinski; U.S. Patent 3,667,605; June 6, 1972* for submerged oil leak control utilizes a cup shaped casing means inverted upon or adjacent to the ocean floor and tubular means extending upwardly from the casing means and communicating with the interior thereof through an opening in the top wall thereof. The tubular means is braced by a plurality of cables connected thereto and to anchor means located on the ocean floor.

Figure 25 shows the apparatus in some detail. It includes an inverted substantially cup-shaped container or casing **10** which is preferably fabricated from dense, durable material such as steel or reinforced concrete. The sidewall **11** of the casing is preferably, but not necessarily, cylindrical and the top wall **12** is preferably, but not necessarily dome shaped, or conical in order to induce the flow of oil toward the central opening **13** in the top wall. A flanged collar **14** is rigidly secured, as by welding if the casing is fabricated by steel, to the top wall around the opening. A tubular member **16** is rigidly secured, as by welding, within the collar. Additional tubular members or sections **17** which are preferably, but not necessarily, of the same diameter as the tubular members **16** are arranged in axial alignment with and extend upwardly from the tubular member **16**.

The lowest tubular section **17** is slidably received in to the tubular member **16** by a sleeve **18** which is secured to the upper end of the tubular section. Similarly, each tubular section above the tubular member has a sleeve secured to its upper end for slidably receiving the lower end of the tubular section thereabove. Means such as the stop pin **21** on the sleeve, which extends through a vertically slotted member **22** on the tubular section **17**, may be furnished to limit vertical sliding movement of the tubular section relative to the sleeve, and thereby prevent an inadvertent disconnection between the sleeve and section.

The uppermost tubular section **17A** is preferably closed by a cap **23** and the upper end of the section **17A** is connected to a discharge pipe **24** which may be provided with a conventional valve **26**.

A plurality of anchors **30**, which are preferably fabricated from concrete, are disposed upon the ocean floor **31** around the casing and are spaced radially therefrom. Preferably, the anchors are located at four uniform intervals around the casing, and each anchor has an elongated flexible cable **32** connected thereto by means of a hook or loop **33**. Additional cables may extend between the anchors and other sections **17**, if desired.

FIGURE 25: SUBMERGED OIL LEAK CONTAINER IN PLACE ON OCEAN FLOOR AND DETAILS THEREOF

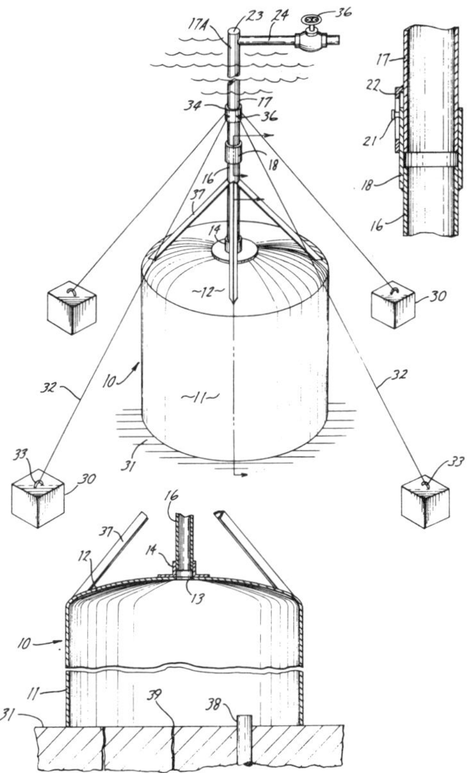

Source: R.O. Zielinski; U.S. Patent 3,667,605; June 6, 1972

An anchor collar **34** is firmly secured to either the tubular member **16** or one of the tubular sections **17** and such anchor collar has a plurality of fastening devices **36** to which the upper ends of the cables **32** are secured. If additional bracing is required, and it may be, two or more rigid braces **37** may be secured to and extend between the top wall and the side of the tubular member **16**. These braces may be fixed in position by welding or the like, and they are preferably heavy-duty angle irons.

When a leak is detected, the casing along with the attached tubular member **16**, may be lowered toward the ocean floor by a suitable hoist of any conventional type. The casing is fabricated from sufficiently dense material that it will readily sink to the ocean floor and any water trapped by such downward movement can easily and freely flow upwardly through the tubular member **16**. If desired, a tubular section **17** can be attached to the tubular member **16** as the upper end of the member **16** approaches the surface of the water. In a similar manner, additional tubular sections **17** can be added, one at a time, as the casing is lowered. When the lower open edge of the casing comes to rest on the ocean floor around and over either the broken casing **38** or a crack **39**, then appropriate anchoring of the tubular sections **17** can be effected.

Either before or after the casing is lowered, the anchors are lowered to the ocean floor by the hoist and positioned around the leak preferably uniformly and equidistantly. The cables **32** will preferably be connected to the anchor collar **34** of the proper tubular section **17** before that particular section is lowered into the ocean. Accordingly, it is only necessary thereafter for a diver to locate the loose ends of the cables and secure them to the loops **33** on the anchors.

As soon as the casing covers the broken casing **38** or crack **39** through which oil is leaking, such oil will be promptly guided up through the casing by its relative buoyancy and then along the domed top wall to the opening. Thereafter, the oil will move upwardly through the opening, the tubular member **16** and tubular sections **17**. When such oil reaches the surface of the water within the uppermost tubular section **17A**, the oil can then be drawn off through the discharge pipe **24** by an appropriate suction or sump pump connected thereto. Such oil can then be delivered to a waiting tanker or barge or, if convenient, piped directly to the shore. Thus, pollution of the water by the oil from the leak is prevented.

WASTEWATER PURIFICATION

The original carboniferous deposits accumulating within, or at the edge of, ancient seas constituted the beginning of petroleum formations. It therefore follows that some part of those ancient seas will be produced with the oil and gas today. This water, which accumulates during the production processing of oil and gas, is frequently great in volume sometimes exceeding 20,000 barrels per day for one field. This water may constitute from less than 10% to greater than 50% of the total fluids produced from a single well.

Depending on several variables, an offshore operator may decide to: (1) Pump the entire oil/water mixture to shore for treatment. (2) Pump the oil/water mixture to shore for treatment following a free-water knockout process. (3) Treat the oil/water mixture on the platform and sell the oil to a pipeline company at the platform. This last choice requires that the oil contain a maximum of 2% water and generally no more than 1% water.

The variables upon which the above decisions are based include:

(1) Distance to shore, 20 to 30 miles is generally the maximum.
(2) The presence and types of solids and emulsions concerned.
(3) The relative specific gravity of the oil and water.
(4) The chemical characteristics of the oil — the paraffin content plays an important role.
(5) Economics of treating on shore versus on the platform.
(6) The percent of water produced with the oil.

Treatment equipment must take into account the fact that the above variables change with time. This requires that adequate safety factors be included in design calculations.

There are two basic types of equipment associated with oil/water separation on offshore production facilities: operation equipment and conditioning equipment.

Operation Equipment: That equipment used in the process of separating well gas from fluids and processing fluids to remove water from the oil. Depending upon the throughput volume in operation equipment, and also upon the presence and types of solids and emulsions concerned, the character of the various emulsifying agents and the relative specific gravity of oil and water, the oil contamination in the separated water will frequently be from 200 to 3,000 ppm or more.

As the water content in the product oil decreases, the oil content in the effluent water increases. Therefore, the 2% maximum (and 1% desired) water content required by the pipeline company has a direct bearing on the amount of oil that must be removed from the discharge, or bleed, water.

Process units that may be classified as operation equipment include the following: emulsion treaters, heater treaters, free-water knockouts and phase separators (gas, water and oil). Prices for this equipment range from $3,000 for a 12,600 gallon free-water knockout separator (4 hour detention time) to $52,000 for a heater treater handling 10,000 barrels per day.

The free-water knockout and phase separators are essentially sedimentation tanks wherein the gas, oil and water separate by gravity under pressure. Valves, located at the phase interfaces, draw off the particular components. The emulsion and heater treaters make use of emulsion breakers (either chemical or heat) to more effectively remove the water phase.

Conditioning Equipment: That equipment used downstream from the operational equipment to treat bleed water prior to disposal overboard. This equipment removes oil from the water. Depending upon the same variables which effect the operation equipment, the effluent from the conditioning equipment will contain from 20 to 200 ppm oil. Close observation is required in order to maintain an effluent in the 20 to 50 ppm oil range. This is not always achieved.

Units which may be classified as conditioning equipment include: gravity separators, oil skimmers, clarifiers, gas flotation cells and coalescers. This conditioning equipment can be installed at a cost of $2 to $3 per barrel or $20,000 to $30,000 for a system designed to process 10,000 barrels per day. This cost is based on the assumption that the equipment would be customized and designed to achieve an effluent of 50 ppm oil. This system would occupy about 300 to 400 square feet and would require three to four months for delivery. The price would double if the equipment were installed after construction of the platform was completed.

Chemical treatment may be used to provide more consistent results in the 50 ppm oil range. This, however, generates a large amount of sludge which is a disposal problem in itself. Operating data indicate 30 to 500 pounds per day ferric chloride would be needed to obtain 50 ppm oil. This would generate four cubic feet of sludge for each pound of chemical used.

Gravity separators, oil skimmers and clarifiers rely on the gravity separation of oil from water. Oil is normally skimmed off the top of these units and water is drawn from the bottom by way of an outside siphon called a gun barrel. These units are normally used upstream from other oil removal equipment to eliminate the bulk of free oil and sedimentation.

Gas flotation units saturate clean water with gas in a pressure tower. This water is then mixed with the water/oil mixture in a flotation chamber in which the gas bubbles expand and carry the oil particles to the surface where they are skimmed off. Clean water is then drawn off and discharged with a portion being recirculated to the gasification pressure tower. These units are capable of removing 90% of the oil from influents ranging as high as one thousand parts per million oil.

In coalescer units the water/oil mixture enters a settling section where heavier solids fall out and free oil rises to the surface. The flow then travels through a graded filter bed or excelsior section where oil wetted particles are filtered out and finely dispersed oil particles are coalesced to particle size sufficient to rise out of the water. After filtering, the water passes to a final settling and surge section.

Proprietary equipment is being developed by several companies which, it is hoped, will reduce the amount of oil in the bleed water, and at the same time reduce the equipment space requirements. To date, this equipment has not been in service offshore but based on existing data, it is felt that effluent concentrations of 10 to 25 ppm oil are achievable. The space requirements for these units range from 50 to 80 square feet, with cost estimates from $20,000 to $30,000 installed offshore.

The offshore oil industry is expanding at a rapid pace and as the industry grows so does its associated pollution problems. Present conditioning equipment is capable of producing effluent bleed water with oil concentrations in the 10 to 200 ppm range. If operated properly, this equipment can discharge effluents which do not produce oil films, sheens or discolorations in the receiving water. The equipment costs about $2 to $3 per barrel per day to install and requires 300 to 400 square feet of valuable platform space.

The more efficient equipment which is currently under development will produce bleed waters with oil concentrations in a consistent 10 to 25 ppm range. These units are needed so that the offshore industry can continue to grow without placing additional loads on an already overburdened environment.

Carmichael-Franklin Process

A process developed by *C.J. Carmichael and A.G. Franklin; U.S. Patent 3,781,201; Dec. 25, 1973* involves separating oil from a mixture of oil and waste water on offshore rigs where the water containing oil cannot merely be allowed to wash overboard to pollute the environment and endanger marine life. The mixture of oil and water is first passed through a primary skimmer tank where oil that quickly separates from water is skimmed off into a blow case and water from the bottom of the skimmer tank is fed to a mixing tank where gypsum and lime are added and the gypsum-oil and excess gypsum and other pollutants are mechanically separated from the water. Figure 26 is a schematic diagram showing one mode of conducting the process.

FIGURE 26: METHOD OF OIL/WATER SEPARATION AT AN OFFSHORE LOCATION

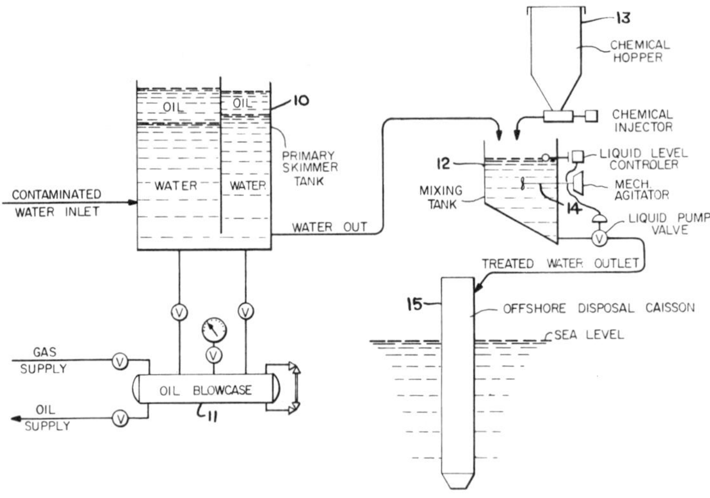

Source: C.J. Carmichael and A.G. Franklin; U.S. Patent 3,781,201; December 25, 1973

The produced water (oil contaminated) is taken into a primary skimmer tank **10** where any free oil that quickly separates from the water is skimmed off into a blow case **11** and eventually returned to the oil facilities on the offshore ring platform. The contaminated water then flows from tank **10** to a mixing tank **12** where the gypsum and lime are introduced from a chemical hopper **13** and mixed with the water by a mechanical mixer **14**. The thus treated water is then released to a disposal caisson **15**. Figure 27 shows an alternative mode of conducting the process.

FIGURE 27: ALTERNATIVE METHOD FOR OIL/WATER SEPARATION AT AN OFFSHORE LOCATION

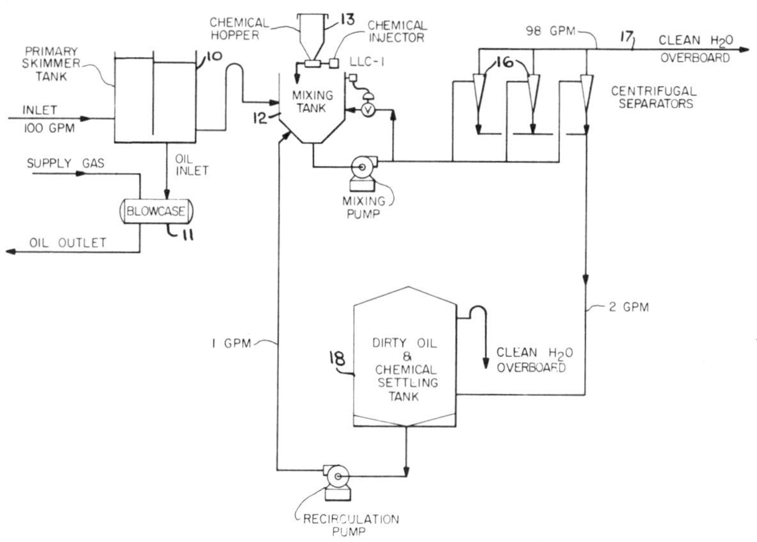

Source: C.J. Carmichael and A.G. Franklin; U.S. Patent 3,781,201; December 25, 1973

The process has the primary skimmer tank **10** and blow case **11**. The blow case **11** is a pressurizable container which collects oil from the skimmer tank **10** and which has a sight glass to tell when the container is full, a gas line to pressurize the container and an oil supply line to remove oil from the blow case as it becomes full. Each of the gas and oil supply lines have valves for control. The water (oil contaminated) is then led to a mixing tank **12** where the gypsum and lime is introduced from a chemical hopper **13** and from there when the water leaves the mixing tank **12** it is pumped through cyclone type separators **16** where all solids (gypsum-oil, excess gypsum) and any other solids are removed. The clean water is then flowed overboard as at **17**. The solids and a small percent of water (about 2 to 3% of the total inlet stream to the cyclone separators **16**) are flowed to a retaining tank **18**.

Since the fluid in this tank is about 90% water and, if given sufficient time, will drop the solids, it will be possible to further remove some water by overflow spill from the tank **18**. When the solids have reached a level in the retaining tank where they no longer have enough retention time to separate from the water and begin to overflow with the water, then the contents of the retaining tank **18** must be drained from the tank and hauled to shore for disposal. If desired, a portion of the liquid-solids in the retaining tank **18** could be recycled back to the chemical mixing tank **15** to use a portion of the unused gypsum. The goal of this process is to insure that water containing no more than 50 ppm of oil is discharged overboard from the rig.

Shell Oil Company Process

A process developed by *H.C.H. Darley and M.C. Place Jr.; U.S. Patent 3,764,008; October 9, 1973; assigned to Shell Oil Company* involves treating a mixture of water, unadsorbed gas and/or liquid hydrocarbons, and oily sand from an operating well. After removing the unadsorbed hydrocarbons, the sand is deoiled by repeatedly entraining it in streams of relatively high speed jets of water, centrifugally separating an overflow of oily water and an underflow of sandy water in which the sand is substantially oil-free, and removing the oil from the overflow to leave a water which is substantially oil-free. Figure 28 shows the essential features of this process.

FIGURE 28: APPARATUS FOR TREATING OIL-WATER-SAND MIXTURES AT WELL HEADS

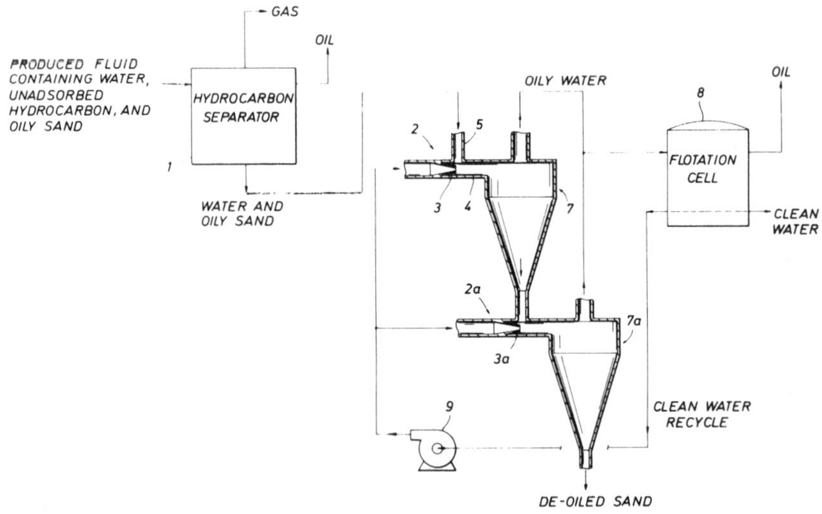

Source: H.C.H. Darley and M.C. Place, Jr.; U.S. Patent 3,764,008; October 9, 1973

In using an apparatus of the type shown in the drawing, a produced fluid containing water, unadsorbed hydrocarbons and oily sand is treated in hydrocarbon separator **1** to remove the gaseous and/or oil phase hydrocarbons that are not adsorbed on the oily sand. Such a hydrocarbon removal can utilize currently available equipment and techniques, such as gravity segregation, hydrocarbon volatilization, emulsion breaking, and the like. The removal of the unadsorbed hydrocarbons leaves a water suspension of oily sand grains.

Where desirable, the sand deoiling steps can be enhanced by contacting the water suspension of oily sand grains with an effective amount of a surfactant for increasing the surface tension of an oil-phase liquid. A particularly suitable surfactant material of this type comprises a polar-group-containing aliphatic solvent solution of a relatively long chain aliphatic alcohol.

The sand in the suspension of water and oily sand is partially deoiled by entraining the suspension in the stream of a relatively high speed jet of water in the Venturi arrangement **2**.

Such a Venturi arrangement can comprise an eductor, or jet pump, type of apparatus. In the illustrated jet-cyclone type of device, water is forced through nozzle **3** to form a relatively high speed jet of water within a relatively large diameter conduit **4**. The slurry of oily sand and water inflows through conduit **5** into conduit **4** where it is entrained in the jet stream. The resultant sand-laden stream of water is separated in the cyclone type of centrifugal device **7**. The present centrifugal separating means can advantageously be a conventional cyclone type of device in which the centrifugal force is generated by the rapid swirling motion of the entering stream of water. The centrifugation removes an underflow stream of sandy water containing at least partially deoiled sand and an overflow stream of oily water-containing oil released from the sand.

This process is, at least in part, premised on a discovery that entraining a water suspension of oily sand in the stream of the relatively high speed jet of water and substantially immediately centrifugally removing a water suspension of at least partially deoiled sand while leaving an oily water is a uniquely efficient way of removing and isolating the oil that was adsorbed on the sand. As a slurry of sand grains is entrained in a stream of water the grain surfaces are subjected to a strong shearing action during the time the water in the stream moves along the grain surfaces until the grains have been accelerated (by the friction of the shearing action) to substantially the speed of the water stream. By immediately centrifuging the sand laden stream, the oily water that is left by the centrifugal removal of the heavier sand grains is substantially immediately removed from the area of agitation so that the oil tends to remain in relatively large droplets rather than being sheared down to a colloidal size.

The sandy underflow containing partially deoiled sand is treated in the same way in at least one additional Venturi jet device **2a** and centrifugal separating device **7a** to subsequently form a sandy water underflow that contains a substantially oil-free sand. As will be apparent to those skilled in the art, the Venturi jet arrangements **2** and **2a** and centrifugal separating devices **7** and **7a** can be separated by conduits, chambers, or the like, as long as substantially all of the partially deoiled sand in the underflow from the centrifugal device **7** is maintained in suspension or is re-entrained to form a sand slurry that is flowed into contact with the relatively high speed jet of water in at least one additional jet device **2a**.

For example, a bank of parallel jet-cyclones (e.g., combination of the Venturi jet arrangements **2** and the centrifugal devices **7**) can discharge their underflows containing partially deoiled sand into one or more chambers from which the sand is re-entrained in water if necessary and is then transported into the jet streams of a second bank of jet cyclones.

The oily water overflow from each such sand grain deoiling stage is flowed into flotation cell **8** in which the oil is removed while leaving a clean water that is substantially oil-free. At least a portion of that clean water is preferably supplied to pump **9** which drives it through the jet nozzles **3** and **3a** of the Venturi devices to provide relatively high speed jets of water.

In general, the flotation cell **8** can be replaced by one or more of a variety of oil-water-separators for removing the oil from an oily water. Such devices can include one or more flotation coalescing, oil-sorbing, gravity segregating, filtering, or the like, elements in various combinations, as long as such devices are adapted to remove substantially all of the oil from an oily water.

Signal Companies Process

A process developed by *J. Duffy; U.S. Patent 3,576,738; April 27, 1971; assigned to The Signal Companies* involves removing residual oil and dissolved sulfides from oil production wastewaters prior to disposal of the wastewater into the ocean, by the injection of controlled amounts of air and soluble nickel catalyst into the wastewater. The mixture is pumped under pressure into a flotation cell or tank where the pressure is released and the air effects oil separation and oxidation of the dissolved hydrogen sulfide.

Figure 29 is a schematic flow diagram of the process. The oil production from water flooding operation is produced from a well (not shown) and pumped through line **10** to settling tank **12**, commonly called a free water knock-out tank. In the settling or free water knock-out tank **12**, the gas products are released and sent to gas processing operations and the oil, with approximately 20 to 30% water therein, is transmitted through line **14** to a heated wash tank **16**. In tank **16** the wastewater is heated and the oil-water mixture separated by settling. The water content of the water is reduced to about 1 to 3% water. Wash tanks such as tank **16**, or heater treaters as gas fired tanks are sometimes called, contain means for heating the wastewater to about 100° to 210°F depending upon the type of crude oil being separated. For example, an 18 gravity crude oil is heated to about 140°F — generally, heavier crudes are heated to higher temperatures.

The oil separated in wash tank **16**, is pumped to shipping tanks or a pipeline for transporting to storage or refinery processing. The wastewater removed in the settling tank **12** is combined with the wastewater from wash tank **16** in line **18** and additional oil removed therefrom in skim tank **20**.

The wash water from the skim tank still has approximately 50 to 100 ppm residual oil which must be removed prior to pumping into the ocean. This water is passed through line **22** to a container vessel or surge tank **24** from which it is pumped by pump **26** into retention tank **28** and flotation cell **30**. Air is injected into the suction side of the pump so that the pressurized wash water can be further cleaned up by the flotation technique. A water-soluble nickel oxidation catalyst is also injected into the wastewater on the suction side of the pump at approximately the point of aeration.

FIGURE 29: PROCESS FOR PURIFICATION OF OIL PRODUCTION WASTEWATER

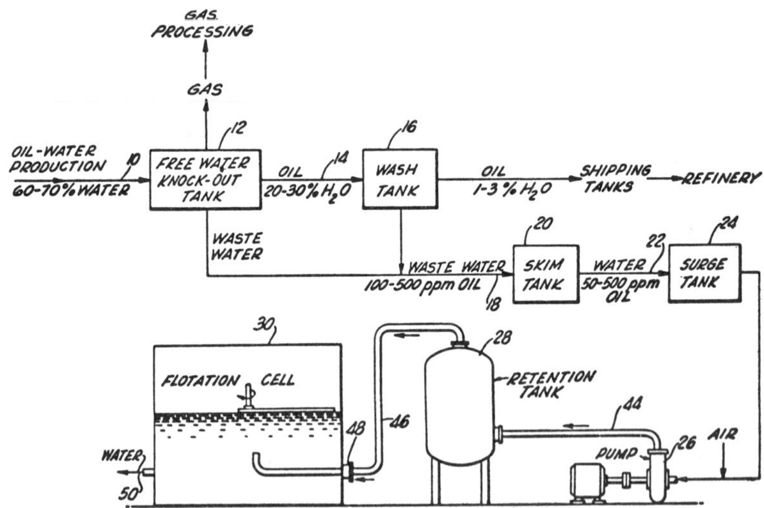

Source: J. Duffy; U.S. Patent 3,576,738; April 27, 1971

PREVENTION AND CONTROL OF SPILLS FROM OIL TRANSPORT

The following are the sources of oil which may be discharged into the sea from tanker operations according to EPA Report PB213,880:

Bilge Oil: Is that oil which collects in the ship's bottom through seepage or leakage.

Slop Oil: Is that oil/water emulsion which is normally collected in an aft center tank as the residue of tank cleaning operations.

Overflows: An overflow of oil usually occurs during the loading or discharge of cargo and the transfer of cargo from one tank to another. This is usually caused by carelessness or inattention.

Ballast Water: This is normally sea water taken into empty cargo tanks to give the vessel stability. The empty cargo tanks usually have a residue of oil in them.

Collision or Groundings: Whenever a loaded tanker vessel is involved in a collision, the result is usually a major discharge of oil into the surrounding waters. Groundings are the most common types of accidents but they do not usually result in major oil spills.

Because of world concern regarding oil pollution of the sea, governments of the world under United Nations sponsorship are meeting to discuss ways of improving tanker operations to reduce or eliminate oil pollution. Four of these proposed improvements are:

Limitation on tank size – this involves limiting tank sizes on all tankers to 30,000 cubic meters. It is in the process of adoption.

Compulsory installation of radar and bridge-to-bridge telephone – both are designed to reduce the occurrence of collisions and to improve navigation. Radar installation is now under discussion by various U.S. agencies. Congress has for approval regulations requiring installation of bridge-to-bridge telephones on all ships operating in navigable waters of the United States.

Establishment of compulsory sealanes – discussions are now being held at the international level on adoption of sealanes. In some restricted waterways (Dover Straits) sealanes have been established.

No deballasting at sea – proposed by the United States at various international conferences. Earliest adoption cannot be before 1975 and will likely be later. This would be the most preferred solution since deballasting, even with utilizing Load on Top, is considered the greatest single source of pollution.

About 98% of all the oil spilled by vessels is from incidents over 1,000 barrels. Most large tanker spills occur within 50 miles of land. Most result from groundings, rammings (the vessel hits a fixed structure), or collisions. Groundings and rammings occur near shore, and collision frequency depends on traffic density, which is highest near shore.

OIL TRANSFER OPERATIONS

The number of spills and quantities lost vary with the type of transport system. About 217,000 miles of petroleum pipelines crisscross the United States transporting 45% of the Nation's annual consumption of petroleum. Pipeline failures accounted for only 3% of the product lost in 1969. An area which has received more attention as a potential cause of spills is those transfer operations where connections are frequently made and broken and where the vessel being loaded or unloaded may be subject to sea motion.

A set of uniform guidelines for the prevention and control of spills in vessel handling at terminals has been prepared by the Technical Advisory Board of the New England Interstate Water Pollution Control Commission, Boston, Massachusetts (January, 1971). The intent of the guidelines is to insure maximum safety from spillage or leakage of oil at terminals and in the transfer of petroleum and petroleum products.

Pretransfer Conference

As part of the guidelines for transfer operations there was specified a pretransfer conference incorporating the following stipulations: No person shall commence or cause to be commenced or consent to the commencement of bulk oil transfer operations unless the following items have been reviewed, agreed upon and complied with by both vessel and shore personnel.

A licensed officer or a licensed tankerman who has full knowledge of the vessel's tanks and cargo handling system shall be on duty or present on the vessel at all times.

A sufficient number of men trained in, and capable of performing competently, the necessary transfer operations, shall be assigned to be constantly on duty both on the vessel and on the pier during cargo transfer operations to keep the transfer operation under constant observation and shall be prepared to take immediate action in case of a malfunction. The State water pollution control agency and the Captain of the Port shall be provided with satisfactory documentory evidence of the training and competency of personnel involved in the transfer operation.

Cargo sequence for loading or discharging products and the proper pipeline for each product has been established.

The handling rate at which oil will be transferred has been established. (Reduced rates are required when commencing transfer, changing the lineup, topping off tanks, or nearing completion of transfer.) The amount of time to be given when the vessel or terminal desires to start, stop, or change the rate of flow has been determined.

A positive communication and signal transfer system shall be established and operating during transfer operations.

The emergency procedures to be followed in order to stop and contain any discharge shall have been established.

Ship and shore personnel responsible for transfer shall be clearly identifiable at all times. Prior to transfer operations terminal and vessel personnel responsible for transfer shall be made known to each other.

Transfer Procedures

The New England Interstate guidelines go on to specify that no person shall transfer or cause to be transferred or consent to the transfer of any oil from any oil carrying vessel to an oil terminal facility or from an oil terminal facility to any oil carrying vessel unless:

> The terminal operator has maintained an inspection and testing program for all active oil handling hoses and equipment owned or used by the terminal in order to detect faulty equipment. Any hoses that are chafed through to the steel reinforcing rings, cut deeply or otherwise found to be defective shall be taken out of service and removed from the dock area.
>
> All hoses through which oil may pass during transfer shall be pressure tested at least semiannually, and immediately after such a test, these hoses shall not be subjected to transfer pressures greater than 80% of that last pressure test, or greater than the rated hose pressure, whichever is less. All hoses belonging to or used by the terminal operator and used in the transfer of petroleum products shall be marked with the terminal facility's name and a hose number. These markings shall be in a color sharply contrasting with the color of the hose and shall be in letters and numbers one and one-half inches high. The terminal operator shall keep a log book of all tests conducted on the individual hoses. This log book shall contain the hose number, test pressure, date of test, place of test and the signature of the person conducting the test. This log book shall be available for inspection by a representative of the State Pollution Control Agency. The same testing requirements apply to hoses belonging to and used by barges that apply to terminals' hoses. Where barge hoses are used at terminals which have no capability to drain the hoses completely after transfer, the outboard end of the barge hoses shall be equipped with butterfly valves for tight shut-off.
>
> A drip pan of sufficient size will be placed at the cargo or bunker manifold and shall be kept in place at all times. It shall be properly positioned and adequately maintained and an adsorbent shall be available in case of overflows to minimize loss of oil. At no time shall the contents of the drip pan be allowed to spill in the water. Contents of drip pans, after removal, shall be disposed of prior to ships departure in a manner acceptable to the State water pollution control agency.
>
> Hoses are supported so as to avoid crushing or excessive strain. Flanges, joints, and hoses shall be checked visually for cracks and wet spots.
>
> Hose handling rigs are of a type which allow adjustment for ship movement and hoses shall be long enough so that they will not be strained by any movement of the ship.
>
> Hose ends are blanked tightly when hoses are moved into position to be connected and also immediately after they are disconnected and drained either into the vessel tanks or into suitable shore receptacles before they are moved away from their connections.
>
> Hoses are not permitted to chafe on the dock or ship or to be in contact with hot surfaces such as steam pipes or to be exposed to other corrosive sources.
>
> Mooring lines are tended to prevent excessive movement of the ship.
>
> The surrounding water shall be inspected frequently during transfer operations. A log of all such inspections shall be kept and signed by the person making the inspection.
>
> No person shall take or discharge ballast to or from any oil carrying vessel unless the scuppers of any such vessel are plugged watertight during the oil transfer or ballasting operation, except on tank vessels using water for deck cooling. However, it will be permissible to remove scupper plugs

as necessary to allow runoff of water provided a vessel crew member stands watch to reclose the scuppers in case of an oil spill. Overboard discharge valves from pump room and engine room bilges shall be tightly closed and sealed with a numbered seal which is to be logged in the ship's log book.

No person shall transfer or cause to be transferred or consent to the transfer of any bulk oil after dark from any oil carrying vessel to a land based oil terminal facility or from a land based oil terminal facility to any oil carrying vessel unless the point of transfer is adequately illuminated.

Vessel-to-Shore Transfer

The New England Interstate guidelines go on to specify that no person shall transfer or cause to be transferred or consent to the transfer of any bulk oil from any oil carrying vessel to a land based oil terminal facility unless:

All cargo risers not intended for use in the transfer are blanked.

Sea valves connected to the cargo piping and stern loading connections are tightly closed and sealed with a numbered seal which is to be logged in the ship's log book.

Lines and valves in the pump rooms and on deck are checked by the ship's master, senior deck officer or deck officer on duty, or licensed tankerman, to see that they are properly set for discharging cargo. An additional check list must be made for the same purposes each time the setting is changed.

Full rate of discharge is not attained until shore lines are proven clear.

A shutoff valve and check valve is located in the dock line to prevent backflow in case of hose or manifold failure.

On completion of transfer operation, hoses or other connecting devices shall be vented, blown down, or sucked out to drain the remaining oil. A drip pan shall be in place when breaking a connection and the end of the hose or other connecting devices shall be blanked off before being moved.

Some of the problems involved in oil transfer from ship-to-shore have also been discussed in EPA Report PB213,880, as they relate to tanker mooring systems. The tankers in the world fleet are growing bigger each year while the ports serving these supertankers are still in the World War II class. Therefore, improved mooring systems are necessary to reduce the potential for spills.

The conventional mooring system uses fixed mooring buoys plus the fore and aft anchors of the vessel to be moored. The connecting hose to the underwater pipeline is connected to a small floating buoy, which is retrieved when the ship is in a fixed position. The known objections to this system are that the vessel cannot remain in this position in rough weather, and when the weather turns bad, it takes too long for the ship to unmoor and lift anchor.

One improved tanker mooring system is the "T"-Jetty mooring system. Some of these jetties are as long as 6,000 feet. The "T" head at the end of the jetty can be constructed to berth as many as eight tankers simultaneously. One of the advantages of this type of installation is that four pipelines can service six to eight berths simultaneously. The disadvantages of this type of installation are:

Extremely expensive to construct.

Ships must stop off-loading during bad weather, and slip moorings.

Approach speed to the installation for large tankers can only be at 1/10

> knot. The problem here is that tankers personnel cannot sense true speed during approach and if the vessel strikes the installation at speeds over $^{3}/_{10}$ of a knot, a large amount of damage can be expected. A new experimental radar system is being tried to overcome this objection.

Another improved method involves the so-called Sea Island Mooring System. This system has the same advantages and disadvantages as the "T"-Jetty Mooring System. In the Sea Island System the oil can either be pumped ashore through underwater pipelines or the oil can be stored on the island to be later trans-shipped to shore installations by smaller tankers and barges.

The currently preferred system is the Single Point Mooring System. While the Single Point Mooring is relatively new in the oil industry (1959), this system is gaining a widespread acceptance throughout the industry as a reasonable cost expense and an effective method to off-load or on-load a tanker. The Single Point Mooring System can handle tankers in much rougher weather than any of the aforementioned systems. One of the great advantages of this type of system is that the wind, high seas and tide forces straining against the ship is only $^{1}/_{6}$ of the forces received at other mooring systems. This is due to the fact that the ship weathervanes around the Single Point Mooring System buoy, bow forward. At the present time, the industry considers the Single Point Mooring System as the next best thing to a harbor mooring.

Shell Oil Company and IMODCO of California are the only two producers of SPM Systems at the present time. However, Standard Oil Company (New Jersey) is in the process of patenting and constructing an SPM System utilizing a single anchor mooring which they believe will be superior to the present system. This SPM System, considered a third generation type, will have an underwater swivel joint. This buoy will have a rigid type conduit from the sea floor to a swivel connection approximately 70 feet below the water surface. When there are sufficient numbers of SPM buoys throughout the world, tankers will convert to bow loading manifolds. The cost of conversion to this type of loading will be approximately $70,000 per tanker.

Shore-to-Vessel Transfers

The New England Interstate guidelines go on to specify that no person shall transfer or cause to be transferred or consent to the transfer of any bulk oil from a land based oil terminal facility to any oil carrying vessel unless:

> All sea valves connected to the cargo piping, stern discharge and ballast discharge valves are closed and sealed with a numbered seal which is to be logged in the ship's log book and with the responsible ship's officer.
>
> All hose riser valves not to be used are closed and blank flanged and all air valves on headers are closed.
>
> Special attention is paid during the topping-off process to the loading rate, the number of tanks open, the danger of air pockets and the inspection of tanks already loading. Notice of the slow down for topping off must be given to shore personnel.
>
> Upon completion of loading, all tank valves and loading valves are closed. After draining, hoses shall be disconnected and hose risers blanked.

Miscellaneous Transfer Precautions

Some other general precautions regarding transfer operations are spelled out in the New England Interstate regulations as follows:

> No vessel while at anchor shall transfer petroleum products while gale warnings (wind velocity 34 knots or more) are in effect or are being displayed by the unit of the Coast Guard having jurisdiction over the

area. Vessel to vessel transfers may only be carried on in anchorage areas designated by the State water pollution control agency after consultation with the United States Coast Guard. This regulation does not apply to the transfer of fuel for a vessel's own use.

Transfer shall cease if a discharge of oil to the waters of the State occurs during such transfer. Notification of the spill shall be made immediately to the Coast Guard and the State water pollution control agency. Transfer may be resumed when, in the judgement of the State water pollution control agency's representative after consultation, if necessary, with the United States Coast Guard or the local Fire Chief, adequate steps have been taken to control the spill and to prevent further spillage.

Transfer of oil by means of a hose through an open hatch is prohibited. An exception will be made only when an emergency arises and this is the only means of moving flammable oil from one vessel compartment to another or of unloading the vessel for the purpose of reducing or preventing fire or pollution hazards, or for preventing foundering, and then, only when all possible precautions to prevent discharge to the waters of the State have been taken.

No person shall discharge exhaust steam containing oil from any coil or other device used to heat oil directly or indirectly into any State or interstate waters unless all oil has been removed from such discharge.

Adequate catchments for leaks and spills must be maintained at the pierheads, and the resulting storm and other drainage will not be discharged unless all oil has been removed. A tight wharf or pier section enclosed by a curb and with drainage discharged to the terminal for separation is recommended.

No terminal operator shall transfer or cause to be transferred or consent to the transfer of any bulk oil from any vessel to a land based oil terminal facility or from a land based oil terminal facility to any oil carrying vessel or from one vessel to another until a sample of the oil to be transferred has been collected, identified by proper labeling, and stored in a place acceptable to the State Pollution Control Agency for a period not to exceed fifteen days. The agency shall determine the information to be provided with each sample and may require chemical analysis of the sample.

Terminal Facilities

The New England Interstate Commission has set forth the following specifications regarding terminal facilities:

All terminals shall be so located and constructed as to provide against flooding by high water and the accidental discharge of oil to water courses.

Any drainage which may contain oil in such amounts as to cause a condition in contravention of water quality standards must first be passed through a properly maintained and adequate oil trap or other removal system. This shall include, but not be limited to, surface drainage from any oil contaminated area of a terminal, or other outdoor area where large volumes of bulk oil are received, stored, or shipped, and exhaust steam or condensate from black oil heaters which is discharged to a watercourse.

Each permanent oil storage tank or battery of tanks must be surrounded by a dike or retaining wall of a capacity not less than 100% of the largest tank inside of the dike, plus 10% of total tank capacity inside the dike to contain spillage and to prevent spillage to the surrounding area. Accumulated drainage shall be removed through a properly maintained and adequate oil trap or other removal system. Dike interiors shall generally be maintained in an oil-free condition.

Security measures should be taken to avoid spillage due to vandalism or unauthorized entry. Such measures may include, but not be limited to fences around property, guards on docks, and locks on dock valves or tank valves when not in use.

Buried tanks and pipelines should be protected from external corrosion by standard techniques of coating, wrapping, cathodic protection, etc. Buried tanks shall be provided with adequate containment or other means to contain subsurface leakage as required by the appropriate State regulatory agency. A schedule should be in effect to pressure test and inspect buried tanks and pipelines at intervals not exceeding five years.

Positive means shall be available to prevent tank overflows during transfer. Procedures shall be established for pretransfer check of receiving tanks and for progress check during transfer. Installation of gauging read-out equipment with high level alarms is recommended.

MacLean Process

A process developed by *G.J. MacLean; U.S. Patent 3,766,739; October 23, 1973* involves a system for containing oil that might be spilled in handling at a marine oil transfer facility where the floating vessel and the mooring are sides of the containment. Two closure devices are provided, one on each side of the hose handling area, between the vessel and the mooring to enclose an area of the water's surface to locally contain a depth of floating oil, wherein the closure sides are movable to compensate for motion of the floating vessel while it is moored. The overall view of such a facility is shown in Figure 30.

FIGURE 30: OIL SPILLAGE ENCLOSURE FOR USE AT DOCKSIDE

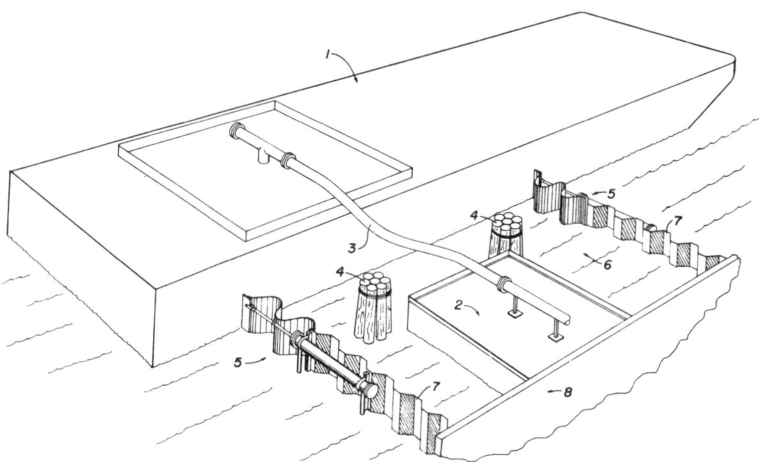

Source: G.J. MacLean; U.S. Patent 3,766,739; October 23, 1973

The barge **1** is shown in position to transfer oil to the unloading dock **2** through a hose **3**. Restraints are attached to bumper dolphins **4** to secure the barge. The closure device **5** is provided on each side of the hose handling area **6**. Sheet piling **7** supports the closure devices and extends to the sea wall or shoreline **8** to enclose the hose handling area.

Wittgenstein Process

A process developed by *G.F. Wittgenstein; U.S. Patent 3,721,270; March 20, 1973* involves a safety installation for preventing pollution by pipelines, particularly those used for transporting liquid hydrocarbons. It is applicable to any pipeline, but particularly to underground or submarine pipelines, whether made of steel, plastic, or any other material.

It may be applied by equipping at least one sector of an existing pipeline or of a pipeline to be built; each equipped sector comprises a fluid-tight jacket of plastic material which surrounds the pipe and which at its ends is sealed on the latter. It is these seals which delimit the sector. The annular space formed between the jacket and the pipe contains a fluid and inserts. The installation also comprises at least one vessel receiving the evacuated flow and a liquid presence detector which gives an alarm and remotely controls operations.

The four essential objects of the device are to ensure reliable prevention of pollution of the environment due to leakage of a hydrocarbon through cracks in the pipe, to evacuate the leakage flow without delay to a vessel, to signal almost instantaneously the existence of a crack, and finally, as soon as the pipe is cracked, to effect remote control of operations by which the dynamic pressure in the pipe is cancelled. It should be emphasized that this scheme is applicable not only to underground pipes but also to submerged pipes.

TANK CLEANING

The inner surface of a cargo tank, if inspected closely, would be found to be rough, uneven and pock-marked with thousands of minute pore openings. The total surface area of the interior of the cargo tanks of a T-2 tanker is approximately 8½ acres.

When a beer glass is filled and emptied a certain amount of the liquid adheres to the side of the container. This is the liquid required to wet the surface of the container. It will vary in amount as a function of its viscosity, temperature, volatility and the roughness and configuration of the container.

In the same way, a certain quantity of oil under fixed conditions is required to wet the surface of the tanks of a tanker. Under average conditions based on long experience, it has been found that this quantity varies from 0.20 to 0.40% of the cargo. A median figure might be 0.30%. This means that if 250,000 barrels of average oil cargo were loaded into a tanker and the tanker immediately pumped out, only 99.7% of the oil would be recovered in the shore tanks and 0.30% or 750 barrels would remain in the ship. The oil remaining in the ship would be found to be adhering to the tank surfaces with a very small portion laying in shallow puddles at the suction bell mouths in the tank and cargo piping.

An operation of great significance in connection with the waste oil resulting from tanker operation is the system used for cleaning or washing down the tanks. Until the early 1930's, the cleaning of tankers was accomplished by long periods of steaming followed by hand washing with streams from fire hoses. There was then developed a system for machine washing of the tanks consisting of opposed revolving nozzles connected to a hose lowered to various levels in the tank to be washed. These nozzles revolve around both a vertical and horizontal axis by the action of a water turbine driven by the washing water at a pressure of about 160 pounds per square inch and a temperature of 170° to 180°F. This is known as the Butterworth System.

According to the New England Interstate Water Pollution Control Commission, steaming and washing of tanks while a vessel is in harbor waters is prohibited, except in those cases of ships equipped with slop holding tanks or except when the vessel is moored at a dock with a facility for receiving all of the discharge from the steaming and washing in a land storage facility.

Esso Research and Engineering Company Process

A process by *J.J. Sheehy and P.A. Sait; U.S. Patent 3,565,252; February 23, 1971; assigned to Esso Research and Engineering Company* provides a system for handling tank washings for oil tankers. Means for separating oil from water are provided to assure that water passing overboard will not have present therein oil in excess of a predetermined concentration.

According to the present process, in a tanker the residue from washing the cargo tanks or the oily ballast water retained on board is sent to a collecting tank. Pumping means consisting of first, a high capacity pump and alternatively, a second low capacity pump, remove the water which has settled to the bottom of the tank. The high capacity pump (or sometimes only the low capacity pump if slower pumping is required) may be used for direct discharge of such settled water overboard from the collecting tank during a process commonly referred to as decanting.

The high or low capacity pump is deactivated by an oil/water interface sensor which substantially reduces the risk of discharging contaminated water during decanting. If the high capacity pump is used first, ordinarily, the low capacity pump subsequently is operated to direct the remaining collected liquid to a separator which separates the oil from the water before the latter is permitted to pass overboard. An oil/water detector and associated conduit means cooperate to shunt mixtures of oil/water back to the settling tank when the separated water contains above a preselected concentration of oil. Figure 31 is a schematic diagram showing the arrangement of equipment involved in the present process.

FIGURE 31: SYSTEM FOR HANDLING TANK WASHINGS AND PREVENTING MARINE POLLUTION

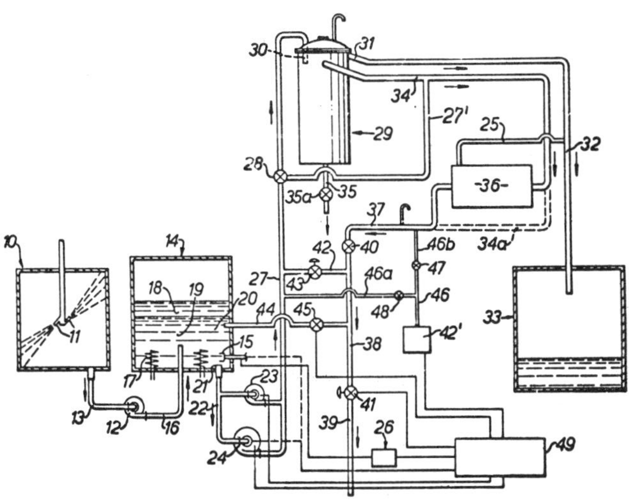

Source: J.J. Sheehy and P.A. Sait; U.S. Patent 3,565,252, February 23, 1971

Referring to the drawing, **10** indicates a tank such as one of several cargo tanks of a tanker being washed by the application of high velocity jet streams of water from a nozzle **11**. A stripping pump **12** removes the tank washings as an oil/water mixture or emulsion (hereafter referred to as a mixture) from the tank by means of conduit **13** directing the tank washings to a collecting tank **14** through conduit **16**. The collecting tank may be equipped with heater coils **17** to assist the gravity separation of oil from water.

Free oil will form a layer **18** at the top of the fluid mass within the collecting tank and a mixture **19**, consisting of water containing a very small quantity of oil in the form of minute droplets will settle to the bottom of the tank. Intermediate the oil layers is an oil-rich transition region **20** wherein the concentration of oil in water increases to a marked degree. Water settling most adjacent to the bottom of the tank will contain very little oil, whereas, the concentration of oil in the water in the transition region will increase as the oil layer **18** is approached.

At the bottom of the tank is an outlet **21** connected to a removal conduit **22**, the latter leading to a low capacity pump **23** and a higher capacity pump **24** which may, under ordinary circumstances, consist respectively of a stripping pump and a main cargo pump aboard the tanker. Pumps **24** and **23** are controlled in part by an oil/water interface sensor **26**.

Leading from each of the respective pumps is a conduit **27** in which a three-way valve **28** is positioned and which leads directly to an oil/water separator **29**. A line **27'** connects the valve **28** directly to line **34** to bypass the separator and to send effluent directly to a coalescer **36** whenever the presence of contaminating oil is insufficient, (e.g., 1,000 parts per million) to require treatment by both the separator and coalescer. The valve is conventional and adapted to be actuated manually, pneumatically, hydraulically or electrically in one of three positions. The separator has an inlet **30** for the admission to the separator of mixtures of oil and water and has an oil outlet **31** connected to the conduit **32** for carrying oil removed from oil/water mixtures from the separator to an oil collecting tank **33**. A line **35** and valve **35a** are provided for draining liquid or for the removal of collected solids from the separator.

Water from which most of the oil has been separated leaves the separator by conduit **34** flowing to a second oil/water separation unit **36**. The latter may operate primarily on a coalescing principle whereby very fine droplets of oil still remaining in the effluent from the separator will be removed so that the amount of oil with respect to the water passing from the unit is below a given concentration. Separated oil will pass through line **25** to conduit **32** and therefrom to recovery tank **33**. Alternatively when the second oil/water separator **36** is not necessary to obtain the desired degree of water purification, it may be omitted and the water from the first separator **29** flows by conduit means **34a** directly to conduit **37**.

On the water outlet side of the coalescer unit **36** there is an outlet system consisting of conduits **37**, **38** and **39** in which are located two-way control valves **40** and **41**. The latter two valves, when open, permit water effluent to flow from the coalescer **36** overboard from the vessel through conduits **38** and **39**. Preferably, valve **41** is pneumatically or hydraulically actuated, whereas valve **40** may be manually actuated.

A conduit **42** with a two-way control valve **43** therein interconnects conduits **27** and **38**. Conduit **44** with two-way valve **45** therein connects conduit **38** with the collecting tank **14**. Valve **45** preferably is adapted to be actuated by pneumatic or hydraulic pressure. Alternatively there may be substituted for two-way valves **41** and **45** a single three-way valve which will shunt the water from conduit **38** either to conduit **39** or conduit **44**. Such three-way valve preferably is adapted to be actuated by pneumatic or hydraulic pressure. In order to assure that water passing overboard from the separator **29** or the coalescer **36** does not contain oil in excess of a preselected concentration, the present process provides means for detecting concentrations of oil above a prescribed maximum.

Accordingly conduits **27** and **37** are connected respectively to the inlet line **46** of an oil-in-water detector **42** by sampling lines **46a** and **46b**. Flow-through sampling lines **46a** and **46b** respectively are controlled by two-way manually actuated valves **48** and **47** so that samples of the water flowing in either conduit **27** or conduit **37** may be sent to the detector **42**. Alternatively there may be substituted for two-way valves **47** and **48** a single three-way valve (not shown) which will pass water samples from either conduit **27** or conduit **37** to the detector **42'**.

The detector **42'** has a radiation source and in a transparent medium such as water will irradiate and cause fluorescence of oil droplets. Through the operation of a photoelectric sensor, an electrical signal is generated proportional to the fluorescence produced which therefore can be calibrated to provide a measure of oil concentration in the water medium.

The signal from the detector **42'** may be sent to a receiver-controller **49** which in response will cause actuation of valves **41** and **45**. Thus, if the detector **42'** senses a concentration of oil in the water passing through sampling line **46a**, greater than a preselected maximum, (e.g., 100 parts per million), the output signal of the detector, by operation of the receiver-controller, will effect actuation of valve **41** to its closed position and of valve **45** to its open position, preventing the flow of liquid overboard and permitting the flow of liquid to the collecting tank.

The interface sensor **26** also generates an electrical signal which again, being transmitted to the receiver-controller will cause it to deactivate either of the pumps **23** or **24** which may be operating to decant the tank. The sensor has a probe **15** defining a gap which will be filled with the fluid in the collecting tank. An ultrasonic signal is sent across the gap which will vary in accordance with the physical characteristics of different liquids in the gap. The sensor may be set to produce no electrical signal output for fluids containing prescribed minimum concentrations of oil and to produce an output signal of 115 volts AC 60 at 10 amps when the physical characteristics of the fluid change drastically, as would occur accompanying a change from water to oil. Thus, the sensor can detect the presence of a relatively large concentration of oil which is present in the transition region and in response can deactivate the pump decanting the tank. The probe of the sensor is installed in the tank at a predetermined height above the outlet so that upon the sensing of a high concentration of oil in the water at the level of the probe the pump decanting the tank is deactivated before an appreciable quantity of oil is drawn into outlet **21**.

Under ordinary circumstances during the washing of respective cargo tanks, the washings will be removed by means of stripping pumps and the washings would be deposited in the collecting tank, which may have a capacity of 2,000 tons. When the tank is relatively filled and after a settling period has passed, for instance about four hours, during which separation of the oil and water may be facilitated by applying heat and by treating the washings with a chemical demulsifier such as Breaxit 7941, the major portion of the fluid contained in the tank will consist of water having less than a prescribed maximum of 100 parts per million of oil. This water is naturally found at the lower levels of the tank and may be removed and sent directly overboard at a relatively high rate without pollution of the seas by oil. This direct passage of settled water is known as decanting.

Accordingly, the first operation or series of similar operations which will be performed using the disclosed system will be decanting. The cargo pump may have a pumping capacity of 2,500 tons per hour, and consequently there is the danger that if the detector **42'** is relied upon to deactivate the cargo pump, relatively large quantities of higher concentration oil in water will have already passed into the conduit system before it is sensed by the detector. Consequently, the interface sensor **26** sensing the imminence or presence of a high concentration of oil in the water at the level of the probe will deactivate the cargo pump before such water of high oil content reaches the outlet of the tank. When the cargo pump has thus been deactivated, valves **43** and **48** may be closed manually and valves **28, 40** and **47** opened. At the same time, the stripping pump will be energized and for the remainder of the operation, liquid removed from the tank will pass to the separator.

The detector will, during the decanting operation sense the presence of oil in the water being pumped through the conduit **27**. If the concentration of oil exceeds the preselected maximum, the detector will send a signal to the receiver controller **49** to effect closing of valve **41** and simultaneous opening of valve **45**. By this means, the liquid, instead of passing through valve **41** and overboard, will be returned through conduit **44** to the collecting tank. At this point cessation of water passing overboard will provide notice that cargo pump **24** should be deactivated manually, that valves **28, 40** and **47** should be opened, and valves **43** and **48** closed.

Stripping pump **23** may then be energized to send further liquid from the tank to the separator at a rate within the capability of the separator to separate the oil and water. Oil removed by the separator will pass to outlet **31** and conduit **32** to the recovery tank. The water effluent, with greatly reduced oil passes by means of conduit **34** to the coalescer **36.** Effluent from the coalescer at less than the prescribed maximum concentration of oil, such as less than 100 parts per million, will pass from the coalescer **36** through the conduit **37** and into conduit **38.** The detector **42** will quickly sense the effluent from the coalescer **36** is clean enough to pass overboard and, as a result, receiver-controller **49** will effect actuation of valve **41** to its open position and of valve **45** to its closed position.

So long as the effluent from the coalescer remains low in oil concentration, valve **41** will remain open and valve **45** closed. However, should a concentration of oil in excess of the prescribed maximum be detected, the detector **42** will again send its electrical signal to receiver-controller **49** which reverses the positions of valves **41** and **45**, thus preventing passage overboard of contaminated water and the return thereof to the collecting tank.

It will be usual for the various cargo tanks to continue to be washed during the ballast leg of the voyage and the washings sent to the collecting tank for a succession of decanting and/or separation runs through the separator and coalescer in accordance with the controls described. Eventually, all of the tank washings will have been sent to the collecting tank and all of the liquid therein shall have passed through the system and the oil which has been separated shall have been collected in the recovery tank. Due to the controls provided and their respective interplay, at no time should water which contains contamination above, for example, 100 per million of oil, be passed overboard.

Gulf Oil Corporation Process

A process developed by *B. Stenström; U.S. Patent 3,722,690; March 27, 1973; assigned to Gulf Oil Corporation* involves a collapsible flotation buoy for skimming an oil layer from water in a tanker. The buoy is provided with mechanical folding means for collapsing the buoy, permitting its insertion into a narrow opening at the top of the tank. The folding means is remotely controlled to permit expansion of the buoy for use when it is inside the tank. This device provides a solution of the problem of cleaning oil tanks, and in this connection, large containers on board tanker ships as well as oil tanks on land are considered.

When a crude oil tanker has discharged its cargo in its home port, the partitions of the cargo tanks have to be cleaned before a fresh load can be taken on. One method for carrying out this cleaning which is common at the present time, consists in spraying the sides of the tanks with water. For this purpose, a suitable number of spray devices are used, which can either be permanently installed in the top of the tank, or temporarily lowered into the tank in connection with the cleaning. Every such device has a spray nozzle which is swingable in at least one plane. The water is ejected at very high pressure and in very large quantities, so that the most effective possible washing away of the oil residues, and rinsing of the tank walls is achieved.

It is also common to add some cleaning agent to the water. As a result of what has just been said, however, the application of such a method makes for a very large volume of a mixture of oil and water and, perhaps, cleaning agent, that is to say, polluted water. It is a known fact that for a long time it has been customary to pump all this liquid into the ocean, making for considerable damage to the environment. On the other hand, the retention of this relatively large amount of water on board, and its discharge only on returning to the loading port, requires the installation on board of suitable storage tanks for this purpose, which implies increased cost, instead. In illustration of the extent of the problem, it can be said that a so-called supertanker, after discharging, has about 1,000 tons of oil residues in its tanks.

However, the difficulty of producing an effective cleaning, and the difficulty of handling the large quantity of polluted cleaning liquid, form aspects of the total problem with which the process is concerned. A third equally important aspect is connected with the fact that the oil residues also involve a risk of explosion. When an oil tanker has been emptied and hence for the most part is full of air, the oil residues on the sides of the tanks give off gases and this results in an air-gas mixture which, at certain concentrations of gas, becomes explosive. These concentrations correspond to an explosive mixture range which has a lower and an upper limit. It follows from this that it is possible to eliminate the risk of explosion by either reducing or increasing the relative gas content. Reduction of the gas content can be done by forced ventilation of the inside of the tank, while a displacement of the mixture ratio to a level about gas contents which fall within the explosive range can be accomplished by evacuating the air in the tank.

This is done in practice in such a way that an inert gas is introduced in place of the air. None of these methods, however, makes for a satisfactory solution of the problem of eliminating the risk of explosion. Aside from the fact that they require installation and operation of a comparatively extensive amount of auxiliary equipment, it will be clear that the protective measures in question, in order to be effective, must be carried out continuously for the whole time, often several weeks, while the tanker is on its return voyage to the loading port. Even an occasional interruption of the ventilation can lead to a rise in the gas content to a dangerous level, and correspondingly, unless the tanks are continuously checked for a sufficient quantity of inert gas, a leakage of air will make the oxygen content of the tank's atmosphere rise from the over-rich range into the danger zone.

Another natural consequence of these conditions is that a breakdown in the ventilation or protective gas installations means an unavoidable moment of risk. In addition, the situation in which such breakdowns can occur or can be feared, are the same in which the risk of fire is greater, for example, on running aground, or in a collision, or when repair work has to be done in the tank, particularly for maintenance of the above mentioned equipment.

It should be pointed out here, that the cursory cleaning of the tank sides achieved by spraying and rinsing according to the above, does not eliminate the risk of explosion. On the contrary, this treatment can trigger an explosion, depending on whether the water is ejected in the form of drops, and with such dynamic energy that the resulting electrostatic charge entrains such powerful field strengths that discharge by sparking occurs. It should be also noted that neither application of the ventilation principle nor introduction of an inert gas in the tank has any cleaning effect; the function is purely protective.

The object of this process is to provide a method for tank cleaning which forms an integral solution of the problem in question, that is to say both protection against risk of explosion and complete cleaning, without difficulties in handling occasioned thereby. More specifically, the present process involves tank cleaning in two stages, the removal of oil residues by collection of oil from an oil layer floating on a ballast of water inside the tank, and a subsequent, conventional cleaning of the tank sides by spraying.

The introduction of water ballast inside a tank, so that as the water level rises a substantial part of the oil residing on the tank sides comes away and forms an oil layer on top of the water, is already known. This layer was then collected with the aid of long gutters inside the upper part of the tank, for example, on the top of a bulkhead. This method, which can thus be characterized as being entirely analogous to the system used in stationary cleaning installations for wastewater, is entirely unsatisfactory for a ship under way. The present process is based on the realization that an effective collection of such an oil layer floating above a ballast of water can be accomplished with the aid of a number of buoy-like devices floating in the mixture of oil and water, with openings at the level of the oil layer, through which the oil can enter, and be carried away from the tank through a hose or pipe.

As mentioned above, the collection operation is designed to be followed by a washing operation. According to a preferred form of execution, however, the spray devices, with the aid of which the second treatment stage is carried out, can also be utilized during the oil collection phase, in such a way that they subject the outer surface of the liquid mixture to an action promoting the movement of the oil toward the buoy or buoys. Figure 32 shows the buoy device used in this process.

FIGURE 32: COLLAPSIBLE BUOY FOR SKIMMING OIL FROM WATER IN SHIPS' TANKS

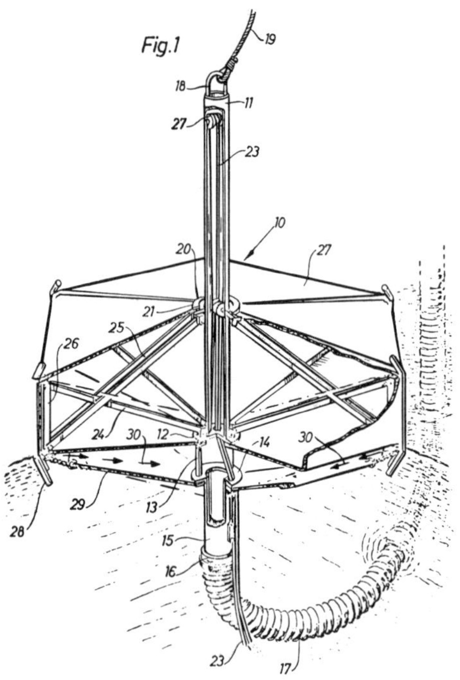

Source: B. Stenström; U.S. Patent 3,722,690; March 27, 1973

It is built around a central pipe **11**, which is attached by its lower end to a bottom plate **12**. The latter is connected in turn, by means of a number of rods **13**, with a flange **14** disposed axially thereunder. To the latter, a downward pointing pipe **15** is welded, and this pipe in turn communicates, by means of a suitable screw, with a hose **17**. At the upper end of pipe **10** there is a lifting eye **18** for a wire **19**.

Somewhat below the middle of pipe **11** there is a ring **20** which is connected to a shaft which extends diametrically through the pipe **11**, and whose two ends emerge through corresponding longitudinal grooves in the pipe. The shaft bears pulleys **21**. At the upper end of pipe **11** there are similar pulleys **27**, and lines **23** run over the pulleys. The latter pass out through the bottom plate **12** and flange **14**, and are so arranged that by drawing them in, it is possible to displace ring **20** between two terminal positions, collapsing the whole device like an umbrella so that it can be lowered through manholes or hatches in the ship's deck.

To bottom plate **12** and ring **20**, bears are fixed for link arms **24, 25** respectively, running in an oblique radial direction outward and upward or downward, respectively. Each of the arms **25** has two parallel legs, between which, corresponding arms **24** are fitted. At their outer ends, the arms are guidably fixed to standards **26**. Together with the latter, the arms form a link mechanism, which in this case consists of six units. The latter are surrounded on all sides by a sheath **27** which is connected watertightly to bottom plate **12**. Thus a floating body is formed, which has a hexagonal shape when seen horizontally. At each of these corners there are staves **28** designed to protect the buoy thus formed.

To the outer, underside of the buoy, a cloth **29** is connected, which can consist of the same material as sheath **27**, and which is tightly connected to flange **14**. On the periphery, however, the latter is not tightly connected to the buoy, but instead there is a number of slots there, or other openings. The space between the bottom of the buoy and the cloth **29** assumes the form of a funnel, and when the buoy is floating on the mixture of water and oil, the oil layer can pass into this funnel as shown by arrows **30**. The oil continues via the center hole in flange **14** and pipe **15** into hose **17** and it is then drawn off therefrom by means of a pump not shown.

As this is taking place, the resulting density of the water mixture rises. Thanks to the fact that the buoy has large cross dimensions in relation to its height and bulk, these variations in density will have a negligible influence on the displacement of the buoy. The oil intake openings will therefore always be at a suitable level, so that for the most part, only oil is sucked in. The exact proportions of the buoy in this respect will, of course be selected according to cases, but it can be mentioned here that practical experiments revealed that in an oil tanker whose tanks, after discharging, were filled with a ballast of water, as high as possible, there was an oil layer about 25 cm thick on top of the water. With a device built as described above, so precise a separation can be achieved that 95 to 98% of this amount of oil was removed.

A process developed by *B.E. Smith; U.S. Patent 3,746,023; July 17, 1973; assigned to Gulf Oil Corporation* is one in which the hold in an oil tanker which has been emptied of its oil cargo is cleaned by first filling the hold with sea water ballast so that oil vapors are forced from the hold and a residual oil layer floats upon the sea water ballast. The floating oil layer is thereupon skimmed or otherwise removed from the aqueous ballast. The hold is then deballasted and washed with a jet of high pressure sea water.

BALLAST DISPOSAL

In addition to accidental discharges, conventional tankers typically discharge oily ballast water and tank washings into the sea before taking on the next cargo of oil. Tankers will normally carry oil from the offshore producing field to a shore terminal. On its return voyage to the field (the noncargo or ballast leg), the tanker must take seawater into its cargo tanks to provide stability. This ballast water mixes with the oil left on the tank walls (called clingage).

Discharge of oily ballast water at or enroute to the field can be avoided if tanks are cleaned at the cargo delivery terminal before the tanker begins its return voyage. The oily washings can be discharged to oil/water separation facilities ashore. The tanker would then load ballast water into clean tanks and head back to the producing field. Shoreside ballast treatment facilities already exist at many refineries and marine terminals. Although expansion of existing facilities, construction of required new facilities, and additional time to clean tanks and pump oily washings ashore would add to the transportation costs of oil, higher oil prices make it more desirable to reclaim oil through separation processes then to pump it into the oceans. Under current international law, tankers are prohibited from discharging any oil into the seas within 50 miles of shore. This rule, largely honored in the breach, would not prevent discharges enroute to OCS sites beyond the 50-mile limit.

The design of a proposed shore facility for the treatment and disposal of ship-generated oily wastes has been presented by the Maritime Administration in Report EIS-AA-73-0949D, Springfield, Virginia, National Technical Information Service (June, 1973).

Ships built with special tanks, separate from cargo tanks and used only for ballast, would also prevent the discharge of oily ballast water. Such segregated ballast systems would avoid the need to mix oil and water, except when cargo tanks are periodically cleaned to remove accumulated clingage or prior to drydocking, generally only for the latter.

Segregated ballast systems incorporating double bottoms also insure against oil spills from groundings, the most polluting form of accident in coastal areas, port entranceways, and harbors. A double bottom uses an outer wall for the hull structure and an inner wall for the structure of the oil tanks. The resulting space permits damage to the outer hull should a grounding occur without necessarily affecting the oil cargo tanks, thus preventing spillage. Double bottoms, however, are not feasible on older ships. According to one school of opinion, double bottoms under some circumstances could jeopardize ship safety should an accident occur.

The costs and effectiveness of segregated ballast systems with and without double bottoms have received extensive study recently in preparations leading to the 1973 Conference on Marine Pollution of the International Maritime Consultative Organization. That Conference produced an international convention which, when ratified, would require tankers of greater than 70,000 tons dead weight to be constructed with segregated ballast capacity while not requiring double bottoms. However, it is likely that smaller tankers will be used to carry OCS oil to shore. Requirement of segregated ballasts on double bottoms, particularly for smaller tankers, has economic implications for tankers operating in international trade. Consideration should be given to present and future design requirements for international vessels prior to establishing design requirements for U.S. vessels.

The U.S. Coast Guard, under the authority of the Ports and Waterways Safety Act, as amended by the Trans-Alaskan Pipeline Act of 1973, is currently in the process of setting standards for the design and construction of tankers in the U.S. coastal trade (which would include tankers used to carry OCS oil to shore).

The Council on Environmental Quality recommends that the Coast Guard require that new tankers in such trade be constructed with segregated ballast capacity preferably with double bottoms when ship safety would not be jeopardized. Existing tankers used to carry OCS oil to shore should be prohibited from discharging oily ballast to the oceans. In addition, the Coast Guard should give serious consideration to requiring new and existing ships to employ advanced accident prevention technologies to improve vessel maneuverability and communications.

The New England Interstate Water Pollution Control Commission has put forth guidelines regarding ballasting as follows: No person shall ballast or cause any oil carrying vessel to be ballasted unless:

> The valves on the lines used are set first; the valves to the tanks to be ballasted are opened second; the necessary valves in the pump rooms, except seacocks, are set third, and then cargo pumps are started before opening seacocks.
>
> All tanks are inspected to see that only the tanks intended are receiving ballast when ballasting is started.
>
> The same attention is given to topping off ballasting as to topping off tanks when loading oil.
>
> When completing the loading of ballast, seacocks are closed before stopping the pumps.

These guidelines do not apply, however, to any vessel whose ballast piping system, ballast pumps, and ballast tanks are wholly independent and not connected to the cargo system.

Until the "load-on-top" system was introduced in recent years, as pointed out by W.M. Kluss of Mobil Shipping Company, Ltd. in *Pollution Prevention*, London, The Institute of Petroleum (1968), the greatest single source of sea pollution was the discharge of oily ballast water into the sea from tankers in normal operations. In the load-on-top system, tanks are washed during the ballast passage in the following manner. The tank washing residues are accumulated in one tank. Most of the clean water in this tank is then carefully drawn off the bottom of the tank and discharged overboard, discharge being halted whenever oil traces appear in the water stream. The tank is allowed to settle, the oil wastes in the tank separate and float to the surface, and additional water is repeatedly withdrawn carefully from the bottom and discharged as before from beneath the floating layer of oil.

Heat may be applied to hasten the separation of oil and water. Some companies occasionally add a demulsifier as well. When all possible water has been withdrawn, the next cargo is loaded on top of the remaining residues in this tank. Usually, this one compartment is segregated from the remainder of the cargo during discharge. Then the segregated material can be directed, as the specific situation dictates, to the fuels processing side of the refinery, to the refinery slop system for ultimate recovery, or mixed with the rest of the cargo being discharged.

In a rolling, pitching vessel, this is no small task. It is estimated that the oil content of the water discharged into the vessel's wake from the load-on-top tank is in the region of 200 ppm, slowly increasing to 400 ppm and finally rising momentarily to 5,000 ppm at the shut-off point. A tanker decanting this residue will not pollute the sea. Even the momentary maximum of 5,000 ppm is only half of 1%. The turbulence in the moving ship's wake would immediately dilute any oil to a tiny fraction of its original concentration. If a vessel experiences rough weather on its way to the loading port and the water in the load-on-top is therefore not reduced to an acceptable level, fresh crude oil can still be loaded on top. But at the discharge terminal the mixture at the bottom of the tank must then be retained on board to be put through another load-on-top cycle.

Shell introduced the load-on-top system in 1962. Other major oil companies soon followed, and today, encouraged by the oil industry, about 75% of the world's crude oil tankers practice the technique. Refineries of the major oil companies accepted load-on-top residue after an experimental period in which they proved to themselves that the residue can be processed through the refinery unit, if it is less than 1% of the total crude cargo, and if the water content of the residue is less than 0.15% of the total cargo. The main problem facing the refinery operator is the removal of salt from the small amount of sea water discharged with the residue. The problem of the tanker operator, then, is to remove as much water as possible from beneath the oil without discharging the oily waste itself.

Members of the oil shipping industry admit that, at present, the best method of measuring the parts per million of oil in an oily water discharge is by the eyeball method. This means that a member of the ship's crew must constantly watch the overboard discharge until black oil is seen with the naked eye. The crew member then signals to the pump man who in turn starts closing the discharge line. This may take from fifteen to thirty seconds. In the meantime, a significant amount of oil is lost overboard.

Chicago Bridge & Iron Company Process

A process developed by *J.S. McCabe; U.S. Patent 3,762,548; October 2, 1973; assigned to Chicago Bridge & Iron Company* involves separating oil from tanker ballast water. Ballast is pumped to underwater disengagement zone operating on water displacement principle. Oil separated from ballast mixture rises to a collection zone from which it is recovered; water escapes from open bottom of disengagement zone to main body of water.

There have been attempts in the past to treat ballast water on board ship prior to its discharge overboard, or to pump the ballast to a land station for treatment and disposal. Neither of these possible solutions, however, is satisfactory. The cost of transferring large volumes of ballast water to a land station for treatment is very high, while treatment of the water on board ship is dangerous because of the risk of explosion. There have in fact been several explosions recorded in large tankers while cleaning operations were being conducted on their tanks.

This process provides a simple but effective way of solving the disposal problem encountered with large amounts of ballast water. Not only does the method prevent contamination of the seas with discharged oil, but it also permits the oil content of the ballast water to be recovered, thereby reducing the economic loss which has heretofore been entailed in discarding ballast water. Figure 33 is a vertical sectional schematic drawing showing ballast water being transferred from a tanker to an underwater separator.

FIGURE 33: UNDERWATER SEPARATOR FOR OIL SEPARATION FROM BALLAST WATER

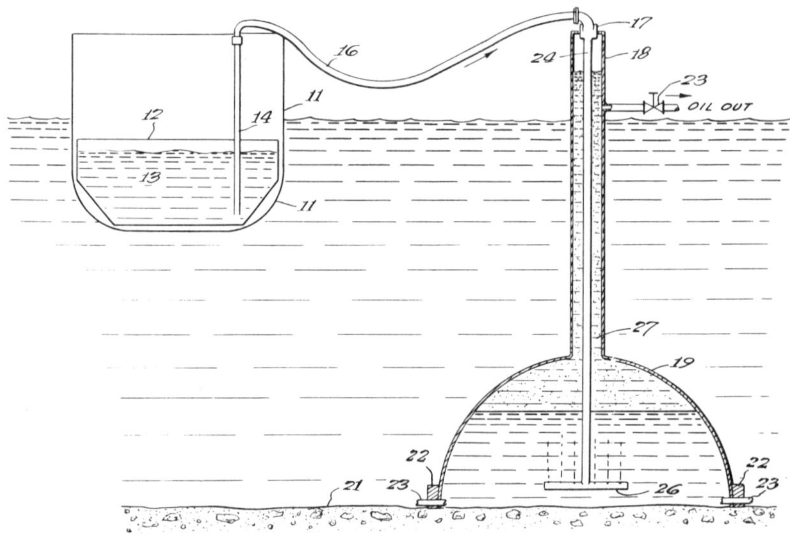

Source: J.S. McCabe; U.S. Patent 3,762,548; October 2, 1973

The hull **11** of a tanker containing therein a ballast tank **12** is used to contain oil or ballast water. Introduced into ballast tank **12** which, in the instance under consideration, contains ballast water **13**, is dip pipe **14** which is connected at its upper end to flexible hose **16** leading to swivel joint **17** at the top of standpipe **18**, which communicates at its lower end with upwardly convex shell **19** which has an open bottom and rests on the floor **21** of the body of water in which the vessel is floating. Shell **19** is held in place on floor **21** by weights **22**, or if necessary or desirable, by additional anchoring means not shown. The interior of shell vessel **19** is open to the main body of water through conduits **23** which permit water to pass freely in and out of the interior of the shell **19**.

Attached to standpipe **18** at a point near its upper end and above the surface of the main body of water is manifold system **23** communicating with the interior of the standpipe. Connected to swivel joint **17** at the top of standpipe **18** is vertical supply conduit **24**, which is attached at its lower end to sparge means **26**. It will be appreciated that the sparge means is not necessary, although desirable, and that the ballast water could be introduced at the top of standpipe **18**. It will be generally found desirable, however, to avoid disturbing the oil phase **27** which accumulates in the standpipe and upper portion of the hollow vessel, by transferring the incoming ballast mixture to the bottom of vessel **19**. Sparge means **26**, by distributing the oil droplets in the ballast mixture, permits these droplets to collect more readily into larger globules and to form the oil phase **27** shown floating in the upper portion of the hollow vessel.

Pfaudler Permutit Processes

A process developed by *A.W. Kingsbury and W.S. Young; U.S. Patent 3,231,091; January 25, 1966; assigned to Pfaudler Permutit, Inc.* involves the separation of heavy, viscous oils such as Bunker "C" fuel oil from ballast water.

Knitted metallic mesh fabric or fine mesh screens have been used for separating oils free from suspended matter from clear water. The fuel oil used in ships may contain residues, and in addition the water used for ballast purposes may be picked up from rivers in which the water contains dirt and other suspended particles. A major problem in the separation of oil from water by wire mesh screens has been the difficulty in cleaning the screens. A deficiency of the wire mesh screens has been that they are not satisfactory in coalescing high viscosity oils. Bunker "C" being a high viscosity oil, having a kinematic viscosity of between 45 and 300 seconds, Saybolt Furol, has not been separable by known coalescing methods on a practical basis.

The process is carried out by passing a mixture of Bunker "C" oil and water through a series of open weave wire mesh screens. Surprisingly, the Bunker "C" oil is largely separated from the water by such operation. In passing through the screens, the oil coalesces and collects in a cohesive mass on the downstream side of the screen. The wire mesh screens are oleophilic in that they retain the oil on their surfaces. A simple draw-off device may be used to remove the coalesced oil from the screen continuously or intermittently. It has been discovered that by using relatively open weave mesh screens, sufficient coalescing action can be obtained to provide satisfactory removal of viscous oil from water. While the open mesh screens tend to clog, they are easily cleaned by a simple process of backwashing. The unexpected simplicity in cleaning is a feature of the process. Figure 34 shows the separator design which may be used in the conduct of the process.

FIGURE 34: SEPARATOR FOR OIL RECOVERY FROM SHIPS' BALLAST WATER

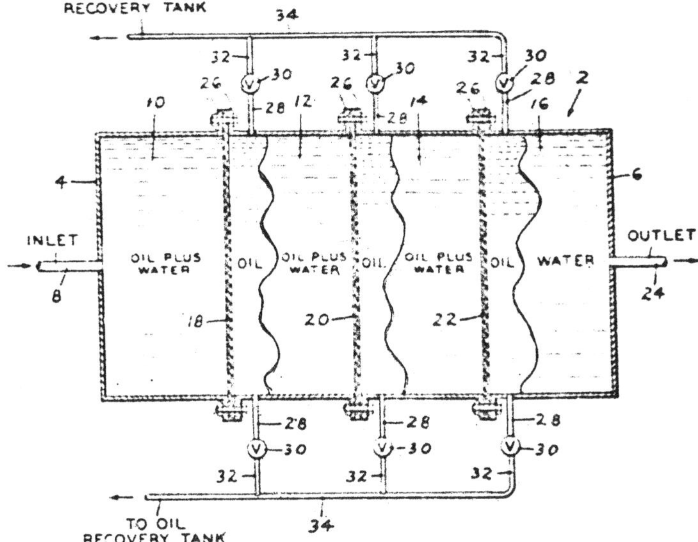

Source: A.W. Kingsbury and W.S. Young; U.S. Patent 3,231,091; January 25, 1966

The separator consists of a shell **2**, which is essentially cylindrical in shape, and contains heads **4** and **6** closing its ends. Head **4** contains an inlet conduit **8** through which oil-laden water is introduced. The shell is divided into chambers **10, 12, 14** and **16** by liquid pervious septa **18, 20** and **22**.

Each septum extends across the separator shell intersecting the liquid flow path which extends from inlet **8** to outlet **24** in head **6**. The septa may contain one or more open mesh screens and may be supported in the shell by any well-known means, such as by flanges **26**. Draw-off conduits **28** are mounted in the shell adjacent to the downstream face of each septum. Valves **30** control the flow from the draw-off conduits **28** through conduits **32** into the collection headers **34**.

In operation, the oil-laden water enters chamber **10** of shell **2** through inlet **8** and passes through septum **18**. A portion of the oil contained in the water is coalesced on the downstream face of septum **18**. The slightly purified water passes across chamber **12** and through septum **20**. Additional oil is coalesced on the downstream face of septum **20** and builds up in chamber **14**. The additionally purified water passes across chamber **14** and through septum **22** where a final stage of oil coalesces on the face of septum **22**. The purified water passes through chamber **16** and leaves the separator through outlet **24**.

The oil which builds up on the downstream faces of the septa **18, 20** and **22** may be withdrawn through draw-off conduits **28**. Valves **30** may be used to control the flow of oil so that the oil may be withdrawn continuously or intermittently. The withdrawn oil passes through conduits **32** into collection headers **34** and is discharged into an oil recovery tank, not shown.

Screen weaves which have been used with good results in the practice of this process include (1) plain weave, where each shute wire passes over and under successive rows of warp wire, and each warp wire passes over and under successive rows of shute wire; (2) twill, which is similar to the plain weave, except that each shute wire successively passes over and under two warp wires, and each warp wire successively passes over and under two shute wires; (3) plain Dutch weave, which has a similar interlacing as plain weave except that the warp wires are heavier and that the shute wires are lighter, and driven close together and crimped at each pass; and (4) twill Dutch weave, which is similar to Dutch weave, except that the warp wires are usually the same size as the shute wires.

A process developed by *W.S. Young; U.S. Patent 3,253,711; May 31, 1966; assigned to Pfaudler Permutit, Inc.* provides an improved method for separating residual fuel oil from ballast water prior to discharge of the ballast water overside. Since the usual deballasting operation is rapid, a flow rate of 600 gallons per minute would not be unusually high, it is important that the separation system be able to operate efficiently at high through-put rates. Moreover, in view of the space limitations imposed by the necessities of maritime design and efficiency, a satisfactory separation system should be compact.

It has been found that when a ballast filled fuel oil tank is pumped out during a deballasting operation, the residual fuel oil is not evenly mixed with the ballast water. On the contrary, the oil concentration will vary from 0 to 100% and the bulk of the oil will come in relatively substantial slugs of almost 100% oil. Except for these slugs of oil, which occur largely near the beginning and ending of the deballasting operation, the pumps are handling relatively clean sea water. The concentration of the oil in the pumped fluid can readily be measured and hence the treatment accorded to the pumped fluid can be selected to meet the oil concentration present. Figure 35 is a schematic portrayal of the apparatus involved in the present process.

A ballast pump **10** draws fluid from the ship ballast tanks through a pipe **11**. The fluid discharged from pump **10** is supplied through a pipe **12** to a three-way motor operated diverting valve **13**. Valve **13** is operated by an electric motor **14** and, depending on the valve setting, may close off pipe **12**, may connect pipe **12** to a pipe **15** or may connect pipe **12** to a pipe **16**.

FIGURE 35: INSTRUMENTED APPARATUS FOR SEPARATING OIL FROM WATER IN HANDLING OILY BALLAST

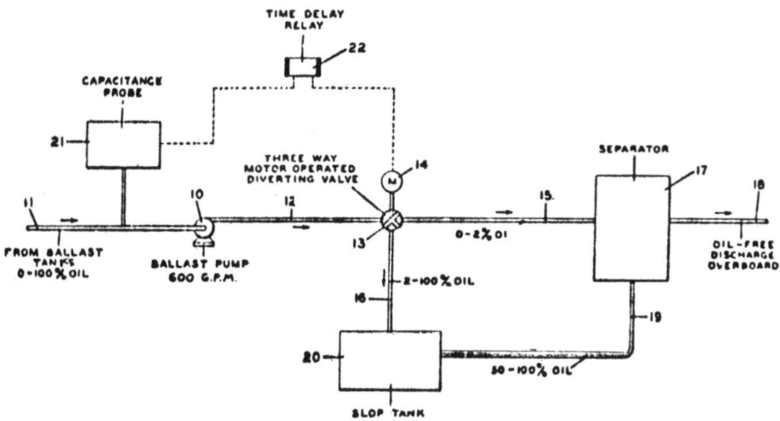

Source: W.S. Young; U.S. Patent 3,253,711; May 31, 1966

Pipe **15** delivers fluid to an oil-water separator **17** which may be of any suitable type. If the residual oil in the ballast tanks is Bunker "C" which is the name commonly given to No. 6 fuel oil, the separator **17** is preferably of the type described in U.S. Patent 3,231,091 described above. The oil-free effluent of separator **17** is discharged overboard through a pipe **18**. The oil separated in separator **17** is supplied through a pipe **19** to a slop tank **20**.

Pipe **16** delivers fluid directly to slop tank **20**. Valve **13** will be set to deliver fluid to pipe **15** and thus to the separator when the oil content of the fluid from the ballast tanks is less than a predetermined amount. The amount will be dependent on the capacity and efficiency of the separator and on the permissible oil content of the water discharged overboard. In the case of Bunker "C" residual oil and a separator of the type described in the preceding process, the valve **13** will usually be set to supply fluid to the separator when the oil content of that fluid is between about 0 and 2%. Where the oil content of the fluid from the ballast tanks exceeds the predetermined amount, e.g., 2%, valve **13** will be set to deliver the fluid from pipe **12** to the slop tank.

As mentioned above, oil in the ballast fluid tends to come in slugs of almost 100% oil, and in the absence of such slugs the ballast fluid is relatively clean sea water which can readily be handled by separator **17** to achieve the desired freedom from oil content for the water discharged overside. Because the oil tends to come in slugs, the presence of such slugs in pipe **11** may conveniently be detected by a capacitance probe **21** which measures continuously the dielectric constant of the fluid in the pipe **11**. Since the dielectric constant of oil is many times less than that of sea water, a slug of oil will be easily detected. Detection of a slug of oil in pipe **11** operates a time delay relay **22** which in turn operates motor **14** to control valve **13**. Time delay relay **22** is provided to insure that the probe **21** has detected a substantial slug or solid stream of oil and has not been affected by a small slug of oil passed in pipe **11**.

Time delay relay **22** similarly provides a time delay on release when detection shifts from oil to water to insure that valve **13** is not operated by a small slug of water in an otherwise solid stream of oil. It has been found that a suitable time delay when oil is detected is of the order of about one second while a suitable time delay when water is detected is of the order of 5 to 10 seconds. In operation, when the fluid in pipe **11** has an oil content less than the predetermined level, e.g., less than about 2%, relay **22** will be in a condition thereof in which valve **13** passes fluid from pipe **12** to pipe **15**. When the oil content of the fluid in pipe **11** exceeds the predetermined level, the capacitance probe will produce an output which, after the set time delay, e.g., one second, will cause relay **22** to shift to another condition thereof in which valve **13** connects pipe **12** to pipe **16**. Valve **13** will be returned to its initial condition when the oil content of the fluid in pipe **11** drops below the predetermined level and after the set time delay, e.g., 5 to 10 seconds.

Since the oil content of the ballast water tends largely to be either well below 2% or close to 100%, great sensitivity in the capacitance probe is not required. Hence the probe may be selected from among many available types. However, the probe may, if desired, be made very sensitive to afford a control operation sensitive to small changes in oil content of the ballast water.

The measurement of the oil content of the ballast water need not be effected by capacitance means, although measurement of the dielectric content is deemed to be the most suitable measurement scheme for this purpose. Other measurement means, e.g., optical inspection, may be used.

The mixture of fuel oil and sea water in slop tank **20** may be disposed of in any desired way. For example, the slop tank may be pumped out into suitable disposal means on land or may be discharged overboard at a later time far out at sea. Or the slop tank mixture may be subjected to a separating operation at a later time when the high through-put requirement of the rapid deballasting operation is not present. In a typical installation the ballast tanks might contain 500,000 gallons of sea water and 3,000 gallons of Bunker "C" oil. For this ballast, the slop tank might be expected to receive about 10,000 gallons of sea water and substantially the entire 3,000 gallons of oil.

TANKER ACCIDENTS

While collisions between vessels or grounding of a vessel on a shoal or reef are the usual tanker accident, it should be noted that all seagoing tankers leak to some extent, as pointed out by H.P. Vind in Report AD 754,746, Springfield, Virginia, National Technical Information Service (December, 1972). Vind's report entitled *Materials for Leak-Proofing Navy Oil Tankers* explores the concept that tankers might be made self-sealing or leakproof.

Some of the promising materials proposed for the purpose are elastomeric sealants for riveted joint construction, rubberized magnetic patches and shingles, gelling and emulsifying agents for thickening oil, rubberized fabric skirts and liners, and reticulated polyurethane foam for retarding oil leaks from tankers. A continued search for new materials is planned.

Abendroth Process

A process developed by *J.C. Abendroth; U.S. Patent 3,772,895; November 20, 1973* is one designed to avoid spillage of a potential contaminant normally fluid at ambient temperatures. A tanker used for transportation of such fluid will be so built and so operated that the cargo will be chilled to a temperature below the point at which it will congeal. Petroleum is an example of such a fluid. Before delivery, the petroleum or other cargo is to be raised to a temperature at which it is again fluid and can be pumped from the storage tanks in which it has been transported. From a method standpoint, a series of tankers can be served by a single high capacity refrigeration device at the port of embarkation, each tanker requiring, therefore, only minimum refrigeration equipment to maintain the paraffin-congealing temperature during transit. Similarly a single high capacity heating system will serve a succession of tankers to restore the contents to fluidity as a port of debarkation.

By way of example, the cargo is assumed to comprise petroleum such as crude oil. Most hydrocarbon oils contain substantial quantities of paraffin and may, therefore, be congealed. In the case of crude oil, the temperature below which the oil cannot be poured will ordinarily range no lower than -70°F. Most are well above that temperature. A text book gives the pour point of Pennsylvania crude oil at 4°F. The pour points given for Oklahoma, Colorado and certain Texas-Louisiana oils are 5°F. Many other crudes will congeal at -15°F or -20°F. Some even congeal at temperatures of 90°F or higher.

If the crude oil cargo is congealed before the carrier leaves port, it will not escape even if the hull is accidentally punctured or broken. The cargo will tend to remain gelatinous or like a chunk of paraffin. Lake or sea water freezing on the oil because of its low temperature would act as insulation and as a secondary hull. Since the refrigeration requirements during transport are greatly reduced after the desired low temperature has once been attained, it is only necessary to provide modest refrigerating capacity on the individual ships.

The method can be practiced without substantial refrigeration if, in lieu thereof, or as a supplement thereto, a congealing component such as paraffin, which is normally congealed at ambient temperature, is introduced into the oil or fuel which constitutes the prime cargo.

Ortiz Process

A system developed by *A.J. Ortiz; U.S. Patent 3,724,662; April 3, 1973* is one in which massive oil spills from tankers are eliminated by confining the oil right at the spill source, before it becomes a slick, by means of a marine pollution control system which includes a hugh plastic bag from which oil is pumped from the bag to a receiving facility. The marine pollution control system, also referred to as MPCS, is a preventative and contingency constituting a complete, self-contained, light and mobile system of men and equipment that can be deployed on very short notice for oil spill containment anywhere in the world. Figure 36 shows such a system in operation at the scene of a tanker accident.

FIGURE 36: MARINE POLLUTION CONTROL SYSTEM FOR REMOVING OIL FROM TANKER HULLS AT POINT OF RUPTURE

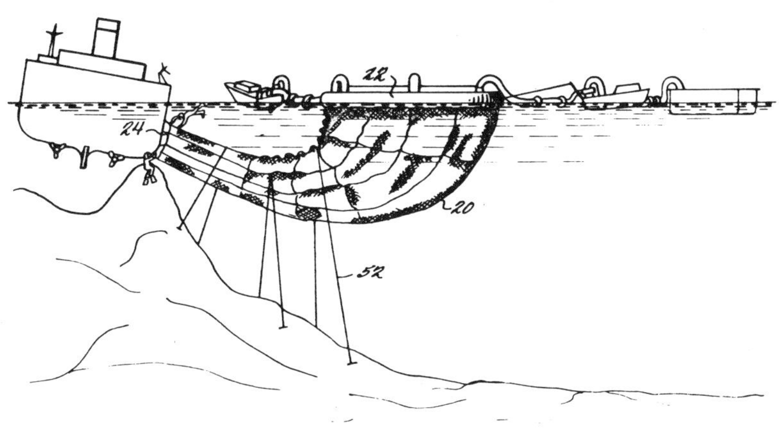

Source: A.J. Ortiz; U.S. Patent 3,724,662, April 3, 1973

The Marine Pollution Control System illustrated encompasses frogmen fitting the mouth of the bag **20** around the source of the oil spill on the ship's hull and in surveillance of the bag as it assumes its inverted tear-drop shape when filled with oil, and on lookout for abnormal conditions, pump suction hoses hooked up to quick release connections on the bag's roof, high capacity positive displacement pumps, aboard pump boats, transferring oil from bag to receiving end, manned boats and pumps, and pump discharge hoses operating on receiving vessel or pipeline ashore with a team captain in direct, closed circuit, electronic communication with every member of the team including frogmen, managing the over-all situation. A helicopter, also under the direction of the team captain would hover over the spill area for the purpose of assisting in the coordination of the salvage work and of rescuing personnel or equipment should the need arise during the operation.

Fire fighting equipment of the chemical type will be provided aboard each boat to guard against the possibility of fire hazard, and a self-contained bag roof CO_2 topping system will be provided on the roof to protect the bag from waterborne slick fires. Thus, the services of the system can be marketed among oil companies or government agencies in the U.S. and abroad for use as an oil pollution control preventive measure as well as a contingency measure. As a preventive measure, it can be kept in readiness in areas of offshore drilling operations, in tanker traffic focal points, and aboard tankers themselves.

This structure includes a roof ring **22** and a mouth ring **24**, all of these forming an integral, one piece flexible structure. An electromagnetic mouth type is used for recovery from a broken tanker or pipelines. In this type, the ring is made of multiple electromagnets in succession forming a flexible ring structure that assumes the shape of the surface of attachment. This electromagnetic ring is energized, as is known in the art, by a self-contained waterproof dry power source (not shown) that is part of the mouth. After placing the mouth, and when control of the salvage situation has been accomplished, power for the electromagnetic mouth ring may come from shipboard equipment, for sustained attachment, but the self-contained power source is designed for a minimum capacity of 24 hours. Bags used for tanker spill containment are fitted with guys **52** around their bodies as shown to allow shaping to keep the body of a bag underwater and preclude any damage should a shipboard fire occur.

Rainey Process

A process developed by *D.E. Rainey; U.S. Patent 3,756,294; September 4, 1973* involves collecting oil at its source of leakage from a container underwater such as a ruptured tanker hull. The apparatus includes an elongated, flexible and impervious conduit having an inverted channel-shaped mouth composed of cushioned sealing material and flexible magnet means for mating engagement with the surface of the oil container surrounding the sides and upper portion of the source of leakage for entrapping the oil and elevated conveying means at the opposed end of the conduit for directing the entrapped oil to a collector. The bottom portion of the mouth remains open to permit entry of water into the conduit to float the entrapped oil and accelerate its passage upwardly through the conduit.

Figure 37 is an overall view of the operation of this process. As shown, a flexible elongated conduit **12** of impervious material, plastic sheet for example, is adapted to extend between a source of underwater oil leakage and an upper storage collector. The source of oil leakage, as illustrated is a rupture or the like **14** in the sidewall **16** of a marine tanker, such as might result from a collision, but it is to be understood that the source of leakage might be in a pipeline or any other kind of oil container.

The entrance mouth **18** of the conduit **12** is of inverted channel or U-shaped configuration and is adapted to be secured in mating engagement with the upper and side wall portions of the container bordering the rupture or other area of leakage. To this end, flexible magnet means of inverted channel of U-shape and a flexible cushioning layer of similar configuration are provided at the mouth **18** of the conduit **12**. The magnet means may be formed from a plurality of individual magnets fastened together by a nonmetallic cable, nylon for example, as a chain or ribbon as shown, or may be formed from a bendable strip of metal or in any other manner to provide flexibility for conforming to the wall curvature of the container. The cushioning layer which may be of sponge rubber or any other suitable material provides sealing contact with the container wall.

FIGURE 37: ALTERNATIVE SCHEME FOR REMOVING OIL FROM TANKER HULL AT POINT OF RUPTURE

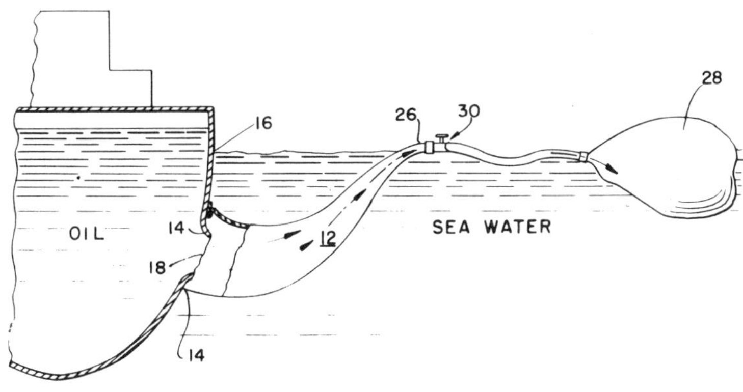

Source: D.E. Rainey; U.S. Patent 3,756,294; September 4, 1973

The discharge opposite end portion **26** of the conduit **12** passes the fluid contents by way of a conveying pipe or hose; and to a storage collector such as the bladder **28** as shown or to another barge or tanker or any other suitable storage collector. Preferably, multivalve means **30** are provided at the discharge end of the conduit alternately to direct fluid flow to selected storage containers. A discharge pump may also be provided but ordinarily is not necessary because of the tendency of the oil to rise in the water medium.

In operation of the apparatus herein illustrated and described after it has been brought to the scene of oil leakage, the conduit mouth **18** is positioned such as by a diver around the source of leakage with the cushioned magnetic upper and side wall portions of the mouth in engagement with the opposed wall surfaces of the leaking container. Manipulation of hogging lines assists in maintaining a mating sealing engagement for entrance of leakage oil into the conduit **12**. The open bottom of the mouth **18** permits entry of water into the conduit to float the entrapped oil and accelerate its passage upwardly through the conduit.

Submersible Systems, Inc. Process

A process developed by *F. McCormick; U.S. Patent 3,491,023; January 20, 1970; assigned to Submersible Systems, Inc.* involves the recovery of oil from sunken or leaking tankers at sea by containment and collection of the oil released therefrom within the circumference of a bubble barrier wall created in a geometric pattern about the location of the stricken vessel.

DRYDOCK OPERATIONS

Preus Process

A process developed by *P. Preus; U.S. Patent 3,786,773; January 22, 1974* is one in which a device is placed on the floor of a drydock to encompass the damaged area of the hull of a ship drydocked therein. The device intercepts the flow of water from the area of the hull during drydocking to filter out hydrocarbons emanating from the damaged hull to preclude contamination of the waters surrounding the drydock and also the drydock itself.

When a damaged ship requires drydocking for inspection and/or repairs and the fuel or cargo tanks have been damaged, suspected or actual pollution of the waters surrounding the drydocks by oil from damaged tanks is inevitable unless definite preventive steps are taken because the great quantities of water expelled from the drydocks drain directly into the adjacent waters and will carry any oil contaminants present or discharged into the drydocks with them.

In the case of collisions or groundings of dry cargo ships, the hazard of oil pollution during subsequent drydocking is great since fuel oil or bunkers is carried in double bottom tanks formed by the bottom hull of the ship and spaced, parallel-planar tank tops or double bottoms disposed above the bottom hull of the ship. Upon grounding or collision, one or more of these double bottom tanks is almost invariably holed if the hull of the ship is penetrated and, upon being holed, that tank is open to the sea. Although there is an initial escape of oil from a holed double bottom tank, the major portion of the oil in that tank, due to its lower density, is pressed against the tank top and remains in the tank as long as there is water beneath the ship to keep it there.

Upon drydocking, the water pressure is relieved and the bunkers are discharged into the drydock. Although a major portion of the bunkers could sometimes be removed from the damaged double bottom tanks by the ship's fuel transfer system, such removal is sometimes inadvisable or impossible due to fear of contamination of fuel transfer system by sea water, because of loss of heating capability necessary to render the bunkers pumpable in the damaged tank or because of the ship's fuel transfer capability. In the cases as described above, oil pollution during drydocking is a serious problem and it is that to which this process is directed.

FIGURE 38: DESIGN OF FLOATING DRYDOCK TO PERMIT CONTAINMENT OF OIL LEAKAGE FROM DAMAGED VESSEL

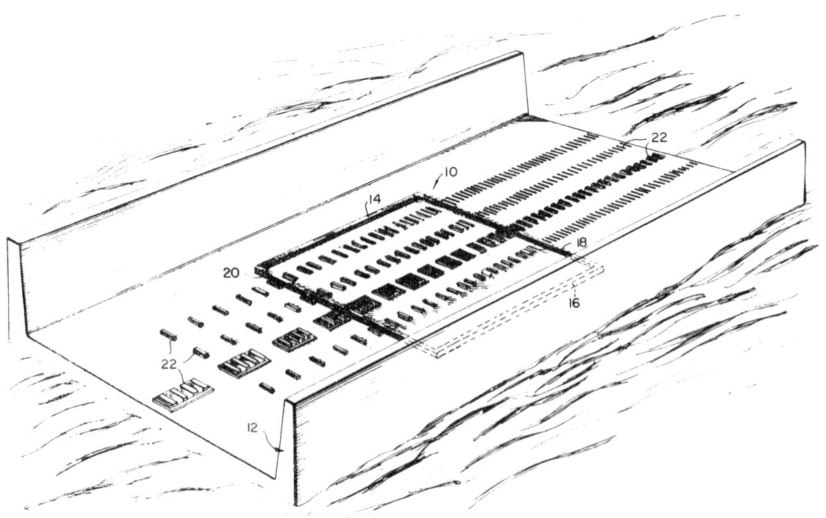

Source: P. Preus; U.S. Patent 3,786,773, January 22, 1974

Figure 38 is a perspective view showing the device of the process mounted on the deck of a floating drydock **12**. The device **10** comprises longitudinal legs **14** and **16** running fore and aft on the starboard and portsides of the drydock and legs **18** and **20** running thwartships the drydock. The legs intersect one another to form a rectangular enclosure beneath what will be the damaged portion of a ship when placed in the drydock **12**. The longitudinal legs **14** and **16** are disposed outboard of what will be the turn of the bilge of a ship when placed in a drydock. Drydock blocks **22** are disposed on the floor of the drydock **12** to support a ship when placed in the drydock in a manner well-known in the art.

The thwartship legs **18** and **20** are of equal height to the drydock blocks **22** at least for the portion thereof coinciding with the bottom of a ship when placed in the drydock such that the legs will be in sealing contact with the ship bottom fore and aft of the damaged portion of the ship. The longitudinal legs **14** and **16** as well as the portions of the legs **18** and **20** outboard of the ship bottom may be the same height as the aforedescribed portions of legs **18** and **20** if the damage to the ship is confined to the hull bottom itself or may be greater in height if the damage extends above the turn of the bilge.

Mesh netting, preferably ¼ inch mesh hardware cloth wire, is attached between the stringers to form perforate sides and a top for the device **10**. A buoyant oil absorbent, water repellant material, having an effective particle dimension sufficient to be retained by the mesh, is disposed in the confines formed by the mesh. Sorbent C, a petroleum absorbent material (Clean Water, Inc.) has proven ideal for the purpose.

In actual use of such a device, as reported in the *U.S. Coast Guard's Environmental Protection Newsletter,* February, 1972 (Vol. 1, No. 1), the M/V Singapore Trader was successfully drydocked without any overboard discharge of bunker fuel although a considerable amount of fuel was discharged onto the drydock floor within the confines of the device.

The primary feature that must be considered in such design is that the oil absorbent material be present in sufficient quantity to absorb the full amount of oil expected to be discharged from the ship. Absorbent C petroleum absorbent material has an absorption capability of about 0.55 gallons of Bunker C per pound of absorbent material and, utilizing the probable capacity of the damaged tanks, the quantity of material needed can readily be determined beforehand.

PREVENTION AND CONTROL OF SPILLS FROM OIL REFINING

In a typical refinery, the large-volume sources of wastewater can be grouped in three categories, as shown in Table 8.

TABLE 8: REFINERY WASTEWATER SOURCES

Source	Flow, gpm	% of Total	Source of Contamination
Cooling water	100 - 6,000	40 - 80	Process leaks, treating chemicals and concentration
Water softener and boiler blowdown	5 - 300	<10	Treatment and concentration
Refinery processes*	20 - 1,200	20	Direct contact with oil and treating chemicals

*Crude oil desalting, overhead oily waste from units such as distillation, cracking, alkylation, treating, etc.

Source: Report PB 213,880

Although the largest percentage of the water discharged from a refinery was used for cooling purposes, the actual amount varies widely from one refinery to another. This depends on whether the cooling system is once-through or recirculating, and to some degree the extent of use of air coolers. Cooling water is normally called non-oily. This is relative, however, since the cooling water can become oily through leaks in heat exchangers. These leaks and treatment chemicals used in the cooling systems are the main sources of contamination in cooling water.

Pollution in a waste stream from water softening and boiler blowdown is relatively minor. Although relatively small volumes of water are involved, the dissolved solids content may be high. Chemicals used to treat the boiler water are the main pollutants. The greatest sources of pollutants in a refinery are the so-called process water or oily water streams. This contamination occurs when the water is used in direct contact with the hydrocarbon process streams. Often this water was used as steam to strip material from process streams, such as sulfides or mercaptans.

Table 9 lists the pollutants and approximate concentrations which may be found in refinery wastewater streams. Visible oil on the surface of wastewater is the most common and troublesome pollutant in refinery wastewater. Concentrations can range from 1 ppm to well over 1,000 ppm during accidental spills.

TABLE 9: UNDESIRABLE COMPONENTS OF REFINERY WASTEWATER

Pollutant	Concentration, ppm
Floating and dissolved oil	1 - >1,000
Suspended solids	---
Dissolved solids	0 - 5,000
Phenol and other dissolved organics	0 - 1,000
Cyanide	0 - 20
Chromate	0 - 60
Organic nitrogen	0 - 50
Phosphate	0 - 60
Sulfides and mercaptans	0 - 100
Caustics and acids	2 - 11 pH
Color and turbidity	---

Source: Report PB 213,880

TREATMENT OF OILY WASTEWATERS

Gulf Oil Corporation Process

A process developed by *J.L. Henning, Jr. and W.J. Robicheaux; U.S. Patent 3,727,765; April 17, 1973; assigned to Gulf Oil Corporation* utilizes a float-supported skimmer for removal of oll from the surface of a body of water. A float-supported skimmer trough is connected to a drain line through which skimmed oil is delivered to a sump and from the sump to a pump for discharging oil from the system. A valve in the discharge line from the pump controls the rate of discharge of oil from the system. An air bubbler line opens into the lower part of the skimmer trough and sends a signal responsive to the depth of oil in the skimmer trough to control the opening of the valve in the discharge line.

Water used in processing at oil refineries frequently becomes contaminated with oil. To avoid discharging oil-in-water effluent from the refinery, much of the water used in refineries is delivered into settling ponds or tanks in which the oil rises to the surface of the water. Skimming apparatus is then used to remove the oil from the surface of the water, and oil-free water is discharged from a low level in the settling pond.

In the usual skimming apparatus a trough is supported at the water level by floats. The oil floating on the water flows over the edge of the trough into the trough and is pumped from the skimmer. If the pumping rate is too high, the level of water in the trough is lowered, which causes the trough and floats to rise and thereby reduce the rate at which oil overlows into the trough. Continued pumping further lowers the liquid level in the trough until the skimmer becomes inoperative. If the pumping rate is lower than the rate at which oil flows into the trough, the trough is flooded and sinks to a lower level because of the increased weight of the liquid in the trough. The device can then become inoperative as a skimmer because excessive amounts of water flow into the trough.

This apparatus permits removal of oil from the surface of settling ponds, or other bodies of water, in which a skimmer trough extends between and is supported by a pair of spaced floats. The skimmer trough is adjustable vertically on the floats to adjust the elevation of the upper edge of the trough relative to the floats, and thereby the initial submergence of the overflow edge. A drain line from the lower end of the trough delivers liquid flowing into the trough to a sump from which the liquid is pumped from the system. A valve in the discharge line from the pump controls the rate of removal of liquid from the skimming apparatus. Control means operating in response to the level of liquid in the trough control the opening of the valve to control the rate of discharge of liquid from the skimming apparatus. The drain line is either flexible or provided with swivel joints to permit vertical movement of the floats in response to fluctuations in the liquid level of the settling pond.

Petrolite Corporation Process

A process developed by *L.C. Waterman; U.S. Patent 3,625,882; December 7, 1971; assigned to Petrolite Corporation* permits the clarification of oil-contaminated water from a desalter or other source, the water being clarified by gas flotation effected in a closed vessel having an upper inclined wall guiding the oily waste material to a collection zone of the vessel without the use of mechanical surface skimmers. The flotation is usually effected at a pressure other than atmospheric. The oily waste material and the separated gas may be further treated for recovery of valuable products, as by being returned to the desalter or to refinery equipment.

Standard Oil Company Process

A process developed by *J.F. Grutsch and R.C. Mallatt; U.S. Patent 3,589,997; June 29, 1971; assigned to Standard Oil Company* involves purifying contaminated wastewater from oil refineries. This system includes a plurality of lagoons in which thrive aquatic microorganisms. As wastewater flows through these lagoons sequentially, an apparatus moves across the lagoons' surfaces introducing air into the surface water, and churning this surface water. The apparatus is equipped with means which direct at least some of the churning water towards the lagoons' bottoms. This aerates the lower strata of the lagoons, keeps sludge stirred up, and provides for control of sludge build-up on the lagoons' bottoms. Establishing underwater currents which have a velocity of about 0.5 foot per second or greater is sufficient to maintain sludge in suspension.

PREVENTION AND CONTROL OF SPILLS OF INDUSTRIAL OILS

TREATMENT OF OILY WASTEWATERS

Mobil Oil Corporation Process

A process developed by *J.T. Jockel; U.S. Patent 3,414,523; December 3, 1968; assigned to Mobil Oil Corporation* involves the disposal of waste rolling mill oil.

Preferably, a waste roll oil emulsion containing, by weight of total emulsion, from about 90 to 95% of water and from about 10 to about 5% of oil and typical oil additives may be disposed of by the steps of (1) adding a mineral acid to the emulsion; (2) subjecting the acid-treated emulsion to centrifuging and removing the oil phase; (3) adding to the remaining water phase an inorganic alkaline substance, while keeping the pH below neutral; (4) passing the water phase through a bed of activated carbon; (5) adding an inorganic alkaline substance, increasing the pH of the material to above neutral; and (6) subjecting the water phase to centrifuge treatment.

Petrolite Corporation Process

A process developed by *W.F. Burns and R.B. Martin; U.S. Patent 3,707,464; Dec. 26, 1972; assigned to Petrolite Corporation* involves clarifying an oil-solids contaminated aqueous stream such as may arise from the operation of a railroad equipment servicing facility. A primary clarifier gravitationally separates the stream into settled solids, a marketable oil product and a clarified water stream. The clarified water stream is treated by creating a flocculation product for removing solids and oil therefrom to produce a clear water stream delivered to a subsequent utilization. Settled solids may be returned to the primary clarifier.

The flocculation product is accumulated in a sludge cleaner to be periodically converted into water and oil phases. The oil phase is returned into the primary clarifier. The water phase is reformed into an oil-free flocculation product and separated from a water filtrate returned into the primary clarifier. Solids from the primary clarifier are accumulated in a sand cleaner to be periodically cleaned to separate oil-free solids from a solids-free water-continuous filtrate returned to the primary clarifier. The reformed flocculation product and oil-free solids are separated in a solids dewatering vessel from a water filtrate returned into the primary clarifier. The relatively water-free solids are then delivered to a subsequent utilization such as land fill.

OIL SOURCE DETECTION, IDENTIFICATION AND MONITORING

DETECTION AND MONITORING

An aerial photography technique has been described in *Chemical Engineering* magazine, May 13, 1974, p. 60, for detecting and recording oil spills on water. The method works well for slicks too faint to be seen with the naked eye. The technique was developed by Eastman Kodak Company at its chemistry research facility in Rochester, New York. In the development of this technique, it was found that polluted and clear water appear white and gray, respectively, when exposed to black and white film in the ultraviolet and blue regions of the spectrum.

Ultrasonic wave attenuation in water with known amounts of fuel oil contaminant was measured. Attenuation was found to vary linearly with concentration of oil when the contaminant level was less than 500 ppm/volume. Investigations were conducted at a frequency of 30 Mhz in distilled water with quiescent flow conditions by F.K. McGrath of the Naval Postgraduate School in Monterey, California, in research on the development of an oil/water pollution monitor. This work has been described in Report AD-747,084, Springfield, Va., National Technical Information Service (June, 1972).

A device to monitor oil in the overboard discharge of ballast water should be an important part of a crude oil tanker pollution control system. Such a device, called the Bailey Oil Content Monitor is manufactured by Bailey Meters and Controls, Ltd. of Croydon, Surrey, England. This device operates on the principle of fluorescence measurement. Fluorescence is a photoluminescent process that involves the absorption of radiation (exciting radiation) followed by the emission of radiation (fluorescence) at a wavelength longer than that of the exciting radiation. An exciting radiation of 3,600 Angstroms is often used, and the resulting fluorescence is in the visible portion of the spectrum. Many organic compounds fluoresce including crude oil. When 3,600 A radiation is used, as in the Bailey Detector, the fluorescence is predominantly from the condensed (3+) ring aromatic compounds.

Evaluation of a Bailey monitor has been reported by J.O. Moreau and J.J. Heigl of Esso Research and Engineering Company, in Report COM-73-10940, Springfield, Va., National Technical Information Service (April, 1973).

A study of laser induced fluorescence to map the extent of an oil slick has been reported by R.M. Measures and M.Bristow of the Institute of Aerospace Studies, University of Toronto in a report for the Canada Center for Remote Sensing of the Department of Energy, Mines and Resources in Ottawa, Ontario. The results have been reported in Report

N72-20479, Springfield, Va., National Technical Information Service (December, 1971). The preliminary results of the study were encouraging and led the authors to predict that a laser fluorosensor could be used for environmental sensing from an aircraft flying at between 1,000 and 2,000 feet on a 24-hour basis.

A method of determining the cost effectiveness of various remote sensing systems for ocean slick detection and classification has been presented by G.C. Gerhard of the University of New Hampshire in Report COM-73-10397, Springfield, Va., National Technical Information Service (April, 1972).

Avco Everett Research Laboratory Technique

A technique developed by *D.A. Leonard and C.A. Chang; U.S. Patent 3,806,727; April 23, 1974; assigned to Avco Everett Research Laboratory* is one whereby the oil pollution content of water is continuously monitored by measuring the oil fluorescence spectrum produced by an ultraviolet light source and by comparing it with the Raman spectrum of water.

Texas Instruments Incorporated Technique

A technique developed by *K. McCormack; U.S. Patent 3,783,284; January 1, 1974; assigned to Texas Instruments Incorporated* can indicate the presence or absence of petroleum products in a water area by utilizing an active infrared source which illuminates the water area which may contain a petroleum product. The reflected infrared radiation is filtered by two filters at two different wavelengths, λ_1 and λ_2. Two infrared detectors produce signals which are proportional to the detected reflected radiation at the wavelengths λ_1 and λ_2. A processing channel is connected to each detector, the processing channels each including a log amplifier, the output of which is coupled to a differencing circuit which produces an output signature signal, $\ln[V(\lambda_1)/V(\lambda_2)]$, which indicates either the presence or absence of the petroleum products in the water area.

SPILL QUANTITY ESTIMATION

Quantity estimation is extremely difficult due to variations in slick thickness over the spill area. For this reason experienced, visual observation will generally provide the most reliable information on the quantity of oil spilled. Table 10 is a guide for estimating the amount of oil on the water's surface.

TABLE 10

Appearance of Slick	Amount of Oil
Barely discernible	25 gal/sq mi
Silvery sheen	50 gal/sq mi
Faint colors	100 gal/sq mi
Bright bands of color	200 gal/sq mi
Dull brown	600 gal/sq mi
Dark brown	1,300 gal/sq mi

Source: Report PB 213,880

SAMPLING

A complete oil sampling program is undertaken for the primary purpose of presenting legal evidence in a court of law. To protect the interests of all parties concerned three basic sampling requirements should be observed: The sample must represent that oil which was spilled; the sample must not be physically or chemically altered by the collection procedure, and the transfer of the sample must be accomplished using a well-defined chain of custody system.

The ideal sampling program is one in which collections are made before and after the spill incident. The very nature of an oil spill, however, generally precludes "before" sampling. The types of samples to be collected depend on the following parameters:

Location of the spill, e.g., offshore, harbor area, inland water;

Uses of the area involved, e.g., drinking water source, shellfish growing area, wildlife refuge, finfish spawning area, recreation, commercial fishing;

Facilities available for analysis, e.g., qualitative, quantitative, none;

Source of spill, e.g., offshore platform, refinery, terminal operation, storm drain, ship in port, ship offshore.

Oil sampling techniques have been reviewed by the staff of the Oil and Hazardous Material Research Section of the Edison Water Quality Laboratory of EPA in Report PB 190,171, Springfield, Va., National Technical Information Service (December, 1969). As noted in that report, sampling of oil in the environment, depending upon the thickness of the slick, can present certain operational problems, most paramount of which is the collection of an adequate volume of sample required for identification by chemical analyses. Several basic dip stick techniques, which are primarily applicable for sampling slicks with a thickness of greater than 2 mm, as well as suggested methods for sampling thin oil slicks are discussed and illustrated. Included in this report are preliminary results on oil entrapment by solid absorbents. Also reported are results of investigations performed by foreign and U.S. scientists, using various types of sampling equipment and materials.

The design of a remote sampler for determining the residual oil content of surface waters has been described by P. Schatzberg and D.F. Jackson of the Naval Ship Research and Development Center in Annapolis, Maryland. It is described in Report AD-760,217, Springfield, Va., National Technical Information Service (November, 1972).

Many precautions must be observed when handling oil samples for analysis since the character of the sample may be affected by a number of common conditions including:

Composition of the container — glass bottles should always be used since plastic containers, with the exception of Teflon, have been found under certain conditions to absorb organic materials from the sample. In some cases, the reverse is also true in that compounds have been dissolved from the plastic containers into the sample itself. This problem also applies to the bottle cap liners; therefore, the portion of the cap that comes in contact with the sample should be made of glass, Teflon, or lined with aluminum foil.

Cleanliness of the container – previously unused glass bottles are preferred. If this is impossible, bottles should be either acid cleaned or washed with a strong detergent and thoroughly rinsed and dried.

Time lapse between sampling and analysis – since the chemical characteristics of most oils, especially the lighter fuel oils, change with time, the time lapse between sampling and analysis should be kept to a minimum. If analysis cannot be completed within 24 hours, samples can be preserved, depending upon the volatility of the oil, by removal of air and exclusion of light. With heavier type oils, such as No. 4 and residual oils, carbon dioxide may be used to displace the air. If dry ice is available, (approximately 0.5 cu in) it may be added to the sample. As soon as the effervescing has stopped, the jar should be sealed. When carbon dioxide or another inert gas is not available, or in those instances where volatile components are present (No. 2 or lighter), the sample can be preserved by carefully filling the bottle to the top with water to displace the air. All samples should be kept under refrigeration until analyses are completed.

Collection of adequate volume of sample – it is desirable to obtain as much of a sample of the oil as possible. It is suggested that 20 ml be considered as the minimum volume of oil needed to perform a series of identification analyses on light oils, e.g., No. 2 and below. For heavier oils a minimum volume of 50 ml is required.

Sampling of oil presents many difficulties not immediately obvious. An oil slick may vary in thickness from several inches down to a monomolecular layer measured in microns (10^{-4} cm). The quantity of sample required is therefore important since such will determine the area of sweep. For example, 5,000 gallons of oil, if assumed to be evenly distributed over one square mile of water, will equate to an oil thickness of 0.0071 mm. If 200 ml of sample is found necessary, then all the oil must be recovered from 28 square meters of open water. If sampling recovery is 50% rather than 100%, the sweep area must be doubled. Table 11 below describes theoretical thickness and area, assuming an even distribution of oil for various magnitude oil spills.

TABLE 11: OIL DISTRIBUTION ON A WATER SURFACE

Spill Area gal/mile^2	Spill Area ml/meter^2	Area (meter^2) Required to Obtain 200 ml/sample	Oil Thickness (mm)
1,000,000	1,430	0.14	1.43
100,000	143	1.4	0.143
10,000	14.30	14	0.0143
5,000	7.15	28	0.0071
1,000	1.43	140	0.00143 = 1.43μ
100	0.143	1,400	0.000143 = 0.143μ

Source: Report PB 213,880

The ideal oil sampling device should be simple to operate, function under diverse conditions, have a few moving parts and not require electrical power, be inexpensive, collect oil rapidly, and not require chemical treatment of sample.

Manual separation is the oldest oil sampling procedure. The method involves collection of oil/water mixture in a container followed by manual separation of the oil and water phases. Collection and separation is continued until about 1 pint of oil has been collected. The procedure is quantitatively inaccurate because of the crude nature of both the collection and separation steps. Collection devices have included a simple pail, a pail fitted with a bottom tap, a sliding plexiglass cylinder whose fall is controlled by a trigger, and a dustpan with a stopcock fitted to the handle. Separation procedure devices have included manual decantation, a separatory funnel and a glass filter funnel fitted with a two-way stopcock.

Another possible sampling technique involves the use of adsorbent materials. Many materials strongly adsorb oil and other hydrocarbons. Such materials include Teflon shavings, straw, polypropylene fiber, rope, glass fiber, paper, and polyurethane foam. Qualitative sampling requires contact of the sorbent with the oil followed by chemical (desorption) or physical (wringing or compression) recovery of the sorbed hydrocarbon. Quantitative sampling requires contact of the sorbent with a known area of liquid surface; this application has been impeded by variable and uncertain adsorption efficiency.

Two applications of the use of textured filter paper have been studied by French scientists, namely, free floating disks and lined cylindrical containers. At film densities of 650 to 2,200 mg/mile^2 (2.6 ml/mile^2 at a density of 0.85) both methods sample quantitatively to ±25% with a single sample.

Adsorption by polyurethane foam appears to be a promising procedure. In practice, a

¼" x 1' sheet is simply dragged across the surface until apparent saturation is reached. The sheet is then passed through a wringer to recover the adsorbed oil. When tested with South Louisiana crude oil, this procedure recovered a sample containing <0.1% H_2O. Infrared analysis revealed no differences between the original oil and the sample recovered.

TAGGING

Methods for tagging oils are also under study. Active tagging requires that an inexpensive, coded material be added to oil. This material must be chemically and physically stable in both oil and oil slicks. It must be readily identifiable by available analytical techniques and it must have no adverse effect on the oil's subsequent use. Some characteristics of various types of active tag materials have been reviewed by L. Melamed of the U.S. Coast Guard in Report AD-761,971, National Technical Information Service (September, 1972).

IDENTIFICATION

Oil analysis involves three levels of laboratory activity: sample preparation and cleanup, basic identification, and final comparison. Identification as a petroleum product is accomplished by comparison of sample characteristics to properties of typical products. Further analysis may be performed to more firmly establish the similarity between environmental samples and reference samples collected from potential sources. At all stages of analysis the judgement of the analyst is crucial. Typical analyses include observation of solubility in organic solvents, infrared spectroscopy, API Gravity, distillation range, gas chromatography, specific metal determinations, viscosity measurement, and sulfur determination. Qualitative analysis is performed to determine the presence (or absence) of a particular type of oil. This analysis may be performed on samples of water, fish and shellfish, and mud. In addition to direct chemical procedures, taste and odor tests may be performed.

It is emphasized that once a spill has occurred, qualitative analysis is useless without a source sample for comparison. That is, a representative oil sample must be collected from the source, such as the ship (each compartment, bilge and ballast), offshore platform, terminal (each tank), refinery (various process sources), and storm sewer (suspected sources). It may also be desirable to collect benthic, phytoplankton and zooplankton samples.

Alternate methods are under study which will detect the presence of oil without the requirement for physical sampling. These methods will use photographic and electronic sensors and include earthbound and airborne instrumentation. Earthbound instrumentation would be deployed in areas of high probability of spills. Airborne instrumentation, on the other hand, would allow occasional spot checks of other areas and effective monitoring of cleanup operations. Ideally, such instrumentation should be capable of detection, identification and of providing aerial and thickness measurements.

Oils and oil/water samples may be analyzed to identify the type of spilled petroleum product or to establish similarity between samples collected from the environment and from suspected sources. Common origin is probable if samples agree in their key characteristics. Analysis of environmental samples involves preliminary cleanup followed by standard analytical procedures. The analytical sequence in Figure 39 involves three levels of activity.

Preliminary cleanup intended to separate the oil from the water and produce an organic phase amenable to analysis.

Identification of spilled material as a specific type of petroleum product; e.g., gasoline, jet fuel, crude oil, etc. Identification is accomplished by analysis according to standard procedure and comparison of results with prescribed characteristics of typical petroleum products. Characteristics of a few such petroleum products are specified in Table 12.

Further analysis more clearly defines the relationship between the environmental sample and reference samples collected from suspected sources. Results of the analyses are used to evaluate similarity.

FIGURE 39: ANALYTICAL SCHEME FOR CRUDE OILS

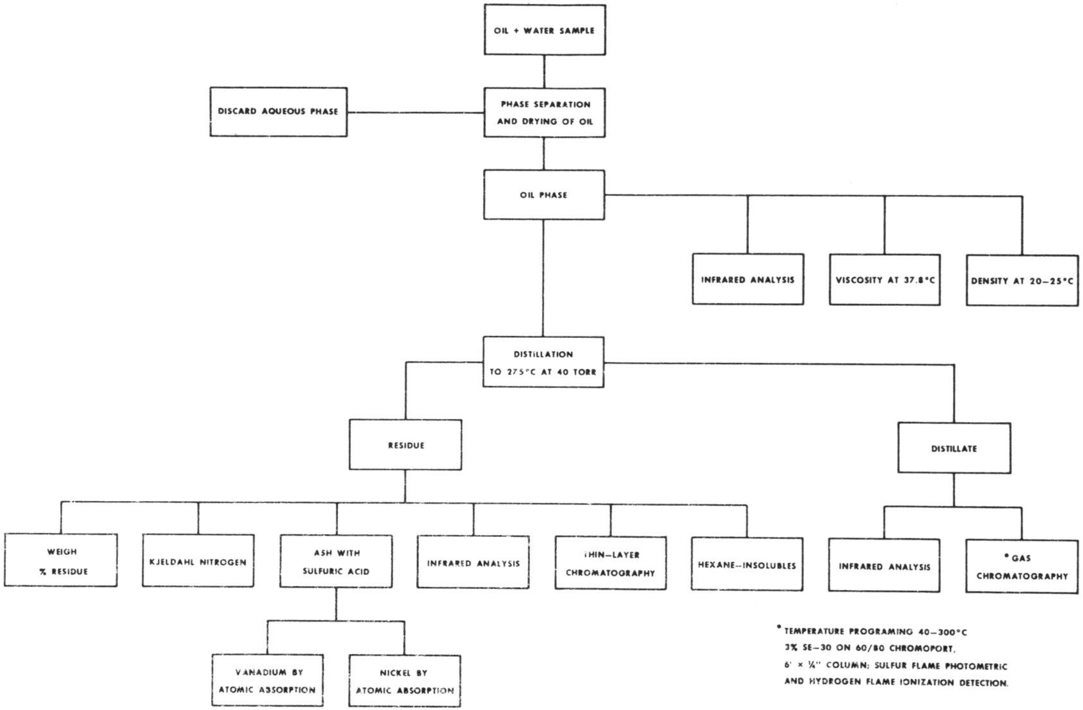

Source: Report PB 213,880

Both the apparent and suspected nature of the sample will influence the actual laboratory procedure. The analyst himself will normally select the procedures necessary to accomplish identification or comparison. The required level of analysis (e.g., identification or identification/comparison) depends on the nature of the problem.

The first step in oil spill analysis is often sample cleanup. Sample cleanup is commonly accomplished by one of three basic methods:

Simple phase separation is generally applicable to distillate fuel oils (Fuel oil Nos. 1, 2, and 4, kerosene, gasoline, etc.), and consists of withdrawing the heavier aqueous phase from the lighter organic phase in a separatory funnel. The organic phase is then dried by passage through anhydrous $CaCl_2$. The dissolved petroleum products may be salted out of the aqueous phase by the addition of Na_2SO_4 to the water.

Drying with centrifugation – When the petroleum product is emulsified with water (as is the case with some crude oils and residual fuel oils), it is sometimes possible to break the emulsion by removal of the water from the matrix with anhydrous calcium chloride, and then centrifuging the sample in a chemical centrifuge to isolate the oil phase as the centrifugate.

Phase separation with solvent addition is accomplished by adding a solvent, such as chloroform to the oil/water mixture. It is sometimes possible to obtain a clean phase separation by solvent addition and to proceed as in simple phase separation to separate the organic phase from the aqueous. The organic phase, after drying with calcium chloride, is heated to evaporate the solvent.

TABLE 12: CHARACTERISTICS OF STANDARD PETROLEUM PRODUCTS

Product	API Gravity (API units)	Kinematic Viscosity (cs)	Distillation Range (IBP - EP)	Comment
High gravity naphtha	45-75		95°-206°F	Sulfur 0.02%
Low gravity naphtha	30-53		160°-410°F	
Gasoline	58-62		96°-408°F	Contains lead, halogens
Jet fuel	40-55		100°-500°F	
Kerosene	40-46		355°-575°F	Narrow API gravity range
Fuel oil #1	>35	1.4- 2.2		
Fuel oil #2	>26	<4.3	370°-675°F (90%)	Sulfur exceeds 0.5%
Fuel oil #4	9-36	5.8-26.4	420°-683°F	
Crude oil	13.5-33.5	2.3-10.5	40°->850°F	Wide dist. range
Fuel oil #6	-2-18	>100	>700°F	

Source: Report PB 213,880

When cleanup has been completed, the separated and dried organic phase is subjected to analysis. Identification is normally accomplished by comparison of sample characteristics to those of typical petroleum products. Characteristics of a few typical products are specified in Table 12. The analyses indicated in Figure 39 should be considered alternatives which the analyst may select to accomplish identification. Only a few analyses may be required in some cases, whereas in other cases the complete analytical scheme may be necessary. It is the responsibility of the analyst to select the combination which will accomplish the most certain identification in the time available. These analytical parameters are:

Solubility in organic solvents – used to differentiate greases and asphalts from other petroleum products, and to distinguish between crude oils and residual fuel oils from different locations. Table 13 gives solubility characteristics for several typical petroleum products.

Specific gravity or API gravity – gravity or density is a distinguishing characteristic of oils. However, since the loss of volatiles, which occurs in the early stages of environmental exposure with volatile distillate fuels and crude oils results in an increase of this parameter, it is of limited value.

Infrared spectroscopy – indicates the relative content of aromatic or carbon-ring-type compounds. May also indicate presence of additives such as silicones. Generally employed to characterize materials less volatile than #2 fuel oil, such as #4 and residual fuels.

Average molecular weight – used to identify low and high boiling products. Not commonly employed with other than pure hydrocarbons.

Distillation range – defined as the temperature difference between high and low boiling compounds in an oil observed during distillation. Actual procedures are specified by ASTM D 86-56, ASTM D 850, and ASTM D 216. Reported as the actual temperatures at which distillation begins and ends, i.e., IBP, initial boiling point and EP, ending boiling point. The boiling range relates to volatility. Although potentially valuable, IBP is useless if the oil has been subjected to weathering before collection due to volatilization of lower boiling components. Typical distillation ranges are presented in Table 12.

Viscosity — a measure of the resistance to flow. May be expressed as Saybolt second units (SSU), the time required for a standard volume of oil to pass through a standard orifice, as specified by ASTM D 445-53T and ASTM D 446-53, or kinematic viscosity at 100°F or 212°F in centistokes (ASTM D 445-65) or in Saybolt Furol units at 122°F. ASTM 2161-63T gives the relationships between the different viscosity units.

Distillation – As a loss of volatile components (if present) occurs on environmental exposure resulting in an increase of the remaining components, these may be put on the same basis (normalized) by distilling all samples to a similar degree. This is accomplished in the laboratory by distillation in a simple distillation apparatus to obtain a distillate boiling point of 275°C at 40 mm Hg pressure. The distillate is analyzed by gas chromatography and the residue for vanadium, nickel, sulfur, and nitrogen content and infrared analysis. Column chromatography may be incorporated in the analysis.

Vanadium – is analyzed according to ASTM D 1548-63.

Nickel – is analyzed either in the final solution from the vanadium procedure by atomic absorption (AAS) or by dissolving 5 grams oil in 100 ml of xylene for determination by AAS.

Sulfur – is analyzed by the procedure described in ASTM D 1552-64, using the Leco induction furnace, or by ASTM D 129-64, utilizing an oxygen combustion bomb. In domestic fuel oils (1967), sulfur values ranged from 0.001 to 0.45% for #1 fuel oil; 0.02 to 1.6% for #2 fuel; 0.2 to 3% for #4 fuel oil; and 0.4 to 4.25% for #6 fuel oils.

Nitrogen – is determined by the Kjeldahl method.

TABLE 13: SOLUBILITY OF PETROLEUM PRODUCTS IN ORGANIC SOLVENTS

	Solvent	
Product	Hexane	Chloroform
Light naphtha	Very soluble	Very soluble
Heavy naphtha	Very soluble	Very soluble
Gasoline	Very soluble	Very soluble
Jet fuel	Very soluble	Very soluble
Kerosene	Very soluble	Very soluble
Cutting oil	Soluble	Soluble
Motor oil	Soluble	Soluble
Paraffin wax	Soluble	Soluble
White petroleum jelly	Partly soluble	Partly soluble
Grease	Insoluble	Soluble
Residual fuel oil	Partly soluble	Soluble
Asphalt	Insoluble	Soluble

Source: Report PB 213,880

As noted in *Instrumentation for Environmental Monitoring – Water*, Report LBL-1, Vol. 2, Berkeley, University of California (February 1, 1973), recently revised standard manuals on the analysis of water and wastewater describe elementary methods for sampling, extracting (if necessary) and weighing oily or greasy (insoluble organic) material which may be present in water or wastewater. Quantitation of such material is important in the treatment of sewage, because oily matter can interfere with bacterial degradation and plug carbon filters or render trickling ponds or sludge digesters ineffective. Quantitation is also essential to determine the levels of oil contamination dispersed throughout the aquatic environment.

Unfortunately, these elementary laboratory methods for measuring the bulk of insoluble organic matter in water and wastewater offer little insight into the identity of the source of an unobserved oily discharge or its probable environmental impact. To determine these aspects, it is necessary to develop information on the composition of the oil, i.e., its constituent organic compounds. Because petroleum in particular is such a gross mixture of organic compounds, extensive detailed analysis of its composition has posed formidable difficulties in the past. In recent years a number of qualitative tests and analytical techniques have appeared in petroleum research and chemistry publications which apply either directly or with modification to oil spill identification.

Because the various techniques employed in oil spill analysis are scattered throughout the technical literature, a standard manual setting forth analytical procedures in a generally accepted, systematized manner is a highly desired goal. As an initial response to this situation, a booklet entitled *Laboratory Guide for the Identification of Petroleum Products* was issued in 1969 by the Federal Water Pollution Control Administration (now the Water Quality Office, EPA).

Besides describing additional sampling techniques, it provides water pollution analysts with a series of specific analytical procedures. The purpose of these procedures is to characterize an unobserved oil spill sufficiently well to identify its probable source when compared with suspect samples. Offering suggested procedures on a trial basis exposes these procedures to the broad, general usage necessary for their acceptance as standardized methods. Although the booklet is incomplete by current standards and regional EPA laboratories already employ alternate approaches, it constitutes a useful initial step toward devising a systematic analysis of oily material in the aquatic environment. Parallel efforts by the American Society for Testing and Materials (ASTM), Subcommittee D-19.10 are underway.

The choice of specific techniques from among the multiplicity of analytical procedures applicable to petroleum and its products depends on factors such as size, sampling methods, personal analytical preferences, and, perhaps most important, the availability of instrumentation. The more elementary bulk properties of an oil spill sample, such as density, viscosity, solubility in selected solvents, ash content, sulfur content, refractive index, distillation range, etc. can be used to characterize an unobserved oil spill for general classification purposes.

Gas chromatography (GC), atomic absorption spectrophotometry (AA), thin-layer chromatography (TLC), and infrared spectroscopy (IR) are examples of moderately sophisticated laboratory instruments which have been applied successfully to oil spill source identification by EPA technical support laboratores, among others. These latter instruments play an even more important role when assessment of the environmental impact of oil spills is the goal, since the questions of "what sort of" and "how much" noxious or toxic organic compounds are present require investigation into the molecular composition of the oily mass.

Because crude petroleum and many of its refined products represent an extremely complex and variable mixture of chemical compounds, mostly organic, no single analytical technique presently available is capable of analyzing this material in toto. However, the still-developing field of GC, in conjunction with ancillary techniques, most closely approaches this ideal; therefore, GC is the most important single technique. The sample requirements for a complex mixture like crude petroleum seldom exceed 10 milligrams for a dual detector GC run. Generally GC analysis is complemented with additional analytical data in order to establish conclusively the identity of an oil spill sample from among several similar suspected samples.

Atomic absorption spectrophotometry (AA), thin-layer chromatography (TLC) and infrared spectroscopy (IR) are currently the most utilized of the sensitive complementary techniques. All four techniques are limited in field applications to laboratory batch-type analysis as opposed to continuous, stream monitoring.

Although perhaps desirable, the present outlook for continuous, remote sensing, oil spill monitors out in the environment is not promising. This is due to a number of factors such as the tendency of oil to float on the surface where it is subject to wind drift, wave action, fluctuation with the water level, and adhesion to solid surfaces contacting it. Devices already suitable for refinery process monitoring become prohibitively expensive when adaptations for general remote environmental monitoring are considered. Airborne oil spill detection devices have been studied. It must be kept in mind, however, that unless the oil slick detected is obviously linked, either through direct, unambiguous contact or strongly suspected recent contact, to the source of the discharge, detailed laboratory comparative analysis between the spill samples and suspect samples will be necessary to identify the source.

A multiparameter oil pollution source identification system has been described by John W. Miller of the Research and Development Department of Phillips Petroleum Company in Report EPA-R2-73-221, Washington, D.C., U.S. Environmental Protection Agency (July, 1973). The feasibility of oil pollution source identification was demonstrated on 80 crude oils from the world's major oil fields. Measurements of 15 diagnostic parameters were made on the 600+°F fraction of the crude oil samples. The parameters are:

- Carbon Isotopic Composition
- Sulfur Content
- Nitrogen Content
- Vanadium Content
- Nickel Content
- Sulfur Isotopic Composition
- Hydrocarbon GLC Profile
- Sulfur GLC Profile
- Saturate Content
- Aromatic Content
- Asphaltic Content
- Odd-Even Predominance Curves
- Carbon Isotopic Composition – Saturates
- Carbon Isotopic Composition – Aromatics
- Carbon Isotopic Composition – Asphaltics

A multivariate statistical procedure was developed and tested to match the 15 parameters of an unknown with those of each 600+°F fraction from the crude oils. All 15 parameters were not required for positive identification; the first six listed above were sufficient to distinguish uniquely among 48 sources.

Outdoor weathering studies demonstrated that of the 15 diagnostic parameters only hydrocarbon GLC profile and saturate, aromatic and asphaltic contents were affected by weathering. The change of the value of these four parameters was dependent upon the length of weathering time (maximum 49 days).

Further, it was found that sulfur isotopic composition was an important identification parameter by virtue of the wide range of its values and its stability to weathering.

ORGANIZATION FOR OIL SPILL CONTROL

One of the best ways to prevent or limit spills is to anticipate them by thorough training of crews and personnel. Workers who are keenly aware of the danger of spills are most likely to be prepared to take immediate and effective action. Anticipatory action should also include the installation, where possible, of equipment to control and stop the flow of oil. The primary responsibility for handling spills should be delegated to one man or to a special committee. Such responsibility should include analysis of potential spill locations; the preparation of plans for dealing with emergencies; procuring, evaluating and making readily available the necessary equipment and materials; and conducting training sessions and drills in cleanup procedures. Lines of responsibility and authority, of course, should be clear. Alert, informed and responsible personnel, thorough training and good equipment are a good recipe for preventing and limiting spills. Chances are the cost of such precautions will be more than repaid by a reduction in lost oil and lower cleanup costs. More important, however, is the protection of our environment from pollution.

The Federal Water Pollution Control Act, as amended (P.L. 91-224, 84 Stat. 93, 1970) directed the preparation of a National Contingency Plan for Oil and Hazardous Materials. This plan, published in the *Federal Register* on June 2, 1970, superseded the National Multiagency Oil and Hazardous Materials Contingency Plan which was approved in September, 1968. The plan provides for a pattern of coordinated and integrated response to major pollution incidents by departments and agencies of the federal government. It establishes a national response team and provides guidelines for the establishment of regional contingency plans and response teams. The plan promotes the coordination and direction of federal, state and local response systems and encourages the development of local government and private capabilities to handle pollution spills.

The objectives of the plan are to develop effective systems for discovering and reporting the existence of a pollution spill, promptly instituting measures to restrict the further spread of the pollutant, to assure that the public health, welfare and national resources are provided adequate protection, application of techniques to clean up and dispose of the collected pollutants, and institution of action to make sure of effective enforcement of existing federal statutes. The plan is effective for all United States navigable waters including inland rivers, the Great Lakes, coastal territorial waters, and the contiguous zone and high seas beyond this zone where there exists a threat to United States waters, shoreface or shelf bottom.

Each of the primary federal agencies has responsibilities established by statute, Executive Order or Presidential Directive, which may bear on the federal response to a pollution spill.

This plan intends to promote the expeditious and harmonious discharge of these resonsibilities through the recognition of authority for action by those agencies having the most appropriate capability to act in each specific situation. The primary federal agencies are the Environmental Protection Agency (EPA), the Departments of Transportation, Defense, Interior, and Health, Education and Welfare.

The plan provides for a National Inter-Agency Committee (NIC) which is the principal instrumentality for plans and policies of the federal response to pollution emergencies. At the Washington level, the plan establishes a National Response Team (NRT) consisting of representatives from the primary agencies. This team acts as an emergency response team to be activated in the event of a pollution spill involving oil or hazardous material which: (a) exceeds the response capability of the region in which it occurs, (b) transects regional boundaries, or (c) involves national security or major hazard to substantial numbers of persons or nationally significant amounts of property. A National Response Center (NRC) in Washington, D.C. is the headquarters site for activities relative to pollution spills. There are established throughout the United States, Regional Response Teams (RRT) that perform functions within the regions similar to that performed by the National Response Team on the National level. Regional Response Centers (RRC), similar to the NRC, are located in each of the predesignated regions.

The plan further provides that in each region there will be established On-Scene Commanders (OSC). The OSC is the single executive agent predesignated by regional plan to coordinate and direct such pollution control activities in each area of the region. The first responsible federal representative to arrive on-scene automatically becomes the OSC until he is officially relieved by the predesignated OSC. The U.S. Coast Guard is to provide for OSCs in areas where they have assigned responsibility which includes the high seas, coastal and contiguous zone waters, coastal and Great Lakes ports and harbors. The EPA will furnish or provide OSCs in other areas.

The actions taken to respond to a spill or pollution incident can be separated into five relatively distinct classes or phases. For descriptive purposes these are:

Phase I - Discovery and Notification
Phase II - Containment and Countermeasures
Phase III - Cleanup and Disposal
Phase IV - Restoration
Phase V - Recovery of Damages and Enforcement

It must be recognized that elements of any one phase may take place concurrently with one or more other phases.

Separate regional plans are to be developed by the EPA and the U.S. Coast Guard for the respective areas of responsibility within each region. All regional plans are to be oriented in accordance with the ten standard federal administration regions. All regional contingency plans will contain, as a minimum, the following items:

(1) a definition of the area covered including the points of change in jurisdiction between EPA and U.S. Coast Guard;
(2) a notification and reporting system beginning with the initial discovery of a spill;
(3) names, addresses and phone numbers of all pertinent federal, state, local and industry personnel involved in the reporting system;
(4) a listing of predesignated OSCs and RRCs;
(5) listing of resources and equipment available in the regional area with names, addresses and phone numbers; and
(6) categorization of water areas by use to establish predetermined cleanup priorities.

THE ON-SCENE COMMANDER

What are the qualifications of a good On-Scene Commander (OSC)? He should be a well trained individual, whose background includes oil pollution research, experience with a number of actual spills of different types of oil, a knowledge of the shipping industry and a good background in law and existing oil pollution legislation.

Basically, the OSC is a decision-making machine, working at least twelve hours per day over a long period of time. To make these decisions in an intelligent manner, he must have the basic background to support his decisions. If a wrong decision is made, the results could range from embarrassment to the federal government, to injury or death to substantial numbers of persons.

A good OSC must have many attributes. First of all, he must have good managerial ability. From utter chaos, he must organize a small army of men and equipment to remove the oil from the water and shoreface. He must be a statistician for he has to provide and direct great amounts of money, equipment, services and manpower. He must have a sense of humor, because no matter how well the cleanup is going, some elements of the public are going to cry for instant and dramatic cleanup. And finally, he must have the ability to walk a tenuous tightrope, to try to please as many of the public, who have been injured by the spill, as possible. For instance: the sunbathers and swimmers want the beaches and surf cleaned at once; the bird lovers want the birds protected at all cost; the vessel owner wants his vessel cleaned of all traces of oil pollution; shore-front property owners want their homes protected and cleaned; the commercial and sports fishermen would rather have the oil come ashore than have it do any damage to marine life; and finally, the oil and shipping industry want the channels kept open at all cost.

Every OSC should be backed up with a fully trained staff. This staff should include, but is not limited to: a public information officer, to handle the countless calls and inquiries from the news media; a stenographer, to keep a running account of all business transacted during any given day; a contracting officer, to handle the myriad details for the hiring of and negotiations with various contractors, purchase request and the recording and handling of petty cash funds. He should also have on his staff technicians to supervise and straw-boss the various contractors and to do general field work during the life of the spill. These technicians should be well versed in oil pollution work. Finally, the OSC should have available to him legal council for consultation when the need arises, and a fully trained chemist, biologist and engineers who have a good background in oil pollution work.

The OSC and his staff should have a suitable headquarters in which to work. Ideally, a Coast Guard Station would provide a suitable headquarters if it were near the scene of the spill. If not, then other quarters must be found. These quarters could be a suite of motel or hotel rooms, a house trailer, a mobile laboratory, as found in the EPA, Water Quality Offices, or a private beach house.

The operations headquarters should be equipped with the following items: wall charts, maps, at least eight telephones, desks, chairs, writing equipment, typewriters and, if possible, a teletypewriter. One of the most important items of equipment in the headquarters is the Commander's Log. The Commander's Log, a permanent bound book, should be maintained from the inception of the spill until the case is closed out. In it there should be recorded, chronologically, a reference to all telephone calls received, meetings held, orders issued, events taking place, personnel changes, visitors received, overflights made, etc. This Log can be an invaluable record of the spill to be later used in court or congressional hearing.

Whenever the OSC and his staff are housed, the first order of business upon arrival is to order phones. At least eight phones should be installed. One of the eight telephone numbers should not be given out for general use, but should be reserved and made known to only a few selected individuals who may have to communicate with the OSC on urgent matters. It is imperative that the OSC have communications with the contractors in the field and with his observers. Small, portable hand-held radios with sufficient range would be the ideal answer. If radio communications are not available, there is another system which can be found in most large cities. This system is called People Beepers. It is a small, belt-clipon radio receiver with a range of approximately 25 miles. To get a contractor or observer in the field, the OSC dials a specified number and asks the operator to have No. A-16 call headquarters. The operator in turn activates a beeper on the radio receiver No. A-16 and transmits the message. The holder of receiver No. A-16 in turn calls the OSC by telephone and communications are established. This system can be rented for a nominal fee. Communications between air observers and headquarters is a necessity, and observation and work vessels should also be able to contact the field headquarters.

A major lack, both for government and for industry, is adequate specialized equipment for dealing with an oil spill. Despite the best efforts of either the responsible party or the Commander, if a major spill should occur at this time, it is not likely to be contained and cleaned up unless it occurs in ideal conditions: close to a cleanup cooperative, on a calm day, with wave heights under three feet.

But response problems are not limited to the availability and capability of equipment; there are also organizational and managerial problems. At every level throughout the response structure, primary and advisory agencies are represented as appropriate. Four primary (Interior, Transportation, Defense and EPA) and five advisory agencies (Commerce; Health, Education and Welfare; Justice; State; and the now defunct Office of Emergency Preparedness) are designated by the plan. This multiplicity of agencies is necessary since relevant responsibilities are fragmented among these agencies. But this fragmentation can be expected to give rise to problems in coordination and continuity. It already has in terms of gaps in response cabability due, at least in part, to a lack of a coordinated research and development (R&D) effort. This effort is supposed to be coordinated by the National Response Team through a multiagency R&D committee.

The rather complex organizational structure appears to be cumbersome and unwieldy. But, except for regional responses to Santa Barbara and the three Gulf accidents, it really has not been tested at the level of a national emergency. The OSCs have been tested but there were complaints that they lacked the expertise needed to perform their assigned responsibilities. The rationale for assigning the Coast Guard OSC responsibilities offshore in Outer Continental Shelf (OCS) operations seems to be based primarily on the expectation that a Coast Guard officer will be familiar with tankers, containment and cleanup. It is not because Coast Guard officers are expected to be familiar with OCS oil operations; and this seems to have been the basis for criticism. In response to these complaints, a memorandum of understanding was signed by Interior and Transportation: it authorizes a representative of USGS to exercise exclusive authority over measures to abate the source of pollution when it is an oil or gas well.

INDUSTRY SPILL CLEANUP COOPERATIVES

In addition to government response teams, industry has established a number of response cooperatives, and most companies apparently have formulated their own contingency plans. Two of the major cooperatives are Clean Seas on the West Coast and Clean Gulf Associates on the Gulf Coast. These are primarily equipment and training cooperatives: that is, the cooperative purchases and maintains containment, cleanup and disposal equipment and will train personnel in how to use it. Members of the cooperative draw on the equipment as needed, using their own personnel to operate it. Permanent staffs are expected to be small. Clean Gulf Associates, for example, employs three marine supervisors and one research engineer full time.

A guide for the formation of oil spill cleanup cooperatives has been prepared by Marine Management Services, Inc. of Washington, D.C., and published by the American Petroleum Institute, Washington, D.C. as of July 1972.

Cooperative programs for the cleanup of oil spills now exist in many harbor, port and other areas in the United States. Where there are a number of them doing business in the same area, oil companies commonly work together to form local cooperatives to contain and clean up oil spills that may occur. Like volunteer fire departments, these groups purchase and pool equipment, establish a system of communications, and adopt emergency plans. Often, these efforts are strengthened by the participation of other interested companies, the local fire and police departments, and the Coast Guard. A survey conducted by the American Petroleum Institute in 1973 showed that ninety such cooperatives were operational on the East, West and Gulf Coasts and on inland rivers and lakes. Some 25 additional cooperatives were under development.

A first step in the formation of a cooperative cleanup program is to identify the companies and government agencies which have an interest in the problem and which can play an active and constructive role in solving it. Those involved can then exchange information on methods of dealing with spills in the specific area and determine the types of equipment and materials that will be needed. It may be found that several firms already have various pieces of containment and cleanup equipment available. If so, it may be possible to arrange for mutual use of the equipment in case of emergency. The purchase of additional equipment, on either an individual or cooperative basis, often is advisable. In any case, equipment should be stored so as to make it quickly available at any hour.

Following study of local needs, inventory of materials, and acquisition of needed equipment, the group should draw up a procedure to combat spills. This would start with an alert system and communications network to inform those who must order out manpower and equipment. The spill must also be reported to the proper governmental agencies. Backup responsibility should be established so that the absence of a key individual will not cripple the effort. Technical assistance should be readily available. The whole point is to have everything planned beforehand to deal quickly and effectively with all types of spills.

RELATIONS WITH GOVERNMENT ORGANIZATIONS

These are the basic federal agencies concerned with a major oil spill: Environmental Protection Agency, U.S. Coast Guard, U.S. Army Corps of Engineers, and U.S. Health, Education and Welfare. However, many other federal agencies may be called on to furnish help during the life of the emergency: they may include, but are not limited to, the following: U.S. Air Force, U.S. Navy, U.S. Army, General Accounting Office, and General Services Administration.

Individual states are encouraged to make commitments to the cleanup of major oil spills. If the state decides that it wants to take complete charge of the spill, it is the policy of the federal government to agree and then monitor the situation. The problem here lies in the fact that very few states have the funds committed by legislation to become involved in a major oil spill cleanup. Interstate agencies, too, are encouraged to make commitments to the cleanup of major oil spills. Interstate agencies can also take complete charge of a major oil spill, but again, the funding limitations applies to these agencies as well as to the individual states. Local governments will rarely have the funds to combat a major oil spill. However, they can be an invaluable source of manpower and equipment.

The academic communities throughout the country can usually be counted on to furnish information and advice on the local environment during a major spill. Quite often, it is found that these institutes have previously made extensive biological studies of local areas, and this can be of invaluable help in comparison with postspill biological surveys.

FINANCING

All cost of cleanup should be borne by the polluter. In most cases, the major oil and shipping companies will shoulder this responsibility without question. If the company agrees to finance the cost of cleanup, they will probably want to take charge of the operations. This is agreeable to the federal government, however, the OSC and his staff should monitor the complete cleanup operation. Should progress of the cleanup not be to the satisfaction of the OSC, he has the authority to make the company stop its operations and have the federal government proceed with the cleanup. All cost borne by the federal government will be reimbursed by the polluter.

The individual state government can assume command of a spill cleanup, if they so desire. But again, the federal government's OSC must monitor all phases of the cleanup. The states in turn can recover expended funds either through the individual state courts or the monies can be recovered through the federal courts. The federal government has established a thirty-five million dollar revolving fund for the cleanup of oil and other hazardous materials. This fund is administered by the U.S. Coast Guard. All monies expended by the federal government in the cleanup process will be recovered from the polluter, either voluntarily or through the federal court system.

As mentioned previously, the OSC should have on his staff a responsible contracting officer. The contracting officer should handle all the details of the contract, payment and the keeping of complete records pertaining to all monies expended during the life of the spill. The OSC and the contracting officer should keep in mind that all paper work involved in cleanup cost will probably end up as evidence in a federal court.

USE OF INDEPENDENT CONTRACTORS

Prior to the Torrey Canyon incident, it was very difficult to find a contractor who had the knowledge of oil spill cleanup or the desire to clean up massive spills of oil. However, since that date, a new industry has evolved — the so-called "third party contractor" for the cleanup of oil spills.

Prior to a major spill, a good OSC should have knowledge of all third party contractors in his area of responsibility. The OSC should know the people he is going to be working with during an emergency. He should know the amount of equipment the contractor has on hand, the type of equipment, and the conditions of the equipment. He should know if the contractor has trained personnel or whether he recruits from the streets. All this knowledge is important to the OSC, for he has to rely heavily on his working contractors. Finally, the OSC should have available to him trained personnel to monitor the work of the various contractors in the field. This is necessary to insure that the federal government receives full value for every dollar spent on the cleanup.

REPORTING REQUIRED

The National Contingency Plan requires that the OSC submit two situation reports daily; one at 0800 hours and one at 2000 hours. These reports are normally submitted by teletype. If any event of dramatic importance occurs at other times of the day, the National Response Team should be notified by teletype at once. When the emergency is over, and the Regional Response Team is disbanded, the OSC should prepare his End of Operations report. This report should contain as much information as possible about the entire life of the spill incident. The Commander's Log will prove invaluable to the OSCs preparation of this report. Once completed, the report should be forwarded to the National Response Team through the Regional Response Team.

According to the New England Interstate Commission for Water Pollution Control, a variety of points should be covered as regards oil spill reporting as follows.

A. In the event of any discharge, the person, firm or corporation responsible for the discharge shall immediately undertake to remove such discharges as required by law. Responsibility for removal shall remain with the person, firm or corporation responsible for the discharge.

B. A telephone report to the Coast Guard and the state water pollution control agency shall be made immediately and shall include: time of spillage, location, type and amount of oil, assistance required, name and telephone number of person calling, and other pertinent information.

C. After removal of an illegal discharge has been completed, and upon request by the state pollution control agency, the terminal operator shall prepare a complete written report of the occurrence and submit such a report to the appropriate state agency within 10 days. If circumstances make a complete report impossible, a partial report shall be submitted. This report shall include at least the following information: date, time and place of discharge; name of terminal, name and owner of vessel or others involved, amount and type of oil discharge; complete description of circumstances causing discharge; complete description of containment and removal operations, including costs of these operations; description and estimate of third party damages; and procedures, methods and precautions instituted to prevent a similar occurrence.

D. Mystery or Unidentified Discharge — Any person sighting oil in the waters of any of the states, not previously reported, shall make a report immediately to the state water pollution control agency and/or the United States Coast Guard, giving such information as might be available. Notification should be prompt and made if any doubt exists as to whether the state pollution control agency is aware of the situation.

OIL CONTAINMENT

Despite efforts to prevent oil spills, oil has been and will continue to be spilled on the water. Once spilled, it becomes incumbent upon the party responsible for the spill to do something about it. If ignored, it will not disappear; it may relocate but inevitably will produce an adverse affect somewhere. Not to report the spill is a violation of federal law; reported, it becomes necessary and advantageous to attempt to clean it up. Pollution control officials have delegated responsibilities in this area. Thus, the only consideration in regard to an oil spill is cleaning it up.

The first priority in a control program is to attempt to limit the spread of the oil mass. Experience has shown that if left to its own devices, oil spreads into thinner and thinner films and breaks apart into smaller patches covering larger areas. The greater the area covered the more difficult and costly the cleanup program becomes. As the oil mass spreads, resources such as municipal, industrial, and agricultural water supply sources; waterfowl, fish and general aquatic flora and fauna; recreational interests, both public and private, as beaches, shorefront properties and homes, marinas, pleasure boats and tourist centers; and shellfish harvesting, to name a few, may be affected. To contain an oil spill within a limited area, oil retention barriers commonly called oil booms have been developed. A satisfactory boom design has to overcome the many forces acting upon it.

Forces are exerted by the oil being contained, and the water in which it is immersed. The boom may be used to encircle an oil slick to prevent its spread, to encircle an oil slick and then compact it so as to decrease the area of coverage and increase film thickness to make recovery easier, or to keep the oil away from sensitive areas. It is an essential tool in any oil pollution control program and generally will be the first piece of equipment placed on scene and the last removed. Considering its importance, an understanding of the way the barrier functions, and the limitations of different designs, is essential.

In order to understand the forces acting on a boom being used to contain oil in the water environment, it is first necessary to understand those forces acting on the oil itself. When oil is spilled on water, it generally tends to spread outward on the water surface forming a thin continuous layer or, depending on conditions, it may tend to accumulate as a slick having some particular thickness.

How it will spread depends on the surface tension of the water, the surface tension of the oil, and the interfacial tension between the oil and water. The tendency to spread is the result of two physical forces; the force of gravity and the surface tension of the water on which the oil has been spilled. The horizontal motion of the oil slick is caused by outward pressure forces in the oil which are a direct result of the gravitational force.

Of the forces acting, gravity and surface tension will tend to increase the spread of an oil film while inertia and viscous forces tend to retard it. The spreading tendency will be increased by waves, wind, and tidal currents. In general, the spread resulting from these random motions will be smaller than those caused by tension and gravity forces. It must also be recognized that the character of oil when spilled on water does change with time, but generally the properties which are important to spreading, namely, density, viscosity and the surface and interfacial tensions, change slowly and, therefore, can generally be predicted.

Another characteristic to be considered is that a globule or droplet of oil will have a predictable rate of rise as it emerges from an underwater position and rises to the water surface. During the rise it is subject to lateral displacement by ocean currents. Large globules of oil (greater than 1 inch in diameter) rise at the rate of approximately 1 fps. Smaller droplets rise at about 1.5 fps. In placing a containment boom to capture oil either emanating from a submerged source, or falling from a substantial height, the rate of rise phenomena has to be considered.

For example, a tanker has ruptured a line on deck and residual fuel oil is running out of the scuppers to the water 20 ft below. The oil is penetrating into the water to a depth of 15 ft. It will take 15 sec to rise up to the water surface. There is an ebb tide and the current immediately off of the anchorage is 1 knot. The oil will return to the surface approximately 30 ft downcurrent of the point of entry. A containment boom placed closer than 30 ft to the vessel would lose much of the oil being spilled.

An appreciation of the above concepts is important in understanding why a boom is necessary and how it may be used. It can provide a basis for the spill control officer to develop his own rules of thumb for predicting the physical size of the slick he is going to have to contend with. Coupled with the rule of thumb that a slick will move generally in the prevailing wind direction and at 3% of the wind velocity and taking into account the effect of tide, current and sea state, where applicable, he can begin formulating the plans for his first line of defense.

A boom is supposed to be capable of retaining oil slicks; concentrating oil slicks so as to increase thickness; acting as a device to move oil across the surface of the water from point to point; and serving as a diversionary or protection barrier to keep oil out. In almost all cases there is a regrettable tendency to overrate these capabilities rather than underrate them. The most satisfactory way of dealing with an oil spill is to contain the oil and then physically remove it as rapidly as possible. The frequent failure of the containment system greatly complicates the physical removal effort. When a barrier is placed in the path of an oil slick, the spread effect is interfered with and a pool of oil, generally much deeper than that which would result from an undisturbed slick, is formed.

The boom's performance is affected by wind, waves, and currents. It must be capable of conforming to the wave profile so as to maintain its freeboard and have sufficient structural strength to withstand the stresses set up by wave and wind action. Sufficient vertical stability is required to overcome the roll effect forces set up by the water current on the fin and wind on the sail to keep the boom from being flattened out on the water surface. Freeboard (sail) must be adequate to prevent oil from being carried over the top of the boom by wind action and choppy wave motions.

Although wind and wave action are important in boom design, water currents are the usual reason for boom failure. Wind is often the controlling factor in moving a slick about on the surface of the water, and wave motion often contributes to breaking up an integral slick into many smaller patches.

The relative current, the resultant of that generated by the water current in the area, be it tidal or river, and the motion of the boom itself, however, will often be the controlling factor in boom failure. Boom failure is defined as a loss of its capability to retain the oil slick. Experimental work has shown that an oil slick being contained by a mechanical barrier will exhibit a shape like that shown in Figure 40.

FIGURE 40: BEHAVIOR OF AN OIL SLICK CONTAINED BEHIND A BOOM

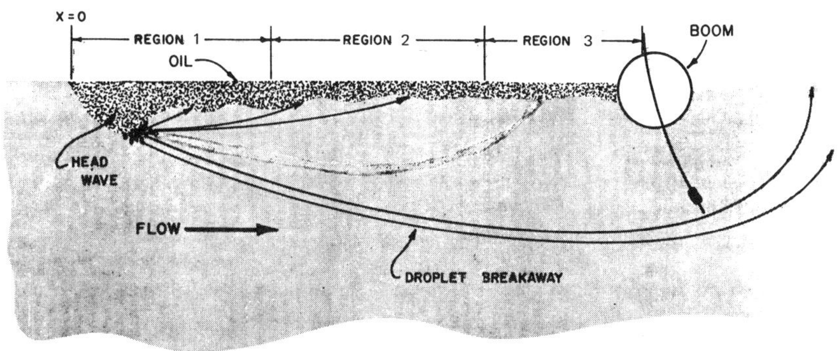

Source: Report PB 213,880

Initial failure will occur when oil droplets break away from the lee side of the head wave. After a critical velocity (v_c) is exceeded, oil droplets will be entrained in the flowing water stream. Unless the droplets have sufficient time to rise through the water and rejoin the slick in Regions 1 or 2, they will be swept under the barrier. Region 1 is the critical area for defining the limits for oil breakaway. The type of configuration which the oil mass assumes in this region is similar to a gravity wave.

For critical conditions (maximum droplet size) v_c will equal 0.62 fps. Below this velocity no droplets will be released from the head wave. Exceeding this velocity does not necessarily mean that oil will be lost under the barrier. It may be redeposited on the slick in Regions 2 or 3. There will be circulation and drag phenomena working in Region 2 as well as the generation of waves at the oil-water interface. These will affect the location of the point of reattachment. In Region 3 which runs from the boom to a point equal to approximately five times the boom skirt depth, any oil droplet entering will be swept under the boom.

The thought should be kept in mind that a mechanical barrier does not have to remain stationary. The relative velocity profile can be changed by allowing the barrier to drift with the oil mass. The failure mechanism can also be modified by instituting skimming action as soon as possible after the slick is boomed. Quite often, however, the amount of skimming capability available will not be adequate to prevent failure without additional action. The failure characteristics can also be modified by the use of ad- or absorbents which will change the thickness profile of the oil in the boom area.

The information having now been presented concerning the factors which should influence boom design and the mechanisms by which they are most likely to fail, what types of barriers are commercially available? What capabilities should the boom have in addition to that most important one of oil retention?

One of the most important is transportability. Cleanup equipment, in general, should be designed for easy transportation from some centralized storage point to the site of a pollution incident. A modular system designed for air transport would be best. Once the equipment is on site, ease of deployment is essential. Regrettably, spills are not in the habit of occurring in ideal weather conditions. The equipment should be designed to provide maximum compatibility with the type of watercraft that are likely to be used in its deployment.

In particular, weight handling, towing, and equipment assembly requirements, should be selected with the capabilities of available vessels in mind. The barrier system design should also be as compatible as possible with the other cleanup equipment available. Such a simple thing as noncompatible connecting hitches can destroy a well laid out spill control program. Last of the major points but by no means least, the containment system should be mutually supporting to the recovery system.

Containing an oil spill without matching provisions for optimum physical removal is winning less than half the battle. Wherever possible, components within the barrier system should be readily adaptable for use with the oil removal system. Oil retained and not removed will end up being oil lost to the environment. The types of equipment available may be broadly categorized into three mechanical-type designs and pneumatic barriers. The mechanical barriers are listed below. The general characteristics of these designs are shown in Figure 41.

(1) Curtain booms which consist of a surface float acting as a barrier above the surface and a subsurface curtain suspended from it. The curtain is flexible along the vertical axis. It may or may not be stabilized by weights to provide greater resistance to distortion by subsurface currents and have a chain or welded wire rope to transfer stress along the barrier.

(2)(3) Light and heavy fence booms have a vertical fence or panel extending both above and below the surface of the water to provide freeboard to counteract wave carryover and draft below the water surface. The flotation assembly is generally bonded to the fence material. The lower edge of the panel is frequently stabilized and strengthened by a cable or chain. The distinction between light and heavy is generally one of size of components and weight.

To prevent the spread of oil slicks over wide areas, many different types of oil retention barriers have been developed. Oil booms may be deployed to encircle oil slicks, thereby reducing the area of coverage and increasing the thickness of slick and providing for feasible recovery, collection and disposal of the waste material. Oil booms may also be installed as a barrier to close off prescribed zones where spills may occur or where oils should not enter.

It must be recognized that boom systems represent only one set of tools, although an important one, in effectively combating oil spills once they do occur. Oil spill booms are almost always deployed in conjunction with other equipment and methods for effective control and cleanup. In the simple case, this will mean a skimming or pickup device, whereas, for the serious spill it may mean any number of methods and techniques used concurrently to combat the situation.

A review of oil containment systems has been published by the Oil Spills Branch of the Edison Water Quality Research Laboratory of EPA in October 1970 and a second edition of this volume published in January 1973. That report categorizes booms as follows:

Commercial floating booms – Some 46 designs are described as listed in Table 14. Systems are listed in alphabetical order so as to indicate no product preference.

Multipurpose booms – These serve both to contain and to remove oils from water surface. They will be discussed in a subsequent section of this book entitled "Combined Containment and Removal Devices."

Improvised booms – In many cases, a commercial boom is not available when an oil pollution incident occurs and maximum use of available materials must be made to provide at least temporary containment or diversion capability until better equipment can be brought to the site. Some 11 types of such booms which have actually been used are listed in Table 15.

Air barriers – These are discussed in the text which follows.

FIGURE 41: TYPICAL BOOM DESIGNS

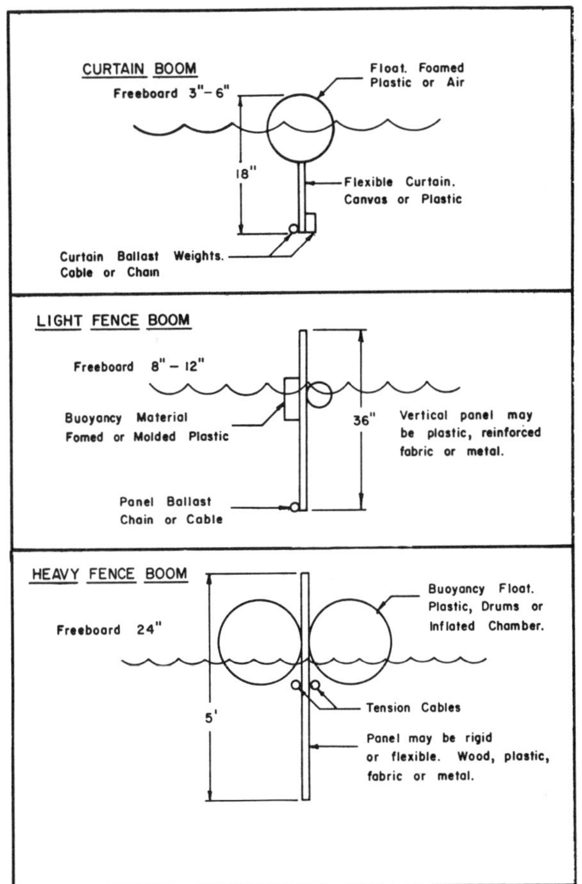

Source: Report PB 213,880

TABLE 14: COMMERCIAL FLOATING BOOM TYPES

(1)	Acme O.K. Corral Boom	(13)	Flexy Oil Boom
(2)	American Marine - New Design II	(14)	Flo-Fence
(3)	Aqua Fence	(15)	Galvaing Floating Booms (France)
(4)	Bennett Offshore Oil Boom	(16)	Gates Boom Hose
(5)	Bennett 3-Foot Harbor Oil Boom	(17)	Jaton Boom
(6)	Bennett 36-Inch Flexi Boom	(18)	Johns-Manville Spillguard Booms
(7)	Bennett 20-Inch Harbor Oil Boom	(19)	Kain Filtration Booms
(8)	Boa Boom	(20)	Marsan Oil Barrier
(9)	Bridgestone Oil Boom	(21)	MP Boom
(10)	Bristol Aircraft Co. Boom (Great Britain)	(22)	Muehleisen Boom
(11)	BP Boom	(23)	Oscarseal: Hover-Platform
(12)	Clean Water, Inc. Inflatable Filter Boom	(24)	Oscarseal - Steel Boom

(continued)

TABLE 14: (continued)

(25) PCR-36 Oil Containment Boom
(26) Pirelli Oil Spill Boom
(27) Quincy Adams Standard Oil Boom
(28) Red Eel Oil Boom
(29) Retainer Seawall
(30) Reynolds Aluminum Oil Boom
(31) Reynolds Aluminum Oil Boom (15-inch)
(32) Sea Boom
(33) Sea Curtain
(34) Sea Fence
(35) Sealdboom
(36) Sea Skirt
(37) 6-12 Boom
(38) Slickbar-Mark V
(39) Slickbar-Mark VI
(40) Slikguard Oil Slick Barrier
(41) SNV Floating Barrier - Uginox
(42) SOS Booms (Sweden)
(43) T-T Oil Boom (Norway)
(44) U.S. Coast Guard Boom
(45) Warne Booms (Great Britain)
(46) Water Pollution Controls Boom

TABLE 15: TYPES OF IMPROVISED BOOMS

Type of Boom	Location Where it was Used
Cork filled boom	Norfolk, Virginia
Cork-float boom	Port Hueneme, California
Fire hose boom	Quiescent waters
Puerto Rican boom	Ocean Eagle oil spill
Rubber bladder boom	Helford River, Great Britain
Rubber tire boom	Torrey Canyon oil spill
Steel pipe boom	Philadelphia, Pennsylvania
U.S. Navy boom	Long Beach, California; Chevron spill, 1970
Wooden float boom	Pearl Harbor, Hawaii
Wooden timber boom	Quiescent waters
Wooden V-boom	Peros Gyiroc, France

Source: EPA, Edison, N.J. Report, *Oil Spill Containment Systems,* (January 1973)

The following is a narrative presentation illustrating one possible use of oil booms for the control of an oil spill. It does not specifically represent any singular pollution incident but rather reflects techniques which met with at least some degree of success in actual incidents.

A vessel with a cargo of crude oil struck a submerged object in the navigation channel approximately 4 miles from shore. The forward center and starboard tanks were holed and approximately 1,000 barrels of a fairly heavy crude oil lost before leakage was controlled. At present the sea is calm and the current immediately off of the channel where the oil was lost is approximately 2 knots. The wind is blowing steadily from the south at about 9 knots and will move the oil slick northerly toward a beach area.

Making use of information previously presented, it is determined that oil will begin reaching the beach area in about 12 hours. Considering the time involved, it is likely that at least part of the slick will reach the shoreface area and that some procedures will have to be implemented for its protection. It is necessary to get equipment out to the site of the slick as soon as possible to remove as much as possible before it does reach the beach area. (See Figure 41.)

Two skimmers are available, one of 400 gpm and one of 200 gpm capacity. Work boats, fishing boats, various types of absorbers, and 1,500 ft of light fence boom, and 1,000 ft of pneumatic barrier. It is determined that the best defense barrier will not be capable of holding the oil offshore in a fixed position while it is pumped off the water. It has also been determined, based on the relative wind velocity, current situation, and the type of spread to be expected from this particular grade of oil, that without some type of containment, a thin film of oil will come ashore within 12 to 14 hr.

FIGURE 41: TYPICAL EXAMPLE OF A GENERAL POLLUTION SITUATION

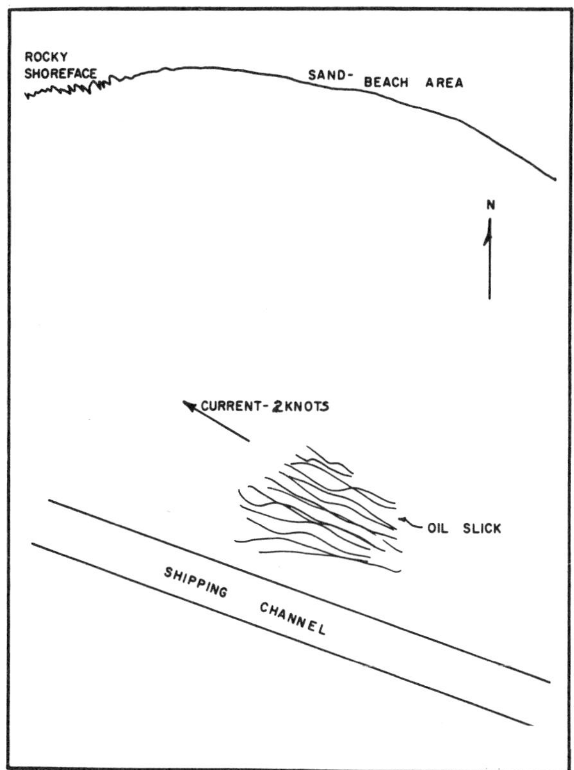

Source: Report PB 213,880

The following is one approach that might be followed. Put the larger skimmer on a work boat along with 900 ft of the fence type barrier and proceed out to the site of the slick. Set up the equipment as shown in Figure 42. The skimmer should be positioned so as to work in the pool of oil collected by the barrier. Note that one end of the boom is held close alongside the work boat and skimmer with a small boat out at the other end.

The oil will collect along the wind side of the boom and then be diverted along the boom to the skimmer. The skimmer should be operated at capacity with the oil-water mix being discharged to some type of storage facility for oil-water separation. The following information should be noted. There must be an oil tight connection between the skimming unit and the boom at **a** and the boom and the work boat at **b**. The position of the upwind end of the boom and the work boat itself must be changed quickly as the wind shifts. The flow of the slick must continue into the area between the boom and the pump boat.

The angle between the wind direction and the boom should not be greater than 20°. With more angle the boom will tend to develop a belly in its shape and oil will be lost under the boom. The work boat and the boat holding the leading edge of the boom are drifting at a speed which has previously been determined will not allow oil to be lost under the barrier. This speed will be adjusted based on on-site conditions.

FIGURE 42: MAIN SLICK RECOVERY

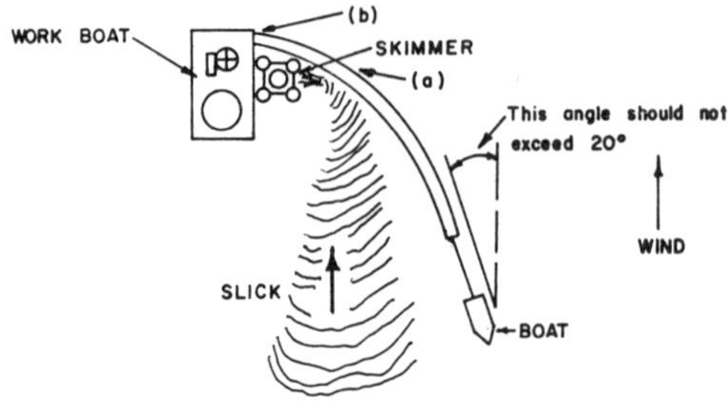

Source: Report PB 213,880

No comments will be made about the type of skimmer, oil-water separation system or storage system. These are obviously a very important concern and will control the speed at which the recovery system will move with the slick to prevent loss.

There are two other problems to contend with: (1) it is obvious from the calculations that it will not be possible to remove all of the oil from the water surface before it gets into the beach area; and (2) the slick is not remaining as one integral mass and portions of it are breaking away and being left behind in the bay.

Next, consider handling those portions of the slick being left behind in the bay. Two small fishing boats and a contractor's work boat having flat deck space in the bow, have been contracted for and the remaining fence boom and the small skimmer mounted on them. These can be arranged as shown in Figure 43.

FIGURE 43: TOWING A FUNNEL BOOM ARRANGEMENT

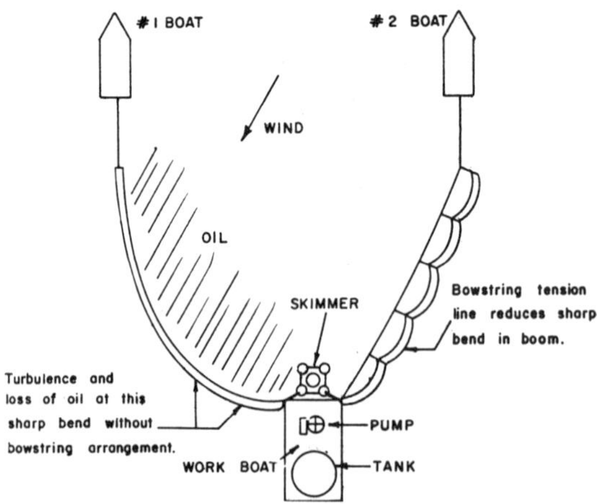

Source: Report PB 213,880

The bow string tension line attached to No. 2 boat reduces the sharp bend that would otherwise exist in the boom and eliminates turbulence and oil loss at the point where the boom would ordinarily bend when it joins the work barge as with No. 1 boat. With the boom system so deployed, this work group begins chasing those slicks left behind and using the small skimmer, pumps them to appropriate storage.

The operating speed of this group is controlled by the structural strength of the boom in question which has been considerably increased with the bow string arrangement, and the thickness of oil being recovered at any one time. Remember that the wind should be used to advantage whenever possible to assist in diverting the oil down the funnel. Note that radio communication among the three boats is essential for proper functioning of this system.

At this point it is necessary to consider setting up some type of protection at the beach area. As all that is left is the air barrier, this will have to be implemented. The air barrier should be placed on the bottom at a distance from shore that will provide it with a minimum of 10 ft of water at low tide. The air supply can come from a diesel powered air compressor mounted on a barge or convenient headland if available. Taking a look at the area, the most difficult area to clean is going to be the rocky section on the northwest side of the beach. The air barrier should be set up so as to divert the oil from the rocky section toward the beach area.

A physical absorber should be placed along the tidal zone of the beach area to absorb as much of the oil as possible to prevent penetration into the beach sand. A float with some type of anchoring system has been established a short distance from the end of the air barrier so that the work barge and skimmer, when they arrive at this point, will have somewhere to attach their fence boom. Initially, the oil that has gotten away from the two skimmers working in the bay will be diverted toward the beach. When the main skimming apparatus arrives on scene, the oil will be diverted along the air barrier to the fixed mechanical barrier to the skimmer (see Figure 44).

FIGURE 44: SHOREFACE PROTECTION SCHEME

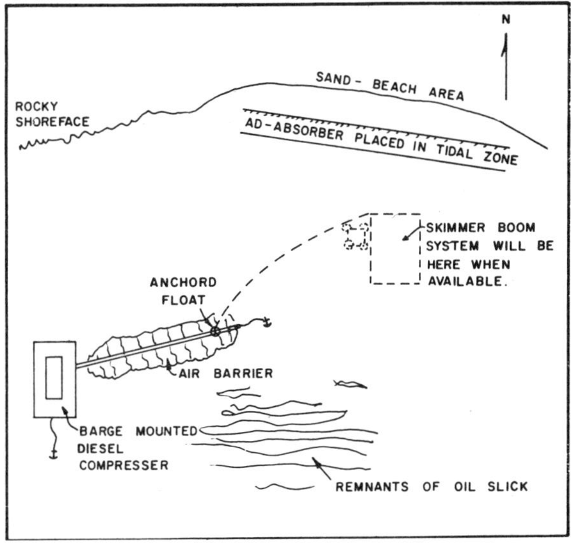

Source: Report PB 213,880

This is a simplified discussion of boom placement which does not attempt to take into account all of the variables which might be encountered. It is cited as an example of some of the procedures which could be implemented. It must be emphasized that getting the boom out and corralling the oil is only the beginning. The very difficult task of removing the material from the environment and its ultimate disposal has to be as well considered and planned out as the boom purchasing and implementation program.

There would have been nothing more useless, for example, than taking all of the barrier that was available in the incident cited above and drawing it across the beach area and then sitting back and watching the oil pile up along its face to finally run under and/or over it to contaminate the beach area. The best designed equipment, if not properly used, will not do the job.

AIR BARRIERS

A bubble, or air barrier consists of submerged, perforated tubing from which compressed air is released. The result is a rising curtain of bubbles which produces an upswelling on the water surface above the tubing. The upswelling produces a raised surface behind which the oil can be trapped. One advantage of bubble barriers is that they do not impede the movement of vessels. But they work successfully only so long as the free current of the water and the wind do not overcome the forces set up by the compressed air. Too, bubble barriers are rather costly to install and maintain.

A discussion of oil spill containment in general with particular attention to air barrier design and construction has been presented by J.B. Herbich of Texas A & M University in Coastal and Ocean Engineering Report No. 150 (March 1972).

The air barrier (air curtain) represents, in certain cases, an alternative method of containing or confining spilled oils within a given area, and also preventing floating materials from entry into high-use waterfront areas. A perforated pipe is laid on the river or harbor bottom and compressed air is forced through the line. The air delivery pipe may also be suspended at some depth above the river bottom depending upon local conditions.

A curtain of rising air bubbles produces an upward current which upon reaching the free surface, spreads laterally in opposite directions. The artificially induced currents generated by the air curtain are designed to counter the normal surface currents prevalent in the body of water, thus confining floating pollutants within the desired boundaries (see Figure 45). The air barrier has distinct advantages over the physical oil booms since vessel traffic may pass through the control area without hindrance; however, it also has certain disadvantages. The Standard Oil Company (New Jersey) reports in an *Oil Spill Cleanup Manual* (1969) that:

> "The capacity of a pneumatic barrier is limited by both environmental and economic considerations. The natural current of the water effects the rising air plume, by causing it to lean over. If the plume is leaned over more than 30° from the position it would assume if no current were acting, the plume will break up and the overall effectiveness is diminished. This problem can be overcome by increasing the velocity of the rising air bubbles. However, for a given nozzle there is a critical velocity above which additional increases in air volume have little effect on the magnitude of the surface generated current; i.e., the efficiency of the barrier is greatly reduced. The efficiency of the barrier improves in proportion to water depth; shallow waters require greater volumes of air."

In conclusion, air barriers have been demonstrated to be successful in quiescent waters. However, more work is required on determining the influence of orbital wave motion on the performance of an air barrier. Generally the barrier is made from a pipe, 1 inch i.d. or less, with a series of orifices drilled through the pipe wall 1/16 inch in diameter or less. Air is supplied by either an air compressor or blower.

FIGURE 45. AIR BARRIER EFFECT

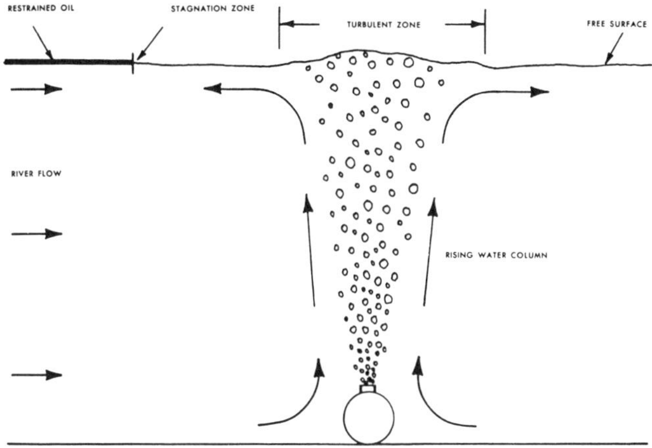

Source: EPA Report *Oil Spill Containment Systems* (January 1973)

As an example of capacities required, a 600 cfm compressor is capable of supplying adequate air to 100 ft of 1 inch diameter pipe having 1/16 inch diameter orifices positioned 40 ft below the water surface. As the air bubbles exit from the pipe and rise to the water surface, they impart momentum to the water. This causes vertical water flow which becomes horizontal at the water surface. The surface current generated retains the oil. Figure 46 illustrates the circulation pattern developed.

FIGURE 46: CIRCULATION PATTERN AND VELOCITY PROFILES IN AIR BARRIER

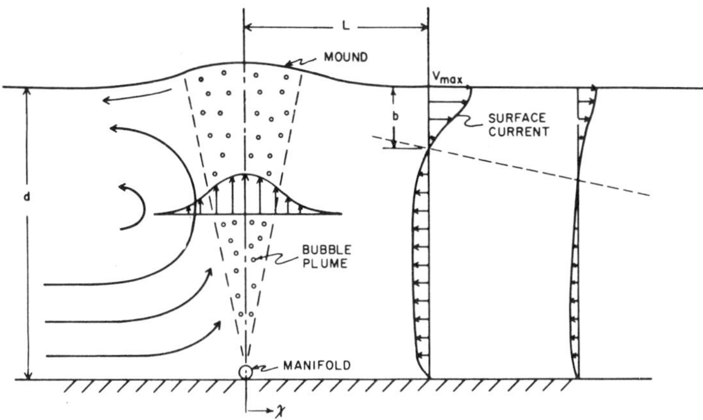

Source: Report PB 213,880

In calm waters the maximum surface current (v_{max}) created is related to the volume flow rate of air per unit length of pipe.

$$v_{max} = k(gQ)$$

v_{max} = maximum generated surface current ft/sec

k = a constant

g = acceleration due to gravity 32 ft/sec^2

Q = volume flow rate/unit length of manifold

The surface current reaches a maximum at a distance from the center line of the barrier varying between d and d/2 where d is the distance between the undisturbed water surface and the location of the manifold (see Figure 46). It then decreases approximately proportional to $1/\sqrt{x}$ where x is the horizontal distance from the manifold. The effective depth of the surface current **b** is d/4 at x = 1 and increases linearly for a larger x.

Water currents will cause a distortion in the plume (see Figure 47). This distortion creates a flow pattern which will allow some of the oil to disperse through the bubble screen. The greater the current, the more severe will be the problem. As with a mechanical barrier, oil drops are lost from the bottom of the slick; in this case, because of the turbulent pattern produced.

FIGURE 47: CIRCULATION PATTERN UPSTREAM OF AN AIR BARRIER IN A CURRENT

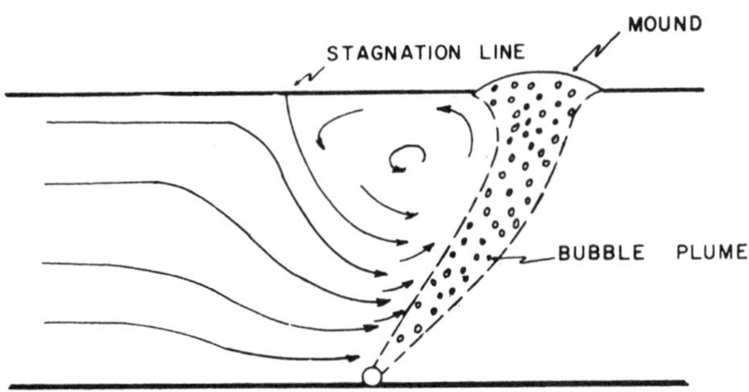

Source: Report PB 213,880

Atlas Copco Design

A pneumatic barrier design developed by Atlas Copco Inc. is described in Report COM-74-10212, Springfield, Va., National Technical Information Service (November 1973). It involves an air pipe submerged at a 10 to 12 ft depth to produce a curtain of bubbles by flow from perforations in the pipe.

It can be designed for specific applications and it is claimed that a permanent installation can be activated in one minute. Such barriers are usually permanently installed in the water around a ship's berth. It is stated in the reference cited above that such barriers are currently in use at European and American oil terminals. The conclusion, however, was that not enough information was available to consider this design as a candidate for offshore terminal use.

Fennelly Design

A design developed by *R.P. Fennelly; U.S. Patent 3,744,254; July 10, 1973* is an air or bubble barrier which can be used to contain and deflect aqueous pollutants, such as oil, flotsam and the like. The air or bubble barrier is confined within its desired line of travel by placing a mesh material between at least one point adjacent the generation point of the gas below the surface of the water and a point which is adjacent the surface of the water. The mesh material can be ordinary wire screening used in household screens. The use of the mesh material insures that the bubble barrier will follow a desired line of travel largely unaffected by currents or eddies which would normally disperse the bubble stream to an undesirable extent.

Figure 48 shows, at the top, the apparatus in position to contain an oil slick that has leaked from a barge in an aqueous environment containing rather strong current as indicated by the arrows. The lower left view in the figure shows the mesh protector in use in a situation containing a variety of current and eddies. The lower right hand view shows the mesh protector in use in a steady unidirectional current.

In order to contain the bubble barrier in a desired line of travel in aqueous environments containing currents or eddies, this process contemplates providing a mesh material **13** to confine the bubble as it rises within a desired line of travel. Where the currents and eddies are in a variety of directions, it is necessary to apply the mesh material on both sides of the outlet **24** in conduit **11** from which the bubbles are generated.

FIGURE 48: PLAN AND SECTIONAL ELEVATION DETAILS OF BUBBLE BARRIER APPARATUS

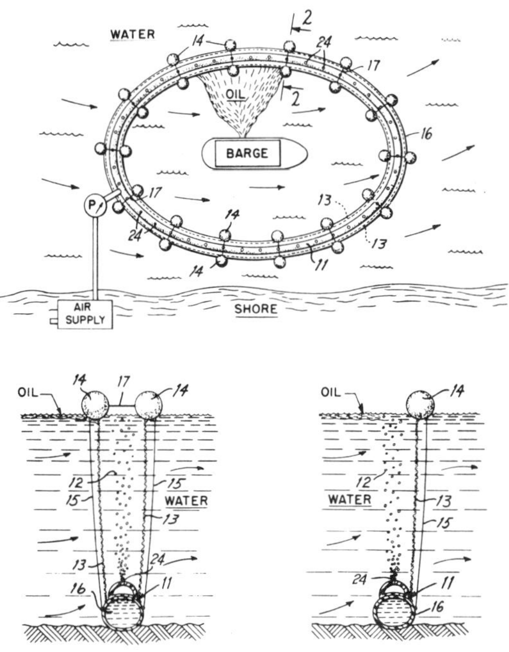

Source: R.P. Fennelly; U.S. Patent 3,744,254; July 10, 1973

The upward travel of the bubbles insures that they will strike the mesh material or screening at an angle which is not substantially normal to the plane in which it lies at the point of impact thereby insuring that the bubbles will not pass through the openings in the mesh material but will rather rebound and remain within the corridor formed by the two portions of mesh material. Each mesh material can be attached at one of its ends to the conduit **11** so that it is adjacent the point of gas generation. Its other end is placed adjacent the surface of the water, preferably by attachment to a float **14**.

Guy wires **15** or restraining members link the floats with the ballast member **16** attached to the gas conduit **11** to insure that the position of the floats will not vary too much from the position occupied by the gas generation means. If desired, the position of the floats relative to one another can be stabilized by including another guy wire **17** as a restraining member. In environments in which the current is generally in one direction, only one piece of mesh material is needed. The presence of the current will insure that the bubbles in their rebounding travel towards the surface and will be held in very close proximity to the mesh material.

Harmstorf Design

According to the EPA Report on *Oil Spill Containment Systems,* Edison, N.J. (January 1973), the Harmstorf Pneumatic Barrier has been commercially available for many years with extensive use in many locations throughout Western Europe and Northern Africa. Harmstorf employs a small diameter plastic pipe(s) properly weighted and placed on the harbor bottom. It is believed that the Harmstorf curtain produces a relatively small diameter bubble under small orifice conditions with relatively high pressure in the air delivery line. Harmstorf's air barriers have generally been limited to harbor areas where surface currents do not exceed 0.8 to 1.0 fps. The cost of this barrier is approximately $30.00 per foot.

The EPA information was stated to be based on product bulletins received from the Harmstorf Corporation, Hamburg, Germany as well as on materials received from Spearin, Preston, and Burrows, Inc., United States representative for Harmstorf Corporation in New York City (1968). However, other information reported in Maritime Administration Report COM-74-10212, Springfield, Virginia, National Technical Information Service (November 1973) indicates that the Harmstorf barrier is designed to operate in 3 ft waves, 15 knot winds and 0.6 knots current. A 2,000 ft barrier was designed to contain 3,400 barrels of oil of 0.75 to 0.92 specific gravity. It was further stated that two men and one vessel were required to deploy a 2,000 ft barrier.

H. Grunau; U.S. Patent 3,651,646; February 28, 1972; assigned to Harmstorf Corporation, Germany has developed a system for the confinement of pollutants on water surfaces until collected or chemically dispersed. The system effects its confining action by generating a curtain or barrier of air bubbles which at the surface of the water form a series of overlapping aerated water hills capable of blocking the passage of a pollutant, such as oil film, therethrough.

The system is designed to enable it to provide a continuous barrier of aerated water hills over long periods of use, even though it may be installed on harbor and waterway bottoms where silty conditions prevail. This is accomplished by a pipe for supplying compressed gas to an outlet, comprising a nozzle plate having an orifice of a size to allow a metered amount of air to pass, with a check valve located downstream of the nozzle plate.

Figure 49 shows an air barrier system using the Harmstorf nozzle as well as details of the nozzle construction. The nozzle **20** is composed of a nozzle body **25** having a length approximating the thickness of the conduit wall and made of any suitable material such as plastic, metal, or synthetic rubber compound. The nozzle body is provided with an external screw thread **25** which threadedly engages a matching internal screw thread **27** provided in the wall of its associated opening in the conduit section **14'** on the sea bottom. The head **28** of the nozzle body is provided with slots **29** for receiving the tool used to screw the nozzle body into position on the conduit section.

FIGURE 49: HARMSTORF AIR BARRIER SYSTEM

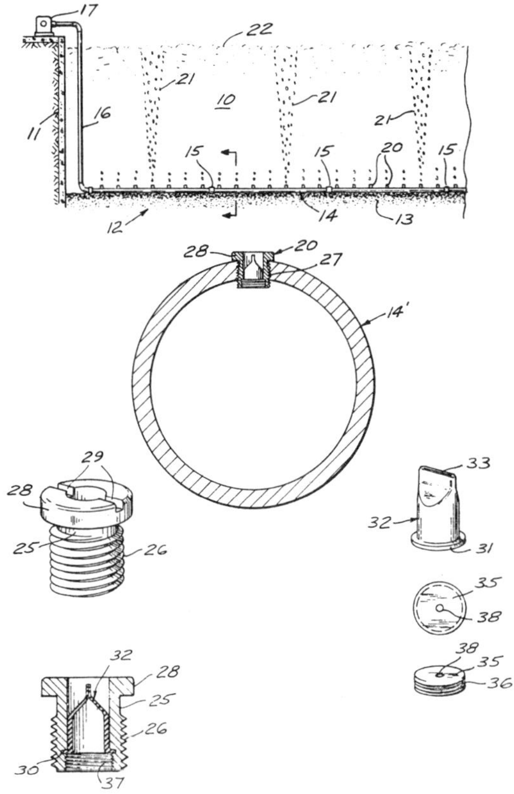

Source: H. Grunau; U.S. Patent 3,651,646; February 28, 1972

Spaced approximately three-quarters of the length of the nozzle body from the head end is an internal annular groove **30** in which is seated an external flange **31** provided on the lower end of a tongue sealed check valve **32**. This valve has a height less than the height of the nozzle body above the groove **30** so that it is entirely contained within the body. The tongue slit or resilient flaps **33** at the top of the check valve, the body of the valve and the flange **31** are made of an integral piece of rubber or other suitable material.

The rubber check valve is preferably inserted into the nozzle body through the bottom of the latter until the valve flange **31** is engaged in the nozzle body groove **30**. The check valve is then locked in position by a nozzle plate **35** provided with an external screw thread **36** which is in threaded engagement with a matching internal screw thread **37** provided in the bottom end of the nozzle body below the groove **30.**

The nozzle plate is provided with an orifice **38** through which air escapes from the conduit. It will be understood that the nozzles are each completely assembled before being screwed into the internal threads **27** provided in the conduit openings. It will be understood from the foregoing, that when the system is in operation compressed air supplied to the conduit

14 on the sea bottom from the air compressor **17** escapes at each nozzle in the conduit. The air escaping at each nozzle initially passes the orifice opening **38** in the nozzle plate **36**. The opening **38** is of a predetermined diameter and controls the volume of air being emitted by the nozzle.

The compressed air emitted through orifice **38** passes through the body of the check valve and due to its excess of pressure over the hydrostatic pressure, forces apart the tongue slot **33** at the top of the body, and thence forms the bubbles which rise in a column **21** through the water and produce the water hills at the water surface. When the compressed air supply is shut off, pressure within the conduit is equalized to the hydrostatic pressure by escape of air through the nozzles, and further reduced by cooling of the air within the conduit.

When the hydrostatic pressure exceeds the internal pressure at the valve tops **33** of the nozzles, this pressure closes the tongue slots, thereby preventing the flow of water and solid particles back into the conduit. The hydrostatic pressure also forces the check valve flanges **31** against the inner sides of the body grooves **30** and the upper ends of the joints between the threads **36, 37** thereby sealing the nozzles completely against the inflow of the solid particles in suspension therethrough and into the conduit **14.**

In the figure, **10** indicates generally a harbor basin bordered on one side with a quay wall **11** and having a harbor bed **12** on which is a deposit **13** of organic silt, finely divided clay or the like. Laid on the deposit is a conduit **14** composed of a plurality of conduit sections or lengths **14'** connected together by couplings **15.** The conduit is laid in a manner to isolate the area occupied by the oil or other contaminant film to be confined and removed. Thus, the conduit may extend across the basin, or may be given any desired form on the harbor bed to effect the intended confinement of the oil film.

The conduit is connected at one end by a section **16** to an air compressor **17** mounted on the quay wall. It will be understood that if an oil film on an open area of water is to be contained, the air compressor will be mounted on a vessel and the connecting conduit **16** is preferably flexible. It will also be understood that instead of laying the conduit on the bed or floor of the water area in which the oil film is located, such conduit may be supported in suspended relation from surface floats.

The sections of conduit may be made of any suitable material such as steel, cast iron, rubber or synthetic rubber, plastic, etc., and preferably has a relatively thick wall. In the duct sections **14'**, the wall thickness is approximately ½ inch. Provided along the upper edge portion of each section **14'** is a longitudinally extending series of spaced air nozzles generally designated **20.** The nozzles are spaced apart so that when the columns **21** of air bubbles arising therefrom reach the water surface they form a series of overlapping aerated water hills **22** which provide a continuous band or barrier having a configuration corresponding to the manner in which the conduit is laid on the water floor **12.** These hills well out transversely on either side of their alignment and prevent the oil film from passing through the pneumatic barrier.

Hoult-Fay Design

A design developed by *D.P. Hoult and J.A. Fay; U.S. Patent 3,593,526; July 20, 1971* is a submerged, segmented, pneumatic boom arranged in zigzig, accordion-pleated configuration. Experiments have revealed that, in a sizeable wave field, the hydrodynamic mode of operation of such a pneumatic boom is very different than it is in calm water. In contrast, in a wave field, the pneumatic boom must operate in a regime of strong turbulent mixing, due to the orbital water movement created by the wave action.

To understand the proposed mode of operation of a pneumatic boom in accordance with this process, consider a tube of length L, submerged to a depth d below the water surface. This tube contains air under pressure which flows out of the tube through numerous small holes, forming bubbles. Let the volume flow rate of air through the tube and out the holes by Q. If the bubbles formed are small enough, and air-water mixture is formed which is slightly less dense than the surrounding water, and hence rises to the surface due to the action of buoyancy, and hence spreads outward, making a surface current. This maximum surface current, U, depends on Q as (g is the acceleration due to gravity):

$$U \sim \left(\frac{Qg}{L}\right)^{1/3}$$

When such a pipe is placed in the open sea, which has waves of height a above and below the means surface (S), at a depth very much deeper than a, $d/a > 1$, the resulting rising column of fluid, generated by the air-water mixture formed near the pipe, rises much as it does in still water, and has a width near the surface, which is much greater than the amplitude of the orbital motion of a fluid particle in the wave field, where it splits into two surface currents which are used to control the spread of oil.

However, experimental work shows that this mode of operation is more wasteful of power the deeper the pipe is placed, because the resulting surface current is independent of the depth of the pipe, whereas the power required to pump the air increases as the hydrostatic pressure increases, such that the power requirements become excessive. On the other hand, when such a pipe is placed at a depth just below the surface, say $d/a \approx 2$, the width of the plume is determined by the amplitude of the motion of a fluid particle in the wave field. Thus the buoyant mixture of air and water is spread over a larger volume than it would be in still water, and hence the column rises at a slower rate than it would in still water.

It has been found that, for a given wave field, there is an optimum depth, for least power consumption, at which the pipe should be placed. This depth, expressed as a ratio of d/a, ranges between 10 and 5, and is preferably about 8. At such depth, the maximum surface current generated is as large as it would be if the pipe were very far below the surface.

Suppose the oil is spreading at a velocity V, due to a combination of wind, waves and surface tension. These effects in the open sea cause V to range from about ⅓ to 1 ft/sec. To stop the oil from spreading, U must exceed V at a distance a (all in the direction of wave or current movement) from the boom, because otherwise the orbital motion W of the waves in such direction would carry the oil past the boom. The volume flow required is:

$$Q = \frac{L}{g}\left(\frac{a}{d}\right)^{3/2} 8^3 V^3$$

provided that d/a is less than or equal to the optimum. The flow rate Q, at the optimum depth, is for a 1 ft/sec spreading velocity, Q/L = 1 cu ft/ft of pipe. If the boom were placed at d/a = 4, the required volume flow increased to 4 cu ft/ft. Hence, for a given wave amplitude, doubling the depth of the boom from d/a = 4 to d/a = 8 results in a two-fold savings in power, as although the required air pressure is doubled, the volume flow rate is decreased by four. If the boom is placed deeper than the optimum, the required Q is independent of depth, but the pressure and hence the power increased linearly with increasing depth.

However, it often happens that oil is carried on a current of such strength that the power requirements as determined above are excessive for the equipment available. It has also been found that if the boom is segmented and the segments are placed at an angle θ to the oncoming current, in zigzag, accordion pleated formation, only the component of current normal to the boom need to be stopped, at a distance x perpendicular to the boom segment which satisfies the above requirements as to wave induced water movement W, hence the above equation reads:

$$Q = \frac{8^3 L}{g}\left(\frac{a}{d}\right)^{3/2} V^3 \sin^3 \theta$$

A typical value of θ may be considered to be about 30°. For booms placed at this angle, the volume flow per foot is decreased by a factor of 8 whereas the overall length of the boom is increased by a factor of 2, for a fourfold savings in power. The practical minimum angle is that at which the orbital motion of the wave would carry oil from one apex of the accordion formation to another, i.e., $\sin \theta_{min} = a/L$, where L is the segment length. A typical value would be 6° for 10 ft high waves, and L = 100 ft. The maximum angle is determined by the maximum power available. The power required per foot of boom is Qp/L, p being the pressure in the pipe. For a given depth, p is fixed. For a given streaming velocity, V, the maximum angle is given by (using the maximum Q available):

$$Q_{max} = \frac{8^3 L}{g} \left(\frac{a}{d}\right)^{3/2} V^3 \sin^3 \theta_{max}$$

Figure 50 then shows a form of pneumatic boom apparatus especially useful for oil slick containment in the open sea in the presence of waves and currents. Its basic elements consist of a plurality of linearly extending hollow pipe segments **12, 14, 16,** interconnected by flexible tubing elements **13, 15, 17** and having apertures **18** preferably arranged in a helical configuration, providing an extended air bubble source.

FIGURE 50: PLAN AND ELEVATION OF PNEUMATIC BOOM DESIGN

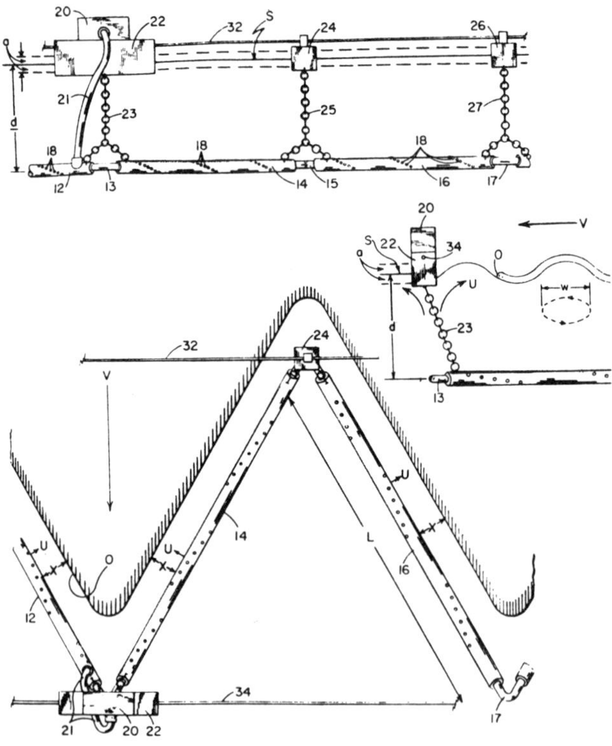

Source: D.P. Hoult and J.A. Fay; U.S. Patent 3,593,526; July 20, 1971

A suitable air compressor **20**, connected to pipe segment **12** through line **21**, provides air compressed at a suitable pressure so that streams of bubbles are produced at pipe apertures **18**, the helical configuration of the apertures also serving to assure the removal of water from bubble producing pipes **12, 14, 16.**

A plurality of floats **22, 24** and **26** having adjustable chains **23, 25, 27**, respectively, are provided for suspending pipes **12, 14, 16** at the predetermined distance d, while the distance between alternate floats is controlled by control cable **32** extending between one set of alternate floats and by control cable **34** extending between the other set of alternate floats to establish the desired zigzag accordion pleated configuration of the segments comprising pipes **12, 14, 16.** Suitable moorings may be provided as desired.

In operation, the apparatus of the drawings is adjusted to establish aperture pipes **12, 14, 16** to the desired predetermined depth (d) by means of adjustable chains **23, 25, 27** and to the desired zigzag accordion pleated configuration by means of control cables **32, 34.** The latter is generally in the range of 20° to 120° included angle between adjoining segments of pipes **12, 14, 16,** depending on the amount of current and power available as discussed above, as well as on the wave caused water movement which determines the minimum distance between adjoining apices of the segments of pipe.

A typical valus is about 60°. Compressor **20** is then started to produce bubbles at apertures **18,** which bubbles rise and cause water movement U, which will oppose the flow V of the oil slick and cause it to stop along its edge O, at distance x from the vertical pipe projection sufficient to prevent wave induced water movement W from permitting the oil edge O from reaching the vertical pipe projection.

Rath Design

A device developed by *E. Rath; U.S. Patent 3,631,984; January 4, 1972* involves airflow through a conduit which conduit hovers on the liquid by means of the airflow and creates a trough-like seal about the contaminating liquid and contains the same or cooperates with the conduit to remove the same by a trough-like skimming effect.

Figure 51 shows the essential features of this process device. In its simplest form, this method of containing a liquid within a larger body of liquid comprises the formation of a duct for air and fluid wherein the action of the air causes the duct itself, together with a fluid containment, to form a dike-like arrangement. This dike-like arrangement is shown to comprise a flexible material such as plastic or the like formed in the general configuration indicated, comprising a basic portion **12** which forms a trough-like arrangement **19** having a pair of shoulders **13** and **15** with depending edges **14** and **16** and lower portions **17** and **18,** all formed in one continuous piece of indeterminate length.

Air ducts **24** and **25** are provided in the area of the shoulders as indicated. These air ducts are supplied with air from a portable air-generating unit consisting of a suitable motor and air blower or the like **20** and interconnected by hoses **21** and **22** with a regulating valve **23** in one or more of the lines if desired. The details of such an air pressure creating unit are well-known in the trade and are not exemplified here.

The air ducts will be formed of suitable material to distribute air either generally throughout the configuration of the shape or at suitable intervals by independent openings and closures (not shown). A series of air skirts such as described in U.S. Patent 3,420,330, and indicated as **17A** and **18A**, and in the general configuration **50** are positioned along the lower edges **17** and **18.**

In activation, the air supply is activated and a length of the material is rolled out across whatever fluid is the basic fluid. When the air is not being activated the relative position of the unit will be as indicated by the regular waterline. When activated, however, the air skirts as well as the pressure of the air in general will raise the unit so that it floats, or hovers, at a higher elevation and just above the waterline.

FIGURE 51: DETAILS OF OSCARSEAL TYPE AIR BARRIER

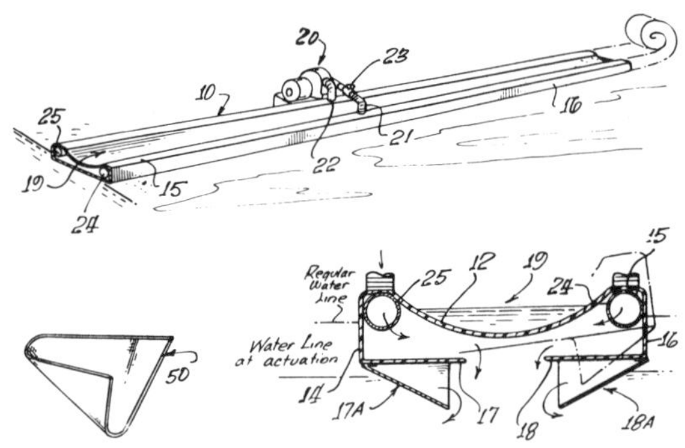

Source: E. Rath; U.S. Patent 3,631,984; January 4, 1972

The air skirts themselves will be touching the water but the basic unit will be above the water. Water which will be normally entrapped within the trough **19** will act as a sort of ballast for the unit and preserve its stability. Oil or the like leaking into the basic area, such as a bay or the like, will thus be entrapped behind the unit **10.** Air will be escaping constantly under the skirts and will create a trough-like turbulence at the edge of the unit, for example, along edge **16**, which will entrap the oil and hold it within the confines and behind the barrier thus created.

According to the EPA, Edison, N.J. Laboratory Report, *Oil Containment Barriers* (January 1973), this design has found proprietary application under the name of Oscarseal Hover Platform. The Oscarseal Hover Platform consists of a series of interconnected captured air floats. The standard floats are fabricated from steel plate in lengths of 40 ft. The air skirt is attached to one side of the float which will form the inner circle. A hinged plate is attached to the opposite side of the air skirt acting as a depth shield and keel. Extension curtain sections for high sea state and locations may be attached to the bottom edge of the shield. Each float has its own blower air supply system, capable of being powered from a variety of sources.

The linkage of the Oscarseal containment systems consist of two clip-links attached to the steel float on the upper edge of the inner side providing for approximately 3 inches of interval between each two air floats. In addition to this a steel cable is provided to be installed on the top side near the clip-links. It serves both as a safety device for the system, for the service of the embedment anchors, and also for towing purposes.

The air skirt is made of a puncture resistant textile designed and manufactured to withstand the impact of the sea and the air-sea interaction. It is claimed that the skirt material does not absorb water, stays lightweight, is easy to handle even when wet; flash dries, is rot and mildew resistant, and can be stored wet or dry. It is further claimed that it is easy to clean with a bristle brush, soap and water. The Oscarseal containment system was developed by the Rath Company in a joint venture with Morrison-Knudsen Company, Inc.

Submersible Systems Inc. Design

The reader is referred to the earlier mention in this volume of an air barrier device on page 95.

Wilson Industries Design

According to Maritime Administration Report COM-74-10212, Springfield, Virginia, National Technical Information Service, a pneumatic barrier is available from Wilson Industries, Houston, Texas. It is designed to operate in 10 ft significant waves (18 ft maximum) in a 35 knot sustained wind and a 2 knot current with oils of specific gravities from 0.75 to 0.95.

It was stated in the report that model tests were conducted on a 1:25 scale with 0.4 ft model waves (10 ft prototype) and currents of approximately 1.1 knots. An estimated 95% of the oil was retained. The device would be permanently installed at a terminal berth; estimated start-up time is only several minutes. According to the report, such a device would have advantages for use at offshore terminal locations.

CHEMICAL BARRIERS

As noted by the American Petroleum Institute in *Oil Spill Cleanup – A Primer,* Washington, D.C., API (August 1973), certain chemicals, known as surface tension modifiers, inhibit the spread of oil. When relatively small quantities of these chemicals are placed on the water surface next to the floating oil, the oil runs from it, greatly reducing the affected area from its original size. With the oil area reduced and controlled, recovery is simplified.

A spray helicopter can deliver such a chemical at the spill site and apply it very quickly, possibly in a matter of minutes. Almost any power boat is also suitable. Only a small quantity of chemical need be used, applied as a coarse spray on the water at the edge of the spill.

As with booms and other containment methods, a surface tension modifier should be used only if plans have been made to clean up the spilled oil. Since this material only keeps the oil from spreading and since even this effect lasts a matter of hours at most, other equipment is needed to remove the oil from the water surface. As with any chemical, approval for the use of a surface tension modifier must be obtained from the appropriate governmental regulatory agencies.

Allied Chemical Corporation Process

A process developed by *S. Roth; U.S. Patent 3,497,450; February 24, 1970; assigned to Allied Chemical Corporation* involves a method for removing a liquid contaminant, particularly petroleum oil, from the surface of water by covering the surface of the liquid contaminant with a salt water solution and then an aqueous solution of polyvinyl alcohol and alkali metal borate so as to form a skin around the liquid contaminant. The entrapped liquid contaminant can then be removed from the water surface by various methods.

Chevron Research Corporation Process

A scheme developed by *R.L. Ferm; U.S. Patent 3,810,835; May 14, 1974; assigned to Chevron Research Company* involves treating an oil slick to contain it and prevent its uncontrolled spreading comprising applying to open water areas in the vicinity of the oil spill a chemical agent which repulses the oil spill.

By judicious application of the chemical agent, the oil slick can be gathered into a limited area which facilitates cleanup. The chemical agent is selected from the group consisting of N,N-dialkyl amides; n-alkyl and n-alkylene monoethers of ethylene glycol and polyethylene glycol; polyethylene glycol monoesters of n-alkyl acids; and n-alkyl and n-alkylene monoesters of propylene glycol.

Phillips Petroleum Company Process

A process developed by *H.E. Alquist and R.T. Werkman; U.S. Patent 3,770,627; Nov. 6, 1973; assigned to Phillips Petroleum Company* involves containing oil on a water surface and removing the oil from the water surface by fusing a finely particulate polyolefin floating on the surface with the oil to incorporate the oil into a crust-like fused mass which will act to form a boundary against extension of the oil mass and which can be easily skimmed from the water surface.

Tenneco Oil Company Process

A process developed by *R.E. Gilchrist and J.C. Cox; U.S. Patent 3,755,189; August 28, 1973; assigned to Tenneco Oil Company* is one in which a composition of matter suitable for the confinement of oil floating on water consists essentially of a drying oil, a carrier selected from the class consisting of liquid alcohols, ketones and ethers, and a water-insoluble metallic soap catalyst. The composition is used to control oil slicks on water by dispersing the composition on the surface of the slick in an amount sufficient to confine the oil slick and thereby allow removal thereof.

FLOATING BOOMS

A boom is a mechanical barrier designed to stop or direct the flow of oil. Floating booms are in use in harbors and other waterways where transfers of petroleum and petroleum products take place. Booms have four basic parts. Some means of flotation is necessary to keep the entire boom at the correct position on the water and to ensure that the boom has enough freeboard to minimize the effect of waves splashing over the top. There must be a skirt to keep oil from flowing underneath the boom. Weights are usually added to the skirt to keep the barrier perpendicular to the water surface. Finally, there is a longitudinal strength member, which must be strong enough to withstand winds, waves and current.

There are many makes of booms, although all have the same basic design. A flexible boom follows wave action easily, although it may offer problems associated with overall strength, durability, and stability. Semiflexible booms are composed of nonrigid sections joined by hinges. They are relatively compliant in wave action, and they are simple to manufacture, store, and deploy. Rigid booms are among the largest in use today, the rigid sections being joined by hinges to provide flexibility. These are the most cumbersome of the three boom types to handle.

In addition to the many commercially available booms, makeshift booms have sometimes been successfully used. They include inflated fire hose, linked railroad ties, or linked telephone poles. The success of any floating boom depends largely on the conditions under which it is used. All booms work best in calm water. The higher the boom rises above the surface of the water and the deeper its skirt descends below, the more oil it will hold and the greater its chance of success in rough water.

If however, the water current is in excess of approximately 0.7 of a knot, about 1.2 ft/sec, oil may escape under the skirt of the boom. To counter the velocity of the current, the boom can be placed at an angle, thereby diverting the oil to a quieter area. The stronger the current, the more the boom must be angled downstream.

In the presence of waves, both water and oil can splash over the top of a boom if the wave height is at least as great as the freeboard of the boom. For this reason, although they have been successfully used in some situations, most booms are generally unsatisfactory in the open sea or in rough water. Under such conditions, they are also difficult to anchor properly, and they tend to break apart under the severe stresses of the open sea. The general topic of oil containment booms has been reviewed in some detail in two U.S. Government publications which are selectively quoted at some length in the pages which follow. These are *Oil Spill Containment Systems,* Edison, N.J., U.S. Environmental Protection Agency

(January 1973); and *State of the Art Review of Oil Containment Barriers for Use at Off-Shore Terminals,* Galveston, Texas, National Maritime Research Center. This is available as Report COM-74-10212, Springfield, Virginia, National Technical Information Service (November 1973).

Acme Products Design

According to Report COM-74-10212 a barrier of the fabric skirt type is available from Acme Products Co., Tulsa, Oklahoma under the trade name of the OK Corral Boom. It has 12 inch diameter flotation logs sealed in a vinyl/nylon material. A 36 inch weighted skirt hangs down from the flotation logs in an example illustrated in the abovementioned report. This barrier is stated to be suitable for fixed installation and is claimed to be fire resistant. It is said that 2,000 ft of such a barrier can be deployed by two men in one hour; one workboat is required. Recovery, cleaning and storage are reported to be fairly easy. For 2,000 ft of barrier, some 1,350 cubic feet of storage space are required.

The EPA Edison, N.J. Laboratory Report entitled *Oil Spill Containment Systems* (Jan. 1973) gives additional details on the Acme OK Coral Boom based on manufacturer's data. Material of the boom is a nylon fabric coated and impregnated with vinyl in brilliant international yellow for visibility. All seams are thermosealed. The unicellular plastic foam flotation units are encased and sealed in the nylon fabric coated material.

Flotation units are 9 ft in length with 6 inch thermal seams for hinges, and are available in 3, 4 and 6 inch diameters. Skirt lengths are available from 6 to 24 inches. There is a ⅜ inch galvanized continuous chain ballast sealed in the fabric at the trailing edge of the boom skirt. Weights are either nylon coated wire rope finished with eye ends, or steel rods from ⅜ through ⅞ inch.

Booms are available in any lengths desired in increments of 10 to 9 ft flotation sections with 6 inch hinges between sections. Boom ends are finished with reinforced nylon webbing fitted with brass grommets for joining sections for deploying, towing and moving. Boom weight is approximately 1.5 lb/lineal ft, including ⅜ inch steel ballast rods. Figure 52 is a drawing of such a boom.

FIGURE 52: ACME FLOATING BOOM DESIGN

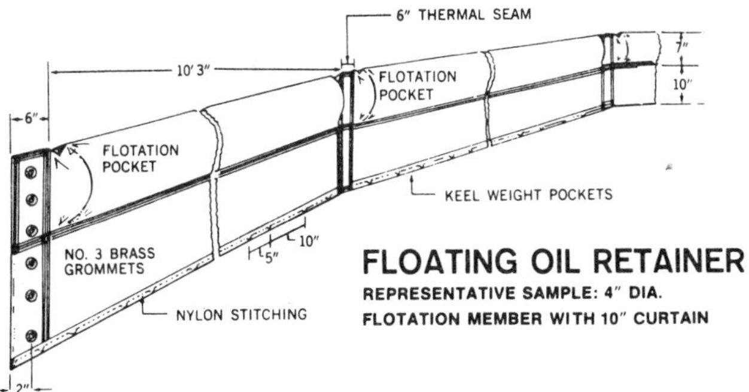

Source: EPA Edison, N.J. Report *Oil Spill Containment Systems* (January 1973)

American Marine Design

The EPA Edison, N.J. Laboratory publication, *Oil Spill Containment Systems* (January 1973) gives details of an American Marine Design II (Optimax) boom. The individual sections come in 100 ft lengths. A choice of two widths is offered, with freeboards of 5½ and 6 inches and drafts of 12 and 24 inches. Both widths use a 6 inch cylindrical flotation, made of Ethafoam, completely sealed into the boom. The fabric is yellow vinyl nylon and is barnacle and algaecide impregnated. The ballast consists of a galvanized chain secured in a pocket on the boom skirt. All metal components are made of galvanized steel or high tensile aluminum.

Section connections are made without tools. Each section comes with foldable 10 ft segments for easy storage. Foldable 5 ft segments are also available at an additional cost of $0.15/ft. The price given for the standard boom having a 12 inch draft is $5.25/ft. This information is stated by EPA to be based on information in a product brochure received in February 1972 from American Marine Inc., Merritt Island, Florida.

Ayers Design

A design developed by *R.R. Ayers; U.S. Patent 3,648,463; March 14, 1972* utilizes an elongate buoyant boom having a foraminous skirt depending therefrom. An impervious shaped section is secured to the bottom of the foraminous skirt to react against water movement under the boom to maintain the lower skirt end depressed in the water. A spreader bar is used to maintain the top and bottom of the skirt spaced a predetermined distance apart.

Figure 53 is an overall isometric view of such a design. In the figure there is shown an apparatus **10** for controlling a spill of liquid pollutant **12** on a body of water **14**. Although the liquid pollutant may comprise any floating liquid, it is typically a liquid hydrocarbon. The apparatus comprises as major components a buoyant boom **16**, a depending skirt **18** and positioning means for maintaining the boom in a given location or for towing the apparatus through the water.

The skirt also comprises a foraminous section which is illustrated as comprising screen mesh but which may be a perforate sheet or the like. The foraminous section allows liquid flow therethrough but collects the oil sorbent particles **36**, such as hay or straw, which have been scattered in the spill area to sorb the liquid pollutant.

The positioning means is illustrated as comprising an upper cable or flexible line **44** connected to the boom by any suitable means (not shown) such as is disclosed in U.S. Patent 3,499,290. The positioning means also comprises a lower cable or flexible line **46.** An important part of the positioning means comprises a spreader bar **48** connected to the flexible lines **44, 46** adjacent the skirt on opposite ends of the boom. The spreader bars act to maintain the upper and lower skirt ends in predetermined spaced apart relation. This lends substantial stability to the apparatus not found in prior art booms. For example, a prototype boom built in accordance with this design has been successfully towed at speeds of 2 fps. The maximum towing speed of conventional booms is about 1.3 fps.

The ends of the flexible lines may be connected to bars **50, 52** which are in turn connected to flexible lines **54, 56** for attachment to anchors for positioning the apparatus in the water or for attachment to a boat for towing the apparatus through the water. The spreader bars are illustrated as spaced from the ends of the boom merely to expose the details of the skirt.

In use, the spreader bars may be secured to the ends of the boom or to the cables **44, 46** adjacent the ends of the boom. When not so secured and the apparatus is subjected to high current velocities, the spreader bars are moved by action of the water adjacent the ends of the boom. Assuming that the apparatus is moored in an area of high current and that the oil sorbent particles **36** are spread on the liquid pollutant upstream from the boom, the liquid pollutant and oil sorbent particles move with the current toward the apparatus. The foraminous section allows water movement past the boom while the float and the imperforate section act to retain the liquid pollutant upstream. The oil sorbent particles may be drawn downwardly under the imperforate section by water movement therepast and are collected by the foraminous section **34.**

FIGURE 53: AYERS FLOATING BOOM DESIGN

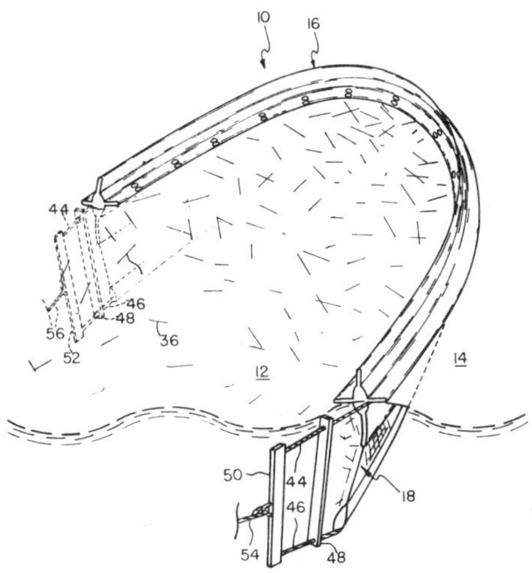

Source: R.R. Ayers; U.S. Patent 3,648,463; March 14, 1972

Barracudaverken AB Design

A design developed by *E.J. Larsson; U.S. Patent 3,757,526; September 11, 1973; assigned to Barracudaverken AB, Sweden* utilizes an oil boom comprising an elongate screen of plastics foil material provided with buoyant bodies and weights to hold the boom in a substantially vertical position when afloat in the water, with a portion of the upper edge portion of the boom above the surface of the water.

The boom comprises two superimposed sheets of plastics foil material which are welded together at a number of sequentially arranged points in the longitudinal direction of the boom to form substantially rectangular closed pockets located on the upper half of the boom in its position of use. Each pocket contains one filling body which extends the pocket to form a buoyant body of requisite buoyancy, the filling body being constructed of corrugated cardboard.

As shown in Figure 54, the boom structure comprises two superimposed sheets of reinforced plastics material **10, 12** obtained by folding double a web of plastics foil. At the upper edge **14** of the foil sheets is inserted a plastics rope **16** beneath which is inserted a number of filling bodies **18,** the position of which is fixed by welding the foil sheets around respective filling bodies such as to enclose the bodies in a closed pocket **20.** Inserted between the pockets are vertically extending plastics rods **22** which are fixed in position by welds made on either side thereof. The purpose of the rods is to stiffen the boom when in use. Each filling body comprises two substantially rectangular plates **24, 26** made of cellular cardboard and bonded together with a suitable adhesive. Each plate consists of two outer layers and an intermediate layer of corrugated cardboard **28, 30.** The plates are of the same thickness and length but have different height such that the lower edges of the plates lie generally edge-to-edge with each other. In this way the lower half of the filling bodies is approximately double the thickness of its upper half so that the portion of the pocket accommodating the section of the body submerged in the water and forming a buoyant body obtains the greater volume, while the larger portion of the upper half of the filling body is only half as thick and is intended to lie at least partially above the water to form a screen for screening-off any oil present on the surface thereof.

The described construction of the floating body formed by the pocket extended by the filling body provides a simple and inexpensive structure with the use of an inexpensive filling material which in addition is light in weight. To provide the desired rigidity of the buoyant bodies, the pleats of the corrugated cardboard plates **24, 26** are arranged with their longitudinal extension at right angles to the longitudinal extension of the boom, as illustrated.

Arranged and securely welded in the lower edge portion of the boom is a further plastics rope **32**, and arranged beneath the rope is a number of reinforcing rods **34** serving as weights to maintain the boom in a vertical position when afloat in the water. The reinforcing rods have approximately the same length as the buoyant bodies, whereby the boom can be folded to form a pack, by folding the boom backwards and forwards at the vertical rods **22.**

A folded boom can be readily stored in a transport box of such construction that the short sides of the box and the lid can be readily removed. The boom can be rapidly drawn from a box of this construction and connected to a motor boat, for example, by means of the ropes **16, 32.** The boom will thereby lie on the surface of the water while being towed and offers but small resistance to the water, thereby enabling the boom to be towed at relatively high speeds. The polyethylene foil used is suitably a 0.3 mm thick, double layer polyethylene foil. The thickness of the cardboard plates, is, for example, 6 mm, so that the largest thickness of respective buoyant bodies is 12 mm.

FIGURE 54: BARRACUDAVERKEN FLOATING BOOM DESIGN

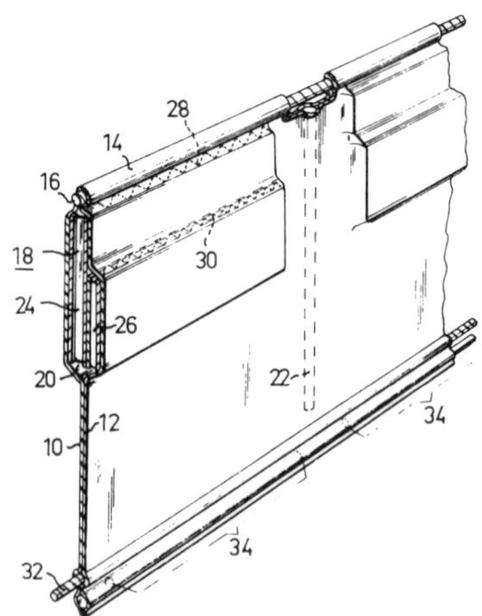

Source: E.J. Larsson; U.S. Patent 3,757,526; September 11, 1973

The ropes suitably have a diameter of approximately 6 mm. The distance between the vertical rods is, for example, 0.9 m and the height of the boom is, for example, 0.7 m. Each boom unit may have a length of 25 to 50 m and comprises a number of sections, each having its respective buoyant body **18.**

Belin Design

A design developed by *M.A. Belin; U.S. Patent 3,739,584; June 19, 1973* utilizes a floating barrier for circumscribing and trapping oil films or like refuse in harbors and other bodies of water which comprises a strip of noncorrodible material such as stainless steel and a pair of buoyant floats removably fitted at intervals on opposite sides of the strap so as to support it vertically in the body of water. Junction means interconnect the floats pairwise and are removably engaged through the strip. A pair of watertight stabilizing compartments is removably secured to opposite sides of an end portion of the strip and are connectible to a hauling or towing craft.

Figure 55 is an isometric view of such a device showing from one side a strip fitted at intervals with floats and at its right hand side end with stabilizing compartments. As illustrated, this floating barrier for circumscribing and trapping oil films or sheets or like refuse, for instance in harbors or ports or in similar sea or large lake stretches, comprises a flexible or distortable strip generally designated by **10** which is advantageously made of a rustless or noncorrodible metal, alloy or similar sheet-like material such as stainless steel the thickness of which may be, for example, from 2 to 5 mm, the width of the strip being, for example, equal to about 1 m while its length may be equal to a unitary measure such as 25 or 50 m.

The strip **10** is provided along its upper and lower edges with reinforcing strips **11, 12** and across its width at suitable intervals with reinforcing strips or braces **13**. Adjacent the rear end of the strip **10** extend transverse strips or bars **14** having holes **15** through which links or bonds such as cables or guys (not shown) may be engaged for interconnecting the strip with a similar strip.

Adjacent the front end of the strip **10** extend likewise reinforcing transverse strips or braces **17** fulfilling a function similar to the strips or braces **13.** Along both flanks of the front end portion of the strip are provided a pair of elongated watertight compartments **18** the general shape of which is rectangular but having on its lower face with a forwardly and upwardly extending sloping wall **19** facilitating penetration of the compartments through the water when the strip is towed, for example, by a tugboat as hereinafter described.

Between each compartment and the contiguous face of the strip is interposed a holding and reinforcing plate **20**. Transverse pins are engaged through holes **22** in the braces **17** and through matching holes in the plates and in the walls defining the enclosures forming the compartments **18.** At their outer ends the transverse pins are held in position by removable clips such as **23.**

The assembly of the front end portion of the strip **10,** plates **20** and bins **18** may be disassembled. The purpose of the watertight compartments **18** is to impart stability and more mechanical resistance to the front end portion of the strip. The latter may be towed by links or cables engaged through suitable holes such as **22**, for example, for being towed by a tugboat to the proper site such as a suitable position in a harbor or along a shore so as to circumscribe and trap an oil film or sheet or more generally a mass of refuse floating in the sea, thereby preventing the same from spreading in noncontrolled fashion.

As suitable intervals along the strip **10** are provided on both sides thereof floats **24** of spheroidal shape as shown. Each float is made up of a pair of shells interconnected along their contiguous rims by a tight welding bead such as **25.** Rearwardly of each float is secured an upright plate **26** extending parallel with the corresponding side face of the strip and rigidly connected with the float by a pair of bracing links **27** which partake of the general rigidity and stability of the floats.

FIGURE 55: BELIN FLOATING BOOM DESIGN

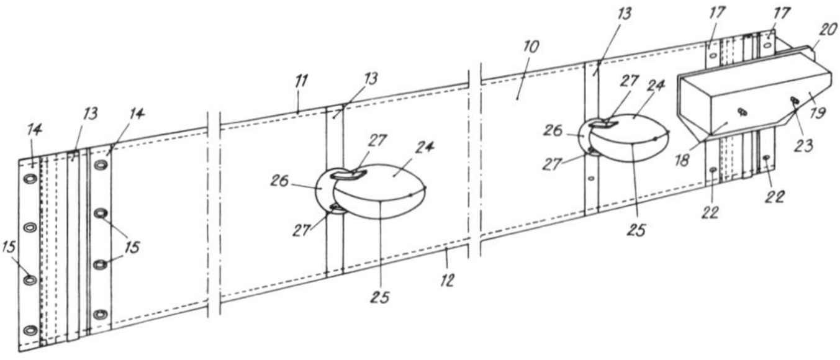

Source: M.A. Belin; U.S. Patent 3,739,584; June 19, 1973

Due to this construction, the compartments and floats may be easily assembled to the strip or disassembled therefrom. Before being used, the strip may be rolled up due to its flexibility and stowed in a limited volume upon a reel which may be, for example, revolubly supported on bearings or headstocks on a ship whence the strap can be unrolled and then equipped with floats and compartments preliminarily to its placement in operative position.

The action of the floats is to hold the strip in vertical or substantially vertical position, thereby enabling the same to perform its duty as a floating barrier. The purpose of the compartments is to impart stability to the strip or barrier particularly when it is dragged or towed to its operative site or position, for example, by being towed by a suitable craft.

Bennett Pollution Controls Ltd. Designs

The Bennett Offshore Oil Boom is shown in Figure 56. The boom is basically a sandwich structure, centermost is the strongest support element, consisting of 3 steel cables, two ¾ inch and one ⅜ inch, running the length of the boom with ¾ inch steel tubing vertical stiffeners spaced 3 ft apart. It has a vinyl covered steel chain link fence over the full length and height of the boom. The freeboard portion is covered with a yellow 22 oz PVC material while the skirt is encased with a polypropylene fiber mechanically entangled and fused by heat. This provides an oleophilic filtration material which passes water but retains oil.

The structure is vertically supported in depth by flotation cylinders attached to the outside of the boom. The cylinders are ethafoam, 9 ft long and 6 inches in diameter, attached by nylon straps. The section lengths are 100 ft with 24 inch freeboard and 48 inch draft, weighing 14 lb/lineal foot. Ballast is effected by lead weights attached to the steel cable at the bottom of the skirt. Such a boom is flexible, follows wave contours, yet is sufficiently stiff to maintain stance in high winds and breaking waves.

According to Maritime Administration Report COM-74-10212, Springfield, Virginia, National Technical Information Service (November 1973), this barrier was designed to operate in 3 ft waves, a 12 ft swell, and current of 2 knots. The design objective was to contain 500,000 cubic feet of oil under the above conditions.

The manufacturer reports that the barrier was tested and found effective in tests on a Louisiana crude under the above conditions plus a 70 knot wind. Tests were also made in 70 knot winds, 6 knot currents and 6 ft waves. Four men can deploy 2,000 ft of this barrier in 20 minutes. Two 600 hp tugs and special storage racks are required. Some 500 ft of barrier can be stored in one rack 12 x 120 ft.

FIGURE 56: BENNETT OFFSHORE OIL BOOM

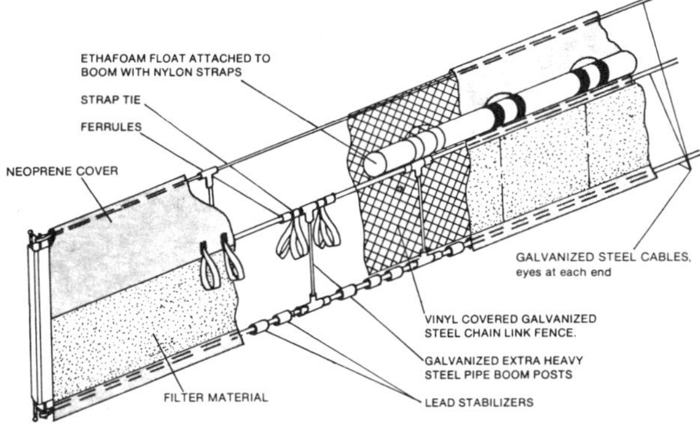

Source: EPA Edison, N.J. Report *Oil Spill Containment Systems* (January 1973)

According to the abovementioned report, recovery, cleaning and storage efforts are estimated to be above average for barriers of this type. There is also a Bennett 36-Inch Flexi Boom design described in a brochure from Bennett Pollution Control Ltd., Vancouver, B.C. received by EPA in January 1972. The construction is shown in Figure 57.

FIGURE 57: 36-INCH FLEXI BOOM

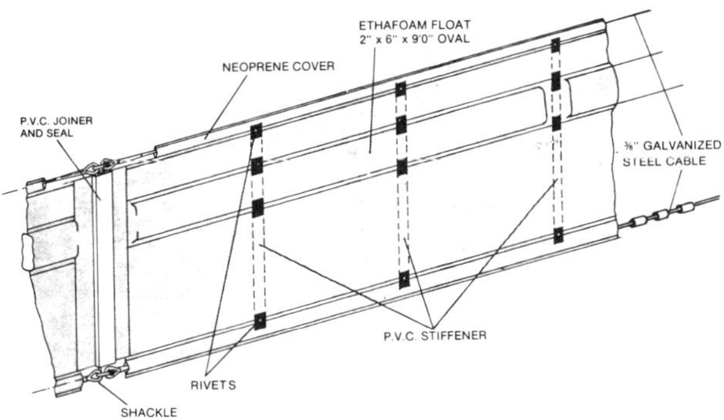

Source: EPA Edison, N.J. Report *Oil Spill Containment Systems* (January 1973)

Individual lengths are 50 ft and have a freeboard of 12 inches and a draft of 24 inches. Sections of oval shaped Ethafoam (2 x 6 inches x 9 ft), set 6 inches apart are used for flotation. The flotation is secured to the boom by ¼ x 1 inch PVC stiffeners, which pass through the flotation at 3 ft intervals.

The boom fabric is a yellow neoprene on nylon material, which covers all boom components. Two ⅜ inch steel cables, encased in the hem at the top and bottom of the boom, are used for connecting sections of booms by means of shackles. A joiner made of PVC is also used for joining and sealing at the connections. Lead weights, secured to the lower steel cable, are used for ballast. One foot of boom weighs 4.5 lb. For harbor oil containment there is a Bennett 3-Foot Harbor Oil Boom as shown in Figure 58. The boom is constructed whereby the freeboard is a barrier and the skirt is a filter. The freeboard cover material is a yellow neoprene on nylon, while the skirt is covered with polypropylene, nonwoven, heat sealed, 4 ply mil cloth. Flotation consists of 6 inch diameter ethafoam rounds 4.6 inches long encased in the freeboard cover material. Vertical stiffeners are $1\frac{1}{8} \times 1\frac{1}{16} \times \frac{1}{4}$ inch PVC I beams. Strength members at the top and bottom of the boom are $\frac{3}{8}$ inch steel cables.

Ballast is achieved by the attachment of lead weights to the steel cable enclosed in the skirt. Section lengths are 50 ft long with a 12 inch freeboard and a 24 inch draft. The boom weighs 3.5 lb per lineal foot. Sections are joined by fabric slid into PVC joiner and the cables shackled together. Finally, there is a Bennett 20-Inch Harbor Oil Boom as shown in Figure 59. Individual lengths are 50 ft and have a freeboard of 8 inches and a draft of 12 inches. Sections of oval shaped ethafoam, 2 inches by 6 inches by 4 feet 6 inches, set 6 inches apart and riveted to the boom fabric, are used for flotation. The boom fabric is a yellow neoprene on nylon material, which covers all boom components. A ¾ inch polypropylene rope, passing through the top hem, and a $\frac{5}{16}$ inch galvanized steel chain (also used for ballast), passing through the bottom hem of the boom, are used for connecting individual sections. Connections are made with shackles and a joiner made of PVC, which also seals the connection. 1 ft of boom weighs 2.7 lb. No price was given by the manufacturer.

FIGURE 58: BENNETT 3 FOOT HARBOR OIL BOOM

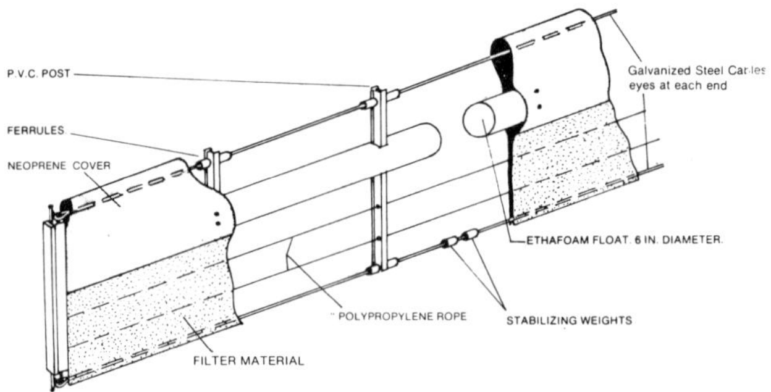

FIGURE 59: BENNETT 20 INCH HARBOR OIL BOOM

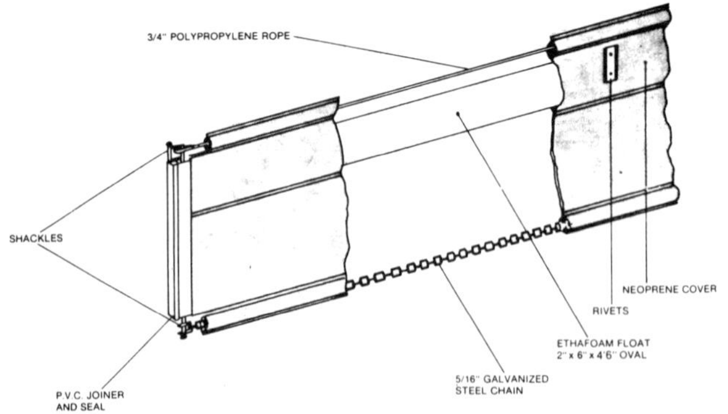

Source: EPA Edison, N.J. Report *Oil Spill Containment Systems* (January 1973)

Blomberg Design

A design developed by *E.G.E. Blomberg; U.S. Patent 3,686,870; August 29, 1972* utilizes a slick confining boom having an elongated flexible body means with upwardly extending portion to which floats are attached supporting the portion above the surface of a body of water and a downwardly extending portion to which weights are attached to extend the second portion below the water surface. A stress relieving rope is connected at spaced apart intervals along the body means by connecting means extending laterally of the body means whereby the boom will retain impurities floating on the water surface.

Figure 60 shows the boom in place as seen from above at an inclined angle and also a cross-section through the boom. The slick confining boom includes a separate body means of long extension consisting of a flexible material as, for example, canvas cloth (fabric). At points along the lower edge portion of the canvas a number of sinking weights and at points along its upper edge portion a number of floats are arranged.

By means of several connecting organs, a number of identically alike boom portions are connected forming a boom of larger size (compound fence), in which the individual boom portions each constitute a section. A tension means, such as a stress relieving rope **5**, extends between the ends of each one of the sections, the rope preferably having a specific weight of less than 1. The water level is indicated by **6**, with the portion of flexible material **1** below water level providing a depending skirt below the surface of a body of water, and the portion of the flexible material above the water level providing an upwardly extending portion above the surface. A terminal and tow wire (line) **7** is connected to the free end of the fence. The fence is connected with the stress relieving rope by means of a tether arrangement, such as supporting wires **13**.

A discharge outlet **43** is arranged in the wall of the fence, through which waste oil and other impurities, which have been collected by means of the fence during its towing, can be taken out to be delivered to removal transportation and/or destruction means which are located behind the fence as seen in its direction of towing.

FIGURE 60: BLOMBERG FLOATING BOOM DESIGN

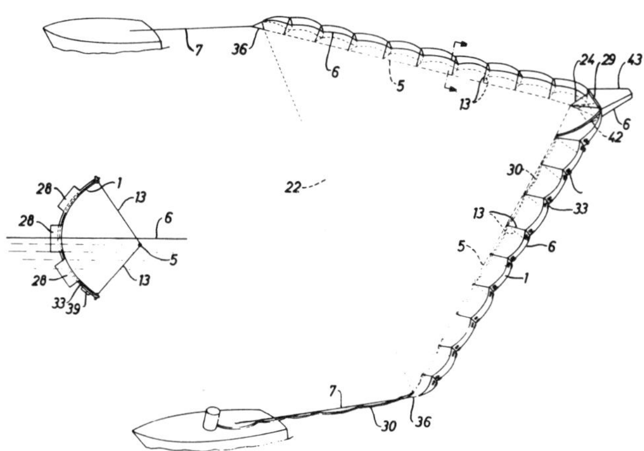

Source: E.G.E. Blomberg; U.S. Patent 3,686,870; August 29, 1972

A distance wire **22** extends between portions of the stress relieving rope located on both sides of the discharge opening. This wire maintains the two portions of the wall of the fence at a desired distance from each other. The tow wires **7** are connected to the free ends of the canvas of the fence and the stress relieving wire by yoke **36**, which maintains the stress relieving rope at a specified distance away from the wall **1** of the fence.

The discharge opening is surrounded by a frame shaped stiffening organ. The frame of the opening is kept in place by means of a number of supporting wires **24**. A number of float bodies **28** are arranged in the longitudinal direction of the fence in interspaced groups. The float bodies entering into each group are vertically spaced in the height direction of the fence in such a manner that at least part of the group is positioned above the surface of the water, when the fence is subjected to normal stresses.

Thereby the part of the group of float bodies, which is above the surface of the water, constitutes a float reserve which can be used when the canvas of the fence is subjected to abnormally great stresses. A protruding bottom frame **42** is carried by the front frame, however, the bottom frame is not necessary for the functioning of the fence. A tube **29** is provided with a number of nozzles directed backwards and is intended for spraying the impurities which have passed through the discharge opening. This tube can be connected by means of a hose **30** to a towboat as shown.

A number of springing laths **33** extending in the cross-direction of the wall of the fence have supporting wires **13** attached to their upper and lower ends by fastening means. The laths are attached to the wall of the fence by fastening means. A number of sinking weights **39** are arranged near the lower edge of the wall of the fence. The sinking weights and the float bodies can be placed near the laths.

Bridgestone Tire Company Limited Design

This design has been covered by the EPA Edison, N.J. Laboratory in its publication *Oil Spill Containment Systems* (January 1973). This boom is a submerging system designed primarily for permanent installation in such places as dolphin and seaberths. It floats when in use and submerges when not needed.

The boom is made of a special compound of rubber as the main material. The skirt is pleated to make it ride more smoothly on the waves. Glass rods are inserted in the skirt as stiffeners. The skirt is weighted with lead weights and flotation is provided by float hoses attached on both sides of the skirt. As there is about 55% surplus buoyancy, even if one of the float hoses becomes damaged, the boom will still float.

The boom is manufactured in two sizes, one approximately 32 inches high with a 12 inch freeboard and a 20 inch skirt; the other approximately 40 inches high with a 16 inch freeboard and a 24 inch skirt. These are approximate sizes as the booms are made with metric measurement specifications. Each boom comes in 65½ ft sections. The smaller boom weighs approximately 7 lb/ft while the larger one weighs 10 lb/ft.

The boom is moored to anchors on the bottom of the waterways. The boom is kept on the surface by air pumped into the flotation hoses with a compressor. The boom is submerged by expelling the air from the flotation hoses. The submerging hose and anchor are attached to the boom on the end furthest from the air input end. This hose and anchor projects into the water below the bottom of the boom. Manufacturer's time estimations for operating the boom are: surfacing (depending on the size compressor used) 98 to 164 feet per minute; and submerging approximately 98.4 feet per minute, according to a product brochure furnished to EPA by Mitsubishi International Corp., New York City, N.Y. in June 1972.

The Martime Administration Report COM-74-10212, Springfield, Virginia, National Technical Information Service (November 1973) quotes data from the Bridgestone Tire Company of America, Gardena, California. They state that the Bridgestone Floating Submerging Oil Fence was designed to operate in 10 ft waves, 29 knot winds and 1 knot currents. It has been tested in 3 ft waves, 15 knot winds and 1 knot currents. They go on to state that this design is used at several major oil terminals.

A design developed by *T. Muramatsu and K. Aramaki; U.S. Patent 3,799,020; December 18, 1973; assigned to Bridgestone Tire Company Limited, Japan* is an immersible oil fence assembly including alternately connected oil fence units and immersible buoys. The oil fence units have tubular float means connected to inflatable bags of the immersible buoys. By inflating and deflating the tubular float means and the inflatable bags, the oil fence is selectively floated and immersed. The buoys are anchored to station the oil fence assembly in position.

Figure 61 is a schematic elevation of this oil fence design showing both the submerged and floating positions. Each oil fence unit **1** consists of an elongated flexible belt member **6**, e.g., an elongated rubber belt, and a flexible tubular float means **7** extending along the longitudinal center line of the belt member. Both the belt member and the tubular float member should be flexible enough to allow the float unit **1** to flex along the profile of water surface. The tubular float means consists of a pair of flexible tubular floats **7** and **7'** secured to the opposite surfaces of the flexible belt member. To ensure that the width direction of the elongated belt member is held substantially vertically across the water surface, opposite longitudinal ends of the belt member are secured to the immersible buoys **2** in vertical posture. Suitable weight members **5** may be embedded in that longitudinal edge of the belt member which is to be held in water.

A permanently floating pilot buoy **9** is disposed in the proximity of that immersible buoy **2** which is connected to one end of the oil fence assembly, and the pilot buoy is connected to the anchors **3** and **3'** of the end buoy **2** through a separate wire rope means **10.** This wire rope means can be a metallic chain too. The wire rope means for the pilot buoy is connected to the wire means **4** of the end buoy of the oil fence assembly. The pilot buoy carries an air valve coupling **11** communicating with the end buoy through a flexible air pipe **8.** This coupling includes an air valve and controls the airflow into and out of the immersible oil fence assembly so that the coupling must always be kept above water surface or must be watertightly closed by a suitable lid so as to selectively be opened for the airflow control.

Assume that the immersible buoys **2** and the tubular float means **7** of the oil fence assembly are deflated and the entire assembly rests on seabed, as shown by phantom lines. A lighter **12** carrying an air source, e.g., an air pump or compressed air tank, berthes the pilot buoy to connect the air source to the air valve coupling for feeding air into the immersible buoys and the tubular float means, through the flexible air pipe. Thereby the buoys and the float means are inflated and when buoyance reaches a certain predetermined level (in excess of the total mass of the oil fence assembly) the entire oil fence assembly moves up to the water surface, so as to define the aforesaid fence line. In this case, the air to be delivered to the assembly through the coupling should be pressurized to the extent necessary for overcoming the hydraulic pressure acting on the tubular float means and the immersible buoys.

FIGURE 61: BRIDGESTONE FLOATING BOOM DESIGN SHOWING SUBMERGED AND FLOATING POSITIONS

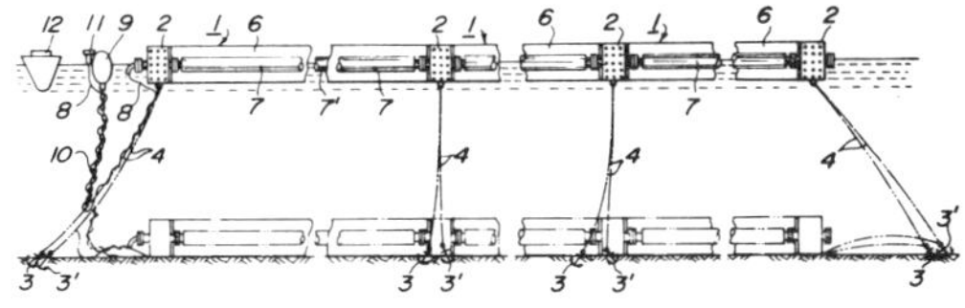

Source: T. Muramatsu and K. Aramaki; U.S. Patent 3,779,020; December 18, 1973

Bristol Aircraft Design

According to the EPA Edison N.J. Laboratory in its report *Oil Spill Containment Systems* (Jan. 1973), in 1967 the Bristol Aircraft Company developed a boom comprised of 10-inch diameter fiberglass pipe onto which strips of marine plywood, 12 and 9-inches wide, were attached to the top and bottom edges of the pipe. The fiberglass pipe is provided in 20-foot lengths with both ends sealed to permit flotation. In the upright position this boom has a height of 31 inches, consisting of the 12-inch strip of plywood serving as freeboard, the 10-inch flotation unit, and the 9-inch plywood board serving as the submersible section. The structure is maintained upright by 28-pound weights attached to the 9-inch plywood strip and positioned at 5-foot intervals along the boom.

Any number of 20-foot sections may be joined together to give the desired length of boom. The assembly is moored to oil drums every 100 feet with support cables tied to marine anchors.

British Petroleum Co. Ltd. Design

A design known as the BP Vicoma Seapack is described in Maritime Commission publication COM-74-10212, Springfield, Va., Nat. Tech. Information Service (Nov. 1973). It is designed to operate in 7-foot waves, 28-knot winds and 1 to 2 knot currents and to contain 1,000 tons of oil per 1,600 feet of barrier deployed. According to the report cited above, theoretical considerations suggest that this barrier will not be effective in strong currents.

The manufacturer reports that in a test, 1,600 feet of barrier served to contain 20 tons of weathered Iranian crude in 7-foot seas, 27-knot winds and 1.2 knot currents. It is stated that two men can deploy 2,000 feet of this barrier in 25 minutes using a single vessel. Recovery, cleaning and storage were estimated to be relatively easy according to the Maritime Commission report cited above. About 125 cu ft of storage are required for 800 feet of barrier.

According to the EPA Edison N.J. publication, *Oil Spill Containment Systems* (Jan. 1973), the boom is a three-tube fabrication in a 30-ounce Butaclor-coated nylon fabric as shown in Figure 62. All joints are hot vulcanized to ensure good retention of mechanical properties even after long storage and to provide good resistance to degradation by mineral oils and detergents.

FIGURE 62: BRITISH PETROLEUM FLOATING BOOM DESIGN

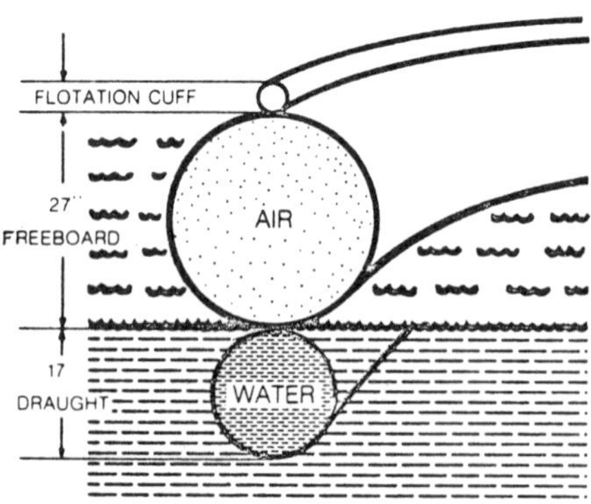

Source: EPA Edison N.J. Report, *Oil Spill Containment Systems* (Jan. 1973).

The three-tube configuration produces a most effective barrier at a very low weight, three pounds per foot. The top tube is a 3-inch flotation cuff. The freeboard tube is 27 inches in diameter while the draft is a 17-inch tube.

The small cuff is inflated with a compressed air cylinder and prevents sinking during laying and recovery operations. The boom is connected at one end to diesel-driven fan and ducted propeller water pump. These simultaneously fill the air and water chambers of the boom respectively. The large volume of the air and water chambers ensures that the boom remains at the air/sea interface and does not suffer loss of barrier efficiency even under severe wave conditions.

Once inflation is completed, the inflation equipment will continue to run for up to eight days maintaining the boom at the correct pressure, sufficient air and water being available to cope with any minor damage that may occur. The manufacturer claims that a typical laying time for the standard 1,600 feet of boom is 20 minutes from the start of the operation. On completion of the operation, the booms can either be recovered at sea or towed inflated to protected water and recovered.

A design developed by *D.H. Desty, L. Bretherick and M.G. Webb; U.S. Patent 3,503,214; March 31, 1970; assigned to The British Petroleum Co. Ltd. & Gordon Low (Plastics) Ltd., England* involves an inflatable barrier which comprises a plurality of air hoses positioned side by side to form, when inflated, a raft which has one or more water ballast chambers attached below. The barrier floats with part below and part above the water surface to impede the passage of floating oil. Preferably the air hoses are graded to give a wedge-shaped raft.

Another design developed by *D.H. Desty, L. Bretherick and M.G. Webb; U.S. Patent 3,635,032; January 18, 1972; assigned to The British Petroleum Company Limited, England* utilizes a floatable oil spillage boom which comprises two arms so as to give a V-configuration. The boom is deployed downwind of a slick which drifts into the apex so that it is concentrated to facilitate collection or destruction. Inflatable booms, e.g., air and water-inflated such as those described above and in U.S. Patent 3,503,214 are particularly suitable.

Brown-Gilbert Design

A design developed by *W.E. Brown and E.E. Gilbert; U.S. Patent 3,708,983; January 9, 1973* utilizes a series of air-retaining structural units connected together in a complete loop so that it will float while surrounding a vessel that may be leaking or spilling oil or some other lighter-than-water fluid. All units are hollow and provided with means for releasing the air so that they will submerge to any desired depth and means for resupplying air so that they can be again raised when needed.

As shown in Figure 63, the apparatus is installed adjacent the pilings **16** of an oil loading dock facility. Preferably, one side of the apparatus is loosely attached to the dock pilings by some guide means such as ring members **18** fixed to one side of the apparatus. Thus, with the apparatus submerged (as shown in the dotted lines), the vessel to be loaded or unloaded **12** can be maneuvered next to the dock directly above the confining apparatus.

A compressor **20** located conveniently on the dock facility supplies air through a line **22** to the connected units **14** of the apparatus and forces water out of them. The apparatus thus becomes buoyant and rises toward the surface while being guided by the ring members around the adjacent pilings. When the apparatus reaches the surface it forms a barrier completely around the vessel and if oil is spilled when loading commences it is confined within the area inside the apparatus.

This apparatus is thus somewhat similar in scope and arrangement to the Bridgestone Tire Co. Floating Submerging Oil Fence described above.

FIGURE 63: BROWN-GILBERT FLOATING BOOM IN SUBMERGED AND FLOATING POSITIONS

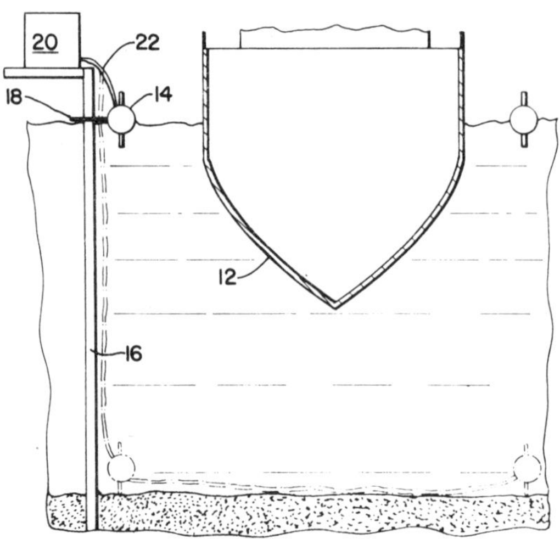

Source: W.E. Brown and E.E. Gilbert; U.S. Patent 3,708,983; January 9, 1973

Cavalieri Design

A design developed by *C. Cavalieri; U.S. Patent 3,565,257; February 23, 1971* employs a barrier for arresting, confining and absorbing oil which consists of a preferably cylindrical body containing a central propylene fiber rope surrounded in succession by a layer of polystyrene (blocks or granules), a layer of a mixture of propylene staples or waste and polystyrene granules, contained in a netting of propylene fiber, and a final layer of propylene staples or waste also contained in a netting. The barrier is floating so as to keep about one-half of its volume above the surface of the water.

As shown in Figure 64, the barrier, which may be shaped as illustrated, namely cross-sectionally circular, or in any other shape desired, consists of a central rope **1** of polypropylene fiber and of a suitable diameter, surrounded by a plurality of rings **2** of polystyrene. These rings may be monolithically built or may consist of granular particles suitably contained in such a manner as to retain a low specific gravity and, therefore, the ability to float.

The rope **1** and the rings **2** are, in turn, completely enclosed in a layer of plastic material **3**, composed of a mixture of staples or waste from fibers of polypropylene and granules of polystyrene, the whole mixture being contained in a net **4** of a polypropylene fiber.

A further peripheric enclosure **5** completes the barrier. Enclosure **5**, of suitable thickness, consists of staples or waste from staples of polypropylene fibers and is contained by an outer netting **6** of polypropylene fiber.

The above described barrier, which may be of any desired diameter and length, has an overall specific gravity such as to allow it to float in the water approximately to the level of the central rope **1**. The relationship between immersed and nonimmersed portions of the barrier is a direct result of the ratio of polystyrene and polypropylene material employed.

FIGURE 64: CAVALIERI FLOATING BOOM DESIGN

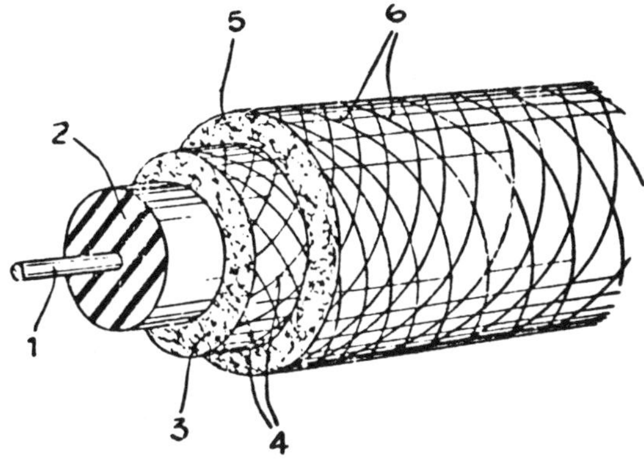

Source: C. Cavalieri; U.S. Patent 3,565,257; February 23, 1971

The barrier may be located along the beaches as a protection against pollutants, or it may be located at a greater distance from the shoreline so as to define an area within which to discharge the polluting products from tankers. In any event, the barrier serves to confine the polluting agents and to prevent them from diffusing on the water surface, because of the barrier's high absorptive capacity.

Cerebro-Dynamics Incorporated Design

A design developed by *A.M. Crucet; U.S. Patent 3,592,006; July 13, 1971; assigned to Cerebro-Dynamics, Incorporated* utilizes a buoyant isolation device which is capable of confining contamination (e.g., an oil slick) present upon the surface of a body of water while conforming in configuration to surface undulations. The device comprises a buoyant elongated flexible barrier which is provided with a plurality of closed fluid chambers. Through the use of a coupling element having a pair of slots which engage the barrier, contamination confined by the barrier may be readily concentrated prior to subsequent removal.

Figure 65 is a perspective view of a portion of the barrier with the barrier floating upon a body of water and confining oil present upon the water surface.

FIGURE 65: CEREBRO-DYNAMICS FLOATING BOOM DESIGN

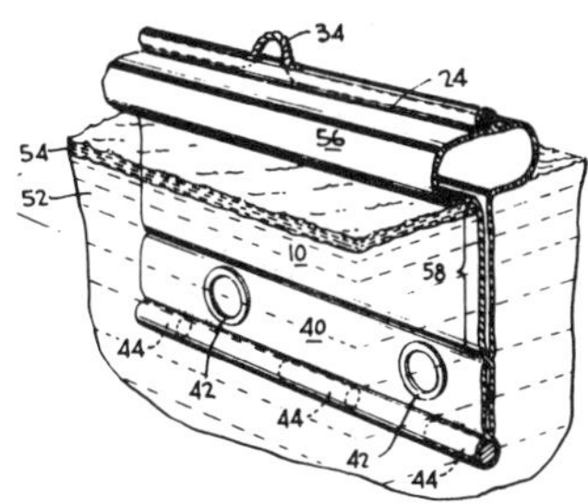

Source: A.M. Crucet; U.S. Patent 3,592,006; July 13, 1971

The buoyant barrier is shown with the barrier floating upon a body of water **52** and confining the spread of oil **54** floating upon the surface. When the barrier is placed in water, the configuration of fluid chamber **10** is modified by water pressure and the air present within its interior is forced upward. The fluid chamber accordingly forms a buoyancy portion **56** and curtain portion **58**. The air present within the buoyancy portion **56** is capable of maintaining the buoyancy of the barrier. Commonly the height of the curtain portion is approximately 2 to 6 times that of the buoyancy portion. The curtain portion is capable of retaining surface and subsurface contamination floating upon or suspended near the water surface. The buoyancy portion remains highly flexible even though the barrier is under the influence of water pressure.

A series of dorsal loops **34** are provided adjacent to the upper portion **24** of the boom. Situated adjacent the bottom portion of the boom optionally may be provided an extended portion **40** which includes a plurality of surge vents **42** and weights **44**.

Chevron Research Company Design

A design developed by *J.A. Sayles; U.S. Patent 3,792,589; February 19, 1974; assigned to Chevron Research Company* utilizes a boom which is constructed of a single sheet of reinforced rubber-like material which is folded along a longitudinal median to form the two sides of it. A tensile stress cable, stress plates, ballast weights, and, in one modification, stiffening members and end plates, are placed between the two sides of the boom in appropriate operating relationship and the sides are then joined together and to at least some of the elements placed between them throughout the area of the sides except for the portions along the top edge of the boom which will form the flotation chambers. However, in the preferred embodiment the sides are vulcanized together and to at least some of the elements enclosed between the sides.

Preferably a series of relatively short inflatable chambers are formed along the top edge of the boom and each chamber has a valve at one end which enables it to be inflated and deflated. The opposite sides of the boom are vulcanized together between flotation chambers to form the chambers as separate air pockets and this construction gives the boom a degree of stiffness in a lateral direction from the top edge to the bottom edge which will function to hold the top edge of the boom above the surface of the water in a region where an air chamber may inadvertently become deflated by being punctured or otherwise damaged.

For some conditions of use, particularly in rougher water, it will be desirable to incorporate stiffening elements in this lateral region between air chambers to insure that a damaged section will remain above the surface of the water and be effective as a barrier.

When the boom is to be retrieved aboard a vessel, or pulled upon a wharf, or otherwise recovered, the successive air chambers are deflated as they approach the retrieving station so that the boom collapses into a flat sheet of material such, as mentioned above, has a smooth, tough surface free of protruding elements, such as bolts, weights, cables, and so forth, throughout the major portion of its length. Hence, the boom can be drawn over hard edges of wharfs or ships without danger of such protruding elements being caught on them and tearing or otherwise damaging the boom structure.

Figure 66 shows some of the constructional elements of this type of boom. As illustrated the sides **10** of the boom **14** are vulcanized together at a location spaced laterally apart from the top edge **18** of the boom and extending for a distance along it parallel to the edge **18** to form an air chamber **20**. The sides also are vulcanized together at spaced locations, as **22, 24** and **26**, along the length of the boom to form separate air chambers **20**, each of several feet length. Each air chamber has its complementary valve means **28** through which the chamber may be inflated with air and through which the air may be released from the chamber to deflate the boom.

The air chambers form the buoyant means which causes the boom to float on the surface

FIGURE 66: CHEVRON FLOATING BOOM DESIGN

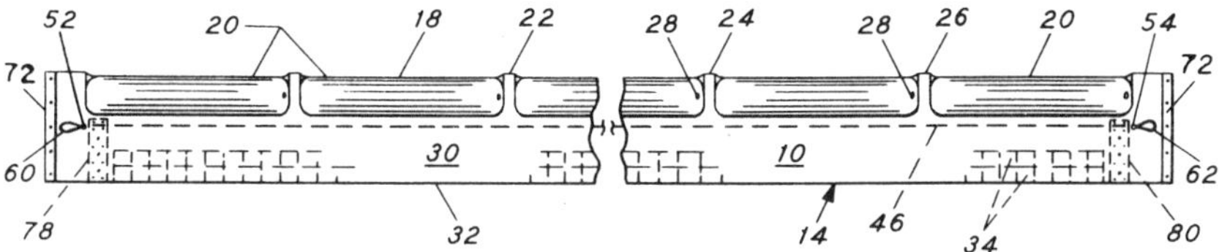

Source: J.A. Sayles; U.S. Patent 3,792,589; February 19, 1974

of a body of water and to extend sufficiently above the surface to form a barrier for contaminants floating on the surface. The sides of the boom which extend below the flotation chambers **20** are vulcanized together and to elements incorporated between them which are described hereinafter to form the skirt **30** of the boom which remains submerged in the water during use.

The bottom edge portion **32** of the skirt **30** has incorporated in it ballast weights **34**. These weights preferably are formed from segments of thin lead sheeting which are distributed along the bottom edge portion of the skirt. Each of the lead segments is spaced apart from the other so that the skirt **30** will be able to bend and flex between the lead segments in the area of the ballast weights.

Although the impregnated fabric of which the boom is made is tough and durable for ordinary handling and shipboard use, it is not designed to withstand directly the high tensile stresses which will be imposed on the boom when it is deployed in the water and exposed to wave action and strong wind and water currents. Therefore, means are incorporated in the boom structure for accepting and distributing the stresses imposed on it in a manner to prevent damage to the fabric of the boom. To this end a tensile cable **46** is placed in the skirt **30** of the boom between and enclosed by the sides **10** thereof which are vulcanized to it.

The cable is located below the air chambers **20** and placed approximately along the longitudinal median of the boom and extends within the skirt longitudinally of the boom to adjacent each end thereof. At a location approaching a respective end of the boom the cable is extended outwardly from between the wall of the skirt **30** through a respective opening **52** and **54** in one side **10** of the boom as illustrated to provide ends of the cable which are accessible from the exterior of the boom. Each terminal end of the cable is formed as an eye or loop **60** and **62**, respectively, to which a similar eye of a corresponding tensile cable of another section of the boom can be connected as by a bolt and nut **64**.

Strips of stiffening material such as strips of synthetic resin which is resistant to the effects of water is secured to each side of each end flap of the section of boom as illustrated at **72**.

It has been found that the tensile cable which, as described above, is embedded in the skirt **30** can, under adverse conditions, be brought to bear against the side of the boom with sufficient force to cause the cable to tear out of the skirt of the boom in a direction normal to the axis of the cable, thus damaging the boom. The boom is particularly exposed to such damage if it snags while it is being towed or is being lifted by a tensile cable out of the water. It is important to provide a construction which will prevent such damage from occurring. To this end a respective stress plate **78** and **80** is placed in the skirt of the boom adjacent each corresponding end portion thereof and the end portions of the tensile cable **46** are securely fastened to a respective stress plate so that the forces which tend to tear the stress cable from the skirt of the boom will be transmitted to the stress plate and distributed over an area of the skirt sufficient to prevent such forces from being concentrated destructively where the tensile cable bears against the side of the boom.

Clean Water Inc. Design

A boom design developed by Clean Water, Inc. of Toms River, N.J. has a tubular air chamber which floats on the water surface and a weighted nylon net skirt which is suspended from the air chamber below the water surface.

According to Maritime Administration report COM-74-10212, Springfield, Va., Nat. Tech. Information Service (Nov. 1973) this Offshore Boom is designed to operate in 8-foot waves, in a 20-foot swell, in 20-knot winds and in 1.5 knot currents. It has been tested in 4-foot waves, 15-knot winds and 1.5 knot currents. Some 220 feet of barrier served to contain 14,000 gallons of #2, #4 and #6 oil according to the above cited report.

Two men can display a 2,000-foot barrier of this design in 3 hours using two 14-foot workboats and an air compressor. Recovery, cleaning and storage of such a barrier are estimated to require an average amount of effort.

Two anchors of 5,000-lb holding capacity are required to moor 1,000 feet of this offshore barrier. As described by the EPA, Edison, J.J. Laboratory in *Oil Spill Containment Systems* (Jan. 1973), this boom is designed for high mobility and quick deployment in moderate sea conditions. The boom requires less than one third its deployed volume for storage and transportation. For example, 1,000 feet of boom can be readily carried and deployed by a truck.

The flotation unit of the boom is fabricated from rubber impregnated nylon material and is vulcanized throughout. The material, originally designed for use in inflatable dunnage bags, is abrasion and puncture resistant. The hardware includes two ⅜-inch top and one ¼-inch bottom galvanized chains. The boom is furnished in 55 foot lengths, with five 10-foot long 12½-inch diameter flotation chambers bifurcated longitudinally to provide two separate compartments for each chamber. The flotation chambers are inflatable to a pressure of about 10 psi.

The boom presents an approximate 15-inch freeboard with an 8-inch impermeable skirt depending (hanging below) below the waterline. A ⅛-inch (18 ounce) nylon net depends from each ⅜-inch chain to form a 2-foot deep filtration skirt beneath the flotation unit on either side of the 8-inch impermeable skirt. Towing bridles and related hardware are available to adapt the boom for sweeping operations, according to a product brochure furnished to EPA by Clean Water, Inc., Toms River, N.J.

Colloid Chemical Co. Design

A boom developed by Colloid Chemical Co., a division of Pollution Control and Research Inc. located in Brockton, Mass. has been designated the Spilldown 360 in Maritime Administration Report COM-74-10212, Springfield, Va., Nat. Tech. Information Service (Nov. 1973).

It is designed to operate in 3-foot seas, 20-knot winds and 2-knot currents. It has been tested under these same conditions and the barrier contained 50 to 100 gallons of 0.85 to 0.97 specific gravity oil in tests with 200 to 500 foot sections. Three or four men can deploy 2,000 feet of this barrier in 15 minutes. An 80 hp boat is required. Recovery, cleaning and storage effort are judged to be average for this type of barrier. As described by the EPA, Edison, N.J. Laboratory in *Oil Spill Containment Systems* (Jan. 1973), the boom is also known as the PCR-36.

The individual sections measure 100 feet in length, are 36 inches wide and weigh approximately 300 pounds. The fabric is a yellow PVC-coated nylon and constructed by stitching, rivets and grommets. Flotation is derived from cylindrical expanded polyethylene floats fastened with bronze chain and toggle connectors. Vertical stiffeners are aluminum rods or fiber glass reinforced epoxy rods enclosed in stitched pockets reinforced top and bottom. The boom has a galvanized chain ballast draped every 12 inches. Sections can be hooked together to make up any desired length.

Denison Design

A device developed by *C.S. Denison; U.S. Patent 3,695,042; October 3, 1972* is a device for containing oil spills in the open sea, comprising a plurality of hanger float structures having a continuous flexible wall or barrier suspended therefrom that encircles and contains an oil spill, the bottom edge of the wall being weighted and submerged, and the hanger float structures being connected with and held in place by spaced anchor float structures to which anchors are attached.

The flexible wall has a height substantially greater than the height of the suspension cable above the surface of the water, and the bottom edge thereof has weights attached thereto at spaced intervals around the circumference thereof. This arrangement results in the lower part of the flexible wall becoming and remaining submerged, even in heavy seas.

The great problem with containment devices previously contemplated has been to keep them in place for prolonged periods, under adverse sea conditions. This problem is solved in the design by the use of a plurality of spaced anchor float structures that are placed outwardly from the hanger floats but which are connected thereto by cables. Each anchor float structure is secured in place by an anchor, placed outwardly therefrom. The result of this arrangement is that the flexible containment wall remains in place even in rough seas, the structure being sufficiently flexible to absorb waves of considerable size without damage.

This structure can be carried to the scene of an oil spill, and thereafter can be quickly erected to confine the oil. Conventional equipment can then be utilized to remove the oil from the surface of the water, after which the containment device can be disassembled and stored for later use. Further, if desired the process can be installed permanently around offshore drilling rigs, to be ready for any spillage emergency. Figure 67 is a transverse section through a portion of the barrier design.

FIGURE 67: DENISON FLOATING BOOM DESIGN

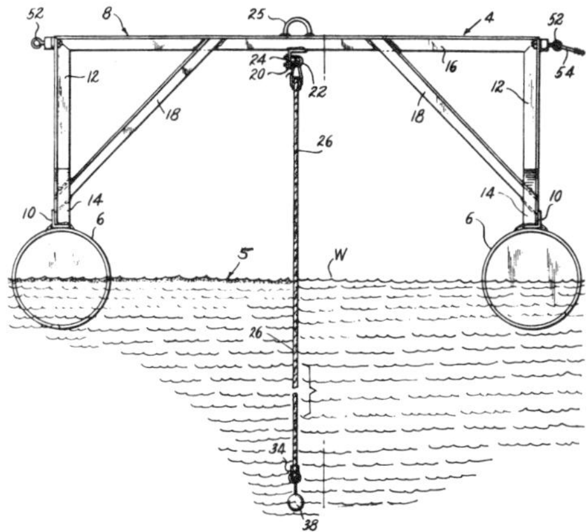

Source: C.S. Denison; U.S. Patent 3,695,042; October 3, 1972

The containment device includes a plurality of hanger float structures or fixtures **4** spaced around the periphery of the oil slick **5**. Each hanger fixture **4** includes a pair of spaced, sealed drums **6** of the 55 gallon or similar size, which drums support a suspension frame **8**. The frame **8** includes a base angle member **10** welded on each drum **6** to extend the length thereof, to each of which is welded the lower end of an upright angle member **12**. The angle members are positioned centrally of the drums and are braced by inclined members **14** welded thereto and to the outer ends of the base angle members.

A bridging angle member **16** is welded to extend between the upper ends of the upright members **12**, and is braced by a pair of inclined angle members **18**. Centrally thereof each bridging member has a short hanger angle member **20** welded to the bottom surface thereof, the hanger members extending perpendicularly to the bridging members and projecting about equally from both sides thereof. A pair of U-bolts **22** is mounted on the opposite ends of the depending vertical flange of each hanger member, the U-bolts on all of the hanger fixtures supporting a continuous suspension cable **24**. Centrally of the top edge of each bridging member is welded a U-shaped hook **25**, for use in handling the hanger fixture. Suspended from the cable is a flexible, continuous wall or barrier **26**, made of waterproof canvas, plastic or some other suitable material.

The flexible wall **26** has a height about twice the height of the suspension cable **24** above the water **W**, and the lower edge **34** thereon is bound and has spaced, grommeted holes therein. Weights **38** are suspended from the grommeted holes **36** and are spaced around the periphery of the wall or barrier **26**. The weights thus serve to submerge the lower half of the barrier or wall, whereby effective containment of the oil slick **S** is obtained.

It will be noted that the weighted wall or barrier hangs like a pendulum from the inverted U-shaped suspension frame **8**, an arrangement which allows the spaced hanger fixtures **4** and the wall or barrier to ride with the waves, even in heavy seas, without being damaged or displaced. Thus, for so long as the hanger fixtures are in place whereby the suspension cable is taut, effective containment of the oil slick **S** is obtained.

The hanger fixtures are connected to anchor fixtures by cables **54** which are attached to extend tautly between the eye-bolts **52** on the outer face of the suspension frames **8** and the anchor fixtures.

Environetics Inc. Design (Boa Boom)

According to Maritime Administration Report COM-74-10212, Springfield, Va., Nat. Tech. Information Service (Nov. 1973), the boom with an 18-inch draft and a 9-inch freeboard is designed to operate in 2-foot seas, 20-knot winds and 4-knot currents. It can contain 60,000 gallons of oil per 2,000 feet of barrier. Three men can deploy 2,000 feet of Boa Boom in 3 hours using one vessel. Some 50 ten pound anchors are required per 1,000 feet of boom.

As described by the EPA, Edison, N.J. Laboratory in *Oil Spill Containment Systems* (Jan. 1973) and as shown in Figure 68, the boom is fabricated of reinforced PVC. It is made up of a self-contained replaceable air bladder for buoyancy and strength. The skirt is nylon reinforced PVC with a ¼-inch chain contained in the hem for ballast, mooring and towing. There are eyelets for mooring, anchoring and towing at each end of the skirt through the ballast chain.

Each section is 50 feet in length with vinyl zipper on each end so that any number of sections may be added. The flotation collar is one foot in diameter and the skirt is 2 feet long. The flotation cover has an air bladder access zipper and double valves for inflation and deflation of the enclosed bladder. The boom is bright yellow in color for maximum visibility. Each 50-foot section costs from $412.50 to $315.00 each, depending on the quantity purchased according to a product brochure furnished EPA by Environetics, Incorporated of Worth, Illinois in April 1972.

FIGURE 68: ENVIRONETICS BOA-BOOM DETAILS

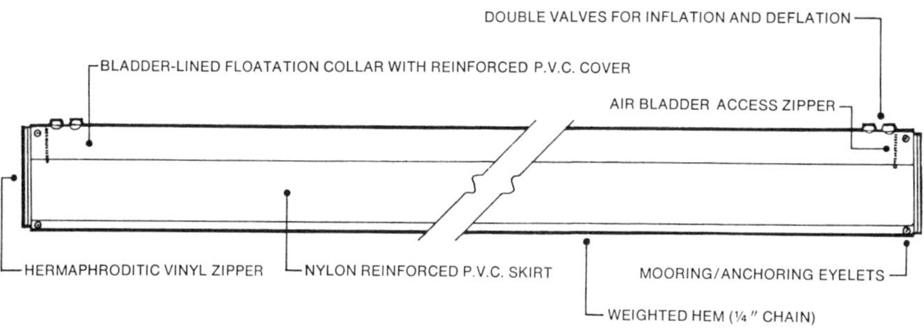

Source: EPA Edison, N.J. Report *Oil Spill Containment Systems* (Jan. 1973).

Esso Production Research Company Design

A design developed by *T.W. Childers; U.S. Patent 3,653,213; April 4, 1972; assigned to Esso Production Research Company* is a floatable plastic barrier, molded on-site in a desired shape and secured to an elongated flexible member, used to contain oil spills in water locations. The flexible member is a cable to which the barrier is bonded directly or attached by clips as the barrier and cable are fed onto the water. The plastic barrier is preferably molded to a 90° "V" shape with the cable formed in or attached to the vertex of the Vee. Vertical drain holes may be punched or drilled at intervals along the length of the barrier to prevent splash from accumulating in the Vee. Mooring lines are attached to the barrier as needed.

Figure 69 shows, from top to bottom, the boom-laying vessel, the boom extrusion apparatus and the boom in service. First the reader is referred to the vessel **10** on which is positioned molding apparatus including an extrusion mold **11**, plastic reservoir tanks **12**, and a control system **13**. A reel **14** is also positioned on vessel **10**. A heavy flexible support cable **15** wound on reel **14** feeds through the extrusion mold in which the cable reinforced plastic barrier, indicated at **16** astern of vessel **10**, is continuously formed.

As seen in the middle view the extrusion mold is supplied with resin and hardener through conduits **20** and **21**, respectively from resin tanks **12**. A control conduit **22** is used to control the molding operation including regulation of the resin mixture and hardening time. As shown, the plastic barrier **16** is preferably formed in a 90° V shape and molded to cable **15** at the vertex of the Vee. The density of the plastic barrier and support cable is sufficient to buoy the barrier so that the wings of the Vee are supported above the surface of the water approximately one foot with approximately one foot thereof submerged.

As shown in the lower view, suitable mooring lines **25** connected to buoys **26** anchored by anchor lines **27** are attached to plastic barrier **16** as needed. These attachments are preferably through a "spring buoy" type system, as shown, to prevent the mooring lines from pulling the barrier under water in strong currents.

Polyurethane foams are preferred as the plastic material due to the inherent toughness and flexibility of this plastic and its resistance to hydrocarbons. However, other plastic foams formed by combining two liquid chemical reactants may be used instead. Plastics of this type are easily handled and form in a few minutes following mixing of the desired ingredients.

FIGURE 69: ESSO PRODUCTION RESEARCH CO. EXTRUDED PLASTIC BOOM – PRODUCTION AND USE

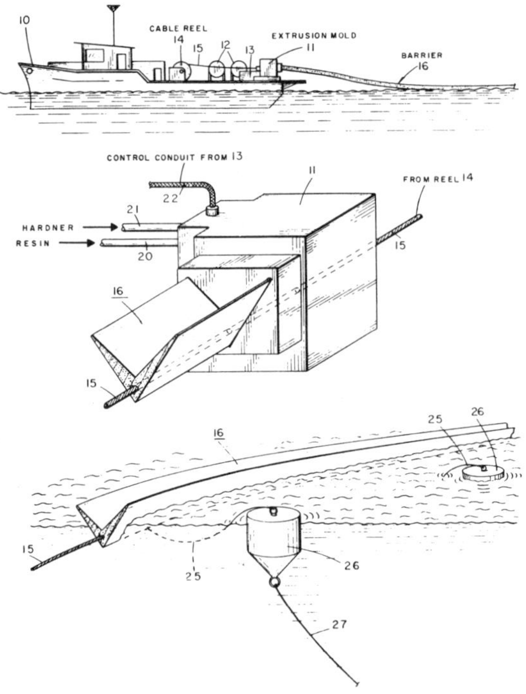

Source: T.W. Childers; U.S. Patent 3,653,213 April 4, 1972

A design developed by *G.R. Cunningham and L.D. Woody, Jr.; U.S. Patent 3,702,657; November 14, 1972; assigned to Esso Production Research Company* is a flow-through pollution containment barrier for the control and removal of potential oil spill hazards in a water environment. It includes a series of cages linked together. Oil sorbent buoyant material, capable of allowing water to pass freely through it while remaining oil wet when contacted by an oil-water mixture is removably arranged in each cage. Liquid impervious material is arranged in the spaces at the joints between the cages to prevent passage of liquid therethrough. Each cage is weighted in order to maintain a predetermined submergence of the cage in a body of water.

A sectional perspective elevation of such a barrier design is shown in Figure 70. There are shown two rectangularly or box-shaped cages **10** formed by upper and lower side rim members **12** and **13**, respectively, solid end panels **14**, wire mesh or screen sides **15** and bottom **16**, center vertically extending slats **17** and horizontally extending upper slats **18**. The cages are connected together by a flexible wireline and socket arrangement **19, 20.** The sockets are connected to end panels **14**. Two flexible membranes **21** and **22** formed of liquid impervious material close off the space between cages **10** to the passage of liquid therethrough.

The interior of each cage **10** is divided into two slots or compartments **25** and **26** which extend the length of the cage. An oil sorbent buoyant bag or pad **27** is positioned in each

FIGURE 70: ALTERNATIVE ESSO PRODUCTION RESEARCH CO. FLOATING BOOM DESIGN

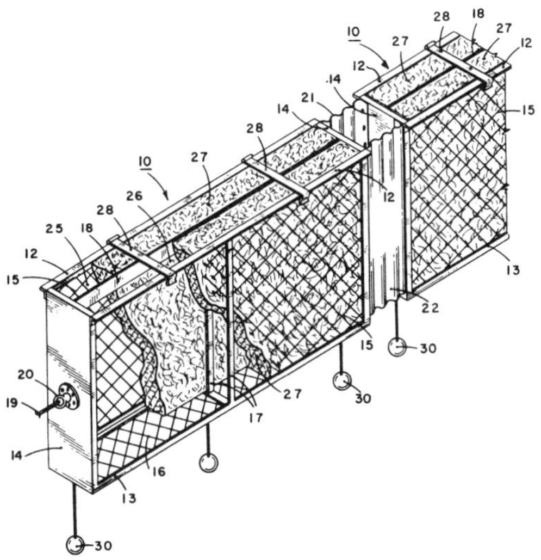

Source: G.R. Cunningham and L.D. Woody, Jr.; U.S. Patent 3,702,657; November 14, 1972

compartment. Hinged snap latches **28** are arranged to connect with the upper rim members **12** to retain the pads in their compartments. Weights **30** maintain desired submergence of the barrier device in the water. The weight requirements could otherwise be satisfied by a skidding system which would also facilitate handling and installation from a land site.

The pads of sorbent material may be removed and replaced by removing the snap latches or other lid arrangement which may be provided to aid in containing the sorbent material. The individual cage dimensions are limited only by the fabrication material sizes. The required submergence, area of exposure and thickness of sorbent material necessary would be designed in accordance with the particular use for the barrier.

Each individual cage consists of a minimum of two compartments oriented such that as one bag of sorbent material is being removed and replaced, the second compartment with its bag of sorbent material remains as an effective flow-through barrier. Sorbent bag size is limited only by the ability of equipment to pick up a soaked bag weight and place the bag on a boat or barge for disposal or reclamation. Sorbent materials suitable for use are those having a cellular structure sufficiently open to allow water to pass freely and yet remain oil wet (oleophilic) when contacted by an oil-water mixture.

Examples of such materials are the generic polymers, such as the polyester, polyethyl and polyurethane foams. Such foam materials have proven durability characteristics in a water environment. The series of cages forms an articulated boom system. The joints between the cages consisting of the pseudouniversal joint arrangement allows a sufficient degree of flexibility in responding to wave or current imposed boom orientations without excessive flexural stressing of the cages. The impervious membrane inserts at the joints makes the joint a nonflow region, forcing flow through the cages.

Additional external and/or internal sorbent material bags or units might be used. The sorbent material would provide the necessary buoyancy. Each unit would be sized sufficiently such that when all but a single bag or unit is removed for a changeout of the sorbent material, sufficient buoyancy would still exist to support the cage assembly.

Fisch Design

A design developed by *P. Fisch; U.S. Patent 3,631,679; January 4, 1972* is a floating loop barrier which can be looped to surround a large water area covered with oil. The loop can be reduced so that the originally thin oil film which cannot escape, will gain more height so that it can be pumped out, separated from the water. The barrier is reduced by having one end passed through an eye at the other end and pulled by a boat. The barrier comprises a flexible steel cable or rope core surrounded by inflated sleeve sections attached in tandem. The barrier can be reeled onto a ship and the inflated sections separated from the rope. The sleeves are deflated and stored on the ship while the rope is wound on a drum for storage on the ship.

Flaviani Design

A design developed by *E. Flaviani; U.S. Patent 3,651,647; March 28, 1972* comprises a variable group of individual watertight floatable barrels rising above the water and affording a barrier against the escape of the pollution material from the area, together with universal joints connecting adjacent barrels of the group to one another and serving to allow the individual barrels to pitch and roll in accordance with the pitch and roll of the water upon which the barrels are floated and thus to maintain the barrier operative despite undue water movement.

Flexy Oil Boom Design

According to the EPA, Edison, N.J. Report, *Oil Spill Containment Systems* (Jan. 1973), the Flexy Oil Boom, made in Canada by Smith-Anderson Co., Ltd., Montreal, Quebec, is constructed of 36-inch wide PVC nylon fabric. When in use, the skirt depth is 24 inches and the freeboard 12 inches. See Figure 71.

Available in 100-foot lengths, the nylon fabric has a tensile strength of 450 pounds per inch and is fitted with aluminum stiffeners for rigidity. Stability is provided and maintained by floats and lead ballast weights spaced two feet on center, and galvanized chain at the skirt bottom running the full length of the section. Weight of a 100-foot section, including floats and ballast, is reported at 250 pounds.

FIGURE 71: FLEXY OIL BOOM

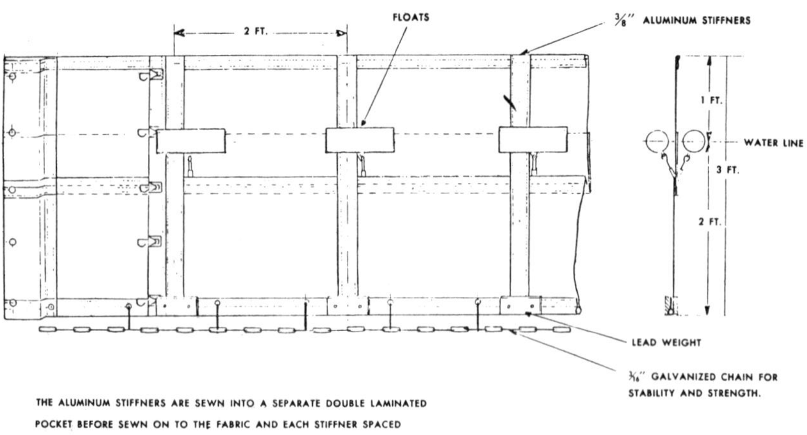

(continued)

FIGURE 71: (continued)

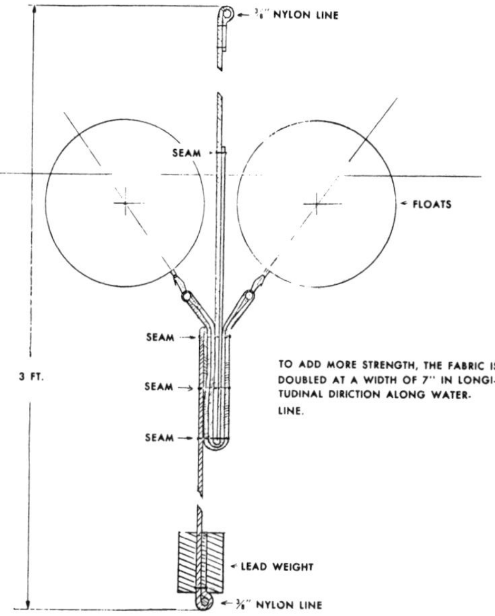

Source: EPA, Edison, N.J. Report *Oil Spill Containment Systems* (January 1973)

For attaching one or more sections together, each end of the boom is equipped with five snap hooks placed two feet from the end, and five matching outlets on the adjoining section; thus providing an overlap of two feet at the junction or connection points. For extra strength, the chain ballast at the skirt base is shackled together. The information was provided to EPA by Hurum Shipping and Trading Co., sole distributors of Flexy Boom in January, 1970.

Flo-Fence Design

This design, made by Logan Diving and Salvage Co. of Jacksonville, Florida is designed to operate in 10-knot winds, 10-foot waves and 1-knot currents according to Maritime Administration Report COM-74-10212, Springfield, Va., Nat. Tech. Info. Services (Nov. 1973). However, no actual test data were available, according to this report.

It was stated that 3 men could deploy 2,000 feet of barrier in 20 minutes using 1 to 2 vessels and a 105 CFM air compressor. Recovery, cleaning and storage of this barrier were pledged to be fairly easy even though no specific data were available. Covered storage, an inert gas atmosphere and humidity control are desired but not required for storage. The Maritime Administration report states that theoretical conditions indicate that the barrier will not be effective in strong currents.

Constructed of fireproof, acid, and oil resistant coated fiber, the barrier is available in 100-foot lengths, weighs 95 pounds per 100-foot section, and can be stored in an area of 6 feet by 5 feet by 3 feet.

Gadd Design

A design developed by *P.S. Gadd; U.S. Patent 3,584,462; June 15, 1971* is an elongated floating boom structure which protectively supports a flexible curtain having float means secured along an upper edge and weight means along a submerged edge, the float means and weight means being connected by slack chains to adjacent portions of the floating boom structure so as to maintain the curtain in an effective operation and working position.

The boom structure mounts protective shielding screens for the curtain, and additionally includes upstanding baffles connectable with float means and/or the floating boom structure. In a modified arrangement, pairs of boom structures support a common curtain therebetween, the curtain being secured along its side edges and being downwardly curved to provide an elongate trough or sluice.

The floating barrier, as generally indicated at **12** in Figure 72, comprises an elongate curtain **13** of suitable material such as one of the available plastics. This curtain is attached in any suitable manner along one edge margin to a suitable float **14** which may be variously constructed of any suitable material, and although wood would be suitable, it is preferred to use a Neoprene tubular hose. This float is of continuous length, and may be formed by interconnecting sections in end-to-end relation. The purpose of the float is to suspend the curtain so as to provide a flexible barrier extending below the water surface.

In order to normally maintain the curtain in an upright position, it is preferred to attach weight means along the other edge of the curtain material. This weight, while it may constitute separate elements, is shown herein as comprising a wire rope **15** which is flexible and therefore adapted to conform longitudinally with the position of the float **14**.

FIGURE 72: GADD FLOATING BOOM DESIGN

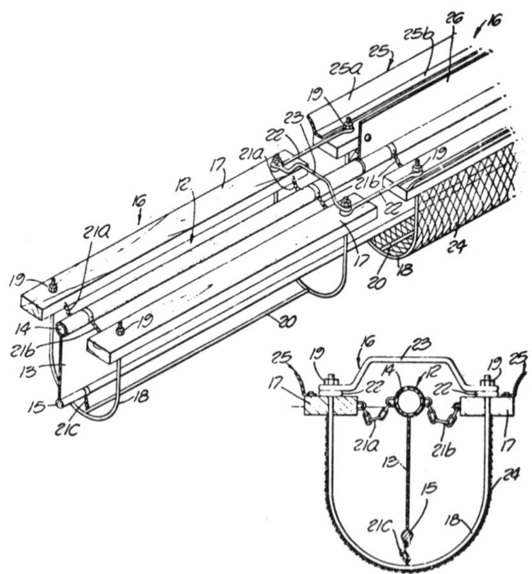

Source: P.S. Gadd; U.S. Patent 3,584,462; June 15, 1971

While the floating barrier as thus far described will function efficiently under normal placid conditions of the water surface, it will be appreciated that under abnormal conditions the floating barrier could be twisted or deformed in a manner to permit the confined materials to pass the barrier. It is a primary feature therefore to provide a stabilizing structure in the form of cradle sections as generally indicated at **16**, which are arranged to be flexibly joined in end-to-end relation for articulated movement which will permit the sections to conform to the longitudinal direction taken by the float **14**.

Each of the cradle sections **16** comprises a pair of boom floats **17—17** of timbers or wood planking which are positioned in parallel relation on opposite sides and outwardly spaced from the adjacent float **14**. The boom floats are retained in fixed spaced relation by means of U-frames **18** which have their ends respectively secured to the boom floats as by retaining nuts **19** having threaded engagement with the frame ends. The U-frame extends below the wire rope **15**, and if desired, the successive U-frames of the cradle sections may be interconnected at their bottoms by tie members **20**.

The floating barrier **12** is supported within the associated cradle sections for limited movement, but held in spaced relation within the cradle by means of a plurality of slack chains **21a** and **21b** which connect from the opposite sides of the float **14** to the adjacent ends of the boom floats, and by slack chains **21c** which connect from the wire rope to the underlying portion of the U-frame. Thus, the curtain will have limited movement within the cradle sections, but will be stabilized into an effective position within the cradle structure under wave, wind and other adverse conditions.

The successive cradle sections are flexibly interconnected by connecting links **22** between the spaced ends of the boom floats on each side, these links having end eyelets adapted to loosely engage the ends of the adjacent U-frames of the sections. While the U-frames ordinarily are sufficient to maintain the spacing of the booms **17—17** of each section, additional cross-tie members, as indicated at **23** may be provided between the ends of the U-frames. The cross-tie members are preferably provided with bowed or offset portions between their ends so as to provide increased clearance above the underlying float **14**. With the cradle sections thus interconnected in end-to-end relation, the boom floats provide catwalks on opposite sides of the sections, whereby it is possible for a person to walk along the apparatus and inspect the various elements or make repairs, if necessary.

The U-frames, as shown, also serve the important function of providing a framework structure upon which there is mounted a mesh screen **24** which provides protective shielding for the floating barrier structure **12**. Although not shown, it is contemplated that if desired a similar screen could extend above the floating barrier.

As an additional feature, splash barriers may be mounted on either or both of the boom floats, and floats **14** in order to more effectively contain the foreign matter. These splash barriers may take the form of flexible elongate baffles **25** secured to the boom float, the baffle being shown as having angularly extending side margins **25a** and **25b** so that when one of the margins is connected to the float boom, the other will extend above the float boom and form a baffle. As exemplary of the baffle which is connected with the float **14**, this baffle as indicated at **26** is shown as of elongate rigid plate construction which may be secured in any appropriate manner to the float **14** as by suitable clamping bands or other conventional means.

Galvaing Design

According to Maritime Administration Report COM-74-10212, Springfield, Va., the Galvaing Floating Boom is made by Gamlen, Naintre and Cie. of Clichy, France. This report indicates that theoretical considerations suggest that this barrier will not be effective in strong currents.

According to the EPA, Edison, N.J. Laboratory report *Oil Spill Containment Systems* (Jan. 1973), Galvaing booms are composed of short compact units of rigid floats inserted into plastic-coated fabric panels or PVC-coated asbestos panels. Individual units (consisting of a vertical panel and a float attached to each side of the panel) are coupled together by neoprene rubber or a plastic-coated fabric. To avoid unnecessary strain on the panels and prevent them from being pulled out of shape, a cord or metallic chain is inserted into the top and bottom hems of the panels running the length of the barrier.

Individual units are only 3.2 to 3.9 feet long to provide for maximum flexibility of the boom. A standard section of the Galvaing barrier comprises either 4 or 5 units depending upon the type of barrier. The length of a standard section varies from 16 to 20 feet. Security hooks are provided to link any number of standard sections together for total length of boom desired. The barrier extends approximately 8 inches above water and 8 inches below.

Three types of Galvaing barriers are available: (1) The Emergency Barrier; (2) The Ballast Water Removing Barrier; and (3) The Fire Barrier.

The Emergency Barrier is made up of polyurethane floats attached onto plastic-coated fabric panels and may come equipped with skirt and ballast fixed to the bottom of the panels. The Ballast Water Removing Barrier appears similar to the Emergency Barrier except a special skirt is attached to the bottom of the panels giving a total submersible depth of about 3 feet. When used for the purpose of retaining and separating oils from ballast water, the manufacturer recommends weighting of the skirt. The two types of barriers described above are relatively lightweight, approximating 1.1 pounds per foot without auxiliary skirt and ballast.

The Fire Control Barrier is designed for containing oil and petroleum product spills within harbors and other similar areas, but is also intended for encircling and minimizing the spread of petroleum fires on water. The Fire Barrier is like the Emergency Barrier in principle. Major differences are aluminum alloy floats inserted into PVC-coated asbestos cloth, and the units being coupled by plastic-coated asbestos and galvanized steel chain. The Fire Control Barrier is available in standard and heavy-duty forms. The standard fire barrier has an overall height of 13 inches, weighs approximately 4 pounds per foot, and cost given in 1967 was $16.20/foot – FOB Marseilles, France.

The heavy-duty fire barrier has an overall height of 18 inches, weighs approximately 8 pounds per foot, and 1967 cost quotations were given as $22.40/foot. After use in a fire, considerable repair work is indicated even for the Fire Control Barriers. One to three months delivery time may be expected for Galvaing barriers and/or replacement parts.

The manufacturer recommends that when the Galvaing boom is towed to the site of an oil spill, it should not be pulled at a speed exceeding one knot. It is also suggested that these booms not be deployed against a fast-moving current nor placed directly across a river or waterway. Rather, the boom should be moored at a desirable angle of closure with the shoreline, enabling recovery of oils in the slower-moving downstream areas of the boom.

Gambel Design

A design developed by *C.L. Gambel; U.S. Patent 3,783,622; January 8, 1974* provides a rigid barrier unit for use in assembling an enclosure around a surface area of a body of water. It is provided with adjustable buoyancy and ballasting chambers so that a nearly neutral buoyancy condition can be established with substantially all of the mass of the unit below the turbulence level of a body of water in which the barrier unit is placed. A method of deployment of such units involves a flotation of the units in horizontal attitudes to the area to be enclosed, followed by a ballasting of the units into vertical attitudes so as to extend around an oil or chemical spill area.

Gates Rubber Company Design

According to the EPA, Edison, N.J. Laboratory Report *Oil Spill Containment Systems*

(Jan. 1973), the Gates Boom Hose is a 25-inch (outer diameter) floating cylinder to which is attached a 20-inch deep skirt ballasted by two pounds of lead weight per running foot of skirt. (See Figure 73).

The one-half inch thick hose consists of two piles of nylon tire cord carcass reinforced with spiral wire, with neoprene outer cover and inner high-tensile Buna-S tube. The Gates Boom Hose rides 19 to 20 inches above the water surface and 5 to 6 inches below (excluding skirt). The 1/16-inch thick neoprene-nylon skirt is built into the hose with reinforcing piles, and comes equipped with grommets along the bottom edge of the skirt to attach lead weights and with grommets at the ends of skirt section to splice adjacent lengths together.

The hose assembly is available in lengths of approximately 25 feet. The ends of the hose are reinforced with fiber glass, also containing a built-in aluminum nipple and aluminum round plate welded to the nipple. Adjacent hose sections are fastened together by stainless steel bands over the end plates. It is noted the Gates Hose is a noninflatable boom.

It is reported the assembly is flexible and has high-tensile strength capable of withstanding end loads of 30,000 pounds. When the Gates Boom Hose exceeds 400 feet in length, the manufacturer recommends that additional reinforcement be provided by steel cables attached to the hose nipples. Boom price is quoted at approximately $50 per foot according to information received by EPA from the Gates Rubber Company, Denver, Colorado in November 1969.

According to Maritime Administration Report COM-74-10212, Springfield, Va., Nat. Tech. Information Service (Nov. 1973), theoretical considerations suggest that this barrier will not be effective in strong currents.

FIGURE 73: GATES RUBBER COMPANY FLOATING BOOM DESIGN

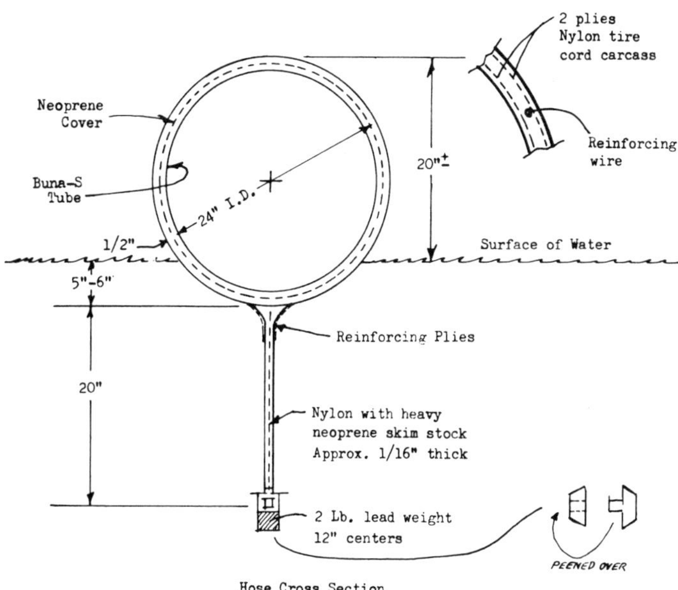

Source: EPA, Edison, N.J. Laboratory Report *Oil Spill Containment Systems* (Jan. 1973)

Grefco Design

A design developed by the Control Products Unit of Grefco, Inc. of Jamesburg, N.J. is designated the Grefco Sorbent Boom, according to Maritime Administration Report COM-74-10212, Springfield, Va., Nat. Tech. Information Service (Nov. 1973).

It is designed to contain 5,000 gallons of oil per 2,000 feet of boom deployed. Two men can deploy 2,000 feet of boom in 2 hours using 2 vessels. It is an 8 inch diameter round burlap "sausage" filled with fiberpearl sorbent. According to the above-cited report, theoretical considerations suggest that this barrier will not be effective in strong currents.

Headrick Design

A hose-type barrier used during the Chevron tanker spill in San Francisco Bay was made by Headrick Industries Inc. of La Canada, California. As described in Maritime Administration Report COM-74-10212, Springfield, Va., Nat. Tech. Information Service (Nov. 1973), this boom contains two floating air-filled cylinders and one submerged water filled tube giving a sort of triangular cross section. They say that theoretical considerations suggest that this design will not be effective in strong currents.

It is designed to operate in any nonbreaking wave and in winds up to 20 knots. In actual tests, 500 feet of barrier has contained 100 barrels of API 30 oil in 5-foot waves, 20-knot winds and 5-knot currents. From 5 to 15 men can deploy 2,000 feet of this barrier in 2 hours using one vessel containing air and water handling equipment. Recovery, cleaning and storage are believed easy but no exact data were available on these points.

Heartness Design

A design developed by *O. Heartness; U.S. Patent 3,701,259; October 31, 1972* is produced by the simultaneous and synchronized unreeling from a vessel of a resilient barrier strip formed with spaced inflatable pockets and a sectioned resilient hose containing a liquid chemical. Each inflatable pocket has an opening formed at the base which is in communication with an injection tube which is removably mounted below each pocket in the lower edge of the barrier strip. Each injection tube contains a chemical powder, reactive with the liquid in the hose to form a gas.

At a predetermined point the injection tube is forced into the hose, the hose is compressed to force the liquid upward into contact with the chemical powder, the gas formed by the reaction inflating the pocket in the plastic barrier. In alternative forms of the device the pocket is inflated by the use of a low boiling point volatile liquid or compressed gas stored in the hose. The barrier strip and hose are unreeled until the oil slick is surrounded, whereafter the ends of the barrier and hose are cut to form an enclosure around the oil spill.

Previous efforts to form barriers around oil spills have involved the separate and cumbersome steps of attaching flotation members to an inflatable barrier; the use of a continuous air tube to produce buoyancy, which tube will collapse if there is a rupture in any part of the air tube; the use of air pumps to inflate the tube after the tube has been positioned on the water resulting in a great loss of time while the oil continues to disperse before the barrier is inflated; the pumping of water and air into a continuous tube to form a partially submerged barrier, which process is slow and does not insure that the water within the tube will remain evenly distributed throughout the tube, thus forming a barrier which may be completely submerged in different areas.

Humble Oil and Refining Company Design

A design developed by Humble is a Bottom Tension Boom as described in Maritime Administration Report COM-74-10212, Springfield, Va., Nat. Tech. Information Service (Nov. 1973). It is designed to survive in 20-foot waves, 2-knot currents and 60-knot winds. It is designed to contain oil in 1.25 knot currents, 12-foot seas and 40-knot winds. A

500-foot prototype has successfully contained oil in 6 to 8 foot maximum waves, 20 knot winds and 1.25 knot currents. This boom was designed to operate in the Santa Barbara Channel. It consists of four basic units:

The Buoyancy Unit – A 4' diameter 13' long high density polyethylene material connected to the skirt by bolts.
Skirt Unit – An 8' wide strip of neoprene conveyor belting.
Lattice Unit – A net of ⅜" polypropylene rope with 12" mesh, which surrounds the float and skirts units, and
Bottom Tension Unit – 2.25" wire rope which serves as ballast and towline unit.

Hurum Shipping and Trading Company, Ltd. Design

A design developed by *R.A. Fossberg; U.S. Patent 3,740,955; June 26, 1973; assigned to Hurum Shipping and Trading Company, Ltd., Canada* is a flexible oil boom which has unique capabilities of being compactly stored and also being extremely stable in heavy seas. The boom includes a curtain wall of sheet material for deploying in a substantially vertical position in the water such that the upper edge is above the water surface and the lower edge is below the water surface.

A plurality of substantially vertical stiffening members are positioned in spaced relationship along the length of the curtain wall, these members being arranged in opposed pairs with the curtain wall sandwiched between. Outrigger members are connected on each side of the boom a short distance below the water line and each outrigger has an inner end pivotally connected to a stiffening member and an outer end having a connector for connecting a float thereto. A restraining member allows the outrigger to swing between a downward retracted position adjacent the stiffening member and an operating position substantially perpendicular to the stiffening member.

Keel members are pivotally connected to the lower ends of the stiffening members at both sides of the curtain wall and these keels are held by restraining members which allow them to swing between an upper retracted position adjacent the curtain wall and an operating position in which they are upwardly and outwardly inclined. The reader is referred to the earlier description of the Flexy Oil Boom whose sole distributors are the Hurum Shipping and Trading Company.

Jaton Design

FIGURE 74: FLOATING OIL RETAINER

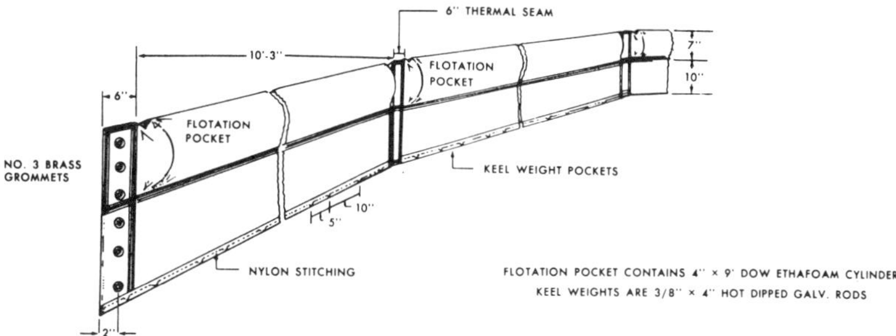

Source: EPA Edison N.J. Report *Oil Spill Containment Systems* (Jan. 1973)

According to the EPA, Edison, N.J. Laboratory Report *Oil Spill Containment Systems* (Jan. 1973), Jaton is the tradename for a floating oil retainer, which is of the float and skirt design and which is available from Centri-Spray Corporation, Centri Clere Filter Division, Livonia, Michigan. The flotation pocket contains a unicellular plastic foam cylinder measuring 4 inches in diameter by 9 feet long. The foam flotation cylinders, in standard nine foot lengths, are thermetically sealed in individual, airtight compartments, (see Figure 74).

The curtain extends down from the underside of the flotation element and is made of vinyl-impregnated nylon. The curtain depth comes in various sizes, 6 to 24 inches. Steel keel weights are stitch-enclosed along the bottom edge of the curtain to span the length of each flotation segment. The keel weights are galvanized steel bars, 4 inches in length and ⅜-inch diameter. Grommets are No. 3 brass spur.

Johns-Manville Corporation Designs

A design developed by *T.O. Bogosian; U.S. Patent 3,685,296; Aug. 22, 1972; assigned to Johns-Manville Corporation* is a buoyant barrier boom which preferably typically comprises a relatively rigid sheet of asbestos rubber of 10 to 20% styrene-butadiene rubber content by weight, having a closed cell sponge attached to both faces of the sheet in a position whereby at least two-thirds of the sheet is below water level when floated in water in an erect position substantially vertical to the surface of the water, the portion of the sheet located substantially below water level being laminated with an additional sheet of the rubber-asbestos material.

FIGURE 75: JOHNS-MANVILLE FLOATING BOOM DESIGN

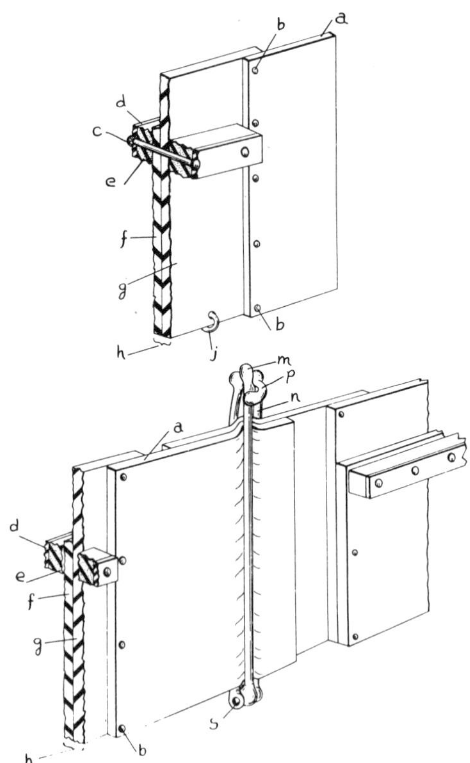

Source: T.O. Bogosian; U.S. Patent 3,685,296; August 22, 1972

In order to provide any additional weight necessary to maintain the structure in an upright position, and in order possibly to provide additional strength, the sheet additionally having attached to each end of the boom a flexible compressed rubber sheet having about a 60 to about 70% content of styrene-butadiene rubber, each of the laminated sheets being about one-eighth inch thick, the sponge material being partially enclosed by the rubber asbestos sheeting of a minor thickness to protect the sponge against abrasion, an anchoring means for restraining the boom in a relatively fixed position in water, and a removable clamp for pinching together two flat and overlapped surfaces of the flexible compressed rubber sheet whereby two or more of the booms may be easily attached in series.

As shown in Figure 75 a rigid, strong oil-resistant rubber-asbestos sheeting **g** has about one-third of the sheet extending above a level of a closed-cell sponge **e** arranged substantially horizontally with the top of the sheet-structure, and having about two-thirds of the sheet located at and below the level of the sponge and laminated to a second asbestos sheet **f** reinforcing and/or adding weight, each of the laminated sheets being about one-eighth of an inch in thickness. Strip **d** is secured by bolt **c** through the laminate **h**. Brads **b** secure the flexible sheet **a** to the barrier sheets. Ring **j** is a convenient place to attach a guide line or anchor. Levers **m** and **n** are hinged at **s** and fastened at **p**, with **n** nested in **m** to clamp the sandwiched flexible rubber sheets of two separate barrier booms, thereby attached in series.

As described by the EPA, Edison, N.J. Laboratory in *Oil Spill Containment Systems* (Jan. 1973), the Johns-Manville Spillguard boom is a proprietary sheet fence design which rides vertically in the water. The boom consists of 10-foot lengths of asbestos rubber sheet. A neoprene flotation liner is cemented firmly to both sides of the boom running the length of the section and enabling the sheets to ride upright at the waterline near the mid-point of the flotation liner. Each of the 10-foot sections is connected to the next section with a 2-ply reinforced rubber hinge. Nine sections, plus the hinges, are combined to form a 100-foot length. The Spillguard boom is available in two sizes:

No. 411 – 4 inches above the water surface and 11 inches below
No. 1224 – 12 inches above the water surface and 24 inches below

The Spillguard boom is furnished in standard lengths of 100-feet and a special connection is provided for joining the standard lengths. Bulkhead connectors are also available which provide a seal between the end of the boom and the stationary vertical member; the connectors allow for free movement of booms with the rise and fall of tides.

The manufacturer reports that the No. 411 boom can be handled by two men in a small outboard-driven boat. The No. 411 boom weighs approximately 3 pounds per foot. The No. 1224 boom, which weighs about 9 pounds per foot, reportedly can be deployed and recovered by four men. During a demonstration in Chesapeake Bay, the company reports that 300-feet of the No. 1224 were connected and set by four men in less than 10 minutes; recovery of the boom required less than 15 minutes hauling aboard a harbor tug.

Costs quoted in 1969 for the Nos. 411 and 1224 oil booms were respectively $7.50/ft and $20.00/ft. The connection for joining standard sections was included in the above costs, whereas bulkhead connectors were quoted at $75 and $125 respectively for the Nos. 411 and 1224 booms.

A large Johns-Manville boom designated the USCG High Seas Boom was designed under contract to the U.S. Coast Guard according to Maritime Administration Report COM-74-10212, Springfield, Va., Nat. Tech. Information Service (Nov. 1973).

It extends 21 inches above the water surface and 27 inches below. It is designed to contain oil in 5-foot seas, 2-knot currents and 20 mile winds. It uses a fabric skirt-type barrier with air-filled flotation bags. It was designed to be air dropped at the scene of a spill. Following drop the barrier deploys from drop package as it is towed around the spill by vessel. The retrieval process consists of taking aboard a surface craft or dock, deflation of air chambers, cleaning and preparation for reuse. Recovery, cleaning and storage judged to require average effort for a barrier of this type but no actual recovery data were available in the report cited above.

Flotation is provided by 12-inch diameter, 6-inch long air-filled tubes which run perpendicular to the skirt. These floats, which are placed every 4 feet along the boom, extend 3 feet out on either side of the skirt. Due to this and the skirt stiffeners, the boom will not lay flat or submerge when working in high currents or towed at speeds or currents in excess of about 10 knots in order to prevent structural damage. The fabric is specially woven to provide more strength at stress points. It is understood that this is the first time that the material was designed for the boom and not vice versa.

The tensile member is a wire rope that runs the length of the boom; however, it is not secured directly to the unit. Wire cords, in groups of three, hold the boom about 2 feet downstream from the wire rope. All towing and mooring of the boom is done directly with the wire rope. This permits the boom to remain flexible during operations and thus it follows the water surface very well.

The flotation tubes, which are interconnected by air hoses, are inflated by compressed gas. As the boom is being withdrawn from its container box, a rip cord on each float opens a valve and the tube is inflated automatically. This makes deployment relatively easy. This information was taken from *Oil Pollution Research Newsletter,* Edison Water Quality Laboratory, September 1971, Volume V.

Another design developed by *T.O. Bogosian; U.S. Patent 3,739,913; June 19, 1973; assigned to Johns-Manville Corporation* utilizes an elongate body of oil absorbing material and flotation material including longitudinal reinforcing or strengthening means whereby a plurality of bodies can be linearly disposed in end-to-end relationship for temporarily fencing oil spills on water for retention and absorption of the oil. The body contents comprise oil absorbing fibers, natural or synthetic or combinations thereof, and may include a flotation material interspersed therewith to aid buoyancy of the body even after saturation of the fibers by oil. Figure 76 shows the construction of such a barrier in some detail.

A closed bag **1** is comprised of a net **2** formed into a cylindrical container. The woof and warp threads **3** and **4** of the netting **2** form openings **6** which permit the water and surface oil to pass freely through the container periphery into the mass **11** of fibers contained in the bag **1**. Netting from materials having good oil-resistant characteristics, such as polypropylene or polyethylene netting, are particularly advantageous for the bag material. Because of their oil resistance, very little of the oil is caught at the surface so that virtually all of it passes through the netting into the mass. The fibrous mass also has a tendency to roll or ball itself into a distinct mass inside the bag leaving the openings relatively free. The ends **7** of the bag may be secured to a rope **8** having its ends **9** and **10** free for dragging the bag through the water.

FIGURE 76: ALTERNATIVE JOHNS-MANVILLE FLOATING BOOM DESIGN

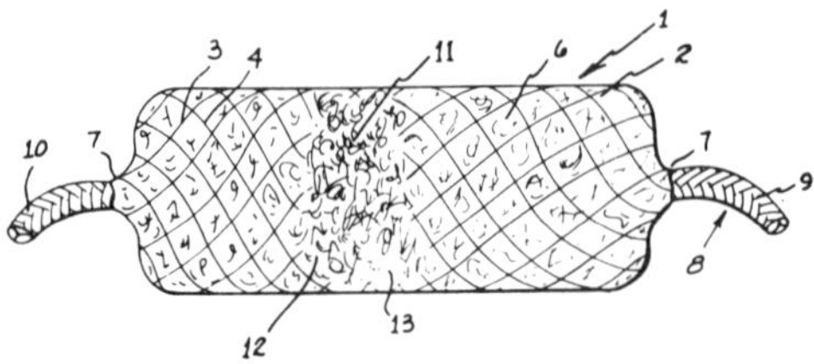

Source: T.O. Bogosian; U.S. Patent 3,739,913; June 19, 1973

An adsorbent material useful*for the mass contained within the netting is reclaimed tire cord known in the trade as "whole-tire-cord-fabric." Unfortunately, because the cord is recovered from old tires, the exact fiber identities are never really known. This fibrous mass is rayon, or nylon, or Dacron fiber, or most likely a blend of the materials, as these materials are the basic cord materials for United States passenger car and truck tires. At times the composition of the mass may comprise fibers **12** having good noncompacting characteristics during oil adsorption. If such is the case, only one type of fiber needs to be used, such as the fibers from whole-tire-cord fabric, even though the exact composition is not known.

However, other materials may be added such as glass fibers **13**, which assist in keeping the adsorbing fibers separated, i.e., fluffy and loose. A composition of about 5 to 10% glass fibers mechanically intermixed with 90 to 95% reclaimed tire cord has proven to be satisfactory.

Kain Design

A design developed by *C.L. Kain; U.S. Patent 3,537,587; November 3, 1970* utilizes a flexible weighted net supported in a vertical position by at least one horizontally elongated flotation unit. A flexible layer of hydrophobic, oleophilic material is attached to the net to generally conform to the position and movement of the net. Water passes through the filter layer while liquid hydrocarbons are blocked.

As described by the EPA Edison, N.J. Laboratory in *Oil Spill Containment Systems* (Jan. 1973), a proprietary design known as the Kain Filtration Boom essentially comprises a deep vertical barrier supported by flotation chambers on each side of the curtain. Approximately one-third of the curtain boom protrudes above the water surface (freeboard) and two-thirds is immersed below the waterline, (see Figure 77). Three types of booms are reported available, including the 3-foot curtain intended principally for use in marinas and for surrounding tankers; the 5-foot curtain for harbor use; and the large 8-foot curtain for offshore sea operations.

FIGURE 77: KAIN FLOATING BOOM DESIGN

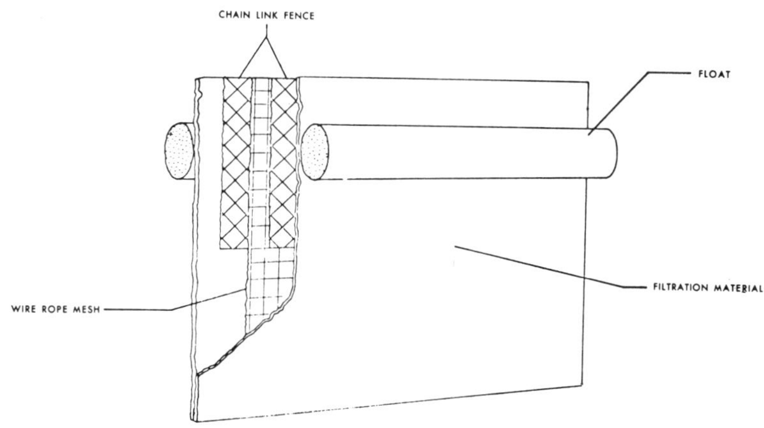

Source: EPA Edison, N.J. Report *Oil Spill Containment Systems* (Jan. 1973)

The vertical section of the boom is basically a sandwich structure. The centerpiece consists of wire mesh or one-half inch steel-cable net (depending upon type of boom) sandwiched on each side by galvanized steel industrial cyclone fencing, and finally encased with filtration fabric comprising the outer sides of the boom. The steel-cable net is woven into 8-inch squares with a longitudinal strength of 400 tons safe working load. The two layers of cyclone fencing serve as a support for the filtration material on each side of the steel-cable net. The filtration fabric is constructed of a polypropylene fiber mechanically enmeshed and heat-fused. The filtration material reportedly allows water to freely pass through in both directions, but precludes the passage of floating oils, etc. A standard section of offshore boom would appear to be 150 feet long.

Flotation chambers providing for proper immersion depth of the curtain are firmly attached to both sides of the curtain by ⅜-inch galvanized wire (through the barrier), thimbles, clips and snaphooks. The flotation cylinders are nylon-vinyl sleeves stuffed with polyurethane bags which in turn are filled with styrofoam pellets. The assembled cylinders for the offshore boom are 14-inches in diameter and manufactured in 18-foot long sections. Tow cables, ropes, grommets, shackles, cables and thimbles represent auxiliary equipment used in completing the Kain Filtration Boom.

The heaviest component of the Kain offshore filtration boom is the curtain element which weighs approximately 1,000 pounds for a 150-foot section. All elements for the offshore boom are stored and shipped together for the standard 150-foot length. The standard length can be reduced into a volume of 800 cubic feet weighing approximately 1,250 pounds. The filtration boom at sea reportedly has considerable freedom of movement and can encompass 360 degrees of travel.

The Kain filtration boom may be purchased outright or leased on a short or long-term basis from Bennett International Services, Woodland Hills, Calif. Purchase costs for 1,000 feet of marina and harbor boom are respectively $18,000 and $23,000 (Nov. 1969 data from EPA); no purchase price is given for the offshore boom. Weekly rental rates for the marina, harbor and offshore booms per 1,000-feet usage are respectively $1,800, $2,300 and $2,900 (prices do not include auxiliary services, and minimum of 2 weeks use for the offshore boom). Other lease plans are also available.

Kingsley Design

A design developed by *W.L. Kingsley; U.S. Patent 3,624,701; November 30, 1971* is a collapsible, continuous, generally annular, curtain floatably supported at the surface of a large body of water, and projecting above and below the surface to retain within the area surrounded by the curtain, oil surfacing from a leak or blowout occurring in a subsea oil well during or after drilling. Guy lines connected with and extending radially outwardly from the curtain have anchoring means at their outer ends for quick anchoring of the curtain around the oil-surfacing area, the lines including yieldable means connected with the portion of the curtain projecting above the surface of the water to enable any section of the curtain to yield radially inwardly under outside wave force thereagainst to permit water to spill into the area enclosed by the curtain but to resist outward movement of such section.

Kleber Design

A design known as the "Balear 311" is also designated as the Kleber design in Maritime Administration Report COM-74-10212, Springfield, Va., Nat. Tech. Information Services (Nov. 1973). Available from M. Jourdan, 6 Avenue Kleber, Paris, France this is a skirt-type barrier with a bottom tension line. It was designed to contain oil in 2 to 5 foot waves, 40 knot winds and 1.55 knot currents.

Two men can deploy 2,000 feet of barrier in seven minutes using one vessel. Recovery, cleaning and storage are judged to be fairly easy. No special storage is needed and 2,000 feet of barrier can be stored in 585 cubic feet of space.

Mack Design

A design developed by *W.T. Mack; U.S. Patent 3,720,062; March 13, 1973* includes an elongate barrier adapted for disposal in the ocean or a large body of inland water for use in confining and collecting oil spills or other pollutants. Much attention has been given lately to oil spills which frequently occur due to a break in an offshore oil well and which may, unless confined and/or collected, pollute or contaminate surrounding areas. Until this time, efforts in this field have been primarily remedial in nature, i.e., after the spill occurs. This device may be used with equal, if not greater, effect as a preventative measure, i.e., before the spill occurs. This design is shown in cross-section in Figure 78.

The barrier **10** is held in a substantially stationary position by means of a plurality of permanent anchors connected by lines **15** to the back side of the barrier. The barrier is made up of sections **10A** which are hingedly connected to one another along the length of the barrier. Each such section is made up of a relatively rigid framework **26** which is triangular in cross section and comprised of horizontally disposed frame members **26A, 26B, 26C** and **26D** along the upper, intermediate and lower edges of the barrier, and upper and lower forward inclined frame members **26E** and **26F**.

The upper frame members are fixedly connected to the intermediate frame members **26B** and **26C** by means of frame members **26E**, frame member **26G** is fixedly connected to frame member **26D**, and the upper ends of the frame members **26F** are pivotally connected to intermediate frame members **26C**. The frame members **26E** and **26F** are held in the inclined positions by means of vertically extending frame members **26C** which are pivotally connected to and extend from the upper frame members **26A** to the lower frame members **26C**.

FIGURE 78: MACK FLOATING BOOM DESIGN

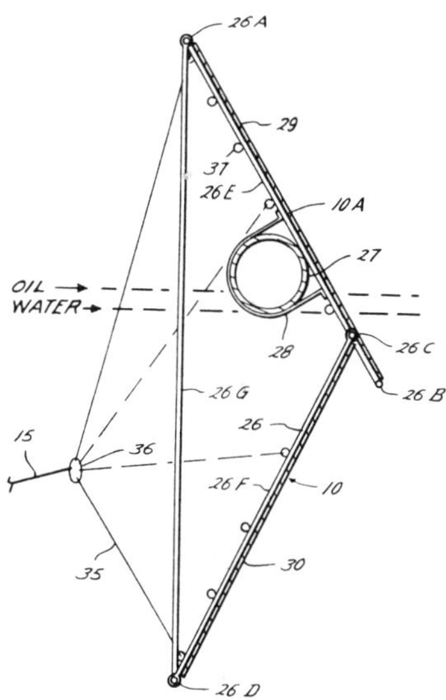

Source: W.T. Mack; U.S. Patent 3,720,062; March 13, 1973

The angularly disposed upper and lower faces on the front side of the framework **26** are covered by rubber belting or other suitable impervious material. Thus, as shown in the drawings, one such covering **29** extends laterally between the ends and vertically between the upper frame members **26A** and intermediate frame members **26** of the upper portion of each barrier section. Another covering **30** extends between the ends and vertically between frame members **26C** and **26D** of each barrier section. The upper edge of the barrier **30** is sealed with respect to the barrier **29**.

A buoyant element **27**, which preferably comprises an elongate, inflatable bladder, is carried on the back side of the framework **26** of each barrier section **10A** by means of one or more straps **28** disposed about the element **27** and secured to the frame members **26E**. The buoyant element is so arranged on the frame that it will, when disposed in a body of water, maintain the frame members **26G** in a substantially vertical position and maintain the frame members **26C** at the lip and preferably the intersection of the frame members **26E** and **26F** at a level beneath the upper surface of the body of water. More particularly, the intersection of the upper and lower faces of the barrier is maintained beneath an oil level above the heavier phase of the body of liquid. Thus, the center of booyancy of the buoyant element **27** is disposed on a level above the intersection and to the rear thereof, and preferably generally vertically above the center of gravity of the frame.

A bridle **35** is connected to the back side of the barrier section **10A** for attachment to the line **15** extending to one of the previously described anchors. More particularly, the upper and lower lines of the bridle connect with an elliptically shaped ring **36**, which in turn is connected to the end of the line **15**.

Marsan Corp. Design

According to the EPA, Edison, N.J. Laboratory Report *Oil Spill Containment Systems* (Jan. 1973), the Marsan Oil Barrier is essentially a plastic curtain with an air or styrofoam filled buoyancy pocket, a stiffened freeboard portion above the pocket, and drop curtain below the buoyancy pocket, (see Figure 79).

The upper few inches of the boom riding above the water surface are described as the stowage fin with grommets located on two-foot centers. Stowage fin stiffeners are provided at each end of a standard barrier section. The buoyant pocket is sewn into the curtain separating it from the stowage fin and drop curtain. The upper sewline contains a dacron tension line running the full length of boom. At the bottom edge of the drop curtain is found a pocket for inserting chain ballast providing vertical orientation of the Marsan barrier within the water. The standard Marsan barrier has a drop curtain 15 inches in depth but 24 and 36-inch curtains are also available.

FIGURE 79: CUTAWAY VIEW SHOWING CONSTRUCTION OF MARSAN OIL BARRIER AND METHOD OF SEGMENTING SECTIONS

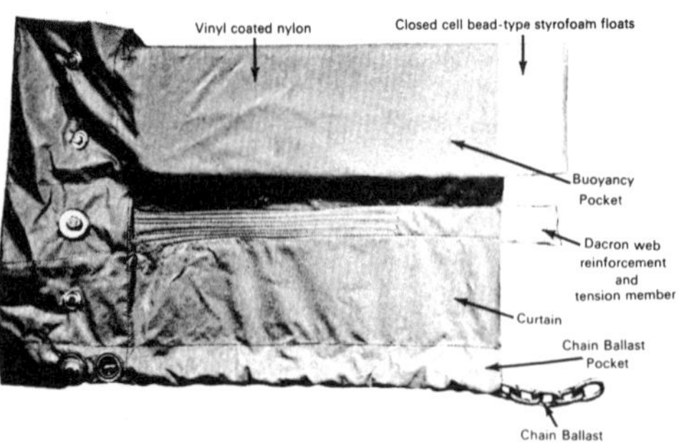

Source: EPA Edison, N.J. Report *Oil Spill Containment Systems* (Jan. 1973)

The barrier is provided in standard lengths of 50 or 100-feet and is constructed of vinyl-covered nylon material, which is fire-resistant and reportedly remains flexible at temperatures below 0°F. The standard boom sections would appear linked together by a series of snap connectors together with special provisions for extending a hose line through the sections when air inflation is used. Self-actuating buoyant air reservoirs, which serve to inflate the hose in the buoyancy pocket, are spaced every 200-feet on the barrier. The cost of the standard Marsan barrier (with 15-inch drop curtain) is $5.95/ft, including actuating air reservoirs. Inflatable barriers with 24 and 36-inch drop curtains are quoted at $6.95/ft according to 1969 EPA data. A Marsan Type II Barrier is described by the Maritime Administration in Report COM-74-10212, Springfield, Va., Nat. Tech. Information Service (Nov. 1973). It is designed to contain oil in 8 to 10 foot waves, 20 knot winds and a 4 knot current and to contain up to 5,000 barrels per 2,000 feet of barrier. It is stated that 2 men can deploy 2,000 feet of barrier in 10 minutes using a 200 hp boat.

Matheson Design

A design developed by *N. Matheson; U.S. Patent 3,710,577; January 16, 1973* utilizes a barrier section comprising a pair of inflatable tubes, one tube being connected intermediate a depending curtain wall and the other tube, the one tube having a smaller cross section than the other tube but sufficient buoyancy for supporting both the curtain and other tube at positions below and above the surface of the water. The one tube serves as a keel, pivotal rotation of the one tube bringing the other tube into floating relation with the surface of the body of water and changing the center of buoyancy to develop corrective forces that maintain the barrier section in an upright stable position.

As shown in Figure 80, barrier sections **10** are supported by the buoyancy of inflatable tube **20**, and tube **21** may be supported entirely or substantially above the surface of the water. This condition of buoyancy changes, however, under extreme winds which blow against the exposed surface of tube **21**. In such event, the tubes will roll over until tube **21** comes into contact with the surface of the water, thereby changing the center of buoyancy of the combined buoyant masses of the tubes and producing a resultant buoyancy force R. Although the curtain may be slightly lifted, the change in location and size of the effective buoyancy mass tends to reestablish the normal flotation of the barrier section. Since the towing force is applied to the upper and lower towing lines **23** and **24** equally, the barrier is maintained in a substantially straight line and extends to a uniform depth from the surface of flotation.

FIGURE 80: MATHESON FLOATING BOOM DESIGN

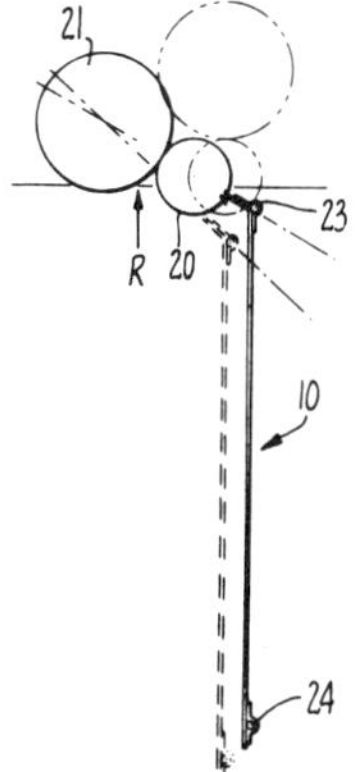

Source: N. Matheson; U.S. Patent 3,710,577; January 16, 1973

Megator Pumps and Compressors Limited Design

F.W. McCombie; U.S. Patent 3,784,264; July 24, 1973; assigned to Megator Pumps and Comprssors Limited, England developed a method for pumping away floating oil slicks comprising a barrier capable of being extended to form an enclosure surrounding the slick. The enclosure is contracted to concentrate the slick and facilitate pumping it away.

Metropolitan Petroleum Co. Inc. Design

As described in the EPA, Edison, N.J. Laboratory Report *Oil Spill Containment Systems* (Jan. 1973), the Metropolitan Petroleum Company of Jersey City, N.J. are manufacturers of the MP Boom, which consists of a 6-inch diameter, closed cell polystyrene bead filled flotation chamber, a 12-inch submerged skirt, and a solid weighted keelson to provide ballast. The flotation material is enclosed in a vinyl envelope for added protection in case of a break in the synthetic fabric covering. This particular boom is recommended for use on sheltered water where wave height is less than 2 feet. Constructed in 6-foot sections, and available in 100-foot lengths, this device is priced at $9.75 per foot (1969 cost data from EPA). Lower prices are available when booms are purchased in lots exceeding 400 feet in length. This same boom, 6-12 Boom, is also available from Worthington Corporation.

Mikkelsen Design

A device developed by *T. Mikkelsen; U.S. Patent 3,499,291; March 10, 1970* comprises a plurality of floatable and foldable flat sections linked together in the form of a zig-zag rail and provided with one or more purse lines arranged to fold the flat sections on being drawn in. In water the boom will provide an upstanding barrier above the surface of the water. The boom can be set up in a ring in the water and by hauling in the purse lines the closed-off area will be decreased. The boom is manufactured of individual plates of porous plastic covered and linked together by a pliable material which is durable to seawater and oil. The purse lines are guided by rings at alternate links. Between adjacent links are connected bands to limit the flattening out of the boom.

Figure 81 is a cross section of boom **10** in which there is inserted plate-shaped floating means **10′** of a foam material. The cover is formed of a double-layered rail **23** of plastic-reinforced glass fiber canvas which surrounds the plate **10′** and which receives a band-shaped ballast material **24** at the lower edge of the section. The plate is not formed over the whole breadth of the boom, but is made so much smaller that an intermediate space is formed between the ballast and floating means. At the side of the section, that is, to the cover **23**, there is adhered a band **25** having a breadth less than the individual length of the section. The connecting bands are of any material sufficiently flexible to permit free folding-together of the boom. The connecting bands serve two purposes. Firstly they limit the boom's levelling out on setting up and before hauling in the purse lines. Further they form a stabilizing element in heavy seas and winds, increasing the boom's sluggishness to wave movement and preventing oil from being washed over the boom.

FIGURE 81: MIKKELSEN FLOATING BOOM DESIGN

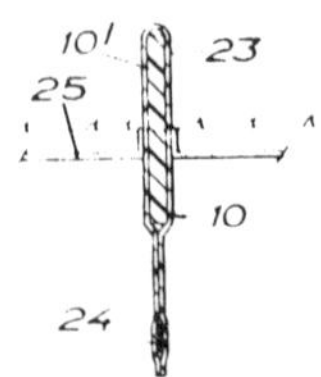

Source: T. Mikkelsen; U.S. Patent 3,499,291; March 10, 1970

M.I.T. Design

A design developed by *J.B. Nugent; U.S. Patent 3,703,084; November 21, 1972; assigned to Massachusetts Institute of Technology* consists of a multiplicity of intercoupling cells comprising floatable material, open at top and bottom, forming an in-depth barrier that is wave conformal, the barrier being appropriately stabilized by weights and being coupled to a tether boom, which, in turn, is connected to moorings. Oil carried over the barrier face by waves is trapped in the cells as is oil carried under the barrier face by current flow.

As shown in Figure 82 the system comprises oil boom **20** coupled via lines **9** to tether boom **10** which is attached to mooring floats **4** via mooring pennants **3**. Oil boom **20** comprises a multiplicity of cells **100** interconnected so as to give the appearance of a honeycomb design. Moorings comprised of anchors **6**, lines **5**, floats **4** and pennants **3** maintain the tether boom in the desired location. The tether boom comprises a series of lightweight plastic floats, 6" in diameter and 2' in length, strung together in a line at spaced intervals. Each of cells **100** comprises three sheets **110** of plastic membrane material arranged in a triangular configuration and coupled together at each vertex of the triangle via floats **120**. Membrane sheets **110** are of reinforced plastic, i.e., woven fabric coated with plastic, approximately 3' x 3' in dimension.

Floats **120** are formed of a plastic such as polyethylene and are cylindrically configured, having a diameter of 3" and a vertical length of 3'. Suspended from the bottom of each of floats **120** is a weight **121** designed to keep the cells essentially in the vertical position relative to the surface of the water. Oil boom **20** is located in the water according to a preferred ratio, i.e., twice as much of cell **100** is submerged as it is above the surface of the water. Vertical stability is of considerable importance to the operating efficiency of the boom system. Besides the weights on floats 120, they are coupled via membrane sheet 110. Additional stability is effected through the use of line pairs **9a** and **9b** coupled at one end to the top and bottom of each of floats 120 and coupled at the other end to tether boom 10, which, in turn, is coupled to mooring floats **4**. The overall effect maintains the oil boom face essentially vertical relative to the surface of the water while also preventing oil boom 20 from being totally submerged due to strong currents and enabling it to be wave conformal.

The improved boom system operates as follows. Where there is both wave motion and current flow, the oil is carried along the surface of the water until it reaches oil boom or barrier **20**. High wave action causes water and oil to spill over the barrier face into the pockets or cells **100**. If there is also a strong current, a certain percentage of the oil is entrained or carried under barrier **20**. As the current flows under the barrier, however, the oil exhibits a tendency to rise up into the pockets of water formed by cells **100**. As the oil is carried along under the barrier, a portion is trapped by each of the succeeding cells. The horizontal velocity of the water towards the top of the cells is considerably less than the current, preferably zero. Because of this velocity condition within the cell and because of the natural tendency of the oil to rise above water, the oil trapped in the cells remains contained therein unaffected by the current flow or by wave motion. The contained oil can then be removed by pumping or other means.

FIGURE 82: M.I.T. FLOATING BOOM DESIGN

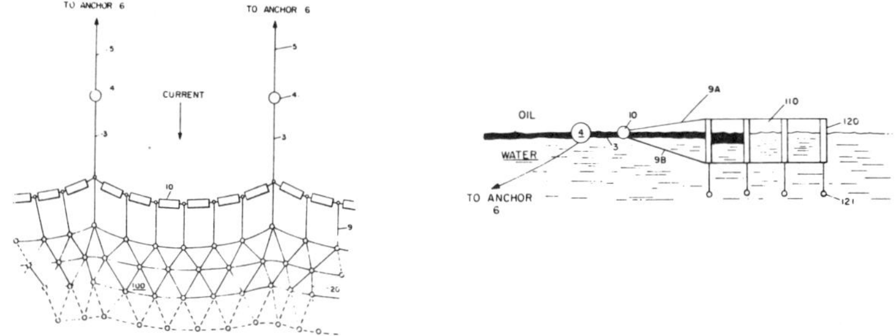

Source: J.B. Nugent; U.S. Patent 3,703,084; November 21, 1972

Muehleisen Design

As described by the EPA, Edison, N.J. Laboratories in *Oil Spill Containment Systems* (Jan. 1973), the boom made by Muehleisen Mfg. Co. of El Cajon, California consists of a series of Ethafoam floats from which is suspended a one-foot barrier shield above water and a two-foot skirt below. Manufactured in 3-foot high by 100-foot lengths, the barrier is fabricated from a vinyl-nylon material, (see Figure 83).

Both ends of each 100-foot section are reinforced by doubling the material to form a pocket to insert an aluminum plate; then punching through this reinforced area for the attachment of grommets to prevent tearouts. At each joint where 1,000-foot sections are joined together, a positive lap type connection (full width) is used to prevent oil from escaping through these areas.

The top edge of the barrier shield is semirigid construction by reinforcement with PVC pipe and rope hem, with the rope inserted to prevent the top edge from drooping while the boom is floating in either a relaxed position or under tension. The bottom skirt leading edge is lead and chain weighted to maintain the boom in an upright position through a series of perpendicular aluminum rods that are inserted in pockets at the folding parts of the boom.

Each section folds at approximately 3 feet 4 inches on center for the full length of these 100-foot sections to enable stowing in an orderly fashion. The bottom leading edge of the skirt has a rope hem with rope insert and a series of grommets for the attachment of a continuous chain. This chain is attached with lap links in such a manner that the chain's primary function is to relieve the stress on the boom during pulling and installation. As the reader will observe this design is similar to the Flexy Oil Boom design discussed earlier.

FIGURE 83: MUEHLEISEN FLOATING BOOM DESIGN

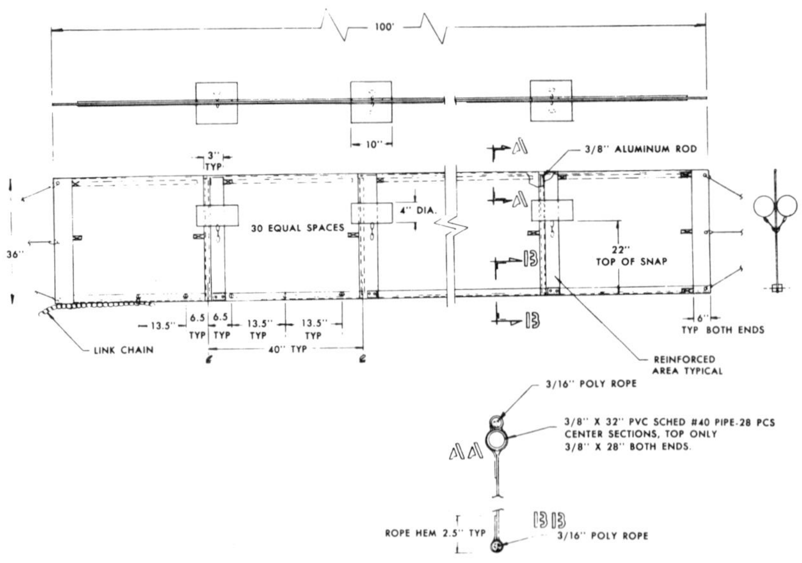

Source: EPA, Edison N.J. Report *Oil Spill Containment Systems* (Jan. 1973)

Murphy Pacific Marine Salvage Company Design

A design developed by *R.K. Thurman; U.S. Patent 3,751,925; August 14, 1973; assigned to Merritt Division of Murphy Pacific Marine Salvage Company* comprises a plurality of interconnected, floating units each supporting a rigid vertical barrier with a part above water and a part submerged. The barriers are interconnected by panels of flexible waterproof material of substantially the same vertical extent as the barriers and both the barriers and the interconnecting panels have flexible, waterproof skirt portions depending below the barriers. Each of the units is secured to a tow line by means of which the boom may be towed to a desired location and there anchored in place.

As shown in the transverse sectional view in Figure 84, this design comprises a vertically disposed barrier member **13** which may be a standard 4 feet by 8 feet piece of marine plywood. Mounted on opposite sides of the barrier member **13** are flotation means such as a plurality of sealed, empty drums **14** and **15**. The front ends of the drums are secured to the barrier member by U-shaped clamps **18** having flanges **18'** bolted to the barrier member as shown.

Secured at the lower edge of the barrier member is a skirt **21** formed by folding a rectangular sheet of flexible textile material such as plasticized canvas substantially in half. The skirt is positioned with the lower end of the barrier member sandwiched between the free ends of the folded sheet as shown. Narrow strips **22** and **23** of marine plywood are placed over the free ends of the sheet and the whole assembly is secured to the barrier member by bolts **24** and nuts **25**. A suitable quantity of ballast such as sand, gravel or scrap pipe is placed within the skirt and the ends are clamped between short strips **27** of marine plywood secured together by bolts **28** and nuts **29**. Preferably, the skirt has a bead **30** formed at the upper end of each fold as shown to prevent it from slipping under the corresponding clamping strip **27**.

The units are secured to towing lines **35** and **36** in such fashion that the towing loads are not applied to the flexible panels interconnecting the units. Thus, the lines pass through eyes **37** and **38**, respectively, on the unit which are formed on the drum clamps **18**. Also, wire clamps **39** are secured to the lines **35** and **36** adjacent the eyes so that for either direction of tow, the tow lines carry the towing loads while the connecting panels **31** between the units remain loose and unstressed. To facilitate towing, sloping bow plane means **40** may be secured to the forward end of the barrier member **13**.

FIGURE 84: MURPHY PACIFIC MARINE SALVAGE COMPANY FLOATING BOOM DESIGN

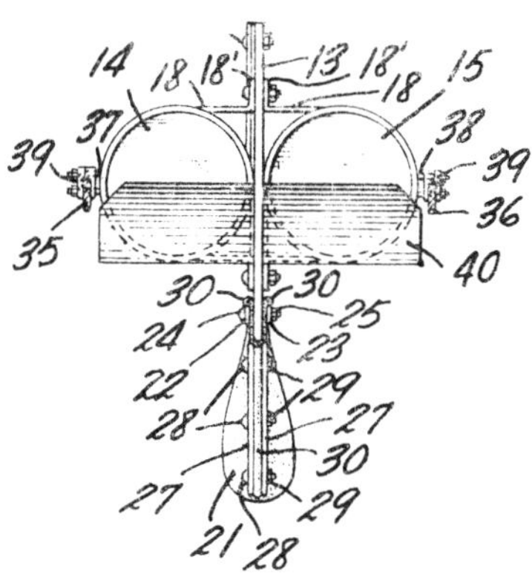

Source: R.K. Thurman; U.S. Patent 3,751,925; August 14, 1973

Newton Design

A design developed by *E.W. Newton; U.S. Patent 3,628,333; December 21, 1971* is a fence structure wherein the fence is assembled from a plurality of modular fence sections to enclose at least part of a selected water surface area, each of the fence sections including vertically elongated strut members having buoyant devices to support them at a selected level in the water, stringers connecting the tops and bottoms of the strut members and plural sets of harness ropes for supporting the modular fence sections in a desired arrangement, together with a cloth web extending above and below the water level supported by the stringers.

The floating fence assembly may be readily transported to a site where lighter than water contaminants may have begun pollution of the surface of a body of water and which may be quickly stabilized in position and moored to effectively contain the contaminants in volume, both over a substantial vertical depth of contaminant as well as over the enclosed area.

A section through such a fence is shown in Figure 85. Each fence section is a 100-foot section and comprises a plurality of horizontally spaced, vertical struts **18** spaced horizontally approximately 10 feet on centers and formed of spring steel tubing to support a fence cloth **32**. The strut is formed of a pair of strut tubes, including a first strut tube **19** which is slightly bowed and a second strut tube **20** which is more predominantly bowed, that is the second strut tube **20** is bowed about a smaller radius, the two strut tubes being interconnected at spaced vertical intervals by horizontal bracing tubes. The vertical struts are of bowed shape so that when the tidal currents are pushing the fence cloth against the concave edge of the strut on the uptide side of the fence (which occurs twice daily), the pressure of the cloth against the strut automatically turns it so that the strength of the strut is always opposing the pressure of the cloth.

The more prominently bowed second strut tube of the plurality of struts making up the fence section have an inflatable float **30** removably secured thereto, for example formed as a vertically elongated bag of neoprene-coated nylon cloth. Each of the curved fence sections are also provided with two sets of inwardly extending fence harness ropes **41** and two sets of outwardly extending fence harness ropes **42** adjacent the opposite ends of the sections, formed for example of ⅜-inch nylon rope, which converge together to inner harness connecting rings **43** and outer harness connecting rings **44**, respectively spaced, for example, about 60 feet from the fence cloth.

The inner harness connecting rings **43'** are designed to be connected to inflatable floats **45A,** which sustain near the water surface the coupling points for positioning grid ropes **46**. Similarly the outer harness rings **44'** are connected to inflatable floats **45c**. An anchor rope **53** is connected to a 60-pound Danforth anchor.

FIGURE 85: NEWTON FLOATING BOOM DESIGN

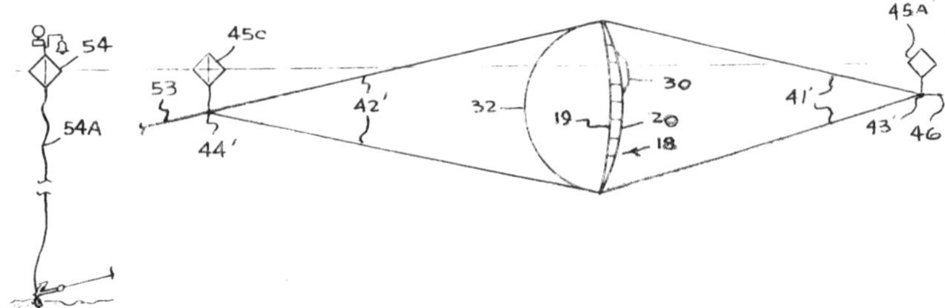

Source: E.W. Newton; U.S. Patent 3,628,333; December 21, 1971

An anchor buoy float **54** is preferably provided for each of the anchors, connected to the crown or fluke end of the anchor by an anchor release rope **54a**. Each anchor buoy is preferably provided with warning flasher and/or a bell mounted on a bracket extending upwardly from the float. The anchor buoy float is provided as a shipping warning marker and also to permit the anchor to be lifted from the region of the flukes and thereby facilitate breaking out of the anchor when it is desired to remove the contaminant fence.

Ocean Science and Engineering, Inc. Design

A design developed by *M. Risin and R.M. Snyder; U.S. Patent 3,597,924; August 10, 1971; assigned to Ocean Science and Engineering, Inc.* is a water-impervious, floating enclosing barrier presenting a substantially rigid vertical surface to the oil with means permitting limited movement of the barrier in the horizontal and vertical planes to compensate for constantly changing wave shapes and forms.

Figure 86 shows the construction of one such barrier using a single sheet of flexible material **42** such as plastic. Equally spaced transverse stiffening rods **44** are secured along the longitudinal length of the barrier. A constant length cable **46** traverses the sheet and is connected to each rod **44**. Bouyancy cells **48** are connected above the cable and ballast as weight members **50** are connected below the cable. This device considers the dynamic geometry of the air/water interface. That interface (the water surface on which the oil slick is carried) is depicted by the line **26**.

FIGURE 86: OCEAN SCIENCE AND ENGINEERING INC. BOOM DESIGN

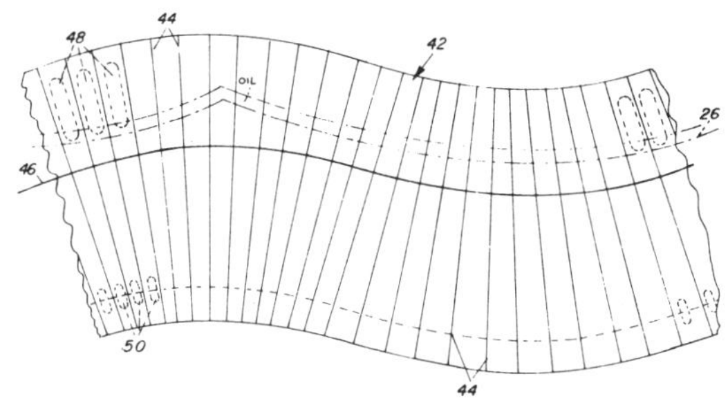

Source: R.M. Snyder and M. Risin; U.S. Patent 3,597,924; August 10, 1971

Ocean Systems, Inc. Design

A design developed by *T.N. Blockwick; U.S. Patent 3,708,982; January 9, 1973; assigned to Ocean Systems, Inc.* as shown in Figure 87 consists of an upper and lower section **20** and **22**, respectively. The upper section **20** extends above the surface of the sea while the lower section **22** represents that part of the structure which protrudes below the surface of the sea and functions concurrently as ballast for the upper section **20** and as the subsurface barrier for the module **10**.

The composition and geometry of the upper and lower section **20** and **22**, respectively, determines the quality of performance for the barrier module **10**. The upper section **20** provides the buoyancy, and functions as the above-surface barrier, i.e., oil barrier, for the module. To perform this operation the upper section **20** should be composed of a high strength but flexible material of low density. Materials such as foam, porous rubber and sponge are suitable, while a polyether based flexible polyurethane foam is most preferred. To maintain its buoyancy the upper section must be water repellant or made to be such.

FIGURE 87: OCEAN SYSTEMS INC. BOOM DESIGN

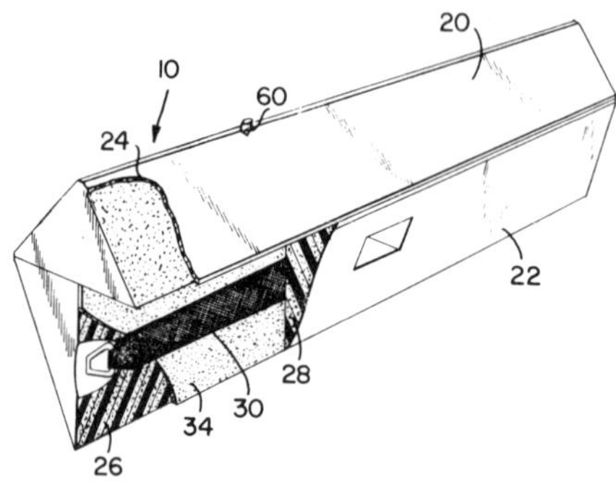

Source: T.N. Blockwick; U.S. Patent 3,708,982; January 9, 1973

Where a porous foam plastic material is used to form the upper section **20** it may be rendered water impervious by sealing the outer periphery. This is preferably accomplished as is shown by applying an appropriate surface coating **24** to prevent fluid from permeating through the porous foam structure.

The type of coating employed and the thickness thereof is not critical as long as the coating will prevent oil from passing through the relatively permeable foam. An elastomer may be used for this purpose. A 30 to 60 mil thin film surface coating of elastomer produces a tough, flexible, seawater and petroleum resistant layer which resists abrasion, puncturing, and chemical degradation. The upper section **20** is bonded to the lower section **22** by any conventional water repellant sealing adhesive preferably an epoxy resin or an elastomer. The sealing adhesive should prevent fluid seepage from the lower section into the upper section.

The lower section of barrier module **10** functions as the submerged fluid barrier while simultaneously providing the static and dynamic stability for the composite structure. The material selected for the lower section must be water absorbent and should otherwise exhibit similar characteristics to the material selected for the upper section such as flexibility and resiliency, and be light in weight.

Once immersed into water the mass of the water-absorbing lower section, which is negligible in a dry state, will immediately increase due to entrapped water, thereby providing the necessary ballast for the upper section. Thus, the lower section which is essentially weightless in water, but which controls and contains a large mass conveniently provided by the surrounding sea, will exhibit in such state a dynamic response characteristic closely simulating the surface characteristics of the sea itself.

The rate at which fluid is absorbed by the lower section is in general proportional to the size and concentration of pores in the structure of the cellular material used in fabricating the lower section as well as the properties of the material itself. Any flexible and permeable foam plastic would be satisfactory, although a reticulated polyether based polyurethane foam is preferred. A reticulated foam can have a relatively large number of pores per linear inch of foam. The foam may be chemically treated to increase its water absorption properties if desired.

The lower section is composed of two symmetrical segments **26** and **28**, respectively, and a relatively thin belt-like member **30** interposed between the segments and aligned in the vertical plane which includes the longitudinal axis of the module **10**. The belt-like member **30** extends along substantially the entire length of the module **10** and is bonded in such position against adjacent faces of segments **26** and **28** by a sealing adhesive **34** such as an epoxy resin or an elastomer. The sealing adhesive **34** not only cements segments **26** and **28** together but represents in combination with belt **30** a partition for preventing fluid from passing from one segment of the lower section into the other. The belt also provides added strength to enable the module **10** to maintain its structural unity under rough sea conditions. The belt is composed of a fabric material having high strength and low stretch properties, such as those exhibited by dacron.

An ancillary advantage in using flexible foam for the upper and lower sections **20** and **22**, respectively, is its ability to be substantially compressed, i.e., reduced in volume for storage and yet when relaxed to expand back to its normal shape. Compression to 25% of original volume and greater has been achieved. This results in ease of transportation and economy of storage. Moreover, foam is also easily handled and can be fabricated into any desired shape. A valve **60** may be incorporated to provide a free flow of air in the upper section **20** of barrier module **10** during packaging and deployment. By this means air may be exhausted during compression and taken in for expansion.

As described by the Maritime Administration in Report COM-74-10212, Springfield, Va., Nat. Tech. Information Service (Nov. 1973) an *Air Deliverable Oil Spill Containment System* is available from Ocean Systems Inc. of Reston, Va. It is designed to operate in 5-foot seas, 20-knot winds and a 1-knot current and to contain 300,000 cubic feet of oil per 2,000 feet of installation. It has been tested in 5-foot waves and a one-knot current.

Pirelli Design

According to the EPA, Edison, N.J. Laboratory Report, *Oil Spill Containment Systems* (Jan. 1973), the Pirelli Oil Spill Boom is available from Grefco, Inc. Jamesburg, N.J. The boom is fabricated of neoprene rubber reinforced with nylon beltstock. It is held upright in water by rigid stiffeners to which polyethylene air floats are fastened. Each individual section is 10 meters long and 90 centimeters wide, providing for 13-inch freeboard and 22-inch draft. The boom is weighted with permanently attached lead ballasts. The flotation units are attached at intervals of 27 inches along the length of the boom, (see Figure 88).

FIGURE 88: PIRELLI FLOATING BOOM DESIGN

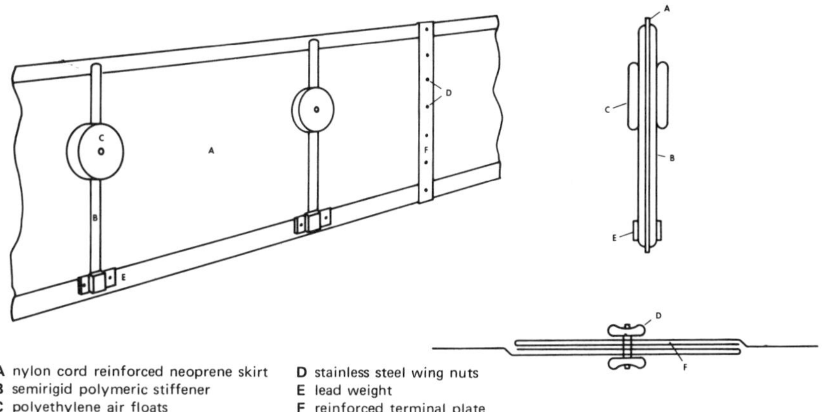

Source: EPA Edison, N.J. Report *Oil Spill Containment Systems* (Jan. 1973)

Lengths of the boom are fastened together by means of stainless steel wing nuts and bolts which pass through reinforced terminal plates. Provision for fastening to posts, workboats or docks are provided at section ends. Nylon rope reinforcement is inserted in the sealed fold of the neoprene fabric at top and bottom of the skirt.

Pneumatiques-Colombes Designs

A design developed by *R.E. Ducrocq; U.S. Patent 3,577,879; May 11, 1971; assigned to Pneumatiques, Caoutchouc Manufacture Et Plastiques Kleber-Colombes, France* is a floating barrier which is constituted by an inflatable enclosure and a skirt, and which, in particular, is possessed of a high degree of longitudinal flexibility which allows it to adjust to the swell of the sea and has sufficient mechanical strength for a tensile stress to be applied at each of its ends in order to draw them together. Figure 89 is a section through such a barrier.

The barrier shown comprises an inflatable enclosure **1** having a tubular form when inflated which acts as a float and constitutes the upper portion of the barrier proper, a flexible skirt **2** and a flexible element **3**, having a substantial resistance to traction and made of the material of which the top portion of the flexible skirt is made. The flexible skirt is formed of two portions **2a** and **2b** which are connected together at the top. Ballast weights are attached (not shown) at the base of skirt **2**.

The elements of the barrier are formed from a sheet of elastomer having a reinforcement **7** constituted by two layers of corded fabric which are superimposed and crossed so that their elements form, with the longitudinal direction of the barrier, equal but opposite angles, of approximately 55°.

This sheet is subsequently bent back on itself on either side of a strip of elastomer which constitutes the flexible reinforcing element **3**. During this arrangement the reinforcement **8** is also placed in position, the reinforcement also being formed of two layers of corded fabric which are crossed and arranged at 55° relatively to the longitudinal direction of the barrier so that the inflatable enclosure is reinforced over its full periphery.

FIGURE 89: PNEUMATIQUES-COLOMBES FLOATING BOOM DESIGN

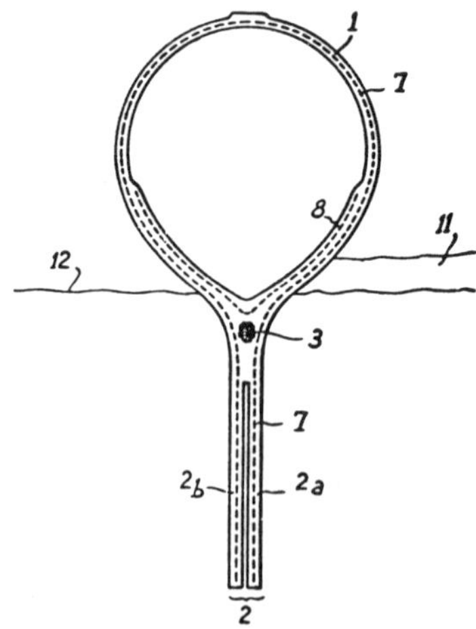

Source: R.E. Ducrocq; U.S. Patent 3,577,879; May 11, 1971

On either side of each barrier element, the elements **3** of high mechanical strength exceed a certain length. This makes it possible to attach them to one another by means of a flexible connection. At these connections, the watertightness of the barrier is ensured by means of a flexible panel which is secured to the two barrier elements and rests upon the skirt **2** of the elements; the panel is located at the side of the layer **11** of hydrocarbons.

A design developed by *R. Ducrocq and C. Moreau; U.S. Patent 3,713,410; Jan. 30, 1973; assigned to Pneumatiques Caouthchouc Manufacture Et Plastiques Kleber-Colombes, France* comprises components consisting of inflatable bags from each of which a plurality of overlapping interlinked panel members are suspended. As shown in Figure 90, this barrier is formed by an assembly of members consisting of an inflatable bag **1**, panels **2**, and a chain **3** attached to the panels **2** by rings **4**. (Only one part of the chain is shown.) The inflatable bag, which contains a valve **8**, is made of a flexible material with a textile or metallic reinforcement of any suitable type, and takes the form of a flat tube when it is not inflated. At its two ends it terminates in a solid part **5** which serves to attach the individual bags together to form the complete barrier. Holes **6** in these end parts **5** are adapted to receive bolts or pins.

Several panels are attached to each inflatable bag. To keep the barrier closed, the panels overlap each other about 20% and their lateral edges are bevelled. For example, if the inflatable bag has a length of 9.6 meters it may be provided with 10 panels having a width of 1.2 meters and overlap for 20% of their width. The panels are sheets of a rubber or plastic material adapted to the particular conditions of use and may comprise a textile or metallic reinforcement. The reinforcing members are selected and positioned so that the panels have substantial rigidity in the vertical direction and a certain flexibility in a transverse direction. Flexible strips **9** encircling the bag and having a length greater than its periphery support the panels which are attached by nuts and bolts **10**. Each panel is supported by only a single strip to avoid imparting rigidity to the inflatable bag. The strips may be made of textile or metallic cloth coated with a suitable plastic or rubber material and preferably are so connected to the balloon as to be able to move slightly if the barrier undergoes substantial deformation.

The barriers are assembled to make a complete barrier by connecting the balloons to each other at their ends **6** and by fastening the chain which extends along the entire length of the completed barrier to the panels by means of the rings passing through a link in the chain and the holes **11** in the bottoms of each two overlapping panels. Each panel has at least one hole in each of its two lower corners which coincide in the overlapped portions and each ring which connects the chain to a panel also connects together two successive panels so that they are mobile with respect to each other. These panels may be attached either to the same bag or to different bags. The method of assembling the panels together and connecting them to the chain permits the panels to swing with respect to each other. The chain acts as a connector and also serves as ballast and traction.

FIGURE 90: ALTERNATIVE PNEUMATIQUES-COLOMBES FLOATING BOOM DESIGN

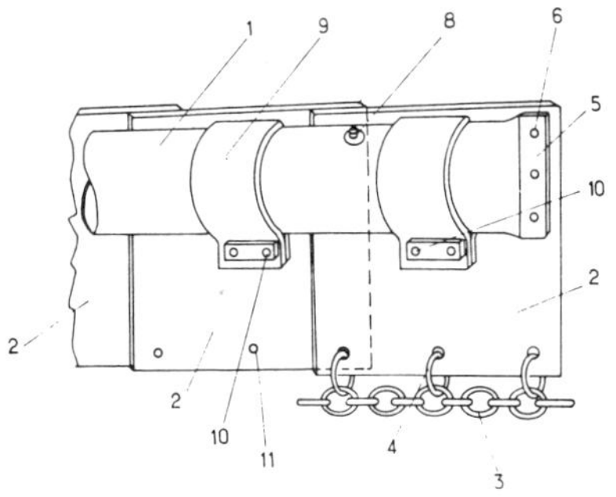

Source: R. Ducrocq and C. Moreau; U.S. Patent 3,713,410; January 30, 1973

Preus Design

A design developed by *P. Preus; U.S. Patent 3,795,315; March 5, 1974* utilizes a technique for controlling waterborne oil slicks where a floating barrier having a fluid pervious skirt is positioned in a controlling position relating to the slick and an oleophilic-hydrophobic fiberous substance is introduced into the slick to absorb the oil and render it impenetrable of the skirt. A material which suitably meets these requirements is a fiberous compound of expanded perlite with clays and fibrous material known as Sorbent Type C.

Preus-Gallagher Design

A design developed by *P. Preus and J.J. Gallagher; U.S. Patent 3,783,621; January 8, 1974* is a barrier for substances floating on water having a flotation member and a liquid pervious and a liquid impervious skirt depending therefrom. The liquid impervious skirt is deflected at currents greater than about one knot, and the oily substances are treated with a particulate oleophilic-hydrophobic substance less dense than water for retention by the liquid pervious skirt.

Floating barriers having liquid impervious skirts have been found to be ideal for containing and controlling substances floating on water where the water is substantially free of currents and wave action. Such barriers find particular utility in the control of oil slicks on bodies of water where the water conditions so permit. Where currents and/or wave action are present in the body of water, however, liquid impervious skirts have proved to be ineffective in retaining, controlling or confining hydrocarbon slicks. Because of the impervious nature of the barrier, the skirt must withstand great hydrodynamic forces without failing structurally or deflecting to a horizontal position, thereby allowing the slick to pass under the barrier. Even where suitable structural strength and ballasting weight are provided to avoid the above failures, it has been found that currents greater than one knot perpendicular to the boom form vortexes which draw floating hydrocarbons under the barrier, thereby effectively circumventing the effectiveness of the barrier.

This barrier overcomes the disadvantages of the prior art by providing a floating boom having a water impervious skirt for containing oil slicks or the like in substantially still water and a water pervious skirt for containing oil slicks on water having wave or current movement therein. Figure 91 shows the elements of construction and operation of this device. With the barrier **10** suspended from flotation collar **12** disposed in a body of water **24** having a quantity of oil **26** floating thereon and relatively little or no current therein, the skirts **14** and **16** depend vertically in substantially parallel relationship and the barrier functions in a manner substantially identical to that of conventional liquid impermeable skirted booms.

FIGURE 91: PREUS-GALLAGHER FLOATING BOOM DESIGN

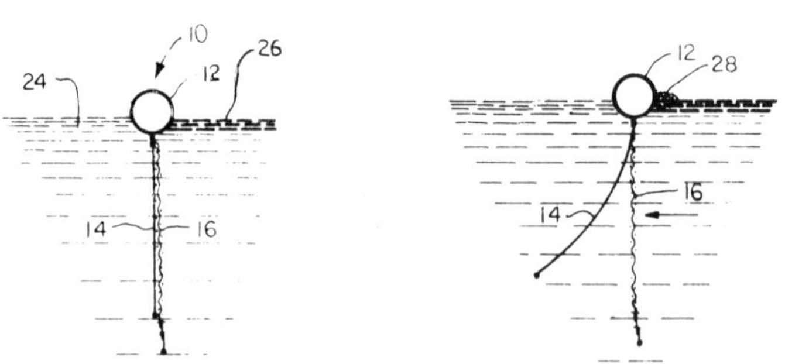

Source: P. Preus and J.J. Gallagher; U.S. Patent 3,783,621; January 8, 1974

Upon the advent of or in the presence of a current much in excess of about one knot in the direction shown by the arrow, the impermeable skirt **14** is deflected with the current as shown and the permeable skirt **16** remains to provide a barrier for the oil **26**. Where the oil is low in viscosity or when the floating substance is some other less viscous material, such as gasoline or the like, the leading edge of the material is preferably thickened up with an oleophilic-hydrophobic material **28**. The deflected skirt **14** assumes a position regardless of the current force, which blocks or otherwise entraps bloom oil traces which may be expressed or otherwise bypass the barrier **10**. In this respect, it is conceivable that the skirt could be fabricated from a mat of the above described oleophilic-hydrophobic material which could be replaced after use.

Preus-Rosendahl Design

A design developed by *P. Preus and C.E. Rosendahl; U.S. Patent 3,579,994; May 25, 1971* provides a barrier for control of waterborne substances having a plurality of units interconnected to one another in end to end relationship. Each unit has articulated flotation chambers and a flexible, depending skirt with permanent ballasting along the lower edge thereof. Means are provided to selectively reef the skirt and water ballasting pockets are formed on the skirt to provide restraint against wind action on the barrier.

Figure 92 shows such a barrier **10** in elevated and transverse section as deployed and inflated for use. The unit comprises a skirt **12** and a plurality of flotation chambers **14a** through **14e**. The unit is formed with a hinge area **16a** through **16d** between each of the chambers **14a** through **14e** and a half hinge area **18a** and **18b** on the free ends of the chambers. The unit can be formed of any flexible material desired, but is preferably formed of a material having at least a heat sealable base, such that the structure may be formed from a single sheet of material at the required places to form independent chambers by turning over and sealing one edge longitudinally and then transversely heat sealing the resultant tubular structure. In lieu of a heat sealable material, adhesive could obviously be used to accomplish similar results if so desired. At least the areas of the material forming the flotation chambers should also be air impervious, and when the unit is used for the control or containment of spills, the entire material should be impervious to liquids.

Each of the chambers **14a** through **14e** is provided with an inflation valve **20** which may be of any type common in the art such, for example, as a flapper valve or the like. Skirt ballast chambers **22** are formed along the lower edge of the skirt **12**, preferably formed by turning over and heat sealing the lower edge thereof. The ballast chambers are filled with a suitable ballast material such, for example, as sand or the like, to provide means to properly deploy the skirt when the unit is in use. Obviously, other ballast means such as individual weights hung from the skirt could be utilized for this purpose in lieu of, or in addition to, the ballast illustrated, if so desired.

Slide fastener connectors **24** are attached to, or formed on, the ends of the skirt to provide means to attach the unit **10** to like units. The slide fasteners may be of any type known in the art such, for example, as metal zippers or the interlocking continuous plastic slide fasteners in common use. Where the unit is to be utilized for control or containment of spills, the fastener should be impervious to liquids and have suitable mechanical strength to withstand stresses imposed by hydrostatic pressure and dynamic forces imposed by wave motion. Obviously, other types of connectors, such as hook and eye or snap connectors used in conjunction with overlapping seals can be utilized for this purpose if so desired.

In order to provide sufficient mechanical strength for the unit **10** in the longitudinal direction, the skirt **12** is provided with a series of longitudinally extending tension absorbing tapes **26**, heat sealed, or otherwise attached to one face thereof. The tapes may be formed of any nonrigid, constant length material having suitable mechanical strength such, for example, as prestressed nylon fabric tape or the like. The ends of the tapes are free from the skirt and are furnished with a connector **28**, such as a calf chain hook or the like. The combined length of the tape **26** and the connectors **28** is less than the effective length of the skirt **12** for purposes to be described later.

FIGURE 92: PREUS-ROSENDAHL FLOATING BOOM DESIGN

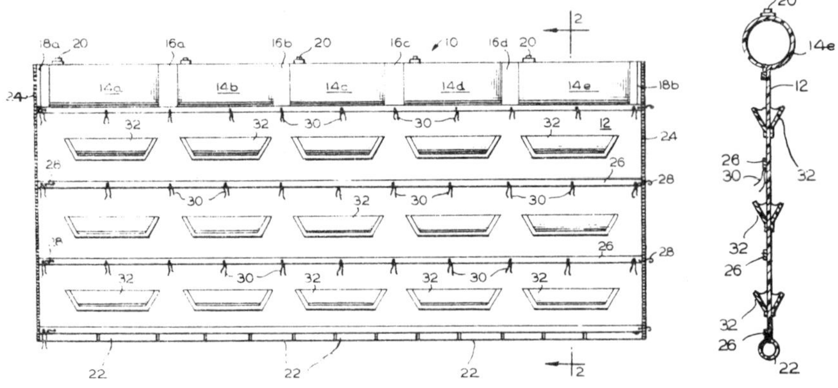

Source: P. Preus and C.E. Rosendahl; U.S. Patent 3,579,994; May 25, 1971

For purposes of reefing the skirt to depths less than the full depth thereof, ties **30** are attached thereto at suitable intervals for shortening the skirt in a manner known in the marine art. A plurality of self-filling, self-bailing pockets **32** are formed on the skirt **12**, preferably by heat sealing material thereto, to provide means to increase the ballasting of the skirt under conditions which will be explained below.

The relative dimensions of the aforedescribed unit are of importance to the process in that the unit is particularly suited for controlling spills on the high seas under most weather conditions and is, furthermore, adapted for handling with a minimum of difficulty. The unit is also particularly suited for compact storage, thereby increasing its availability and utility. In order to achieve these results, the unit must be able to conform to wave action in relatively heavy seas and high wind conditions without losing containment of the spill either over or under the unit. In order to accomplish this, the flotation chambers must have enough buoyancy and shape to provide sufficient freeboard and kiting characteristics to contain either wave carried or airborne spill fluids. For these purposes, it has been found that a flotation chamber between 2 and 5 feet, preferably 3 feet in diameter is required.

Quincy Adams Design

According to the EPA Edison, N.J. Laboratory Report, *Oil Spill Containment Systems* January 1973), a boom is available from the Quincy Adams Marine Basin Inc., Quincy, Mass. The boom was designed to be a permanently installed boom. The boom is made with a one-eighth inch thick rigid polypropylene fin. Flotation is supplied by 2 inch by 4 inch creosote treated fir timbers. It is constructed in 10 foot units. The connections of the units consist of a neoprene membrane and a brass piano hinge with a removable pin. The membrane is brass riveted to the polypropylene fin and to the piano hinge. The ballast is lead weights riveted to the fin. The standard weight of the boom is 4.5 pounds per lineal foot. Large flotation and heavier weights are available depending on tides and usage.

Rath Design

A hybrid (combined bubble curtain and mechanical) design developed by *E. Rath; U.S. Patent 3,665,713; May 30, 1972* utilizes high pressure jets which are operated in a predetermined configuration beneath the surface of the water in such a manner that their activation causes the contamination to be totally contained within a confined area and additional contamination from the same source to be directed by the same area and at the same time form a physical as well as hydraulic barrier to its spreading beyond the confines of the containment area.

As shown in Figure 93, the unit is made up of flotation tubes **11** and **12** which may also function as air supply reservoirs. The device may include a wave guard **31** supported by brace **32**. A drop shield **50** hinged at **53** carries at its lower end pipe **51** containing holes **52** from which air jets emerge. Hose **60** carries compressed air from tube **12** to pipe **51**. The brackets **55** at spaced intervals and pinned by appropriate latching pin **57** as indicated, may be adjusted so as to have the angle of the drop shield at that angle which is deemed most desirable for the particular operation and conditions encountered. The drop shield may thus be pulled completely up for storage and easy transportability and dropped and latched in whatever position may be desired when in use.

FIGURE 93: RATH COMBINED BUBBLE CURTAIN AND MECHANICAL BARRIER OF OSCARSEAL TYPE

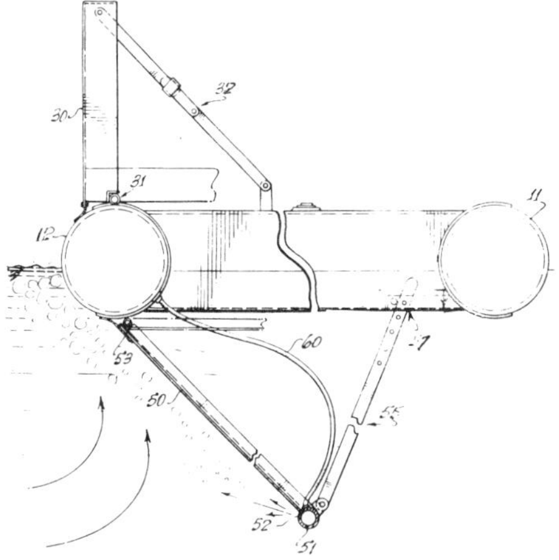

Source: E. Rath; U.S. Patent 3,665,713; May 30, 1972

This seems to be the same device described by the Edison, N.J. Laboratory of EPA in *Oil Spill Containment Systems* (January 1973) under the title of Oscarseal Steel Boom (see Figure 94), which is available from Morrison-Knudsen Co. Inc. of Boise, Idaho. Each individual section of the Oscarseal containment boom measures 40 feet in length, is 7 feet wide and weighs approximately 5,000 pounds. Flotation is derived from two longitudinal steel pipes and boxed cross-members. The forward pipe is to be 14 inches in outside diameter and the rear pipe is 12¾ inches. Both have wall thicknesses of 3/16 inch and are sealed with welded-on end caps. Two 12 x 12 inch boxed flotation members will be spaced 13 feet inward from the ends of the flotation pipes. In addition, two 24 x 12 inch boxed flotation members are located at each end of the float, both of which are installed on a 45 degree slope to give "lift" to the float when being towed.

A 2 foot upright barrier runs the length of the forward 14 inch pipe section. It is constructed with a steel frame and 12 gauge steel face. A rubber seal is attached to the lower edge of this barrier to prevent any passage of an oil slick between the 14 inch flotation

FIGURE 94: OSCARSEAL STEEL BOOM

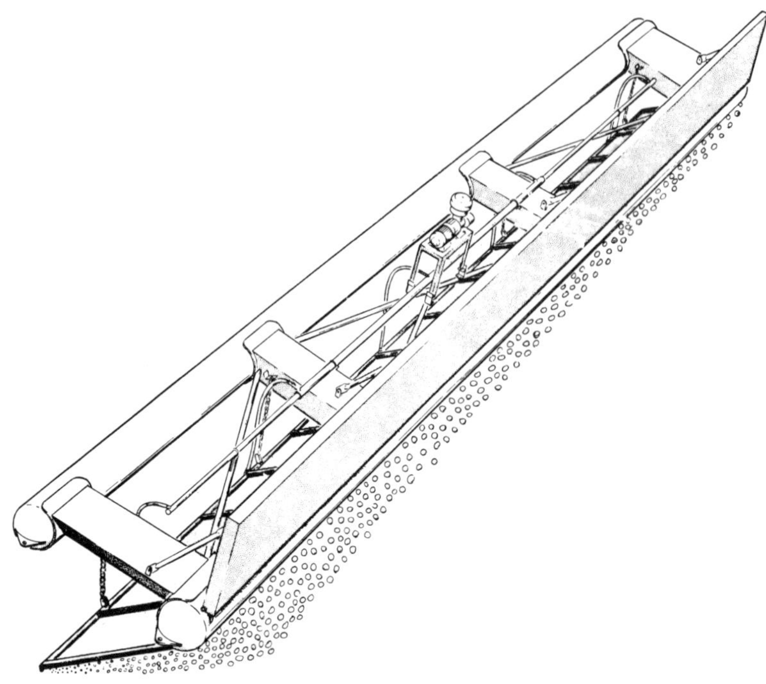

Source: EPA Edison, N.J. Report *Oil Spill Containment Systems* (January 1973)

pipe and the upright barrier. The barrier itself is hinged to the pipe so it can be folded to facilitate shipment or storage. Angling downward beneath the 14 inch flotation pipe is a slope sheet that is 5 feet wide and 40 feet long. It has a welded frame of 1 x 1 x ⅛ inch angle steel and a skin plate of 16 gauge steel. The entire assembly is attached to the 14 inch pipe by hinges, which permit folding for transport or storage, and is also sealed with rubber belting at the point of contact with the flotation pipe. It is suspended beneath the float at a 45 degree angle by adjustable chains.

Running the length of the slope sheet at its low (undersea) end is a 1½ inch diameter pipe manifold, drilled with a total of 160 holes, each ⅛ inch in diameter and spaced on 3 inch centers. From these holes, air is emitted to escape upward along the slope sheet, thus giving a rolling action to the water.

Air is supplied at 6 pounds pressure to each unit by means of a 3 hp electric motor that drives a rotary compressor at 1,800 rpm. The compressor delivers 61 cfm of air to the manifold, and by increasing the speed of the motor to 2,4000 rpm, can deliver 96 cfm. Motor and compressor are mounted on an angle iron frame in the center of the float. Electrical power at 440 volts is required to supply a series of Dynamic Oil Boom units. This can be accomplished by a diesel powered generating unit aboard a service barge or tug.

Each 40 foot float unit is equipped with 1 inch pad eyes and 1 inch shackles for coupling to another unit, thus enabling the boom to be extended to any length. Wire straps may be used on the backside connections in order to permit formation of a desired arc around an oil slick.

According to Maritime Administration Report COM-74-10212, Springfield, Va., National Technical Information Service (November 1973), the Oscarseal design is available from TRC Development Inc. of La Jolla, California. This device is designed to operate in 8 foot seas, 25 to 40 knot winds and 3 knot currents.

Reynolds Submarine Services Corporation Design

A design developed by *A.L. Markel and J.R.R. Harter; U.S. Patent 3,731,491; May 8, 1973; assigned to Reynolds Submarine Services Corporation* is an oil containment boom comprising an elongated strip of corrugated metal which is formed from a plurality of sections secured together end-to-end. The corrugated metal is disposed substantially vertically in a body of water and floated therein with suitable flotation means. In one form, the flotation means comprises a strip of plastic material adhesively secured to the corrugated metal. In another form, the previously described flotation means is supplemented with the aid of outrigger or stabilizer floats which extend generally laterally outwardly from the corrugated metal strip. Weights may be attached to the corrugated metal strip to vary the freeboard height of the floating boom.

A single strip of plastic material provides the required flotation to give the fence-like boom about one foot of freeboard and provides a reserve buoyancy of approximately 100%. The fence-like boom may be constructed in continuous sheets 100 feet or longer by 4 feet or greater in width. The freeboard may be varied up to one-half of the vertical height of the boom for heavy sea conditions.

By way of a specific example, one boom made in accordance with this process weighed approximately 2.5 pounds per lineal foot. For storage, 100 linear feet of the boom was compactly received on a spool 45 inches by 4 feet in diameter. The boom may be launched by unspooling, or in quick deployment operations, by dropping it from an aircraft or vessel in a compact, coiled cylinder. When floating, a girth strap is released, the fence or boom unwinds itself and is ready for use.

The bottom of the fence-like boom is weighted to provide a good metacentric height and stability in a seaway. For use in rough sea conditions, small lightweight stabilizer or outrigger floats may be bolted to the boom to provide further reserve flotation and a righting moment arm to each side of the boom. The stabilizer or outrigger floats may be attached at regular intervals such as 6 foot intervals. The side of the boom not containing the plastic flotation means would normally be the side of the boom exposed to the oil slick. Therefore, there is provided a smooth surface for easy cleaning when operations are finished. In order to attach one section of the boom to another, flexible fabric reinforced rubber couplings may be utilized or the sections may be bolted together.

Figure 95 shows the unreeling of an oil containment boom of this design (upper view) as well as the deployment of the boom behind a small boat (lower view). The boom **10** may be introduced into a body of water from a spool means **14**. The latter is supported at the stern of a ship indicated generally at **16**. The spool means is supported by a suitable holder means **18** and is rotatable on axle **20**. In the lower view the boom **10** has been deployed and is being pulled by a suitable cable means attached to the end of the boom so as to contain an oil slick or the like on the surface of the body of water. Oil containment booms made in accordance with this technique have been towed at speeds up to 4 knots per hour.

A plastic floating means **22** is suitably attached to the sheet of corrugated aluminum **12** as it is being unwound from spool **14** such as by means of a contact cement in order to secure the plastic floating means to the corrugated aluminum sheet. The corrugations are illustrated at **30** for the corrugated aluminum sheet material **12**. These corrugations are oriented vertically in order to provide floating flexibility in both the vertical and horizontal planes for the boom. The flexibility afforded by the corrugations provides sea-keeping compliance and increases the boom's effective strength. The increased surface area of the corrugation tends to have an adhesive effect causing oil to adhere to it at the water line.

FIGURE 95: REYNOLDS FLOATING BOOM DESIGN

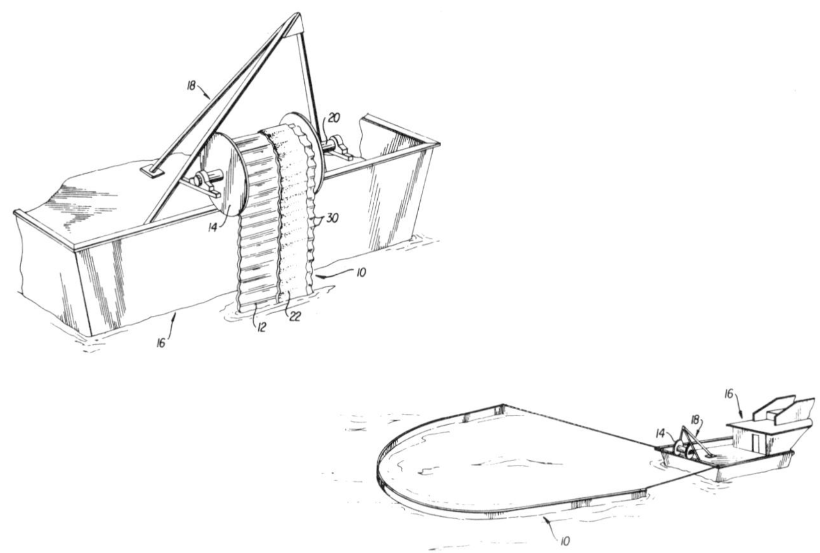

Source: A.L. Markel and J.R.R. Harter; U.S. Patent 3,731,491; May 8, 1973

As described in the EPA Edison, N.J. Report, *Oil Spill Containment Systems* (January 1973), the Reynolds Boom is constructed of corrugated aluminum (3003 H14). The corrugations are oriented vertically to provide flexibility in the vertical and horizontal planes. Flotation is provided by a single strip of plastic alloy material which affords 7 inch freeboard and a 21 inch draft for the 28 inch boom (see Figure 96).

FIGURE 96: REYNOLDS ALUMINUM OIL BOOM DETAIL

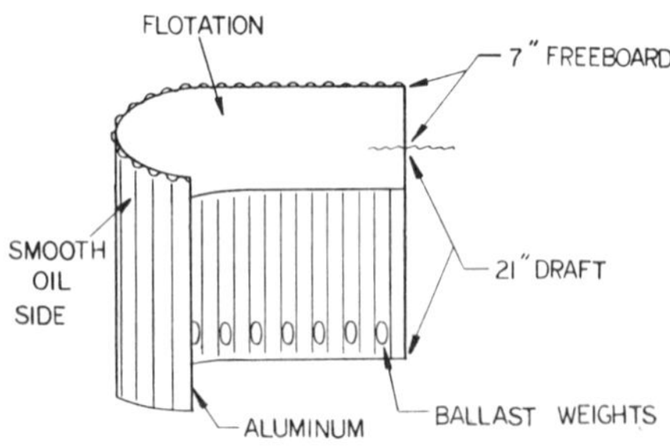

Source: EPA Edison, N.J. Report, *Oil Spill Containment Systems* (January 1973)

The barrier weighs approximately 2.6 pounds per lineal foot and may be reeled on to a cylindrical spool 40 inches in diameter. The bottom of the fence is weighted to provide good metacentric height and stability in a seaway. The side of the fence without flotation would normally be the oil side of the boom. This provides a smooth surface for easier cleaning when operations are finished. One hundred foot sections are attached with bolts to provide for booms of greater length. The fence uses no longitudinal cables or other strength members. The tensile strength of the metal sheet is well in excess of a ton. All metallic components of the containment device are galvanically compatible for long life in salt water environment.

According to Maritime Administration Report COM-74-10212, Springfield Va., National Technical Information Service (November 1973), the 28 inch Reynolds boom is designed to operate in 10 to 12 foot seas, 1 knot current and 20 to 40 knot winds. It is also designed to contain 500,000 gallons of oil per 2,000 feet of barrier. It has been tested in 3 to 5 foot waves, a 0.5 knot current and 20 knot winds in the Atlantic Ocean. Four men can deploy 2,000 feet of this barrier in one hour using two vessels. An anchor of 3,000 pound holding capacity is required to man 1,000 feet of barrier.

There is a smaller inshore version of the 28 inch boom for rivers, harbors, industrial ponds and waterways. It is lighter in weight and smaller (15 inches) for ease of handling and is for use in waters where wind conditions are not severe. It is constructed of continuous sheets 100 feet long of corrugated aluminum alloy (3003-H14) weighing 1 pound per linear foot. The corrugations are oriented vertically to provide flexibility in the vertical and horizontal planes.

Flotation is provided by a single strip of plastic alloy material on one side of the upper portion of the boom to afford a 5 inch freeboard and a 10 inch draft. The side of the boom without the flotation would normally be the oil side of the boom. This provides a smooth surface for easier cleaning when operations are finished. The bottom of the barrier has ballast weight attached to provide height and stability in the water. The sections can be attached with bolts to provide for booms of greater lengths. The barrier uses no longitudinal cables or other strength members. All metallic components of the device are galvanically compatible for long life in salt water environment. Three 100 foot sections may be stored on a spool arrangement 48 inches by 53.5 inches by 52 inches in height. The spool is designed for easy deployment of the boom.

Rupnick Design

A design developed by *E.J. Rupnick; U.S. Patent 3,635,347; January 18, 1972* is one in which the dispersion of pollutants having specific gravities less than that of water and floating on a body of water are controlled by encircling the pollutant body with a floating reservoir wall having an open top and open bottom. The reservoir wall comprises an elongated length of flexible water impervious material having disposed along its upper end a plurality of inflatable flotation cells inflatable through a common duct running the length thereof and having disposed along its lower end ballast means for maintaining the lower end of the wall beneath the surface of the pollutant and body of water.

The ends of the reservoir wall are joined together in sealing relationship by compressively engaging inflated cells disposed along the ends of the reservoir wall. Pollutants issuing into a body of water can be directed into the interior of the enclosed reservoir formed by the reservoir wall by a flexible or inflexible conduit secured at one end around the source of pollution with the opposite end leading to the interior of the reservoir. The reservoir wall is held in a predetermined location by means of anchors resting on the floor of the body of water and connected by lead lines to the lower end of the reservoir wall.

Figure 97 is an overall view of such a design which includes an outer reservoir **10** and an inner reservoir **20** supported in spaced relation from the outer reservoir by guy wires on lines **1**. Leading from the source of pollution to the interior of inner reservoir **20** is a conduit **2** which directs the pollutant from its discharge point into the interior of inner reservoir **20**. An outer duct or ducts in the inner reservoir connect to line **3** leading to a transport, tanker or other vessel **4** for removing the pollutants as necessary from the reservoir system. A series of anchors **5** having guy wires **6** leading to the lower end of the outer reservoir wall are positioned at predetermined intervals around the periphery of the outer reservoir wall to hold the inner and outer reservoirs in a predetermined location.

FIGURE 97: RUPNICK FLOATING BOOM DESIGN

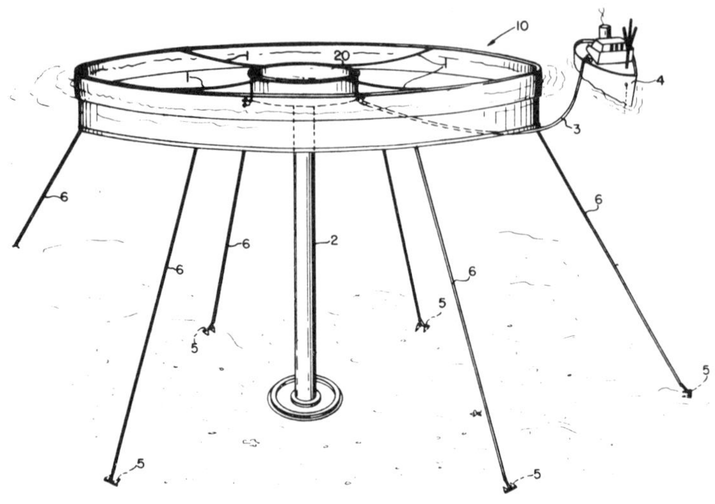

Source: E.J. Rupnick; U.S. Patent 3,635,347; January 18, 1972

In more detail, the outer reservoir wall of reservoir **10** is made of a flexible water impervious material. Materials which may be used include rubber, polyethylene film, polypropylene film, or other synthetic suitable materials. Typically a heavy gauge material which will not easily puncture and does not become brittle after extended periods of use is preferred. The outer reservoir wall has a series of flotation cells disposed along the upper length dimension thereof, the flotation cells inflatable with air or other gas through a common filler duct. Ballast means are attached to the end of the reservoir wall opposite the flotation cells to stabilize the reservoir wall in a vertical position and maintain the lower portion of the reservoir wall beneath the surface of the body of water and pollutant. The ballast may be a series of weights about which the lower portion of the reservoir wall is enclosed and seamed, or may be a series of weights clipped through reinforced eyelets in the lower edge of the reservoir. Any number of ways of attaching the ballast to the lower edge of the reservoir wall are possible.

Saavedra Design

A design developed by *M. Saavedra; U.S. Patent 3,645,099; February 29, 1972* comprises a panel structure defined by an elongate rectangular sheet of pliable material of double thickness that has means on each end thereof for removably engaging the ends of adjacently disposed panels, and each panel including two oppositely disposed longitudinally extending buoyant members that are situated at intermediate positions between the top and bottom of the sheet. The buoyant members when floating on a body of water serve to support the sheet at such a position that a substantial section of the sheet projects above the buoyant bodies to act as a barrier to prevent the escape of oil.

Figure 98 is an overall view of the design. An oil slick **A** is shown surrounded by an endless floating wall **B** that extends above the surface of the water **C** a substantial distance. The wall is defined by a number of elongate rectangular panel structures **D** that may be removably connected end to end by means **E**. Each panel structure is buoyantly supported by two longitudinally extending floats **F** secured thereto. Each panel structure includes a pliable rectangular sheet of double thickness that has at least one intermediate section **12** formed into an accordion-like configuration. Each intermediate section is situated between two sections **14** of the sheet with section **12** permitting independent upward and downward movement of the two sections **14** due to wave action. The two sections **14** on each side of a section **12** have doubled over, longitudinal strips **16** of a pliable material secured thereto, the doubled over strips serving as elongate envelopes in which the two cylindrical elongate floats **F** are disposed, the floats preferably formed of plastic tubing that have closed ends.

FIGURE 98: SAAVEDRA FLOATING BOOM DESIGN

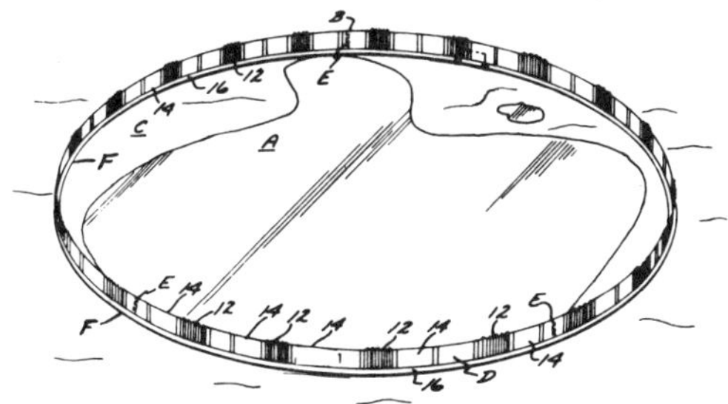

Source: M. Saavedra; U.S. Patent 3,645,099; February 29, 1972

Sawyer-Tower Products Inc. Design

A design developed by *W.A. Reilly; U.S. Patent 3,548,599; December 22, 1970; assigned to Sawyer-Tower Products, Inc.* provides a barrier for intercepting surface spread of oil in a body of water which includes a semiflexible buoyant section and a semiflexible submerged stabilizing section which holds the upper surface of an elongated trough-shaped deck above the normal water level, from which liquid sloshing over one side of the barrier into the trough can be drained out of one or both ends of the trough. A series of the barriers float in enclosing array around a source of oil leakage.

Figure 99 shows an offshore drilling rig platform with the device in place. The device is greatly exaggerated in size for purposes of illustration and is shown partly in section. The barrier is heavy duty (6 ounce, 2 x 2) nylon fabric impregnated or coated with an oil and salt water resistant neoprene or vinyl polymer, giving a fabric weight of about 28 ounces per square yard. This is fabricated with heat seals or stitching or both to form five connected pockets. Two circular pockets at the top are filled with long cylindrical bodies **12** of buoyant material, preferably polyurethane foam. They are joined with an intervening multiple layer web forming with the adjacent portions of the bodies, a trough shaped semiflexible deck **21**. Suspended below the two elements is a triangular shaped body of fabric sealed to the bottoms of the cylindrical elements forming a third pocket packed with wedge-shaped, solid, dimensionally stable, heavy waste reclaimed rubber blocks **24**, in sections of convenient handling length, acting as a weighting and spacing material. Suspended centrally of the barrier at the bottom below a multiple layer vertical web of the fabric covering is a fourth fabric pocket containing a three inch neoprene oil hose **27**. The hose is fairly stiff but has a certain amount of flexibility.

FIGURE 99: SAWYER-TOWER PRODUCTS INC. FLOATING BOOM DESIGN

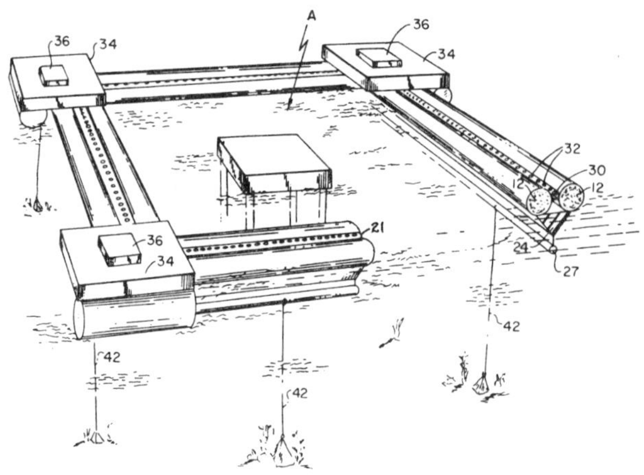

Source: W.A. Reilly; U.S. Patent 3,548,599; December 22, 1970

Laid inbetween the cylindrical elements on the web is another neoprene oil hose **30** which has a row of perforations as shown at **32**, so that the ends thereof can be coupled at corner rafts **34** to pumping mechanism **36** for pumping surface water which sloshes over one or both of the cylindrical elements **18** out of and draining the trough formed therebetween. The perforations may face downwardly or in other directions or may be placed in spiral or other configuration around the hose **30**. In operation the lower hose **27** is open-ended and fills with water when the barrier is floating and acts as a rudder to resist snaking of the barrier and helps to hold it in a straight line to prevent the oil from going underneath the barrier.

In operation then, when strung as an enclosure about a source of oil leakage, the barrier maintains a level and a straightness due to the resistance to bending both in the vertical and horizontal directions. As a result the material does not bounce out of the water and waves seldom lift it sufficiently to allow passage of oil underneath the subterranean portion of the device. Sloshing over the top is cared for by the pumping operation through the pipe **30**.

It has been found that the barrier can be best made in 100 foot lengths with one foot diameter elements **18**, but these of course can be connected together to form further lengths between pumping rafts. Moreover, it is contemplated that in extreme cases two of them can be positioned in side-by-side parallel relation to give double protection. The buoyant material may be in continuous lengths or sectional. As can be seen, the barrier may be transported in knocked-down condition and be finally assembled at or near the water site, and anchored as required as indicated by the anchors **42**, connected to the rafts or to the barrier or both.

Scandinavian Oil Services Design

As described by the EPA Edison, N.J. Laboratory in *Oil Spill Containment Systems* (January 1973), SOS Booms, manufactured in Sweden and distributed in the United States, are available in two types: the inflatable emergency boom and the permanent boom of glass-fiber reinforced polyester.

The inflatable emergency boom, suggested for temporary or intermediate use, consists of a circular float section 6 inches in diameter made of PVC plastic, a 14 inch skirt extending downward from the float, and chain ballast provided in the bottom hem of the skirt. This bottom chain is also used in towing the boom. The SOS emergency boom is manufactured in standard lengths of approximately 80 feet. Adjacent sections are joined by vertically slit rigid PVC tubes and hooks for linking the respective ballast chains.

It is reported that the inflatable emergency boom retains air for long periods of time and that these booms have been in the water for more than two months without need for refilling. The boom may be inflated by means of a motor-driven air compressor, compressed air bottles, etc. Equipment is also available for simultaneously inflating more than one section of the boom at a time. A standard 80 foot length of inflatable emergency boom weighs approximately 29 pounds (unit weight of 0.4 pound per foot) and had a purchase price in 1967 of $350, East Coast U.S. (i.e., $4.40 per foot).

The SOS permanent boom consists of a 7 inch diameter fiberglass tube filled with urethane foam. Attached to the flotation tube is a nylon reinforced, PVC coated skirt 18 inches in depth. The SOS permanent boom, available in 10 foot sections, comes equipped with chain ballast fitted into the bottom edge of the skirt. A price of $16.50 per foot, delivered East Coast, U.S., has been reported. These booms, manufactured by Skandinavisk Oljeservice AB of Goteborg, Sweden, are marketed in the U.S. by Surface Separator Systems, Inc. of Baltimore, Maryland.

According to Maritime Administration Report COM-74-10212, Springfield, Va., National Technical Information Services (November 1973), there is also an SOS Flema Offshore Boom. This boom has 18 inch draft and 16 inch freeboard, and consists of two hose sections made of nylon-reinforced oil resistant PVC. The two hoses are arranged one above another in a vertical plane; the upper one is air filled to provide flotation and the bottom one is water filled to provide ballasting. A 328 foot barrier section of this design has been tested in 2 foot waves and a 1.2 knot current with light fuel oil. Recovery, cleaning and storage of this SOS offshore boom are judged to be relatively easy.

Sea Boom Design

According to the EPA Edison, N.J. Laboratory Report, *Oil Spill Containment Systems* (Jan. 1973), there are available from Submarine Engineering Associates Inc. of Cohasset, Mass. two basic types of sea booms— the permanently floating model, 3PF, and the submersible model, 3SU. They are designed and constructed to remain in the water constantly. Each boom can be utilized independently as a complete barrier or in conjunction with each other.

In a typical installation, the permanently floating version and the submersible boom completely enclose a terminal. The portion of the terminal for docking and undocking would use the submersible boom. One man, normally the dock foreman, would raise and lower the boom by turning two valves to allow ship movement. Both booms are in 24 feet section lengths with 12 inches freeboard and 24 inches draft. The permanently floating barrier weighs 13 pounds per foot, while the submersible version weighs 15 pounds per foot.

The manufacturer lists the materials used in the construction of the boom as Sea Rubber, Searethane and Sea Coat, all trade names. The boom is black with white and international orange striping. The manufacturer claims the booms are designed for continual immersions and heavy abrasion operations for periods in excess of two years.

Sea Curtain Design

The Sea Curtain, manufactured by Kepner Plastics Fabricators, Inc., Torrance, California, is a relatively large plastic-rubberized oil spill barrier consisting of an upper air or foam filled flotation chamber and a lower hanging skirt. The foam filled flotation consits of polystyrene beads enclosed in an inner liner. The Sea Curtain is available in standard 20 foot lengths and in four different sizes and types as follows (see Figure 100).

Type A — Heavy-duty, offshore, ocean service boom with foam filled float 20 inches in diameter and with a 30 inch skirt suspended below. Approximate price is between $10 and $15 per foot.

Type B — Heavy-duty harbor and channel boom with foam filled float 12 inches in diameter and with an 11 inch skirt below. Approximate price is given as $6 to $9 per foot.

Type C — An emergency service boom with lined-inflatable float 19 inches in diameter and a 32 inch skirt. Price range is between $4 and $6 per foot.

Type D — Light-duty boom with 12 inch inflatable float and an 11 inch skirt. Price range is reported as $2 to $4 per foot.

FIGURE 100: SEA CURTAIN FLOATING BOOM DESIGNS

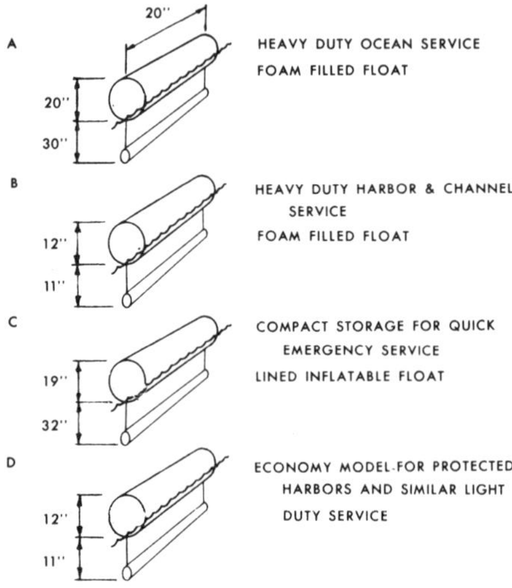

Source: EPA Edison, N.J. Report, *Oil Spill Containment Systems* (January 1973)

The Sea Curtain is provided with ballast in the form of a chain running through the bottom of the skirt. Recently, according to the EPA Edison, N.J. Laboratory Report, *Oil Spill Containment Systems* (January 1973), Kepner Plastics has started design of booms containing chain ballast on both sides of the skirt with up to ⅜ inch cable being used on the heavy-duty ocean boom. It is possible to add chain ballast onto the offshore boom until the unit weight approaches and even exceeds 10 pounds per running foot of boom. Extremely heavy ballast may be best for severe offshore winds and waves, but until additional data are secured, the company suggests for offshore booms that lesser amounts of ballast in the range of 5 to 7 pounds per foot be employed for ease in handling, towing and installation. The Kepner heavy-duty offshore boom weighs approximately 2 pounds per foot without ballast. The offshore boom with heavy ballast is believed to be in the cost range of $15 to $17 per foot.

Standard fittings at the ends of each section of the Sea Curtain are reportedly similar to that provided on the Slickbar Boom (given later in this report) although the company is known to be developing new joints providing tighter and more secure connections between sections. The foam filled boom is apparently recommended by Kepner Plastics over the inflatable-type curtain for durability and longer service upon open waters.

Sea Fence Design

The Sea Fence is an oil spill boom in the experimental-development stage expected soon to be in commercial production by the Aluminum Company of America (ALCOA). The Sea Fence is made up of rigid vertical sheets of aluminum held together by steel cable, with foamed plastic material for flotation, and neoprene joints providing a flexible seal between the aluminum panels. The boom is fireproof and reportedly capable of being stored onto reels. A prototype model has been tested. However, no further information is currently available concerning the Sea Fence, according ot the EPA Edison, N.J. publication *Oil Spill Containment Systems* (January 1973).

Sealdboom Design

The Sealdboom, manufactured by Uniroyal, Inc., Providence, Rhode Island, is a thin-wall barrier made of nylon-rubber. Closed-cell polyethylene foam support floats provided on each side of the curtain are fully enclosed within the barrier. Vertical stiffeners, which are elastomer-coated spring steel strips, are located 21 inches on center. Available in 30 foot sections, the barrier provides 12 inch freeboard and a 24 inch draft below the waterline. The curtain is fabricated from Paracril coated nylon and is reported to be abrasion resistant and vulcanized similar to an automobile tire for added strength. The boom, which weighs 3.8 pounds per foot, uses lead weights spaced 21 inches apart for ballast. Cost of the Uniroyal barrier is reported as $12.00 per foot (October 1969 price).

Sea Skirt Design

The system consists generally of lengths of inflatable tubing provided with weighted skirts, the widths of which can be changed as conditions require. The tube and integral skirt sections can be stored on a reel in a deflated condition. When needed, the units are inflated as they are being unreeled for deployment into the water. Additional skirt sections can be added during the unreeling and inflating step, or they can be added later. Inflation to a pressure of 0.25 to 0.50 psi is reportedly sufficient to provide buoyancy and still allow flexibility to permit conformance with the water surface (see Figure 101).

FIGURE 101: SEA SKIRT FLOATING BOOM DESIGN

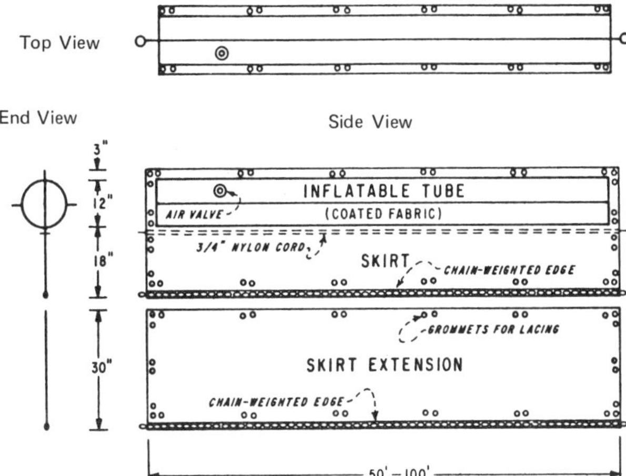

Source: EPA Edison, N.J. Report, *Oil Spill Containment Systems* (January 1973)

When deploying, three anchor points are utilized to form an elongated diamond shape, with the long apex being in the downwind or leeward direction. A windward opening can be provided to permit ingress and egress of boats. The windward apexes are held relatively stable by anchor lines, but the leeward apex is free-floating, so that it will not submerge and allow oil to spill over. An oil-collecting sump is attached to the leeward apex to permit separation and pumping out of the oil. If the wind direction changes appreciably, either the anchors can be moved or the sump can be moved from one apex to another. Under circumstances where the 12 inch inflated section is not sufficient, due either to rough seas or a thick accumulation of oil, it is possible to increase the height of the above water barrier by pyramiding the inflated tube units. No costs are available on this oil boom. This boom is manufactured by Core Laboratories, Box 10185, Dallas, Texas 75201.

Shell Oil Company Design

A design developed by *R.R. Ayers, P.E. Titus and J.R. Hanson; U.S. Patent 3,664,504; May 23, 1972; assigned to Shell Oil Company* employs a technique for deploying a floatable barrier where the barrier is initially collapsed in a storage location provided by a container having means allowing escape of the barrier from the storage location upon sinking of the container and means for sinking the container. As the container is sunk, the barrier floats out of the storage location and may be deployed merely by uncollapsing the same. Figure 102 shows the floatable barrier collapsed in its sunken container (bottom view) and the escape of the collapsed boom from the sunken container (upper view).

A major component of the apparatus **10** is a container **14** having the barrier **12** collapsed therein. The container is sinkable allowing escape of the barrier therefrom which may then be uncollapsed, as by pulling on the ends thereof. The container **14** comprises a receptacle **16** of any convenient shape, but which is illustrated as generally rectangular. The receptacle includes a barrier-receiving compartment **18**, means **20** allowing escape of the barrier **12** from the compartment **18** and means **22** for sinking the container **14**. In the embodiment illustrated, the escape allowing means **20** merely comprises the open top or side of the receptacle **16**. It will be apparent to those skilled in the art that this escape allowing means may be a movable closure or the like for the top or a side of the receptacle.

The sinking means **22** is illustrated as comprising openings in the receptacle **16** below the water line thereof to allow ingress of water thereinto. In the illustrated embodiment, the receptacle should be more dense than water so that it sinks when placed therein. It will be apparent that suitable weights and the like may be attached to the container to ensure sinking thereof.

It is normally desirable to have the container **14** sink when placed in the water. Under some circumstances, however, it is desirable to have the container initially buoyant. For example, it may be desirable to tow the container through the water to the area of an oil spill or it may be desirable to tether the container in a crucial area, as near an unloading pier or near the mouth of a water course carrying affluent water. In these circumstances. the sinking means **22** may comprise an opening below the water line of the container, a normally closed closure associated therewith and remotely controlled means for opening the closure. It will be apparent to those skilled in the art that this arrangement may be carried out in any number of ways.

The container **14** also comprises a plurality of lifting lugs **24** of any suitable type to enable the apparatus **10** to be lifted and moved about with conventional equipment. The container also preferably comprises a lip **26** around the open top thereof providing strength for the receptacle and for purposes more fully explained below. After the container has been sunk to allow escape of the barrier **12**, it may be abandoned if desired. It is, however, preferred to raise the container for subsequent reuse and/or for use in the spill controlling operation. The simplest recovery means is a buoy and line connected to the container whereby the container may be located and hoisted aboard a vessel. It is preferred,

FIGURE 102: SHELL DEPLOYABLE FLOATING BOOM DESIGN

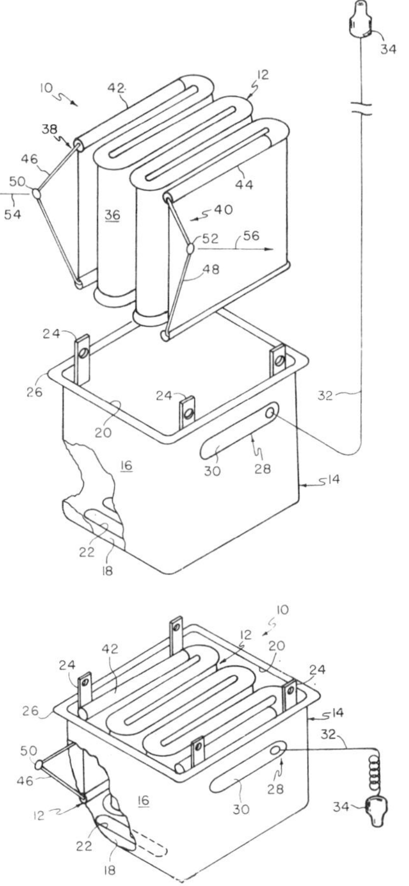

Source: R.R. Ayers, P.E. Titus and J.R. Hanson; U.S. Patent 3,664,504; May 23, 1972

however, to buoyantly raise the container to the water surface. To this end, there is provided a buoyant recovery means **28** comprising an inflatable member **30** secured to opposite sides of the receptacle **16** adjacent the lip **26**. It will be seen that the lip provides substantial protection for the inflatable member to prevent puncture thereof when in the collapsed condition.

Inflation of the member **30** is preferably accomplished by the use of a pressurized gas cartridge actuated in any convenient manner. For example, means may be provided responsive to water pressure to pierce the cartridge when the container sinks below a predetermined depth, e.g., 10 feet. In the alternative, the buoyant recovery means **28** may comprise a force actuable plunger for piercing the cartridge in response to a predetermined force existing in the line **32** connecting the force actuating mechanism to a buoy **34**. When the container sinks sufficiently to tension the line **32**, tension in this line may release a spring actuated plunger to pierce the cartridge.

The boom or barrier may be of any type capable of being collapsed for storage in the compartment **18**. Preferably, the barrier **12** is folded in accordion fashion. The barrier may include a plurality of solid buoyant floats or may comprise an inflatable boom of the type shown generally in U.S. Patent 3,494,132. In any event, the barrier comprises a foldable float section **36** and means **38,40** on the ends of the float section **36** for uncollapsing or unfolding the section. Each of the uncollapsing means **38,40** comprises a stiffening section **42,44** secured to the float section and about one full fold length long, and a bridle **46,48** having a connection **50,52** thereon. The connections may comprise an eye for attachment by a hook to a towing vessel.

The apparatus **10** may be placed in the spill area by throwing the device overboard from a vessel or by dropping it from an aircraft or helicopter. As water flows through the openings **22**, the receptacle **16** fills with water causing the container **14** to sink. If the barrier **12** is of the type comprising solid floats, it substantially immediately begins to escape through the open top of the receptacle. If the barrier is of the inflatable type, suitable means may be provided to partially inflate the same to ensure removal from the container rather than wedging therein. Complete inflation may await escape of an inflatable barrier from the container. The connections **50,52** are then attached to separate vessels to impart uncollapsing or unfolding forces to the barrier as represented by the force vectors **54, 56**. In the alternative, the connections may be attached to a pair of lines leading from opposite sides of a single vessel in the barrier towed slowly through the water to unfold the same rearwardly into a draped or arcuate configuration. The barrier may then be used in a conventional manner to contain the areal extent of the pollutant spill or may be used to facilitate removal thereof.

Recovery of the container is effected by inflation of the members **30** which are designed to impart a buoyant force to the receptacle **16** sufficient to float the container **14**. The container may then be removed from the water by securing a line to one or more of the lifting lugs **24** and hauling the container aboard the vessel.

Slickbar, Inc. Design

The firm of Slickbar, Inc. of Saugatuck, Conn. makes Slickbar Mark V and Slickbar Mark VI booms. The Slickbar Mark V boom was designed especially for use in high seas and rough and choppy water. It is 36 inches wide with a continuous fin of polyester woven monofilament fabric. The material is impregnated with international orange PVC and treated with UV inhibitor. The material is designed to be flexible in the horizontal plane and "rigidized" in the vertical plane. This feature eliminates the need for vertical stiffeners. Flotation is provided by floats of yellow, hard-skinned polyurethane foam attached to each side of the fin. There are two pair of floats per 10 feet of the fin. The placement of the floats afford a 12 inch above surface and 24 inch below surface fin (see Figure 103).

A reinforcing cable of ⅜ inch stainless steel is attached to the fin and placed below the floats. All loads imposed on the boom are transferred to the cable. The combination of specially designed fabric and stainless steel cable give the boom a tensile strength in an assembled condition during a static load test in excess of 10,000 pounds. Ballast is provided by lead weights, hardened with antimony, riveted to the bottom of the fin, 2.1 pounds per foot of the boom. End sets on the boom sections are made of extruded aluminum. Connections are made without tools or any additional pieces. To disconnect, just pull a pin and the boom is detached. The Slickbar Mark V sells for $12.00 per foot (1972 price).

According to Maritime Administration Report COM-74-10212, Springfield, Va., National Technical Information Service (November 1973), the Slickbar Mark V is designed to contain oil in 25 foot waves, 40 knot winds and 1.3 knot currents. This report quotes the manufacturer to the effect that he has over one million feet of this barrier in use, inshore and offshore.

The Slickbar Mark VI consists of individual floats made of yellow, cylindrical shaped, solid polyethylene, 50 inches long and tapered on both ends. Floats come in 4 inch or

FIGURE 103: SLICKBAR – MARK V FLOATING BOOM

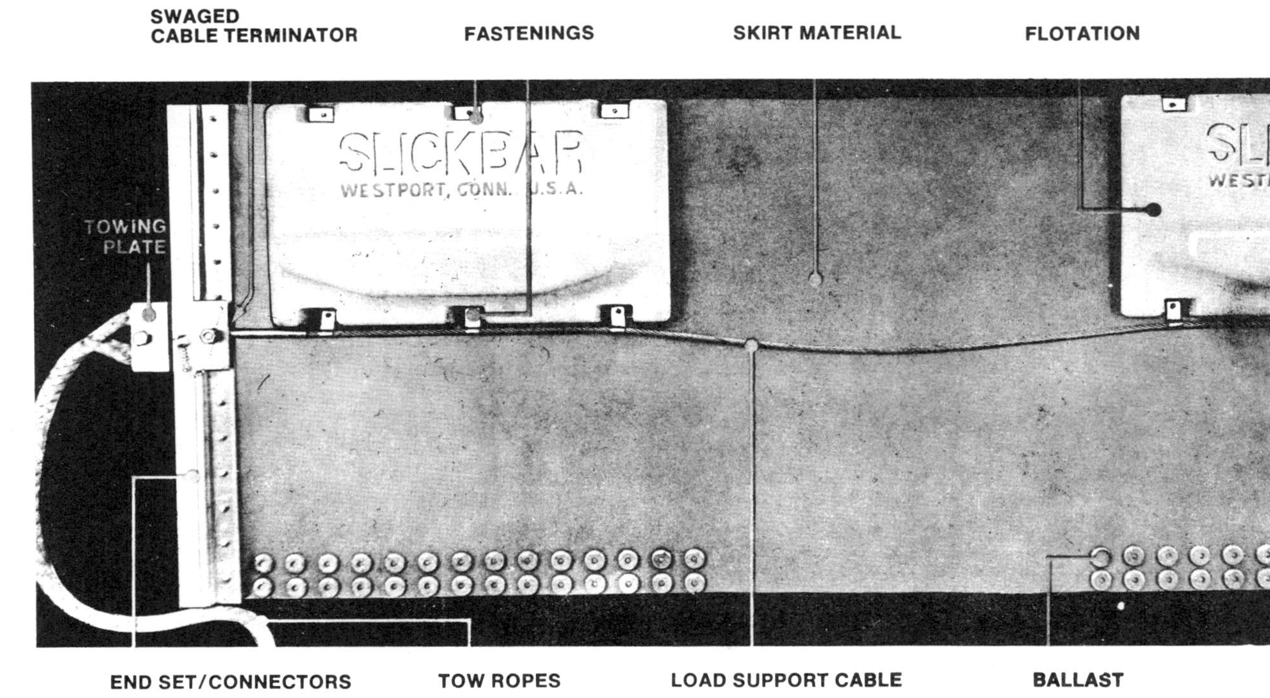

Source: EPA Edison, N.J. Report, *Oil Spill Containment Systems* (January 1973)

FIGURE 104: SLICKBAR – MARK VI FLOATING BOOM

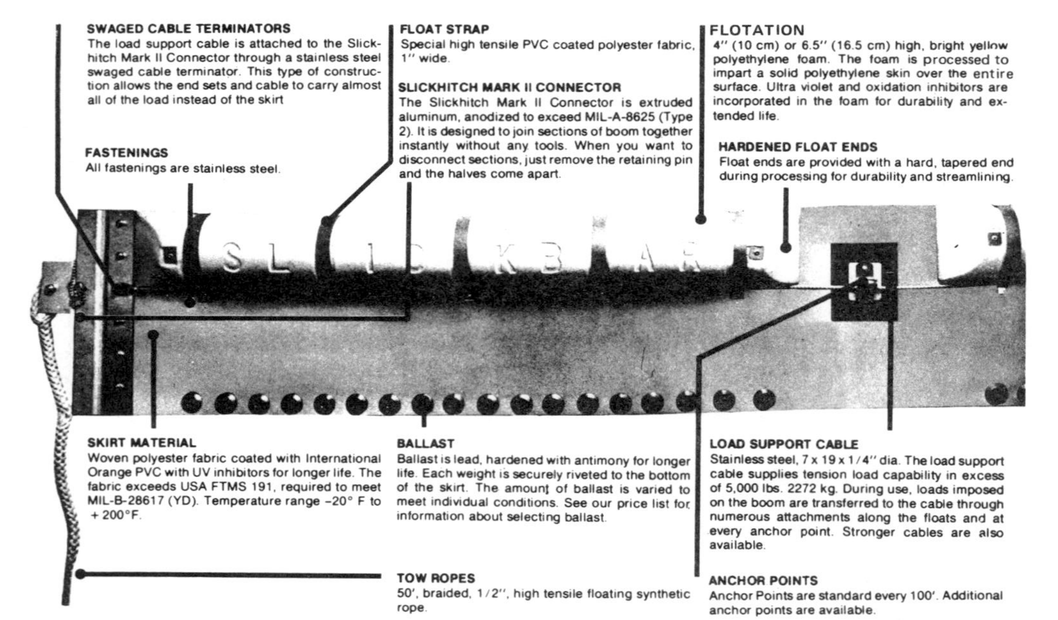

Source: EPA Edison, N.J. Report, *Oil Spill Containment Systems* (January 1973)

6 inch diameters and are secured to the boom skirt by vinyl coated polyester straps (see Figure 104). A woven polyester fabric coated with international orange PVC is used for the skirt. The skirt is available in 6, 8, 10 and 12 inch depths. The 12 inch depth is used only with the 6 inch diameter floats. Individual lead weights, fastened to the bottom of the skirt, furnish ballast. Weights come in several sizes to meet the need for the current velocity where the boom will be used. A load support cable made of ¼ inch stainless steel is placed below the floats and runs the entire length of the boom. Both ends of the cable are provided with eyes. Boom sections are connected by pairs of aluminum interlocking plates called the Mark II Slickhitch. Boom sections may be ordered in lengths from over 200 feet to less than 50 feet.

The boom varies in price according to the size of the floats, depth of the skirt and size of ballast. Prices range from $4.25 per foot to $10.90 per foot. Booms ordered over 200 feet in length come equipped with end connectors, while with booms under 200 feet, end connectors are additional. Also, booms under 100 feet are priced 10 or 20% higher according to the EPA Edison, N.J. Laboratory in *Oil Spill Containment Systems* (January 1973).

Slikguard Design

According to the EPA Edison, N.J. Laboratory Report, *Oil Spill Containment Systems* (January 1973), the Slikguard barrier is available from Slikguard Inc., Division of Alfred G. Peterson & Sons Inc. of Avon, Massachusetts. The individual sections are 50 feet in length with an overall width of 36 inches. The freeboard varies from 9 to 12 inches and the draft is 25 inches, plus or minus 3 inches. The fabric is a yellow nylon based material with a coating of PVC, treated with an antifungal and antibacterial agent. Individual segments of foam, 18 inches in length, are used for flotation. Each segment is protectively sheathed. The segments are fastened 6 inches from the top edge and spaced every 24 inches on both sides of the boom (see Figure 105).

FIGURE 105: SLIKGUARD OIL SLICK BARRIER

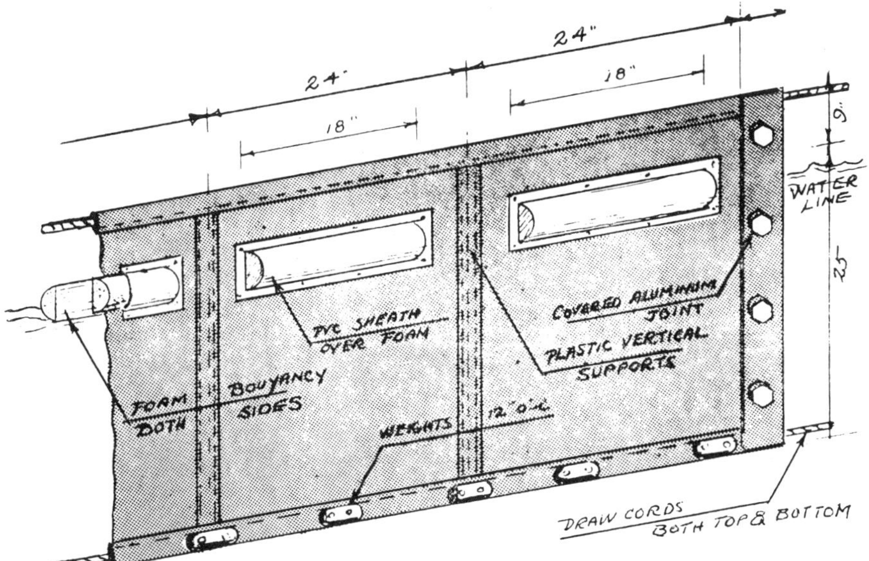

Source: EPA Edison, N.J. Report, *Oil Spill Containment Systems* (January 1973)

The ballast has two draw cords that are passed through the sewn hem at the top and bottom. The draw cords can be used for towing and securing the end of the boom to a pier or bulkhead. Sectional end connections are made by bolting the provided end aluminum plates. The boom is constructed of noncorrosive materials and sells for $8.00 per foot (1972 prices).

Smith Design

A design developed by *M.F. Smith; U.S. Patent 3,638,430; February 1, 1972* provides rugged high strength, fire resistant floating booms each incorporating a continuous flexible fin suspended at and extending below the surface of a body of water, supported by short floats formed of highly fire resistant material, such as foamed aluminum blocks, spaced apart longitudinally along the upper edge of the fin to permit flexing and accordion folding of the structure and to provide flexible articulating movement with surface waves. All parts of each spill control boom are formed of highly fire resistant and high strength materials substantially impervious to impact, bending or snagging between vessels, docks, pilings and similar structures, and optionally incorporating resilient fender materials protecting the hulls of adjacent vessels and barges as well as docks and pilings from impact damage, while retaining spilled petroleum products or other floating materials discharged from a dockside vessel in the immediate vicinity of the spill, and preventing the movement of such spilled material under docks and other structures.

Figure 106 shows this boom design in place adjacent to a pier and also shows a sectional view of the boom as well as a detail showing the boom flexing in use. The high strength, fire resistant booms **10** each incorporate a continuous, flat, relatively flexible fin **11** forming a fence or curtain positioned vertically with a major portion of its vertical height submerged under water, and a small upper portion of its height retained above water. The fin is supported by a series of floats **12** secured along its upper exposed edge. Both the fin and the floats are formed of highly fire resistant materials, preferably capable of withstanding temperatures of up to 1000°F for periods as long as one hour, in order to provide ample time for the spreading of foam or the release of carbon dioxide or similar fire suffocating agents to cover any flammable floating gasoline, petroleum products or the like spilled adjacent to the spill control boom.

Fin **11** is preferably fabricated of a knitted wire mesh centrally enclosed between outer layers of flexible elastomer materials having high heat-resisting characteristics. Various polyamide materials having high heat resistant properties may also be employed to form each outer layer supporting embedded in one face thereof, or encapsulating and enclosing therebetween, the flexible knitted metallic wire mesh. Other alternative materials from which the continuous flexible fin may be formed include Thermanol woven carbon filament fabric, asbestos fabrics bonded with neoprene layers, and glass fiber fabrics similarly bonded with neoprene layers. However, neoprene laminates have exhibited some tendency to swell and delaminate from supported fabrics unless laminated edges are thoroughly sealed during fabrication.

The floats may be formed in any convenient length **L**, and float lengths between 12 and 18 inches have been found to provide excellent conformity with vertical crests and troughs of surface waves, allowing the boom to flex vertically as well as laterally while performing its spill control functions without interruption, as shown. The continuous fin having a core of knitted metallic mesh is well adapted to permit such vertical and lateral flexing at points between adjacent floats, where the boom easily adjusts itself to differences in the vertical height of the water surface formed by advancing troughs and crests of surface waves.

The maximum distance **D** between floats **12** is preferably two-thirds or less of the normal diameter of piles or supporting vertical members beside which boom **10** is normally deployed to minimize the risks of snagging the boom between adjacent floats on such pilings or similar structures. As a further precaution against snagging the floats on such pilings, the floats may be provided with tapered ends formed by sharply raked end surfaces of the

FIGURE 106: SMITH FLOATING BOOM DESIGN

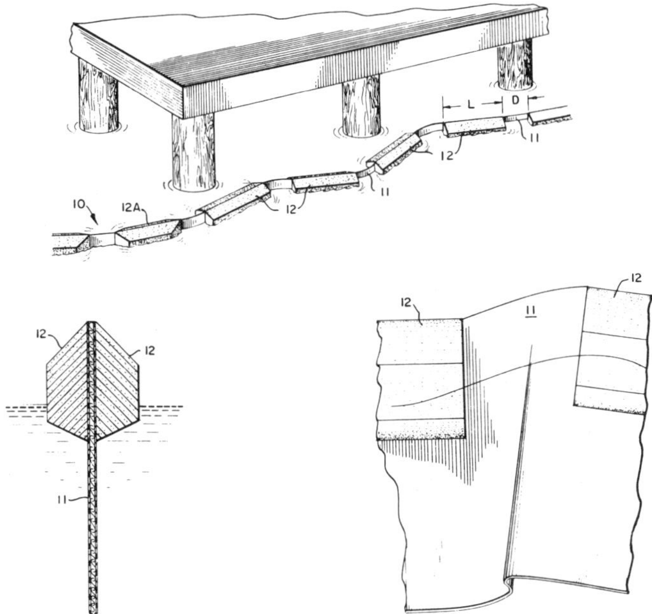

Source: M.F. Smith; U.S. Patent 3,638,430; February 1, 1972

float blocks as shown on floats **12A** illustrated at the left end of the boom. Floats **12** are preferably made of highly temperature resistant material, such as foamed aluminum blocks having a closed cell foam structure, and are secured to the fin by any convenient means such as straps, rivets, bolts, adhesives or the like. In place of foamed aluminum blocks, blocks of other foamed metals, ceramics and high temperature resistant compounds may be employed as floats.

Smith-Blair Design

A design developed by *M.F. Smith and R.M. Blair; U.S. Patent 3,563,036; February 16, 1971* provides accordion-folding floating booms for confining spilled oil or other floating material incorporating a thin continuous flexible fin positioned vertically and provided with numerous, short, inflatable balloon-like float pockets mounted along its upper edge. The float pockets are all deflatable and collapsible for compact accordion folded stowage of the boom in limited volumes of space for storage, shipment and delivery to the site by water transport or by airdrop. Automatic inflation of succssive inflatable float pockets upon unfolding deployment of the booms is achieved by individual pressure sources actuated by the deployment process, providing inflation pressure to produce fully inflated expansion of the float pockets for buoyant flotation of the boom structure.

Compressed gas charge cylinders triggered by unfolding of the boom supply the desired inflation pressure. Alternatively, chemical reactants enclosed in adjacent enclosures are mixed together upon unfolding deployment of the boom to produce sufficient amounts of gaseous reaction product to provide inflation pressures required for each buoyant balloon-like float pocket.

The concept of the Smith-Blair design is somewhat similar to that of the Heartness design described earlier; the reader is invited to refer to that earlier text for the concept of successive inflation of floats by chemical means when a boom is deployed.

Spandau Design

A design developed by *H.D. Spandau; U.S. Patent 3,641,771; February 15, 1972* is one in which a series of gas inflated bodies are secured end-to-end to form a buoyant toroidal barrier which confines oil floating within the area circumscribed by the barrier. Each of the bodies includes a coupling means which secures adjacent bodies together in a leakproof seal and also acts as a harness structure for anchoring the composite toroidal body in place. Water and/or oil may be placed inside the inflated bodies to act as ballast and to provide storage for oil.

Stamford Australia Pty. Ltd. Design

A design developed by *P.Y. Williams, K.C. Williams and C.E. Heath; U.S. Patent 3,744,253; July 10, 1973; assigned to Stamford Australia Pty. Ltd., Australia* is intended to restrain the passage past the boom of oil spilt onto water. The boom comprises a plurality of alignable interconnectable floats, a continuous screen barrier extending through, between, above and below the floats with the part below the floats formed of two walls spread apart at the bottom to provide an open bottom water ballast chamber, props and stays to support the screen barrier erected.

Figure 107 is a perspective view of a boom in operative position, the leading float being partly in section, and the tow being omitted. Each float **1** is divided longitudinally into two equal parts **2** and **3** each incorporating buoyancy chambers **4** and on the inner edge coamings, whereby the two parts are joined together by rivets after the screen barrier is inserted between the parts. The screen barrier, an impervious flexible sheet, comprises a top fin **7** and two integral bottom walls **8** and **9**. It is secured between the float parts **2** and **3** juxtaposed the bottom wall **8** and **9** by rivets. The fin and the bottom walls are held erected by wire stays **20** at the respective ends of the float. The stays are secured in a manner whereby they can be released to permit the fin and the walls to be collapsed onto the float for stowage or transport. The ends of the fin and the walls project beyond the float a sufficient distance to permit maneuverability of a boom.

FIGURE 107: STAMFORD AUSTRALIA PTY. LTD. FLOATING BOOM DESIGN

Source: P.Y.Williams, K.C. Williams and C.E. Heath; U.S. Patent 3,744,253; July 10, 1973

When the boom is subjected to the forces of wind and/or wave and/or current particularly athwart the structure, the open bottom ballast chambers formed by the several walls **8** and **9** and the water therein constitutes a restraining force against any tendency of the boom to heel or lift. In each float assembly the relative angular relationship between the walls and the underface of the float is such that heeling action, due to relative movement of water, acting on the walls creates counteracting forces to the float buoyancy to maintain the boom stable. It is preferred to space the walls apart at 60 degrees.

Surface Separator Systems Design

Designs known as Warner Booms, Type F and T are made by Surface Separator Systems of Baltimore, Maryland, according to Maritime Administration Report COM-74-10212, Springfield, Va., National Technical Information Service (November 1973). This boom has a 16 inch diameter air cylinder, neoprene sealed and a 22 inch neoprene skirt hanging therefrom and weighted at the bottom.

Susquehanna Corporation Design

A design developed by *T.E. Sladek, J.E. Palmer and M.F. Steele; U.S. Patent 3,638,429; February 1, 1972; assigned to The Susquehanna Corporation* is a floating barrier having upper and lower pivotally connected sections. The upper section is buoyant and thus seeks the water surface, and the lower section is of neutral buoyancy so as to provide ballast and to remain below the surface of the water. A restraint strap is connected to the upper and lower sections to limit the relative movement therebetween. In operation, when there is an upward movement of the water surface, the upper section pivots upwardly to seek the water surface and the lower section remains below the water surface, thereby preventing the floating material from moving over or under the barrier. The restraint strap limits the movement of the upper section from a substantially horizontal position adjacent the lower section to a substantially vertical position when subjected to wave motion.

Figure 108 shows the barrier in use when subjected to wave action. When the barrier segment **12** is subjected to current or wind forces in the direction shown by the arrows, the upper section **16**, because of its buoyancy, pivots upwardly about hinge **20** to follow the rising level of the floating material **F** and separates from the lower section **18** which remains in its normal position below the water surface because of its neutral buoyancy.

FIGURE 108: SUSQUEHANNA CORPORATION FLOATING BOOM DESIGN

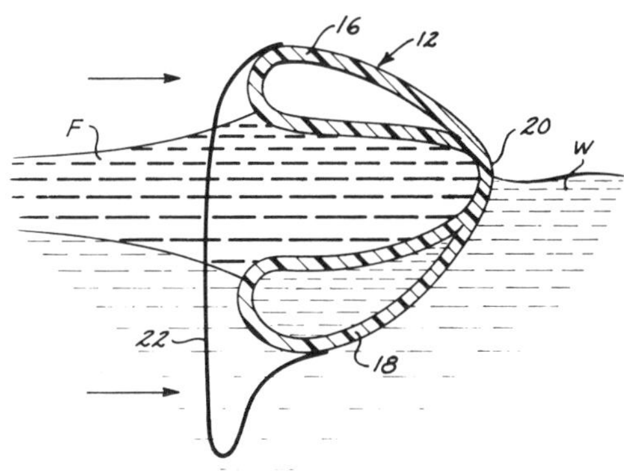

Source: T.E. Sladek, J.E. Palmer and M.F. Steele; U.S. Patent 3,638,429; February 1, 1972

In view of the spacing between the upper and lower sections, they are able to contain and trap therebetween the increased accumulation of floating material caused by the wind or current forces. It will be seen, therefore, that this floating barrier segment readily and automatically adjusts to wind and current forces to prevent the escape of any floating material over or under it. The strap **22** acts to prevent excessive movement of the upper section.

Trelleborg Rubber Company, Inc. Design

A design known as the Red Eel Boom, available from Trelleborg Rubber Company, Inc., New Rochelle, New York, can be assembled on site during oil spill incidents. Sections ready for assembly include a 164 foot length of 23½ inch wide PVC (two ply), shaped cellular plastic floats, plastic bags for sinkers, seven battens for each 164 foot section and metal hank couplings to attach each section. Price for each 164 foot section is quoted at $426.40, or $2.60 per foot (1970 prices).

A Universal Boom design available from Trelleborg Rubber Company, Inc. is cited in the Maritime Administration Report COM-74-10212, Springfield, Va., National Technical Information Service (November 1973). It has 14 inch freeboard and 27 inch draft, and is said to be for use in heavy sea and wind.

Uniroyal, Inc. Design

A design developed by *W. Juodis and G.J. Gauch; U.S. Patent 3,685,297; August 22, 1972; assigned to Uniroyal, Inc.* is one in which detachably connected floats form a barrier on the surface of a liquid for confining materials, i.e., pollutants floating thereon. A plurality of adjacent floats are longitudinally arranged in an end-to-end relation. The end wall of at least one of the floats has a movable portion which is biased in a longitudinal direction toward the corresponding end wall of the next adjacent float so as to form a seal between adjacent floats. Each of the opposed end walls includes a portion adjacent the movable portion thereof for operatively securing together the floats, and for preventing the movable portions from moving apart further than a predetermined distance so as to maintain the movable portions in sealing engagement.

Figure 109 shows a longitudinal elevation and a sectional cross elevation of such a design. Each float **10** has a cylindrical body portion **10a** which supports the system on water surface **19**. Cable **20** joins the float and provides considerable tensile strength to assist the system in resisting the forces generated by the waves of a rough sea. Cable **24** supports a vertically downwardly extending sheet **26**. The upper marginal edge portion **26a** of the sheet is interposed between the body portion of the float and the cable. In response to inflating the float, the exterior surface presses against the upper marginal edge portion of the sheet, in turn pressing the latter against the cable **24** so as to squeeze the upper marginal edge portion of the sheet in a liquidtight manner between the cable and the exterior surface of the float member.

The upper marginal edge portion of the sheet is folded over the cable and is cemented or otherwise suitably connected against itself to form a sleeve or overlap of about 3 inches in width through which the cable extends. The sheet extends in a vertical upward direction between adjacent floats, extending through and sealed within the annular region thereof, thereby preventing pollutants from passing between adjacent floats. The lower marginal edge portion of the sheet is folded over a bottom cable **28** and is cemented against itself to form a sleeve or overlapped portion **26c** of about 3 inches in width through which the bottom cable extends. Preferably, the bottom cable is made of steel and is about 1 inch in diameter. The overlapped portion may extend along the entire length of the sheet and may have a plurality of spaced holes extending completely therethrough for bolting ballast weights **30** thereto. The sheet is thus held in a substantially taut condition by the ballast weights so as to extend vertically downwardly into the body of liquid.

The sheet is made up of individual sheet sections **26d** and **26e**, each associated with eleven float members, i.e., extending along approximately the length of eleven floats. The length

FIGURE 109: UNIROYAL FLOATING BOOM DESIGN

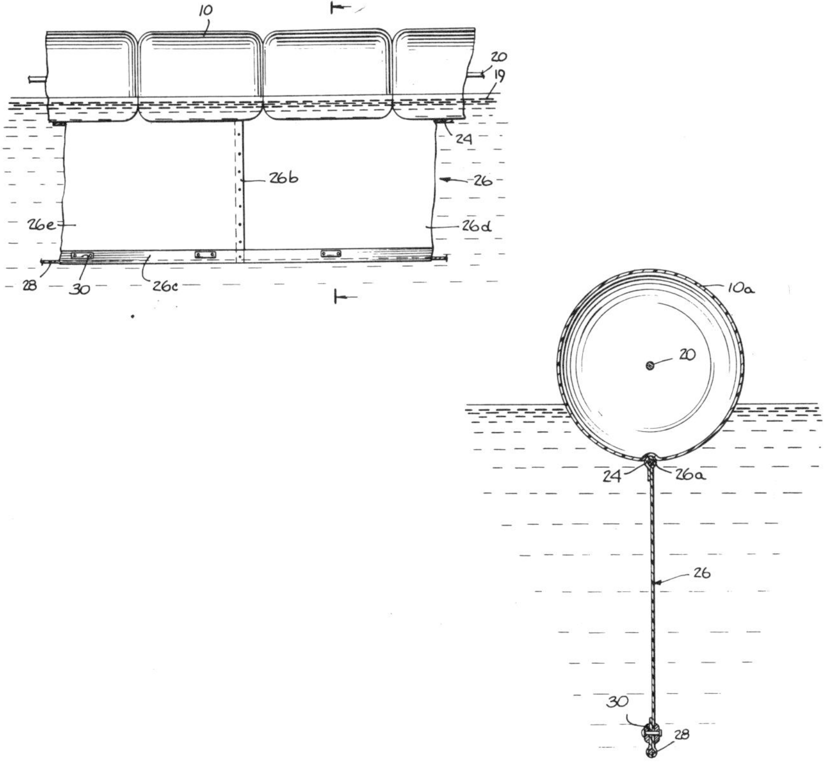

Source: W. Juodis and G.J. Gauch; U.S. Patent 3,685,297; August 22, 1972

of the individual sheet sections may vary for ease of manufacturing and handling. Each of such sheet sections has opposed vertical marginal edge portions **26b** preferably overlapped and fastened together in the region spaced from the interface between adjacent float members, i.e., about 18 inches therefrom. This vertical joint, formed by overlapping the vertical marginal edge portions, occurs at about every eleventh float member and forms a substantially fluidtight seal. The sheet located at the underside of the float members and supported thereby is made up of a series of flexible waterproof panels which may be of plastic coated or rubber coated fabric. Enough of these panels are used so that they extend the entire length of the apparatus. The vertical height of the sheet is chosen to provide an effective barrier to pollutants floating on the surface of the liquid **19**. A plurality of ballast weights secured to the overlapped lower marginal edge portion **26c** of the sheet which maintain the sheet in a relatively stable vertical downward direction.

In order to tow the apparatus through the liquid, a bridle may be fastened to the accessible extremities of one of the end float members, the corresponding end of the lower marginal edge portion of the sheet, and the corresponding end of the bottom cable **28**. Similarly, to keep the apparatus in a predetermined location at the site of the pollution, an anchoring device may be fastened to the bottom cable. The float apparatus may thus be assembled and inflated at a convenient location and subsequently towed in its assembled and inflated condition to the site of the pollution where it may be anchored in place.

U.S. Coast Guard Design

The reader is referred to the earlier discussion under the Johns-Manville heading where the "USCG High Seas Boom" is described.

Versatech Corp. Designs

A design developed by *J.E. Manuel; U.S. Patent 3,686,869; August 29, 1972; assigned to Versatech Corporation* comprises an elongated sheet of flexible material and a plurality of resiliently collapsible air chambers secured to the sheet and arranged in two series each extending along a different side edge portion thereof, the arrangement being such that the barrier can be flattened and wound on a reel for storage and transport and unwound from the reel and deployed on the surface of a body of water when in use. When the barrier is deployed, the inflated air chambers afford buoyant support and the sheet extends in trough-like fashion to define a water chamber. Figure 110 shows transverse sectional views of the device collapsed for transport (above) and deployed on a water surface (below).

FIGURE 110: VERSATECH CORP. FLOATING BOOM DESIGN

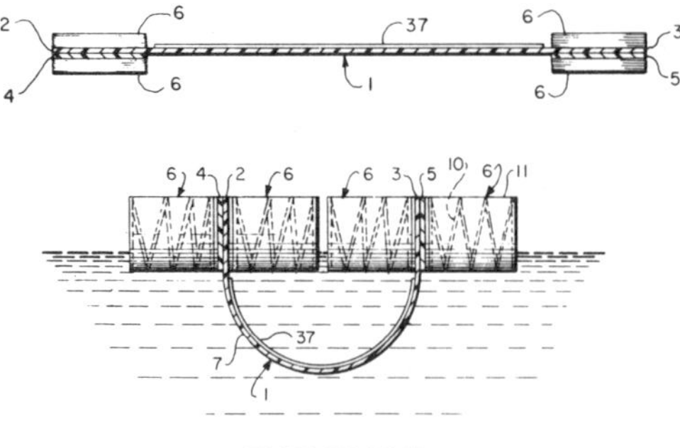

Source: J.E. Manuel; U.S. Patent 3,686,869; August 29, 1972

The barrier comprises an elongated flexible water impervious sheet **1** having parallel side edge portions **2** and **3** to which reinforcing strips **4** and **5**, respectively, are secured throughout the length of the sheet. A plurality of resiliently collapsible air chambers indicated generally at **6** are provided, air chambers **6** being arranged in pairs with each pair being secured to one of the edge portions **2** and **3**. The air chambers **6** of each pair carried by edge portion **2** of sheet **1** are opposed to each other, one end of one air chamber **6** being secured directly to edge portion **2** and projecting therefrom, the like end of the other air chamber being secured directly to strip **4** and projecting therefrom, and the pairs of air chambers being spaced along edge portion **2** throughout essentially the entire length of the edge portion so that edge portion **2** thus carries two series of the air chambers. The air chambers **6** carried by edge portion **3** are arranged in the same fashion just described for those carried by edge portion **2**.

The main body of sheet **1** extends as a greatly elongated rectangle. The combination of sheet **1** and air chambers **6** can be flattened and wound on a reel, the winding operation causing the air chambers **6** to collapse to the condition shown so that a relatively great length of the barrier can be wound on a single reel. When the barrier is unwound from the reel for deployment onto the body of water, air chambers **6** expand under the influence of springs **10** to inflated condition as soon as the restraint resulting from the wound configuration is removed.

The transverse width of sheet **1** between edge portions **2** and **3** is substantial. Accordingly, when the barrier has been deployed, the main body portion of the sheet can depend in trough-like fashion from the inflated air chambers so that body portion defines a water chamber **7** for stabilizing the buoy. Water chamber **7** can be filled by pumping water into the trough as the buoy is deployed or simply by immersing the barrier briefly in the body of water during deployment, such immersion being accomplished, for example, by running the barrier under an inflated buoyant roller disposed on the surface of the body of water and carried by the deploying vessel on which the reel is mounted.

The flat springs **37** extend transversely of sheet **1** and serve to assure that the main body portion of the sheet will assume the arcuate configuration seen in Figure 110 when the barrier is deployed. Sheet **1** and walls **11** can be fabricated from various types of commercially available fluid impervious sheet materials including extruded polymeric materials and composite fabrics. Extruded polyvinyl chloride sheet with a thickness on the order of 0.006 to 0.01 inch is particularly useful.

Another Versatech Corporation design of collapsible nature known as the Aqua Fence is described in EPA, Edison, N.J. *Oil Spill Containment Systems* (January 1973). A conceptual scale model of the Fence reportedly has been constructed and tested satisfactorily in simulated wind and sea conditions. The Aqua Fence system is conceived as a helicopter transportable package which can be flown to the site of an oil slick. The package will consist of the machinery and sufficient raw materials for fabricating a continuous Aqua Fence of optimum length. Additional raw materials can be stationed at designated supply points. The operation of fabricating Aqua Fence will consist of automatically feeding two rolls of plastic sheet through forming machinery which folds and joins the sheets into a continuous row of "pockets". Flotation will be provided by inflating the pockets with air during the sealing operation.

Warne Designs

As described by the EPA, Edison, N.J. Report in *Oil Spill Containment Systems* (January 1973), Warne Booms are manufactured by William Warne and Co., Ltd., England, and distributed in the United States by Surface Separator Systems, Inc. of Baltimore, Maryland. The Warne Booms are heavy duty booms made of fabric reinforced synthetic rubber and consisting of an inflatable or plastic filled flotation tube, fabric skirt, and pocket at the bottom edge of the skirt for enclosing chain ballast. Four types of Warne Booms are currently available as shown in Figure 111.

FIGURE 111: WARNE FLOATING BOOM DESIGNS

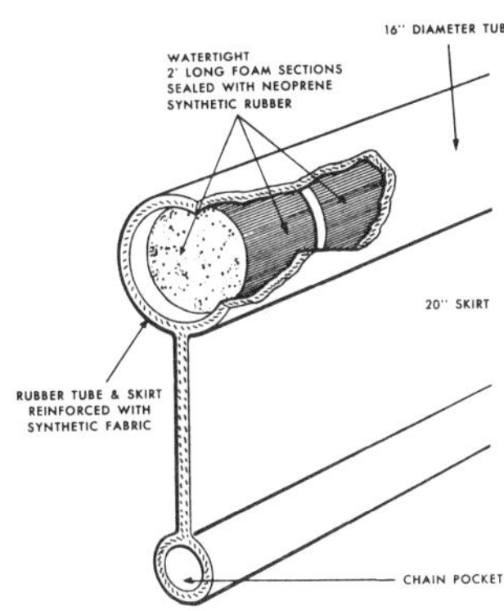

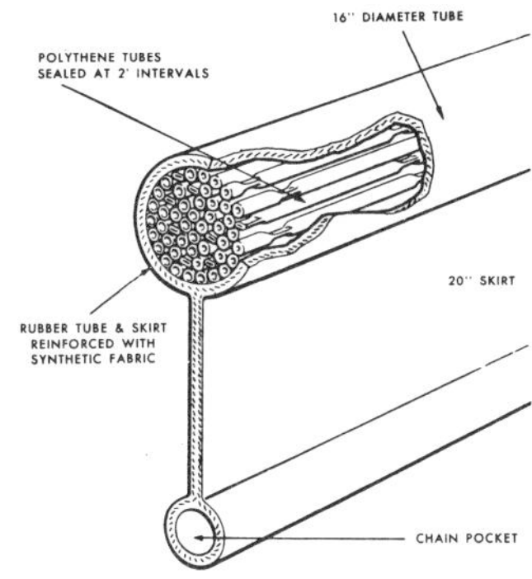

Foam Filled Boom (Type F)

Polythene Tube Filled Boom (Type T)

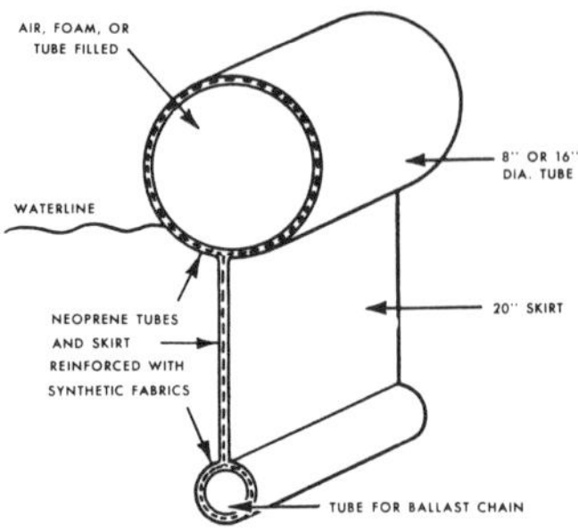

Inflatable Oil Spillage Boom (Type Y)
Rising and Sinking Boom (Type Z)

Source: EPA, Edison, N.J. Report *Oil Spill Containment System* (January 1973)

The rising and sinking boom (Type Z) is primarily intended for permanent installation at off load and on load terminal facilities and cutout berths where there is heavy vessel traffic. This boom has an 8 inch diameter inflatable chamber, 20 inch skirt and 3 inch tube at the bottom of the curtain for containing chain ballast. Standard sections are 25 feet and 50 feet long, and adjacent sections are connected by pin hinge joints. A permanent air supply is required for this installation. An air line runs the full length of the boom and release valves are incorporated into each boom section to avoid over pressurizing. Although the main line air pressure is maintained around 40 psi, the pressure within the flotation chamber approximates only 1.5 psi. Prices quoted in 1969 for the Type Z boom were $440 for the 25 foot length and $750 for the 50 foot length, (i.e., $15.00 to $17.40 per foot).

The inflatable oil spillage boom (Type Y) is similar to the rising and sinking boom. Whereas this boom is available with either an 8 inch or 16 inch flotation chamber, the skirt, the 3 inch weighing tube and other appurtenances and connections are identical to those employed for the Type Z boom. The inflatable barriers may be rolled up when not in use, or the hinged joints permit the boom to be folded concertina-wise in an enclosed area of the harbor away from vessel traffic. Spring loaded nonreturn air valves connecting adjoining sections of the boom ensure that the entire length of barrier will not lose buoyancy if one section fails; yet the same air system permits the complete boom to be inflated from a common point.

The inflatable Warne booms have been used for controlling the spread of petroleum fires, and in some cases, water hoses have been added across the top of the boom making it possible to spray a water curtain over the entire length of the barrier. Recent price quotations for the Type Y Warne boom are given as follows: 8 inch float, 25 foot section, $445; 8 inch float, 50 foot section, $755; 16 inch float, 25 foot section, $585; 16 inch float, 50 foot section, $940. The above price range is $15.10 to $23.40 per foot.

The polythene tube filled boom (Type T) is different from the inflatable booms in that the flotation chamber is tightly filled with 2 inch diameter polythene tubes which are sealed or crimped at 2 foot intervals and at the ends of the boom section. The Type T boom is intended for use in strong tidal conditions and rough waters. The Type T boom is similar in outer configuration but reportedly more flexible and durable than the inflatable barriers. The polythene tube filled boom, constructed of synthetic rubber and reinforced with plastic fabric, has a 20 inch weighted skirt and is available with either 8 inch or 16 inch flotation chambers, and in standard lengths of 25 and 50 feet. Because of the nature of the Type T boom, it is less vulnerable to failure than inflatable booms used under the same conditions. This boom is relatively expensive and prices for East Coast, U.S. delivery are as follows: 8 inch float, 25 foot section, $595; 8 inch float, 50 foot section, $975; 16 inch float, 25 foot section, $900; 16 inch float, 50 foot section, $1,525. The above price range is $19.50 to $36.00 per foot.

Except for the difference in the type of flotation chamber, the neoprene coated, plastic foam filled boom (Type F) is essentially the same including its application potential as the Type F boom. Prices for this boom are in the same general area as the Type T barrier. The Warne booms are relatively heavy and their unfit weights are in the range of 6.0 to 8.5 pounds per foot. The majority of weight is centered in the ⅝ inch chain ballast inserted in the bottom of the curtain of most booms. This chain has a unit weight of 4.3 pounds per foot.

The company also specifies that any type of boom must receive proper and continuous maintenance to operate satisfactorily over long periods and for its permanent installation the company recommends weekly inspection and servicing if necessary. If an oil boom is held within the water for considerable time, maintenance and routine cleaning of the boom become very important. Without cleaning it has been shown that accumulated marine growths can greatly limit the effectiveness of a boom besides causing serious damage to the boom itself.

Washburn Design

A design developed by *F.R. Washburn; U.S. Patent 3,673,804; July 4, 1972* provides a portable, flexible, floating firewall having a rectangular galvanized metal body and four essentially identical flotation packets, two disposed on each side of the metal body. The flotation packets are formed of Styrofoam and encased in a plastic container which is covered with a layer of asbestos cloth on all exposed portions of the container. A plurality of steel bands secure the flotation packet to the body and an outwardly projecting rigid shelf is mounted immediately above the flotation packet to hold the flotation packet in place. Two horizontally directed cables are anchored near each end of the body on both sides thereof. The cables are adjustable in length in order to control the degree of bending which will be allowed the body. Ballast weights removably connected by cables to the bottom edges of the body continuously urge the body in the upright position in the water. Figure 112 shows the design in longitudinal elevation and in somewhat enlarged transverse section.

The side of the boom which will be exposed to heat from a fire has an insulation on it consisting of Johns-Manville Permatone "S" Flex beard. There are two flotation packets **2** on each side of a unit, one side being encased with asbestos cloth and the other left uncovered as far as fireproofing is concerned. Each unit is held upright in the water via ballast blocks **11** and **10** of either the 20 lb or 10 lb weight, respectively. Ballast cables **12** and reinforcement cables **8** are preferably aircraft cables. Ballast cables are a minimum one eighth inch, whereas those for the reinforcement vary from one eighth to three eighths inch in accordance with the demand of the purchaser. Each flotation packet is mounted beneath a rigid shelf and held in place by three quarter inch steel bands **18** bolted to the unit in such a manner that the same bolt will extend to the flotation packet on the opposite side of the unit as well.

The rigid shelf **4** is secured to the main body by an L shaped bracket **3**. Each unit is equipped with handles **6** on both sides of the unit and on both ends. Each unit has "S" bolts **9** on the lower edge to which the ends of the ballast cables are connected via snap hooks.

Directly beneath the shelf **4** is a piece of the Permatone "S" fireproofing **5** to protect the piece of ⅜ inch plywood against which the flotation packet will press in order to maintain a maximum lift without placing any strain on the steel bands **18** which are used merely to retain the packet closely against the unit. The main body **1** of each unit consists of a galvanized 18 gauge metal body measuring 3 x 10 feet. Each flotation packet is 6 x 15 x 24 inches and made of a block shaped member of Styrofoam **21** which is first encased in a plastic cover **20**, then covered by a waterproof ducking (not shown), then covered in asbestos cloth **19** on the fireproof side. No asbestos is necessary on the side not to be fireproofed.

FIGURE 112: WASHBURN FLOATING BOOM DESIGN

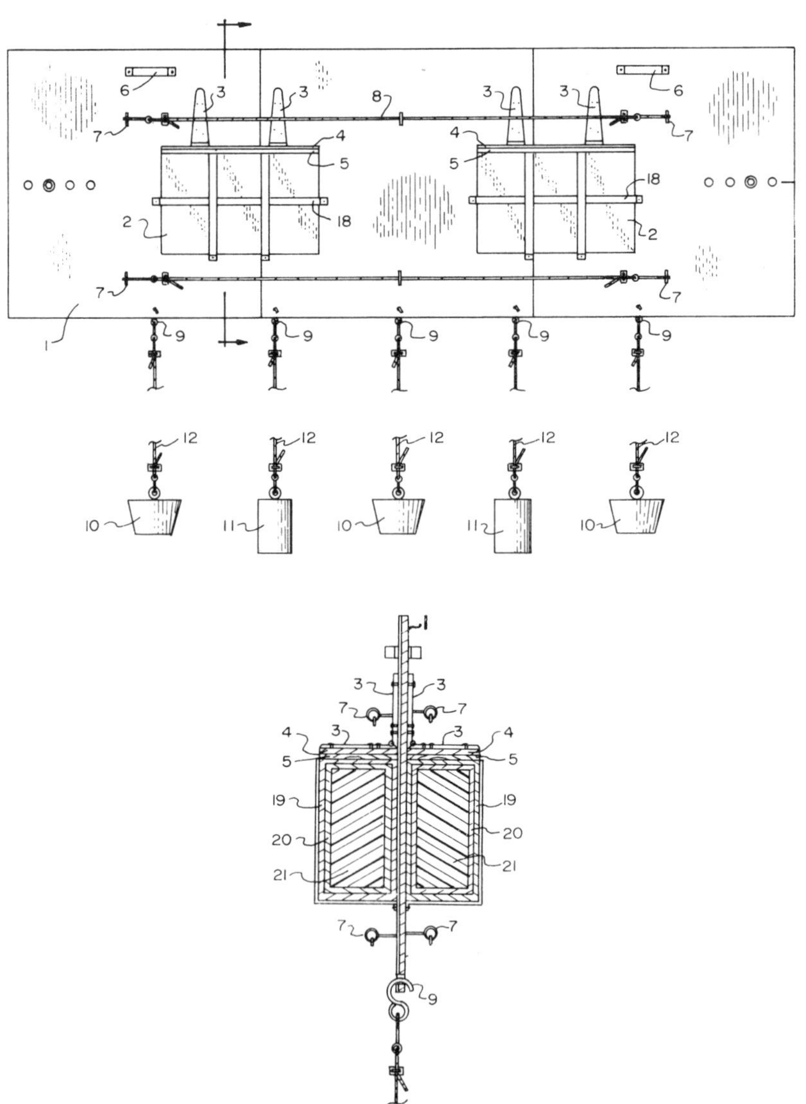

Source: F.R. Washburn; U.S. Patent 3,673,804; July 4, 1972

Water Pollution Control Inc. Design

According to the EPA, Edison, N.J. Report *Oil Spill Containment Systems* (January 1973), Water Pollution Control Inc. of Fort Lee, N.J. has developed a 48 inch diameter inflatable boom device which is stored as a thin plastic "ribbon", attached on a reel, and designed for compact stowage on ships, helicopters and drilling rigs. The boom can be retrieved after use and eliminates the need for skirt attachment.

Deployment around the spill periphery is made from a helicopter or small craft using a reel to unload an evacuated ribbon from an anchored and marked point of beginning. The purpose of stringing the material flat is to make possible the accomodation of several thousand feet of containment equipment on a spool of reasonable diameter and weight. After making turns for a closure back to the original point (around buoyant floats), inflation of the evacuated tube begins at a single point into a valve.

Inflation of the ribbon is accomplished by use of a small pump discharging through a venturi, taking sea water into its suction, and air into the venturi inlet. Gauges provide the unit volume ratio (air to water) desired to meet the sea condition and determine what portion of the tube will serve as the underflow skirt or surface trap. This design eliminates the need for a skirt attachment. Inflation of 4,000 linear feet of 48 inch diameter tubing reportedly can be accomplished in approximately 30 minutes with a pump delivering a one-to-one air to water ratio through the venturi. The tubing can be modular and therefore is capable of being expanded or contracted as need requires. After the oil has been removed and the need for containment is over, the ribbon can be retrieved.

Wilson Industries Design

According to Maritime Administration report COM-74-10212, Springfield, Virginia, Nat. Tech. Information Service (November 1973), a "Low Tension Barrier" is offered by Wilson Industries of Houston, Texas. It was developed under U.S. Coast Guard contract as a high seas barrier. It is a skirt type barrier with a complex tension line arrangement consisting of a main tension member and multiple bridles to a barrier skirt. It is designed to operate in 5 foot waves, 20 knot winds and 2 knot currents.

Worthington Corporation Design

The 6-12 Boom, available from the Pioneer Products Division, Worthington Corporation, Livingston, New Jersey, is the same boom that is offered by Metropolitan Petroleum Co. In addition, Worthington also has available the "Acorn" barrier, which has a flotation chamber of approximately 16 inches and a skirt of approximately 15 inches. This barrier, which is claimed to be applicable for open sea conditions, weighs 4.4 pounds per foot.

Zaugg Design

A design developed by *R.E. Zaugg; U.S. Patent 3,613,377; October 19, 1971* comprises multiple chambers joined to form, when floating, a triangular cluster of flexible bag-like tubes including a ballast chamber partly filled with water and virtually immersed beneath the other chambers so that it lies mostly submerged, a main chamber partially inflated with air and rising above the water surface like a large continuous pillow and a support chamber more firmly inflated with air and lying above the top of the ballast chamber and behind the main air chamber to help support the latter when the wind tries to beat it flat on the water surface. The design includes a means for manipulating the ends of the chambers including inflating and sealing them and means for storing and reeling out the barrier to whatever length of it is required under actual working conditions.

This will provide a buoyant barrier which is continuous when deployed and is highly flexible under all conditions of use so that it can faithfully follow the surface contour of a liquid, especially among wave troughs and crests so that none of the surface borne floating material can escape the barrier's confinement by passing either over it or under it.

COMBINED CONTAINMENT AND REMOVAL DEVICES

Air Preheater Company Inc. Design

A design developed by *L. Bakker; U.S. Patent 3,628,665; December 21, 1971; assigned to The Air Preheater Company Inc.* involves an apparatus for restraining an oil slick floating freely on the surface of a body of water whereby it is at all times under control and not free to spread with movement of water to contaminate the surrounding areas. The oil slick controlling apparatus contains oil-water separation means whereby oil may be removed from the oil slick for further utilization.

The present device comprises essentially an oil-water separator which is included within a containment ring in the form of an annular boom floating on the surface of the water and surrounding a continuing source of contaminating oil. The oil-water separator may be made to occupy only one or several of many articulated sections which comprise the containment ring or annular boom around the oil slick. Thus the size and effectiveness of the apparatus may be varied to meet the existing need by the addition to or the deletion of one or more sections. Sections of the containment ring not used to house the oil-water separation means may contain storage tanks for oil as it is removed from the water.

Figure 113 shows the present device in operation (top view), shows the linkages between individual units (middle view) and shows the oil-water separator incorporated in some units (lower view). In the drawing an oil slick **10** is shown as floating on the surface of the body of water adjacent an offshore oil drilling platform **14**. Although the source of oil polluting the surrounding water is indicated as coming from a subsurface break in the well adjacent the drilling platform, a wrecked oil tanker, an industrial leakage of oil, or any other source of leaking oil would be an equally suitable source of contaminating oil for use of this containment device.

The containment device comprises essentially a multiplicity of individual tanks **16** in an end-to-end arrangement and pivotally linked together to form an endless containment ring extending around the oil slick. The exact linkage used to join adjacent tanks together is of little significance inasmuch as various types of pivotal linkage could be satisfactorily used, an example of which is shown in the middle view of the drawing wherein a pin **18** merely links the apertured extensions **22** of each segment together to form a pivoted joint.

Depending upon the size of the oil slick and the amount of oil to be separated from the water, a number of tanks of the containment ring are interspersed by tanks other than tanks **24** housing an oil-water separator. Each of the tanks housing an oil-water separator has an inlet duct **26** for the oil-water emulsion that may be deployed toward the greatest concentration of oil-water emulsion of the oil slick, an outlet **28** for water which has been separated from the oil, and an outlet port **32** for oil which has been separated from the water.

The oil which has been separated from the water may be pumped directly through port **32** to an independent collection tank, or it may be stored in one of the adjacent tanks **16** of the containment ring until such time as it may be collected and removed therefrom. Thus oil, after separation from the water, is exhausted through port **32** and connecting duct **36** to an adjacent storage tank **16** having suitable porting connections. Simultaneously, water separated from the oil is ejected through an outlet port **28** directly to the clean water outside the oil slick and separated therefrom by the containment ring.

In order that the oil-water separator will float upright in the water at all times, flotation tanks **42** are secured to each side of the segment housing an oil-water separator. Moreover, each segment housing an oil-water separator has an anchor **45** secured thereto by a cable **47** that provides stability to each oil-water separator and the intervening oil storage tanks **16**. If the containment apparatus is to be located in basically still or calm water areas, the anchors may be replaced by guy wires or lines to the shore or other stabilized areas. Electric power for the various pumps of an oil-water separation segment of the oil containment ring may be provided by any suitable source that is housed on the drilling platform **14**, in a section of the containment ring itself, or if available, brought in from an outside source. On occasion the source of oil is other than an offshore well and prevailing winds or currents force the oil slick emanating therefrom to float in a single direction downwind from its source. In such an instance the oil containment boom need be formed only in semicircular form on a single side of the oil source to receive the entire flow.

FIGURE 113: AIR PREHEATER CO. SPILL CONTAINMENT AND REMOVAL SYSTEM

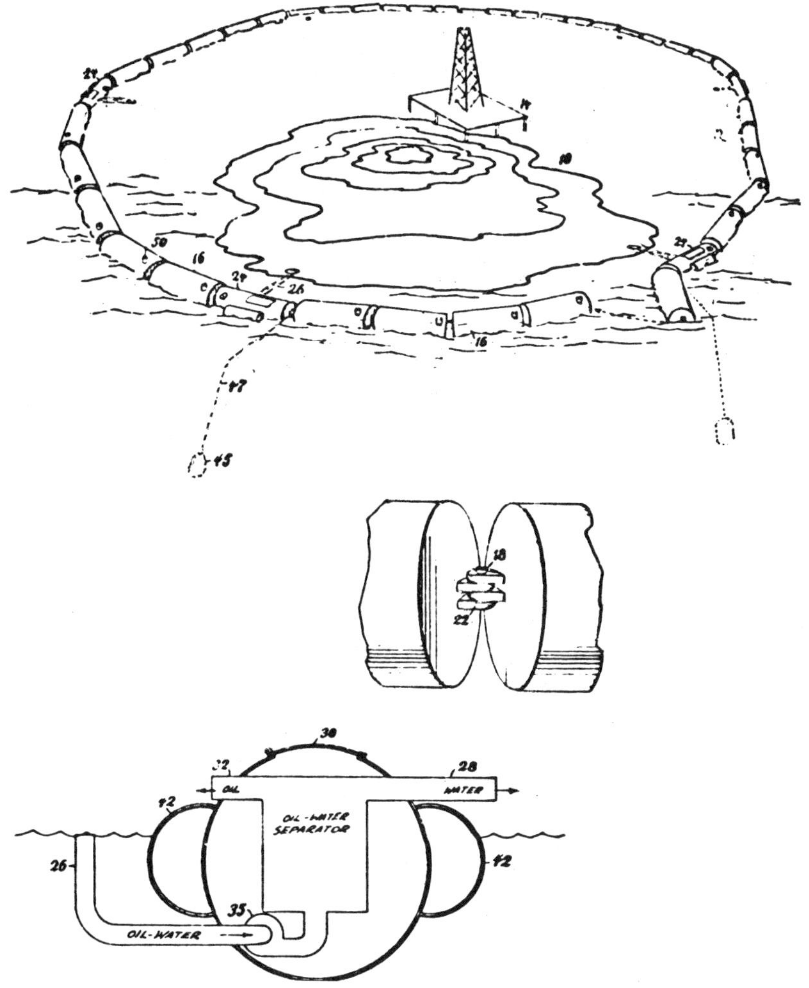

Source: L. Bakker; U.S. Patent 3,628,665; December 21, 1971

Selected tanks may be provided with a quick or easily actuated release **50** whereby they may be moved aside to provide an opening that permits a floating tank, a barge or other work ship to come into the area enclosed by the oil containment boom **16** in the manner shown in the Figure.

The tank sections may be assembled or linked together on shore or at a staging area and then towed into position by a suitable tugboat or other vessel. Otherwise the separate units may be delivered via shipboard and assembled at the point of use. Whether the segment tanks **16** that comprise the oil slick containment barrier are linked together so that they extend completely or partially around the source is dependent upon such variables as the amount of oil that is leaking into the water, the prevailing winds and currents, and the likelihood of contamination in nearby areas. For massive or continuous leakage an array of two or more concentric containment rings may be adapted to encircle the floating oil slick.

The tanks **24** in which the oil-water separators are housed are preferably formed of welded steel plate having an access door **30** on the top through which are supplied the various elements which comprise the oil-water in accordance with U.S. Patent 3,468,421. Each oil-water separator tank **24** is additionally provided with flotation tanks **42** and an anchor **45** held by anchor line **47** whereby the tank housing the oil-water separator is maintained relatively stable in all types of weather and sea conditions.

The tanks **16** linking the oil-water separator tanks **24** are elements of the oil containment boom that may be used as storage tanks for the oil removed from the oil slick or simply as links in the boom that encircles the oil slick. Connections **50** in each tank permit the supply or exhaust of oil to a convenient storage area. The tanks **16** may be of standard metallic plate construction, or they may be formed from rubberized fabric that enables them to be transported in deflated condition on a barge or the open deck of a ship, and upon inflation serving as links in the containment boom.

In operation the oil-water emulsion of the oil slick is drawn inward through inlet duct **26** by pump **35** to an oil-water separator of the type represented by U.S. Patent 3,468,421. The oil removed therefrom is then directed outward through duct **32** to a storage tank while substantially pure water is exhausted through exhaust port **28** to the space outside the containment boom.

Anti-Pollution Inc. Design

A design developed by *C.C. Cloutier; U.S. Patent 3,613,891; October 19, 1971; assigned to Anti-Pollution Inc.* involves a flexible boom which confines the liquid so that it can be removed by a scoop unit. The scoop unit has a plurality of paddles which cooperate with a bottom plate to enclose and seal off a portion of the oil slick so that the oil and water in the sealed off portion will separate in layers.

The bottom plate has a plurality of apertures which permit the water and a small amount of the oil in the sealed off portion to flow therethrough, the water flowing back into the body of water while the small amount of oil is trapped between the surface of the body of water and the bottom plate. This portion of trapped oil acts as a check valve to permit the water enclosed on subsequent passes of the paddles to pass through the holes in the bottom plate and to prevent the oil picked up on these subsequent passes from flowing through the apertures. After separation of the oil and water, the oil is carried by the paddles to a sump for removal to a storage area.

The boom, which is secured to the scoop unit, encircles the oil slick and can be winched inwardly so that the encircling area can be made smaller, thereby drawing the oil layer into the scoop unit. The boom has an inflatable portion to provide buoyancy and a depending weighted skirt which prevents wave action of the body of water from allowing any oil to pass under the boom and escape.

Figure 114 shows the boom in place with respect to the oil recovery device and also longitudinal and transverse sections of the boom employed. The numeral **10** generally indicates a liquid removing apparatus for removing undesired surface layers from bodies of water. Although the preferred example illustrated here shows the liquid removing apparatus **12** being towed by a barge **10**, it will be apparent that the apparatus and barge can be combined into a single unit with the liquid removing apparatus serving as the aft portion of the barge. The numeral **14** illustrates a scoop unit to which is secured a floating boom **16**, the boom being extensible to encompass the liquid layer which is to be removed. The boom is designed to guide the surface layer towards the scoop unit which is then activated to remove this material from the water and deposit it in a suitable sump from which it may be pumped into a suitable storage tank for recovery purposes.

FIGURE 114: ANTI-POLLUTION INC. SPILL CONTAINMENT AND REMOVAL SYSTEM

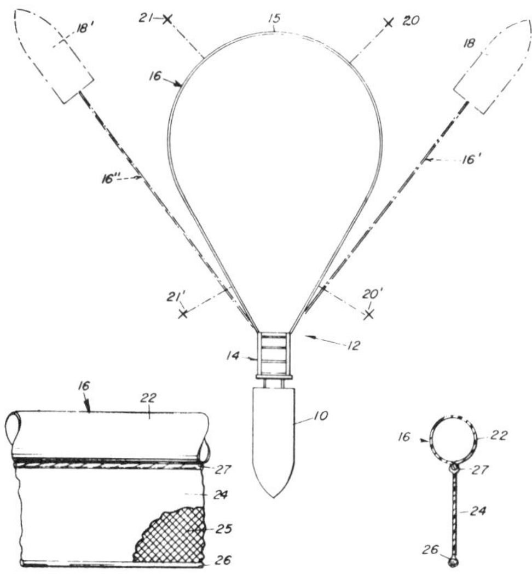

Source: C.C. Cloutier; U.S. Patent 3,613,891; October 19, 1971

There are several configurations and methods of utilizing the boom **16** to gather the surface layer, and where this layer is an oil slick the particular method used will generally depend on the type of source which is causing the oil slick or the area over which the oil slick is dispersed. For example, if the oil slick has not dispersed over too large an area and the cause of the oil spillage has been eliminated, the most appropriate way of using the boom may be to encircle the oil slick and confine it to a particular area. In this particular example, when the oil slick is so confined to a particular area, one end of the boom can be fixedly secured to one side of the scoop unit and the other end of the boom may be attached to a take-up or winching mechanism so that the boom may be moved inwardly, thereby causing the enclosed area to become smaller and forcing the oil into the scoop unit.

On the other hand, if the slick is dispersed over a wide area, it may be more practical to divide the boom **16** into two parts, opening the closed circle at, e.g., point **15**, and forming **16′** and **16″** each being fixed between one side of the scoop unit and corresponding pulling barges **18** and **18′** as shown in phantom. The pulling barges **18** can tow the scoop unit **14** and the barge **10** towards the slick so that the oil is forced into the scoop unit, or if the wind is blowing the oil slick towards barge **10**, pulling barges **18** and **18′** can be anchored, or the corresponding ends of the booms can be anchored allowing the wind to force the oil into the scoop unit.

The method of use wherein the boom encircles the oil slick might also be used where there is a continuous flow from an oil source such as a leaking oil well or a sinking oil tanker. The boom **16** would then confine the source to a relatively small area so that it would not dissipate. The boom can be anchored as shown in phantom at numerals **20, 20', 21** and **21'** thereby keeping the oil slick encircled and stationary until the source of spillage is eliminated. When using the boom in this anchored example, a constant flow will force the oil layer into the scoop unit for removal. In addition, it has been found that during operation of the unit, a current is produced by the unit itself which also helps to draw the oil into the unit.

In order to permit use of a winch apparatus for taking up the boom **16**, as mentioned above, the boom will preferably be made of a light, flexible material such as plastic which is substantially fire resistant and punctureproof and capable of being inflated and deflated. The boom can be similar to any of those disclosed in the prior art, but U.S. Patent 2,682,151 to Simpson et al shows a particularly suitable example. The boom as disclosed in Simpson has two portions, an inflatable and deflatable hose or tubular portion **22** and a depending skirt portion **24** integral therewith.

Preferably, the boom will be a unitary member with a steel mesh wire **25** formed integrally therewith to give it strength. A cable **26** may be carried at the lower edge of the depending skirt **24** and a cable **27** carried at the juncture between the hose portion **22** and the skirt portion **24**. These cables serve as weights to hold the skirt down and to give the boom longitudinal strength so that it may be towed anchored and rolled up on a winch mechanism. The boom can be made in sections, if desired, for easy handling and this would be a preferable example if the unit is used in the open sea, where the size of the unit and boom would have to be relatively larger than a unit which is to be used in calm water. The steel mesh embedded in the plastic of the boom is particularly necessary in a device designed for use in open or rough water, the mesh serving to distribute stresses caused by towing and anchoring the boom at spaced points. Also, the mesh protects the boom from being cut or otherwise damaged by contact with drilling rigs, ships, and the like, or by floating debris.

Bridgestone Tire Company Limited Design

A system developed by *T. Muramatsu, K. Aramaki and Y. Kondo; U.S. Patent 3,771,662 November 13, 1973; assigned to Bridgestone Tire Company Limited, Japan* involves sweeping oil films on the water surface into a mobile U-shaped oil fence line. An oil collecting zone is formed in the U-shaped oil fence line where all the oil films swept thereby are collected to form a comparatively thick oil film so that the oil is efficiently removed from that zone.

According to this scheme as shown in Figure 115, an oil recovery system **1** for recovering spilled oil **11** floating on a water surface is pulled or drawn by two boats **2** and **2a**. The oil recovery system includes a linear oil fence assembly having an elongated belt member **3** which is preferably flexible. Being drawn by the two boats, the belt member is bent into a U shape. Two end floats **1a** are provided at the opposite longitudinal ends of the elongated belt member to facilitate the connection of the oil recovery system to the boats, for instance, through ropes **1b**.

A machine boat **2b** is connected to the rear end of the oil recovery system relative to the moving direction of the boats. The machine boat carries various machines to actuate the oil recovery system, which include an air compressor **22** for inflating the tubular floats **5a** and **5b**, as shown in the sectional detail of the boom, for keeping the elongated belt member afloat, a section pump **18** for sucking the spilled oil through a suction hose **19**, an oil-water separator **20**, and an oil tank **21** for storing oil recovered from the water surface.

In order to collect the spilled oil at the closed end of the U-shaped belt member and to accumulate the oil in the form of a comparatively thick oil film, one or more bridging belt members are provided so as to bridge the two legs of the U-shaped flexible belt member. Two bridging belt members **4** and **4a** are used which are disposed substantially in parallel to the central portion of the elongated belt member connecting the two leg portions of the U shape. The construction of the bridging belt members is identical to that of the elongated belt member of the oil fence assembly.

A float means is secured to the belt member, which float means includes a pair of flexible and inflatable tubes **5a** and **5b** bonded to opposite surfaces of the belt by fastening straps **7**. The structure of the float means **5** is not restricted to such tubes **5a** and **5b**. Upon inflation of the tubes, the belt member floats. To keep the plane of the belt member substantially vertical across the water surface, a plurality of reinforcing bars **8**, each carrying a weight member at one end thereof, are fastened to or embedded in the belt member. Thus, that edge of the belt member wherein weight members are embedded sinks in water, and the tubular float means **5** and the weight members act to hold the width direction of the belt member substantially upright.

The belt member is preferably made of rubber or other water resistant elastomer material. The belt member and the flexible tubular float means **5** are required to be highly flexible in a direction perpendicular to the water surface, when afloat, so that the oil fence line formed thereby flexes in accordance with the profile of water surface, such as different wave profiles. Thereby the belt member acts to keep oil or other sea drifts on one side of the belt member even when the shape of the water surface profile varies.

For certain applications, it is preferable to form a stream of air bubbles directed from the belt member to the inside of the water surface area surrounded by the U-shaped oil fence line. Such air bubble flow may accelerate the accumulation of the oil particles into a thick film. An air tube **9** having a plurality of air nozzles **10** is secured to the belt member along its weighted edge. A suitable air bubble guide **9a** is mounted on the belt member so as to direct the air bubbles in the aforesaid desired direction, at a position immediately above the air nozzles of the air tube. To facilitate the drawing or traction of the oil recovery system **1**, a tension member **24**, e.g., a chain, may be hung from the lower edge of the flexible belt member **3**.

FIGURE 115: BRIDGESTONE TIRE COMPANY SPILL CONTAINMENT AND REMOVAL SYSTEM

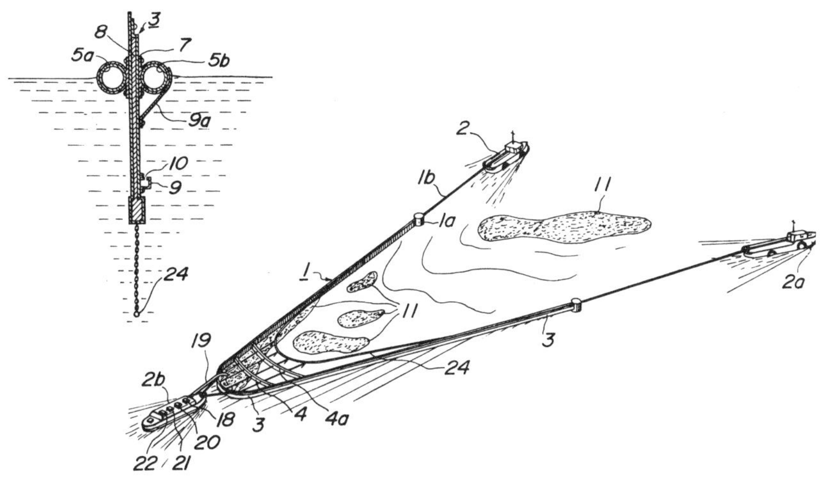

Source: T. Muramatsu, K. Aramaki and Y. Kondo; U.S. Patent 3,771,662; Nov. 13, 1973

The British Petroleum Company Limited and Gordon Low (Plastics) Limited Design

A design developed by *D.H. Desty, L. Bretherick and M.G. Webb; U.S. Patent 3,503,508; March 31, 1970; assigned to The British Petroleum Company Limited and Gordon Low (Plastics) Limited, England* involves barriers which have one or more skimming chambers which run lengthwise along the barrier and which have perforations which connect to water level when the barrier is inflated and floating. Pumping out the contents of a skimming chamber causes them to be replaced by material from the surface of the water and this enables oil floating on the water to be recovered.

As shown in Figure 116, the perforated tubular skin **26** contains a barrier comprising hoses **10** and **11**. When the two hoses are inflated, the skin forms a membrane which is stretched taut between them. The perforations **27** are situated at the level of the "waist" of the figure-of-eight so that they are at the water line when the barrier is afloat. A flexible flap **28** (of polythene or material of similar density and properties) is attached below the perforations.

FIGURE 116: BRITISH PETROLEUM-GORDON LOW SPILL CONTAINMENT AND REMOVAL DEVICE

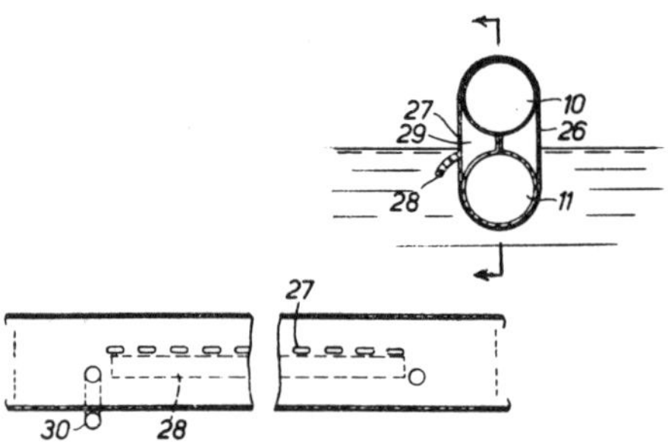

Source: D.H. Desty, L. Bretherick and M.G. Webb; U.S. Patent 3,503,508; March 31, 1970

This arrangement creates a skimming chamber **29** which runs the whole length of the barrier and which is maintained in the open position by the pressure in the hoses. When, during use, the contents of the skimming chamber are removed by pumping at the outlet **30** (more than one such outlet may be provided if desired), the contents are replaced by fresh material which enters through the perforations. The fresh material may be air, oil, water or any mixture of these depending upon the position of the barrier due to wave action. However, the flexible flap has a density between that of crude oil and that of seawater and therefore it tends to remain at the oil/water interface. This encourages the preferential skimming of oil, i.e., of material from above the flap.

In use, the barrier can be used to surround a patch of oil spilt on the water and remove the spillage at its periphery via the outlets. As the quantity of oil is reduced, the length of the barrier can be shortened so that the oil is enclosed within a reduced area and consequently advantageously increased in thickness. Alternatively, the barrier can be used to recover elongated narrow oil slicks by "trawling". In this case the barrier is attached to vessels which sail along the length of the oil slick with the oil slick between the two vessels. Towing causes the barrier to adopt a curved configuration and as the oil enters through the perforations it tends to move to and accumulate at the apex of the curve. The oil can therefore be recovered by pumping from the vicinity of this point.

Damberger Design

A design developed by *C. Damberger; U.S. Patent 3,726,406; April 10, 1973;* involves a towed boom used in conjunction with a water/oil separator. Figure 117 shows the boom in plan view at the top, the fence elements in the center and the separator at the bottom.

FIGURE 117: DAMBERGER OIL SPILL CONTAINMENT AND REMOVAL SYSTEM

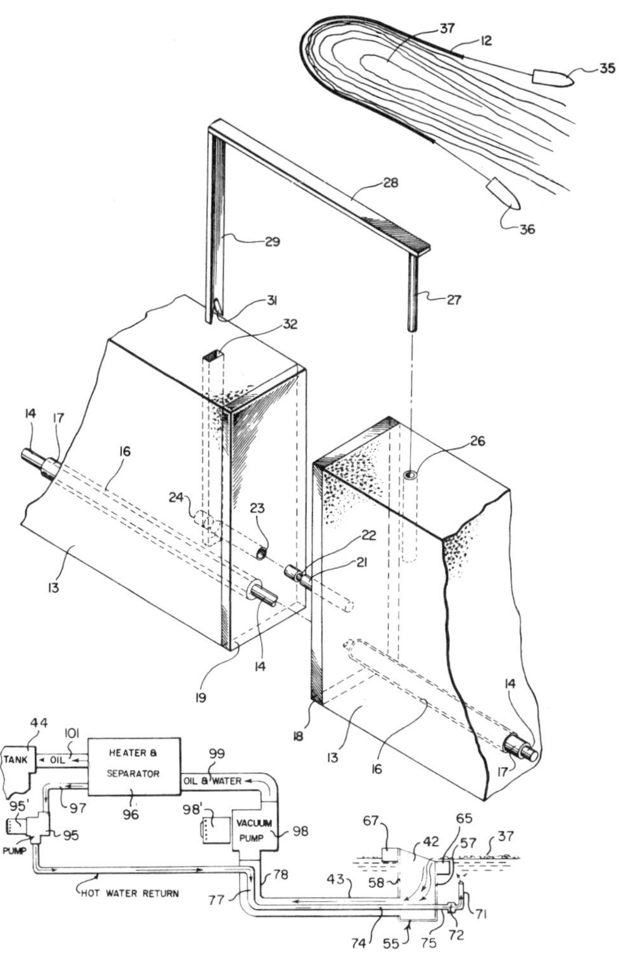

Source: C. Damberger; U.S. Patent 3,726,406; April 10, 1973

With the fence **12** assembled and with the ends of the cable connected to boats **35** and **36**, an oil slick **37** can be concentrated and dragged to a collecting location where it can be removed by the skimmer device to be described. The oil slick fence is made up of a series of floatable oil resistant blocks **13** which are joined together by a steel cable **14** that is extended through longitudinally extending holes **16** and coated with a fiberglass coating **17**. These blocks are preferably made of floatable plastic material and at the end of the blocks are steel plates **18** and **19** at their respective opposite ends. The plate **18** carries a male locking pin **21** that extends perpendicularly therefrom and has a locking groove **22** at the outer end. This pin mates with a hole **23** extending through the plate **19** and beyond into the body of the plastic block as indicated at **24**.

To lock the blocks together in end-to-end relationship while being assembled on cable **14**, two holes are extended vertically from the upper surface of the block and downwardly. One hole is round in section as best seen at **26** and is adapted to accommodate a swivel pin **27** carried on one end of a pin locker handle **28** and on the opposite end of the handle is a depending pin locker **29** extending downwardly and parallel to the pin **27**, the lower end of which has a V-shaped locking notch **31**. This pin locker **29** is of rectangular section and extends downwardly into a hole **32** of similar section that intersects with the hole **24** for receiveing the pin **21** for engagement with the groove **22** to thereby hold the blocks together upon the steel cable **14**. It will thus be apparent that with the pin locker **28** inserted into the top of the blocks **13** for the locking engagement of the pin **21** that the blocks **13** will be held against outward displacement from one another along the cable **14**.

At a time when it is desired to separate the blocks and to provide an opening in the fence **12** and therebetween, the pin locker handle **28** is pulled upwardly and the pins **27** and **29** will be disengaged from the upper surfaces of the blocks. If it is desired to maintain a separation of a greater distance of the blocks **13** from one another, this can be effected by using a pin locker device with a longer handle than the handle shown at **28** or even have an adjustable handle for different lengths. If desired, instead of having the pin **21** and the pin holes **23** and **24**, the blocks can be locked together by merely inserting the pins **27** and **29** into top holes **26** and **32**, both of which can be round as desired. However, by providing a pin connection **21** the blocks are held more appropriately against twisting with one another and without placing undue strain or bending of the cable **14** between the blocks.

The skimming and separating apparatus incorporates a hot water supply to thin the slick at the collection point. The hot water and oil hoses **77** and **78** extending from the oil skimmer **55** are connected to operate the oil skimmer. The hot water hose **77** is connected to a water pump **95** which takes hot water from a heater and separator **96** by a pipe **97** and delivers the hot water to the inner hose **77** and pipes **74** and **75** via fitting **72** to the spray head **71** to direct the hot water against the underside of the oil slick **37** to render it more effective and so that it will more readily pass over the float **67** and into the skimmer body **55** having forward and rearward faces **57** and **58**.

The large hose **78** is connected to a vacuum pump **98** that is connected by a pipe **99** to the heater and separator so that the oil slick once collected in the skimmer body is delivered to the separator **96** wherein the water in the oil slick is heated so that hot water is taken from the heater and separator by the water pump **95** and the separated oil is delivered from the heater and separator **96** through a pipe **101** to the storage tanker **44**. The respective pumps **95** and **98** have their respective drive motors **95'** and **98'**.

It will thus be seen that as the slick is drawn by the floating fence **12** toward the skimmer device **55** that the oil slick can flow over the float **65** through opening **42** into the skimmer body and thus be taken through the apparatus via conduit **43** for delivery to a storage tank **44** wherein it will have been collected from the water surface and upon being reconditioned made available for actual use without the oil having been lost to the sea or having polluted the water thereof and spoiled the shores and sea bird life.

Davidson-Cole Design

A design developed by *W.M. Davidson and H.W. Cole, Jr.; U.S. Patent 3,710,943; January 16, 1973* is one in which a flexible, inflatable, elongated barrier useful for the containment, separation and recovery of oil spilled on water is constructed in the form of a tunnel inside of which there is a continuous passage for oil that enters the barrier beneath the water surface on the upstream side and leaves the barrier on either end of the tunnel. The barrier is weighted at the bottom by suitable ballast means and buoyed at the top by long, continuous air chambers. Lengthwise cables are attached along the top and bottom of the barrier for towing and control purposes. The barrier can be submerged during emergency conditions and is constructed for roll-up on a reel either as a continuous single element or multiplicity of sections. Figure 118 shows transverse sections of this barrier-separator under conditions where it is located at the wave crest and the wave trough.

FIGURE 118: DAVIDSON-COLE OIL SPILL CONTAINMENT AND REMOVAL BOOM

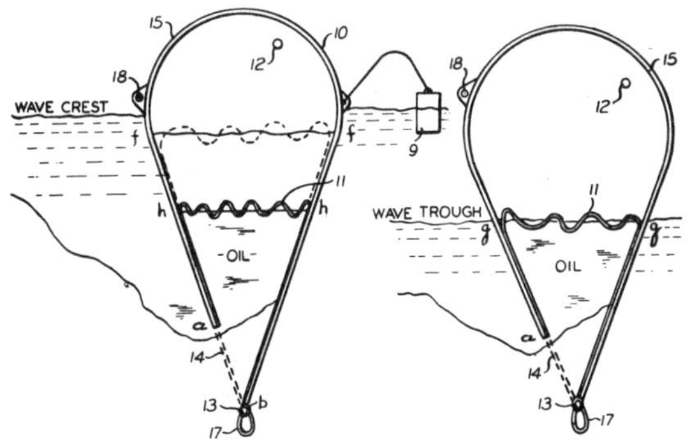

Source: W.M. Davidson and H.W. Cole, Jr.; U.S. Patent 3,710,943; January 16, 1973

As illustrated, the inflatable chamber of the barrier may be simply constructed by bonding together two flexible sheets of plastic or rubber material, such as PVC, polypropylene, neoprene, or other similar film material. Elongate, flexible sheets **10** and **11**, of equal width and length and rectangular in shape, are placed flat against each other and then bonded together to form a long, inflatable chamber. Valve **12** is attached to this chamber to provide for inflation with gas, such as air. Elongate, perforate, flexible sheet **14** is bonded to sheets **10** and **11** to form a collection chamber with open ends and with elongate inlet in communication with the chamber.

An alternate method of construction is to extend perforate sheet **14** to continuously envelop the exterior of the barrier, both edges of perforate sheet **14** being joined; use of this construction avoids the need for cable attachments directly to the air chamber. Harness **15** can be added at regular intervals along the length of barrier sections to support the loads from cable **18**. Cable **13** and cable **18** attached to harness **15** are used to interconnect barrier sections and to control the position of the barrier in relation to drag forces applied against the barrier.

Air pressure inside the air chamber causes sheets **10** and **11** to separate and assume a loosely arranged and horizontally disposed configuration. Outer elongate sheet **10** conforms flexibly to internal air pressure and to local changes in external water pressure, its shape being alterable by adjustment of ballast **17** and cables **13** and **18**. Elongate sheet **11** is forced against the surface of the water and oil inside of the collection chamber, its function being to complete the air chamber and thereby provide variable volume to the air chamber responsive to wave action. Each of the barrier sections can be inflated separately and its pressure adjusted independently from other fence sections; alternatively, a common air supply can be connected to all barrier sections simply by interconnecting the air supply lines attached to valves **12.** The use of perforate sheet **14** permits water and oil to enter inside the collection chamber while excluding any heavy debris.

Deepsea Ventures, Inc. Designs

A design developed by *J.P. Latimer; U.S. Patent 3,565,254; February 23, 1971; assigned to Deepsea Ventures, Inc.* is shown in Figure 119.

FIGURE 119: DEEPSEA VENTURES, INC. OIL SPILL CLEANUP DEVICE

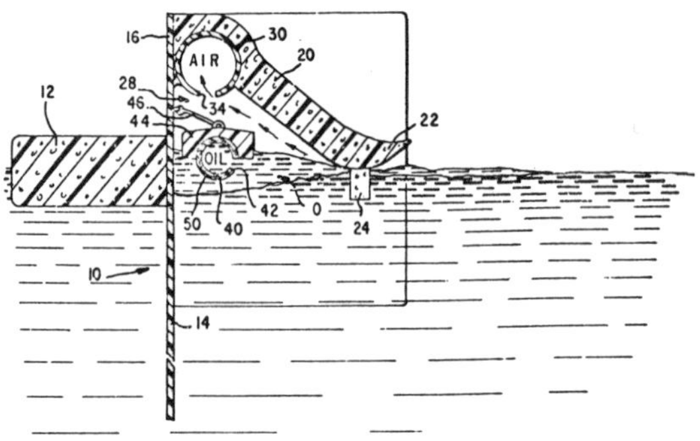

10 Boom assembly
12 Plastic foam buoyancy means for boom
14 Skirt
16 Upward extending portion of boom
20 Foot of oil removal device
22 Upward directed portion of foot
24 Plastic foam buoyancy means for foot
28 Air space
30 Tubular vacuum pipe
34 Holes spaced in tubular vacuum pipe
40 Oil pipe
42 Slot in oil pipe
44 Buoyant means of support for oil pipe
46 Flexible connectors
50 Flexible electrical elements for when oil heating is needed

Source: J.P. Latimer; U.S. Patent 3,565,254; February 23, 1971

A design developed by *C. Garland, J.J. Victory and J.P. Latimer; U.S. Patent 3,666,098; May 30, 1972; assigned to Deepsea Ventures, Inc.* is shown in Figure 120. The figure shows, at the top, a vessel **20** at dockside surrounded by an oil spill **24**. The center of the figure shows a venturi aspirator for oil removal. The bottom view is a transverse section of the floating barrier collector. In the upper view, the buoyant confining and collecting boom **10** is represented as connected to eductor **12**, having input **16** and discharge conduit **14**, so as to flow oil into the system as an assistance in withdrawing the confined oil slick through the boom. The oil is passed to removal truck **18**. In the middle view, eductor **12** is illustrated as including a plurality of jet nozzles **25** and **26** which direct the liquid motive flow from tank truck **18** through the discharge conduit **14** from nozzle **22** so as to entrain the ingested oil slick. In practical applications of this collector, it is an advantage to use the motive liquid flow from the storage tank rather than seawater. This arrangement uses recirculated suction liquid rather than external seawater, which would require twice the storage tank holding capacity.

As shown in the lower transverse sectional view, boom **10** includes a plurality of laterally or horizontally aligned and graduated perforations **46** which engage the oil slick for ingestion, and their height may be adjusted by varying the size of the buoyant floats which surround the boom at intervals. The flexibly constructed air skirt **36** may depend on either side and be offset with respect to perforations **46**, skirt **36** being connected to the boom top by means of bolt **38** and nut **40** or the like. Weights **52** and **54** may be presented at the ends of the skirts so as to hold the skirt as a shield within the oil slick interface. Additional buoyant floats may encircle the boom and be held in place by clasps. In this view, seawater is indicated by **34** and oil by **24**. An air space within the collector is indicated by **42** and the body of the boom tube by **44**. The devices **48** and **50** are buoyant floats.

FIGURE 120: ALTERNATIVE DEEPSEA VENTURES INC. OIL SPILL CONTAINMENT AND REMOVAL DEVICE

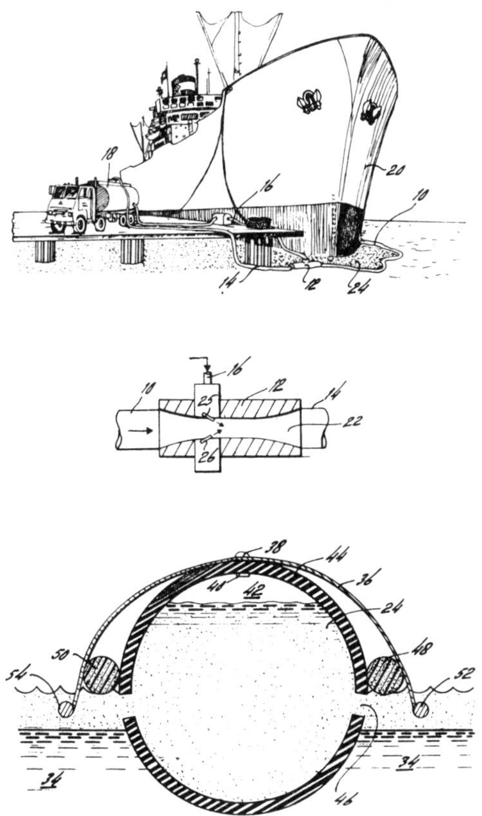

Source: C. Garland, J.J. Victory and J.P. Latimer; U.S. Patent 3,666,098; May 30, 1972

Johns-Manville Corporation Design

A design developed by *W.G. Ekdahl; U.S. Patent 3,747,760; July 24, 1973; assigned to Johns-Manville Corporation* involves containing the oil in a barrier to form a confined oil slick having a leading edge. Due to the flow of the water, a head-wave is formed along the leading edge of the confined oil slick and oil removal apparatus is located at the head-wave to remove oil from the head-wave.

Field tests have shown that oil contained in a barrier will form a head-wave at its leading edge when there is relative movement between the barrier and the water on which the oil is floating. A head-wave is a pool of oil along the leading edge of the confined oil slick which extends deeper into the water than the remainder of the confined oil slick. The depth of the head-wave is typically twice the depth of the remaining portion of the confined oil slick and can be four times the depth of the remaining portion of the confined oil slick. Figure 121 shows a plan view of an oil barrier with oil removal apparatus placed adjacent the ends of a head-wave and also an elevational side view with portions broken away to show the formation of the head-wave at the leading edge of the oil slick.

FIGURE 121: JOHNS-MANVILLE OIL SPILL CONTAINMENT AND REMOVAL SYSTEM

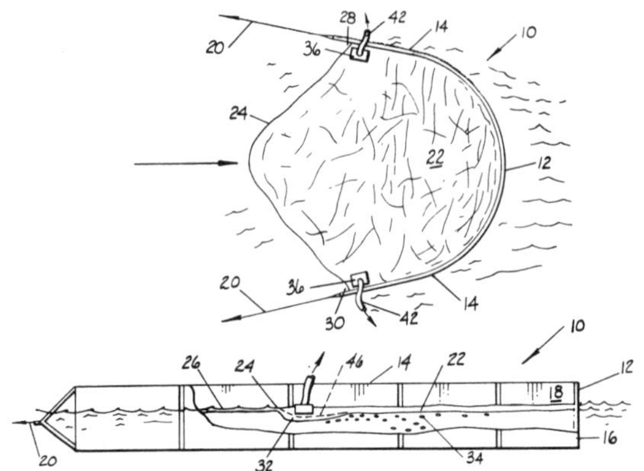

Source: W.G. Ekdahl; U.S. Patent 3,747,760; July 24, 1973

A conventional barrier **10** having an apex or bucket portion **12** joining sidewall portions **14** is used to confine the oil. The barrier is designed to float in the water with a lower portion **16** of the barrier submerged and an upper portion **18** of the barrier extending above the surface of the water. Tow lines or anchoring lines **20** are provided at each end of the barrier to secure the barrier to towing vessels or suitable anchorages (not shown).

The direction of flow of the water relative to the barrier is shown by arrows. The flow of the water in this direction can be caused by any number of factors with the direction of flow being the important aspect and not the cause for the flow in that direction. For example, the water could be caused to flow in the given direction relative to the barrier by a current and/or by towing the barrier through an oil slick. However, as a result of this relative flow between the water and the barrier, an oil slick **22** builds up within the confines of the barrier.

The oil slick has a leading edge **24** where the slick meets open water or a relatively thin layer of oil **26** being carried by the water into the barrier. In plan view, the leading edge of the confined oil slick generally assumes a parabolic or convex configuration between points **28** and **30** of the barrier sidewalls. Along the leading edge a head-wave **32** of oil is formed. The head-wave extends along the entire length of the leading edge meeting the barrier at points **28** and **30**. The oil-water interface when viewing the head-wave in transverse cross section is generally convex with the thickness or depth of the oil slick at this point being substantially greater (two to four times thicker) than the remaining portion of the oil slick intermediate the head-wave and the bucket portion **12** of the barrier.

In transverse cross section, the leading edge of the head-wave is blunter than the trailing edge of the head-wave which gradually merges with the remaining portion of the oil slick. The width of the head-wave is independent of the length of the head-wave or the distance between the sidewalls **14** and the amount of oil contained within the barrier but is dependent on the relative velocity of the water with respect to the barrier and the thickness of the oil slick. The head-wave is substantially the same thickness or depth along its entire length between points **28** and **30** where the head-wave meets the sidewalls of the barrier.

Globules of oil **34** are also shown. These globules ordinarily break away or are torn away from the head-wave by the scrubbing action of the water which then carries the globules beneath the barrier. However, when recovering the oil according to this method, the loss of these globules beneath the barrier is minimized. According to this process, conventional oil removal apparatus **36** is placed at each end of the head-wave **32** adjacent the barrier sidewalls **14** as shown.

Oil is pumped from the surface of the water through inlets of the apparatus. Thereafter, the oil-water mixture is discharged through hoses **42** to barges or other vessels (not shown) having means to store and/or dispose of the oil removed from the water surface including, when necessary, means to separate oil from any recovered water.

The oil removal apparatus **36** should have a combined oil removal capacity at least sufficient to remove the oil at a rate to overcome the forces tending to drive the oil downward and out from beneath the barrier. Preferably, the combined capacity of the oil removal apparatus is equal to or greater than the rate at which oil is being added to slick **22**. Thus, when the oil recovery apparatus is recovering oil from the head-wave **32**, it reduces the depth of the head-wave, especially adjacent the points of actual recovery. The head-wave then becomes less prominent as shown by phantom line **46** and at least approaches the thickness of the remaining portion of the confined oil slick.

Consequently, the tendency of the water to scrub off globules of oil from the head-wave is minimized by reducing the profile of the oil-water interface and the pressures tending to force the oil down and beneath the barrier where the head-wave meets the barrier are also minimized. By placing the oil removal apparatus adjacent the ends of the head-wave at the barrier, the pressure is relieved and the depth of the head-wave is reduced at the very point where the pressures on the oil and the depth of the head-wave would ordinarily be causing the oil to flow beneath the barrier. Of course, additional oil removal apparatus can be placed intermediate the head-wave and the bucket portion **12** of the barrier. However, the critical locations are adjacent each end of the head-wave adjacent the barrier and to a lesser extent anywhere along the head-wave.

M.I.T. Design

A design developed by *J. B. Nugent; U.S. Patent 3,768,656; October 30, 1973; assigned to Massachusetts Institute of Technology* is a floatable unit for accumulating oil from the surface of water, essentially comprising a framed member having three sides and open at the top and bottom thereof and also having an open end called the bow, this member including an appropriately located ramp and baffles. The accumulator is inserted into a line of booms at appropriate intervals. A wave with oil on its surface enters the accumulator, surges up over the ramp, and the oil is trapped in the unit for later removal.

Figure 122 is a top view of the device showing a pair of separators located in a boom system. The oil accumulator **20** is a framed member constituting a box having two sides **20A** and **20B**, open at the top and bottom. The stern **20C** is enclosed while the bow end is open. Within accumulator **20** near the center (for maximum stability) is tapered ramp **22** coupled to sides **20A** and **20B** and designed so that when a wave carrying oil on its surface enters the bow, it is caused naturally to surge up the ramp and spill over it into the accumulator. The end of the ramp nearest the bow is lower than the other end which, in turn, is higher than the nominal sea level. At the low end of ramp **22** is baffle **26** running horizontally between sides **20A** and **20B** and vertically such that its bottom edge is as deep as sides **20A** and **20B**, i.e., the three surfaces are contiguous. Accumulator **20** is maintained partially submerged by the use of flotation members **24** attached to the sides **20A** and **20B** of unit **20**.

When a moderate size wave with oil on its surface enters the box, it surges up the ramp by reason of its contained energy. It spills over the end of the ramp. Because oil is lighter than water, the oil is trapped and the water is free to discharge through the open bottom of the box. By the addition of side walls **28** on the ramp such that the walls are nearer together at the upper end of the ramp, a jetting action (higher lift) may be attained. This higher lift could be utilized with a higher ramp to prevent backflow of the collected oil when the sea is rough.

As shown, a series of accumulators **20** are inserted into a line of booms at appropriate intervals. For stability, each of the accumulators is several normal wave lengths long and approximately one-quarter as wide and is inserted in the boom so that the longer dimension is approximately normal to the boom.

FIGURE 122: MASSACHUSETTS INSTITUTE OF TECHNOLOGY OIL SPILL CONTAINMENT AND REMOVAL SYSTEM

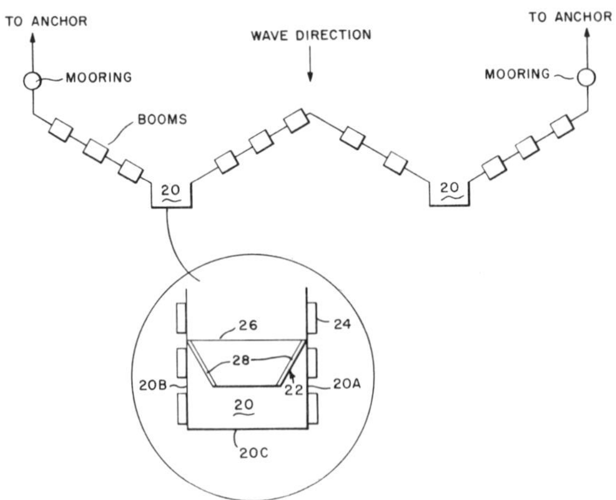

Source: J.B. Nugent; U.S. Patent 3,768,656; October 30, 1973

Mobil Oil Corporation Design

A design developed by *P.C. Dahan; U.S. Patent 3,476,246; November 4, 1969; assigned to Mobil Oil Corporation* utilizes a floatable collar section comprising a main inflatable tube provided with means for skimmnig a floating liquid and a ballast means comprising a weighted skirt located below the inflatable tube. The skimming means can be formed with the main inflatable tube or can be attached thereto. A plurality of collar sections can be attached end-to-end forming a floatable collar to enclose and confine a liquid floating on seawater.

Figure 123 is a cross-sectional view of such a barrier-separator in use. As shown, there is a main inflatable tube **70** to which is attached a skimming means formed of a plurality of inflatable tubes **71, 72, 73, 74** and **75.** The skimming means is formed into a trough having a hook shape when viewed as a vertical cross section. The skimming means is held to the desired shape during use by strap **76.** To the strap and the underside of inflatable tube **73** is attached a skirt **77** which extends the length of the tube and provides an effective barrier to the floating oil **78.** To the bottom of the skirt is attached a tube means **79** adapted to retain weights.

The skimming means is shaped so that inflatable tube **75** provides a barrier to seawater but permits floating oil products to enter the recess **80** formed by the skimming means. A flexible conduit **81** is attached to the exteriors of tubes **72** and **73** and the open end **82** thereof is in communication with the oil within recess **80**. The oil is pumped from the recess through tube **81** to storage means not shown. The tubes **70, 71, 72, 73, 74** and **75** are attached in a manner to prevent leakage from the recess into the seawater. Thus, for example, when the inflatable tubes are made of rubber, the tubes can be attached by vulcanization.

FIGURE 123: MOBIL OIL CORPORATION OIL SPILL CONTAINMENT AND REMOVAL SYSTEM

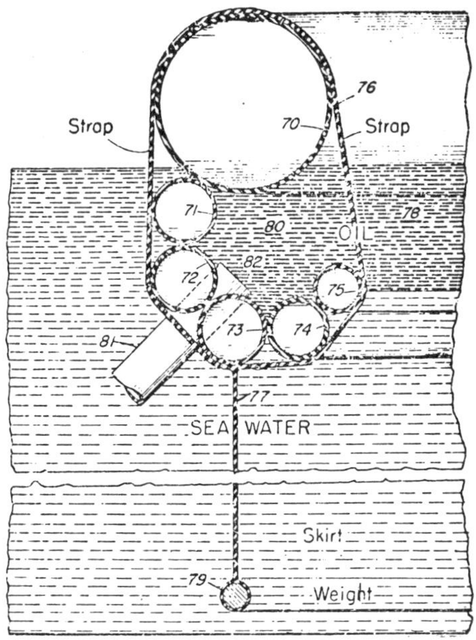

Source: P.C. Dahan; U.S. Patent 3,476,246; November 4, 1969

Ocean Pollution Control, Inc.

A design developed by *H.J. Fitzgerald; U.S. Patent 3,641,770; February 15, 1972; assigned to Ocean Pollution Control, Inc.* for confining and removing oily material on the surface of a body of water consists of a polygonal ring formed by flexible tubular float members inflated to a pressure on the order of 0.5 psi with a weighted skirt depending therefrom and with a transfer pipe having its inlet end removably supported in the leeward corner of the ring at a level within the thickness of the oily accumulation.

Figure 124 is a cross-sectional view of the barrier-oil remover design. As may be seen, each of the tubular float members may be formed of a plurality of, for example, four segments **22**, each of which extend lengthwise of the elongated tubular member around a portion of its circumference. The segments are formed of flexible, impermeable sheet material such as nylon fabric or other strong textile material impregnated with or laminated to neoprene, rubber or similar plastic material. The proximate ends of the adjacent segments are adjoined by marginal portions which are bonded face to face, for example, by cementing, heat sealing or vulcanization to form radially outwardly extending flanges **24**. These flanges are provided with holes, preferably reinforced, for example, by grommets **26** for convenience in attaching the float members to one another and to ancillary parts of the apparatus.

The lower flange **24** preferably encloses an integral cable or rope **27** formed, for example, of nylon or similar material which is provided at each end with a projecting loop or other means of attaching the float members together and for transmitting the tensile forces imposed either by wave action or by towing to or from the point of use without straining and possibly tearing the float members.

Attached to the bottom edge of the lower flange, for example, by boltlike fasteners extending through the grommets or by lacing a line through the aligned grommets is a depending skirt **30** which is formed of a flexible, impermeable sheet material, for example, the same type of material used in fabricating the float members although possibly of lighter gauge. Secured adjacent the lower edge of the skirt is a chain **32** which weights the edge of the skirt to insure that it extends generally vertically downwardly without impairing its flexibility. The width of the skirt exceeds the anticipated maximum thickness of the oily accumulation to prevent the oily material from escaping beneath the bottom edge of the skirt. The skirt may suitably be made integral with the float member, for example, by merely extending the marginal portion of one of the lower segments **22**. However, forming it of a separate sheet removably attached to the float member as shown allows it to be replaced readily in the event of damage or in the event a deeper skirt is desired, for example, because of rough seas.

FIGURE 124: OCEAN POLLUTION CONTROL, INC. OIL SPILL CONTAINMENT AND REMOVAL BOOM

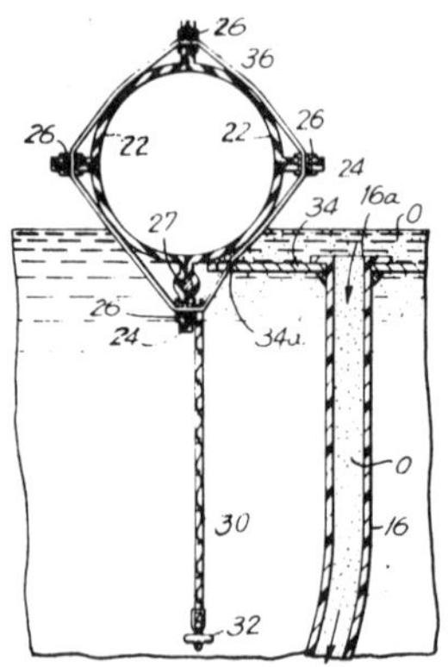

Source: H.J. Fitzgerald; U.S. Patent 3,641,770; February 15, 1972

The buoyancy and flexibility of this assembly is such that the ring will float lightly upon the surface conforming readily to its rapidly undulating and fluctuating contour and effectively entrapping the oily accumulation which will build up on the surface within the ring as indicated at **O**. As shown in the leeward corner of the ring, the pipe **16** is supported on a web **34** formed, for example, of a heavy but flexible fabric material extending obliquely across the corner with its opposite ends secured to the float members **10**, for example, by bands **36** encircling the float members through the grommets and extending through holes **34a** in the ends of the web **34**. The input end **16a** of the pipe **16** is open above the web

34 and is supported at a level slightly below the upper surface of the oil accumulation **O** so that the liquid entering the pipe **16** will consist essentially of the oil material with little if any water.

Ocean Systems, Inc. Design

A design developed by *L.S. Brown, F.A. March, R.P. Bishop and B.C. Gilman; U.S. Patent 3,650,406; March 21, 1972; assigned to Ocean Systems, Inc.* provides a system for collecting and retrieving a liquid of low density from the surface of a body of liquid of higher density which includes a catch basin having a floating weir as one section thereof through which the lower density liquid flows, and means for retrieving the liquid from the interior of the basin. The floating weir consists of a buoyant upper section and a water absorbent lower section representing the ballast for the upper section. Figure 125 shows a plan view of the system in operation and also a detail of one section of the floating weir.

FIGURE 125: OCEAN SYSTEMS, INC. OIL SPILL CONTAINMENT AND REMOVAL SYSTEM

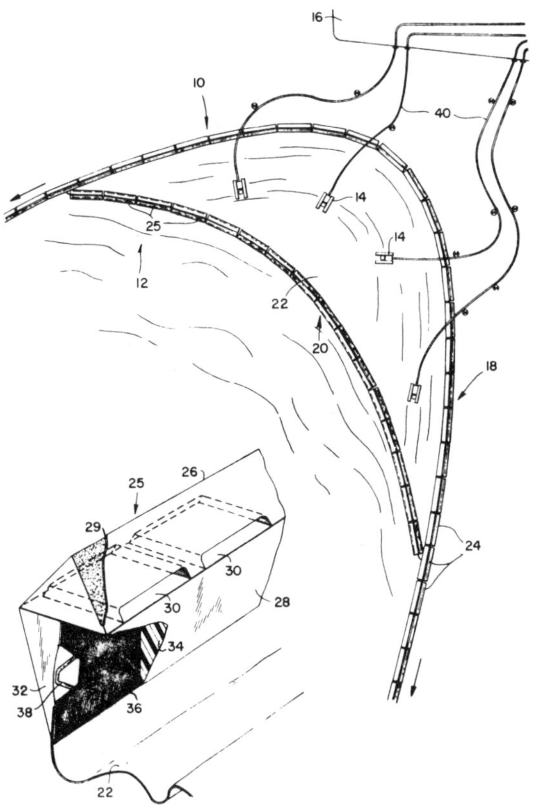

Source: L.S. Brown, F.A. March, R.P. Bishop and B.C. Gilman; U.S. Patent 3,650,406 March 21, 1972

The system as shown is composed of a catch basin **10** which is adapted to be towed in the direction of the oil slick **12** by two tow vessels (not shown). A number of pumps **14** are located in the interior of the basin to deliver oil accumulated in the basin to an oil receiving vessel **16**. The catch basin is made up of a floating oil boom **18**, a floating weir **20** and a reinforced support bottom **22**.

The floating oil boom is defined by a plurality of serially connected liquid confining barrier modules **24**. The barrier comprises a plurality of serially connected modules each defined as a composite structure having a buoyant substantially water impervious upper section and a water absorbing lower section which represents once immersed the sole ballast for the upper section and concomitantly the subsurface barrier for the module. The upper and lower sections of each module are fabricated from polyurethane foam. The lower section is divided into two sections with a thin beltlike member vertically aligned between the segments and extending longitudinally for substantially the entire length of the module. The shape and construction of each barrier module would thus be similar to the construction of the floating weir as explained below.

The floating weir consists of a plurality of serially coupled modules **25** where each module comprises an upper section **26** which is bonded to a lower section **28** by means of any conventional water repellent adhesive preferably an elastomer or epoxy resin. The upper section provides the buoyancy for the module and through apertures **30, 30** controls the oil flow into the basin while the lower section extends below the surface of the sea and functions concurrently as ballast for the upper section and as the subsurface barrier of the module.

Both the upper and lower sections, respectively, are fabricated from a flexible polyurethane foam material with the upper section preferably of a polyether based polyurethane foam and the lower section of a reticulated polyester based polyurethane foam. Other materials such as porous rubber and sponge having low density may also be used. To maintain a state of buoyancy, the upper section is rendered water repellent by coating the surface with a sealant **29** such as an elastomer or an epoxy resin. The bonding water repellent adhesive between the upper and lower sections (not shown) prevents water seepage from the lower section into the upper section. Apertures **30, 30** located in the upper section of each module control the flow of liquid from one side of the module to the opposite side thereof and into the interior of the basin as will be explained hereinafter.

The lower section of the module is divided into two segments **32** and **34**, respectively. A thin beltlike member **36** is vertically interposed between the two segments and extends longitudinally along substantially the entire length of the module. The belt **36** is bonded in such position against the two segments by means of a sealing adhesive (not shown) such as, for example, an epoxy resin or an elastomer. The sealing adhesive not only cements the two segments against the belt and to each other but represents in combination with the belt a partition for preventing any oil from passing between the segments **32** and **34**, respectively. The belt is preferably composed of a fabric material such as dacron to provide structural rigidity and limited elasticity. A conventional coupling element **38** is attached to each longitudinal end of the belt for coupling the modules to each other in abutting relationship to seal the joined ends. The joined ends may be otherwise sealed in any conventional manner.

The trapped seawater in the lower section provides the stability for each module resulting in excellent sea conformance characteristics for the weir. For compatibility and to optimize overall performance, it is preferred that the floating boom be constructed in a manner similar to that described for the weir. The cross-sectional geometry of each module is preferably that of a nonsymmetric rhombus composed of two equilateral triangles having a common base with the bottom triangle having approximately twice the height of the upper triangle.

The basin is completed by connecting the bottom support member **22** to both the weir and floating boom. The preferred method of attachment is to sew the bottom support member to the belt as shown at the interface between the mated segments **32** and **34**. An alternative approach might be merely to cement the bottom member in place between segments **32** and **34**, respectively, of lower section **28**. Attachment to the floating boom will depend upon the chosen construction for the boom. Where the boom is constructed as shown except for passages **30, 30**, attachment may be made by sewing the bottom support to the belt as described above with reference to the weir. The preferred material for the bottom support member is a nylon reinforced rubber sheeting which is flexible and has high tensile strength. The material should also be unaffected by sunlight, seawater and aromatic hydrocarbons.

In operation, two tow boats will be connected in the vicinity of the oil spill to the opposite ends of the floating boom for movement in the direction of the oil spill as indicated by the arrows. The boom and weir will assume a parabolic contour in response to water currents and/or tow movement. Oil is trapped by the boom and is funneled toward the weir where part of the piled up oil will pass through passages **30, 30** into the interior of the catch basin from which it may be simultaneously removed by pumping means **14** through discharge lines **40** and into the receiving vessel **16**. A grating (not shown) may be placed in front of each mouth **30** of the weir to prevent flotsam from clogging the opening. The water saturated lower section of each module of the weir tends to keep the openings **30, 30** at the same position relative to the oil-water interface. Thus, a high proportion of oil to water flows into the catch basin.

M.F. Smith Designs

A design developed by *M.F. Smith; U.S. Patent 3,539,013; November 10, 1970* utilizes an oil-absorbing boom for the purpose of collecting and removing from water thin films of oil. The boom is made up of an elongated flat tubular sleeve of polymer netting enclosing within itself a plurality of flat elongated slabs or bats of "picker-lap" fibrous polymer material such as blown polypropylene film arranged end to end within the tubular sleeve and sufficiently spaced apart to permit accordion folding of the sleeve at fold lines between adjacent bats. A tension bearing rope or cable is positioned within the tubular sleeve alongside the successive plurality of absorbent bats to reinforce the structure for carrying its own weight or impact loads placing it in tension between its ends.

The design includes means for separating the absorbed oil from the absorbing boom which may then be prepared for reuse, to be redeployed as required. Figure 126 is a schematic plan view of a maritime oil transfer terminal showing a tanker near a dock surrounded by an oil boom with an oil collection boom of the type deployed downstream to collect any escaping oil as well as an enlarged fragmentary elevation view schematically showing pinch roll apparatus on shore positioned to draw in the boom and separate collected oil from it by compressive pinch roll action.

Thus, a tanker **20** is shown in oil transfer position beside a dock **21** surrounded by an oil spill boom **22**. Arrows **23** show the direction of the prevailing current passing the installation and the compressible collection boom **10** is shown deployed extending partway across the waterway from one shore toward the opposite shore at a point downstream from the tanker positioned to receive oil carried downstream by the current **23**.

As shown, after an oil film has been brought into contact with the collection boom by the prevailing current and after all or the major portion of the oil film carried by the current has been caught and collected in the interstitial spaces between the loosely matted fibers of the bats along the length of the boom, the boom may be withdrawn from the waterway by such means as the pinch roll oil separation apparatus illustrated schematically.

Pinch rolls **24** and **26** which may be powered to draw between themselves the compressible boom are shown in power driven operation and the boom is moving from the water between the pinch rolls which have the effect of compressing each of the bats in turn as these bats are drawn between the pinch rolls. As a result, a large portion of the collected oil is squeezed out of the compressible bats into a sump **27** beneath the pinch rolls. The squeezed collection boom issuing from the pinch rolls is preferably accordion folded in a storage stack **28** near the shore **29** where it is ready for prompt deployment in the event of another oil spill.

FIGURE 126: M.F. SMITH OIL SPILL CONTAINMENT AND REMOVAL SYSTEM

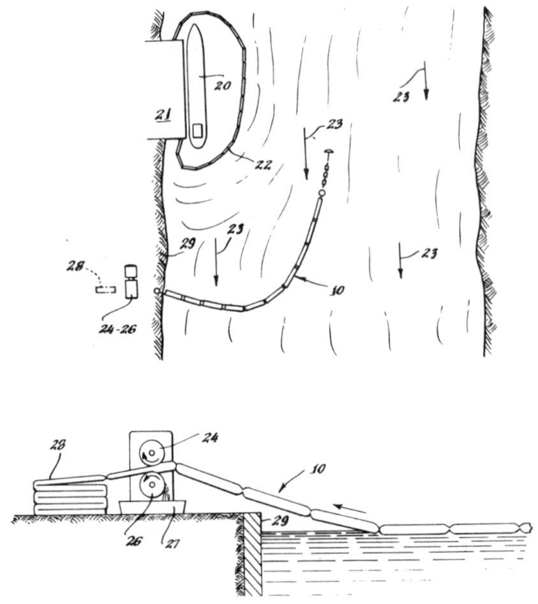

Source: M.F. Smith; U.S. Patent 3,539,013; November 10, 1970

Another design developed by *M.F. Smith; U.S. Patent 3,679,058; July 25, 1972* deals with the construction of various types of absorbent compressible booms for use with various types of compression roll apparatus which remove the oil and return the flexible booms for additional absorption.

One type of boom **10** is shown in fragmentary perspective in the upper part of Figure 127. The net sleeve **11** is preferably formed of similar lightweight polymer fibers which may be heat sealed in net configuration or which may be woven or knotted like a fisherman's net in the form and spacing desired.

The bats **12** are spaced apart longitudinally within the tubular sleeve by a distance of from two to three times their thickness to permit accordion folding in the manner of the accordion folded oil spill boom described in U.S. Patent 3,146,598. Along the inner edge of the bat a tension cable preferably formed of stainless steel wire rope **14** is also laid longitudinally down the central portion of the net beside the bats.

Another type of boom is shown in fragmentary perspective in the lower part of Figure 127. In this case, solid blocks **48** of highly resilient compressible buoyant foams, such as polyethylene foam, may be cemented as abutting strips along the edges of an oil collection slab **39** which is formed of such material as fully reticulated open cell polyurethane foam, thus producing a unitary foam slab assembly **49** which may be directed between pinch rolls in order to express the oil collected in the central open cell oil collecting foam portion **39**.

FIGURE 127: DETAIL OF M.F. SMITH ABSORBENT BOOM CONSTRUCTION FOR OIL CONTAINMENT AND REMOVAL

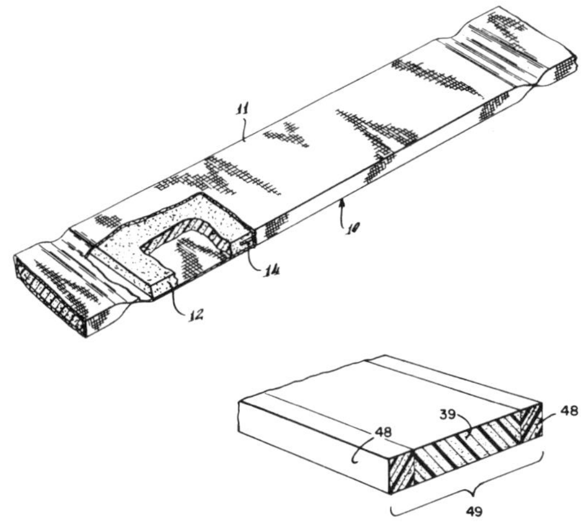

Source: M.F. Smith; U.S. Patent 3,679,058; July 25, 1972

G.J. Trapp Design

A device developed by *G.J. Trapp; U.S. Patent 3,495,561; February 17, 1970* is one in which a floating boom made up of a number of buoyancy chambers flexibly coupled together is towed over the surface, each buoyancy chamber having an aperture, preferably slightly above the water level, through which the surface layer of oil (with a certain amount of water) enters an inner chamber within the buoyancy chamber, and the liquid in the inner chamber is drawn off through a pipe, passed through a centrifuge, the water is discharged and the oil is stored.

Figure 128 shows a plan view of such a boom device containing nozzles for collecting oil and water and also a vertical section on the longitudinal center line of one of the buoyancy chambers making up the boom. As shown, the floating boom is made up of a number of buoyancy chambers **11**, each of which is provided with a tow line **12** to enable it to be towed. The buoyancy chambers are coupled together by means of couplings **13** which are so arranged that relative movement is allowed between the individual buoyancy chambers in all directions except the transverse direction which would allow the buoyancy chambers to vary their separation from each other. This is to allow the boom as a whole to conform to the undulations of the sea. Hinged or flexible shields **14** are coupled between the forward ends of the buoyancy chambers to reduce or prevent the escape of oil between them.

A second line of buoyancy chambers **15** lying behind the front line may be used if desired to ensure that any oil which does escape collection by the front line of buoyancy chambers is picked up, the second line being preferably arranged in staggered relation to the first line so that the center of each buoyancy chamber **15** is behind the gap between two byoyancy chambers **11**. The second line of buoyancy chambers is towed by means of cables **16**. Control lines **28** are coupled to the rear ends of the buoyancy chambers of the second line; but if this second line is not employed, the control lines are coupled to the rear ends of the buoyancy chambers of the front line in place of the cables **16**.

FIGURE 128: G.J. TRAPP OIL CONTAINMENT AND REMOVAL SYSTEM

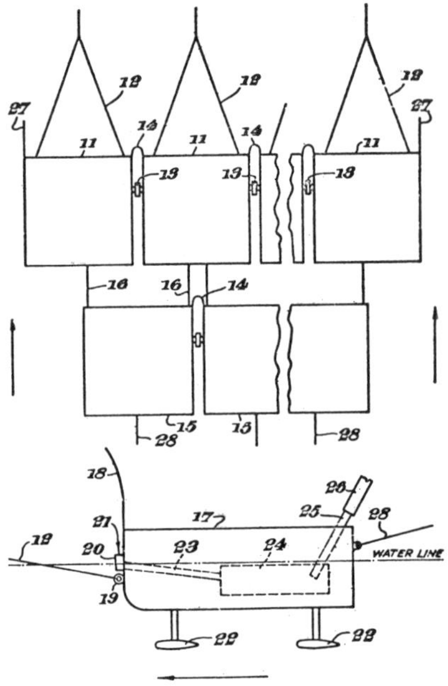

Source: G.J. Trapp; U.S. Patent 3,495,561; February 17, 1970

The boom is intended to be towed slowly through the area over which the oil pollution extends. For this purpose, the tow lines **12** are connected to a beam (not shown) supported horizontally and preferably a little above the surface of the sea with the axis of the beam lying transverse to the center line of a ship. The beam is preferably supported on a superstructure, preferably made of light metals, carried on the deck of the ship which projects for an adequate distance, for example, 20 to 30 feet in front of the bows of the ship.

It would be possible to tow the boom by supporting the beam above the sea behind the stern of the ship; but in that case, not only would the ship have to sail through the polluted area but the passage of the ship and the disturbance created by the propellers at the rear of the ship would disturb the oil film and, to a certain extent, disperse it so that the boom would be less effective. Accordingly, the preferred arrangement is to mount this superstructure at the bow end of the ship. Ideally, the ship would have a square front in place of the normal pointed bow. The superstructure may, if desired, be made retractable.

As shown in the lower sectional view, the chamber comprises an outer sealed body **17** with a curved upwardly projecting shield **18** at the forward end to reduce the possibility of water and oil passing over the top of the unit so that some oil would escape collection. A little below the water line is the anchorage **19** for the tow line. Ideally, the buoyancy chambers are drawn forward without an upward or downward bias to the traction force, but this may be difficult to achieve, and if the beam is held a small distance above the surface of the sea, the angle of the upward tilt of the cable connected to the anchorage need not be very large. The height of the beam may be made variable or it might even be allowed to float. Projecting outwardly from the buoyancy chamber is a lip **20** the top of which is a little above the water line. The lip and the front end of the casing **17** of the chamber define a slot **21** through which the surface oil and a certain amount of seawater are taken into the buoyancy chamber as the chamber is pulled forward through the water. The general shape of the front wall of the buoyancy chamber and the arrangement of the lip and its distance above the water line are such that when the chamber is being towed at the designed speed, the water will escape comparatively easily below the buoyancy chamber while the oil will resist being dragged down below the chamber and the layer of oil on the surface will become locally thickened immediately in front of the lip until it builds up and overlaps the lip and runs into the slot. The description of the action of the lip will not be fully realized in practice, but the arrangement is intended to ensure that the oil/water ratio entering the slot is maintained as high as possible.

In order to secure the maximum efficiency in operation, it is important to maintain the vertical position of the buoyancy chamber as constant as possible in relation to the water surface even when this surface is undulating and heaving. For this purpose hydroplanes **22** may be fitted below each buoyancy chamber.

The mixture of oil and water entering the slot passes along a sealed duct **23** in the buoyancy chamber to a sealed collecting chamber **24** which is on a lower level than the lip. A rigid pipe **25** leads to a connection with a light flexible pipe **26**. Since the water and the oil will pile up to a certain extent in front of the lip due to the fact that the buoyancy chamber is being towed, forwardly projecting guards **27** are preferably fitted to the buoyancy chamber to minimize the loss of oil around the sides thereof.

The flexible pipes **26** from all the buoyancy chambers are supported at intervals by the aforementioned beam or superstructure and lead to suction pumps on the ship. Positive pumps must be used since air may enter the collecting chambers with the oil and water, and the height of the pumps must be limited to the distance through which atmospheric pressure will lift the mixture. From the pumps the mixture is led to centrifuging machinery in which the oil and water are separated. The water is discharged from the stern of the ship while the oil passes to tanks in the ship.

Water Pollution Controls, Inc. Design

A design developed by *J.M. Valdespino; U.S. Patent 3,532,219; October 6, 1970; assigned to Water Pollution Controls, Inc.* involves a portable inflatable apparatus for confining and collecting oil on the surface of water, separating the oil from the water and containing such oil until collected, without the use of mechanical parts.

As shown in Figure 129, for the oil containment barrier there is provided a relatively long lightweight tube **10** which preferably is constructed of plastic or other inexpensive material having high tensile strength. The tube may be of any desired diameter and length although a tube approximately 6 feet in diameter and approximately 4,000 feet long is believed to be satisfactory. The tube preferably has a seal **11** at both ends. When not in use, the tube preferably is wound on a reel and stored in any convenient location ready for rapid transportation to a point of use.

When an offshore drilling well or a tanker becomes disabled and springs a leak, the reel is quickly moved to the site and placed onto a small boat or onto a helicopter (not shown) which then encircles all or a major portion of the oil slick on the surface of the water. As the tube is being unwound from the reel, a flotation member **15** is connected to the tube by straps or other connectors **16** at spaced intervals in such a manner that the tube will extend around the corner of the flotation member. Preferably each of the flotation members **15** has a weight or anchor **17** connected thereto by a chain or cable **18** so that when the flotation member is positioned it will remain in such location.

The opposite ends of the tube may be connected together, as illustrated, by a connector **19** passing through aligned openings, or if a smaller ring is dictated by existing conditions, the opposite ends of the tube can be disposed in overlapping relationship with each other and the ends can be lashed or otherwise secured to an intermediate portion of the tube.

It is contemplated that when a large oil slick is encountered, two or more tubes could be connected in end-to-end relation to provide a larger circle. The introduction of air and water into the tube causes the tube to be expanded so that it will float and form a barrier around the oil slick and will confine the slick to the area within the circle. As illustrated, a prevailing wind indicated by the arrow **33** will cause most of the oil within the area encircled by the tube to move toward one side of the tube.

FIGURE 129: WATER POLLUTION CONTROLS, INC. OIL CONTAINMENT AND REMOVAL SYSTEM

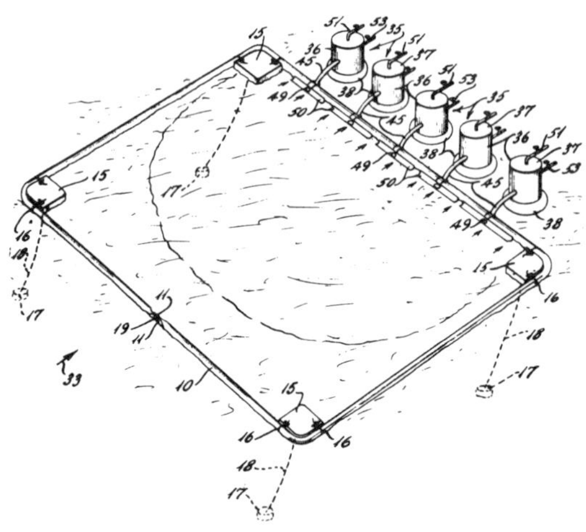

Source: J.M. Valdespino; U.S. Patent 3,532,219; October 6, 1970

At the leeward side of the encirclement a plurality of containers or tanks **35** are provided and each of the tanks includes a generally cylindrical side wall **36** and a top wall **37** connected to the side wall in airtight relationship. The bottom of the tank is open so that seawater can enter and be discharged from the tank. A flotation collar **38** such as an inflatable tube or a ring of foamed material such as Styrofoam or the like is connected to the lower portion of the side wall so that the tank will float with most of its length located above the water level and will remain in substantially upright position.

Each of the tanks has an inlet hose **45** which terminates in a trap **49** on the inner side of the encirclement. The trap is disposed slightly below the water level so that water and oil from within the encirclement will flow into the trap. If desired, a trough **50** can be connected to the trap and adapted to extend lengthwise of the tube and be supported thereby in any desired manner. The trough is adapted to direct oil along the length of the tube to the trap.

In order to cause water and oil to flow from the trap into the tank **35**, a vacuum line **51** having a valve therein is connected to the upper portion of the tank and this line may be connected to the low pressure side of an air pump (not shown). The tank **35** may have an outlet pipe **53** communicating with the interior thereof and extending outwardly through a valve. Normally the valve is closed; however, when sufficient oil has been collected within the tank, a tanker or other vessel can move to a position alongside the tank and connect the outlet pipe **53** to an exhaust pump so that after the valve is open, the oil in the tank can be pumped. As the oil is being removed, the water line within the tank again will rise at least to the level of the outlet pipe.

Worthington Corporation Design

A design developed by *W.C. Smith; U.S. Patent 3,703,811; November 28, 1972; assigned to Worthington Corporation* utilizes an oil boom which comprises a continuous length of tubing made of flexible material, a first weighted fin of flexing material suspended from the tubing along its length thereof and valve means secured to the tubing at predetermined locations along the length thereof for selectively permitting the ingress and egress of fluid to and from the tubing whereby the tubing can be inflated and deflated. Thus, the oil boom may be utilized not only to contain an oil slick floating on water but also as a conduit to store and/or transfer oil which has been removed therefrom.

As shown in Figure 130, the boom section **10** comprises a continuous length of tubing or conduit **12** which may be inflated or deflated. Depending from the tubing is a first fin or skirt **14** which functions much like the keel of a boat in maintaining the boom in a proper vertical position when the boom is in the water generally designated at **16**.

FIGURE 130: WORTHINGTON CORPORATION OIL CONTAINMENT AND REMOVAL BOOM

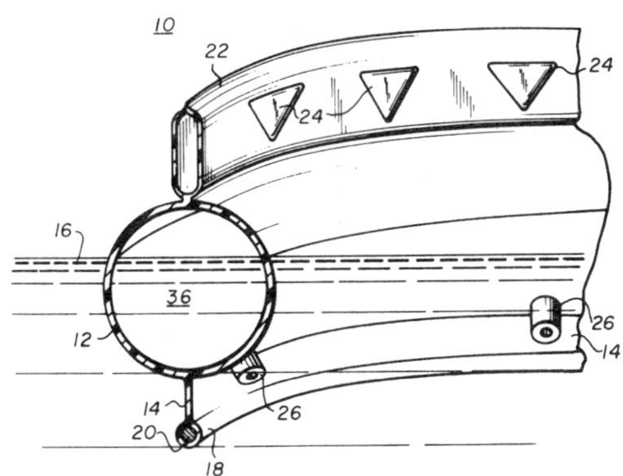

Source: W.C. Smith; U.S. Patent 3,703,811; November 28, 1972

Embedded in the lower border area **18** of the depending fin **14** is a ballast, preferably in the form of a continuous flexible metal cable **20**, which aids in maintaining the boom in its proper upright position in the water. Finally, the boom includes a second fin or skirt **22** upstanding from the tubing along its length thereof. The second skirt further includes flotation means therein. In the example illustrated, this flotation means comprises a plurality of individual flotation pockets **24** provided in the fin **22** during the manufacture of the boom.

To facilitate the ingress and egress of fluids to and from the tube **12**, a plurality of valves **26** are spaced at predetermined locations along the length of the tube. Once the oil slick is surrounded, the end of the tubing containing recovered oil **36** or one of the valves **26** can be connected to the oil recovery means (not shown) and the oil recovered from the slick.

OIL REMOVAL FROM OPEN WATER SURFACES

Once an oil spill has occurred, the most positive approach to protect the environment is to physically remove the oil from the water. This may be accomplished, to a more or less degree, by the use of mechanical pickup devices commonly called skimmers. The numerous types of skimmmers presently under development do not permit a presentation on each particular unit. If the reader desires information on particular devices, the Edison Water Quality Laboratory's May 1970 publication titled, *Oil Skimming Devices,* is recommended.

The broad principle of skimming unwanted material from the surface of a body of liquid, for example, dross from a bath of molten lead or excess fat from the surface of a pan of soup, is, of course, very old. Special problems arise, however, if the body of liquid is very large in extent, and it undulates and heaves as the surface of the sea will do.

The ideal oil skimmer should be designed for: (1) easy handling (2) easy operation (3) low maintenance (4) ability to withstand rough handling (5) versatility to operate in various wave and current situations and (6) ability to skim oil at a high oil to water ratio. The present day skimmers are generally designed to emphasize one or more of the above characteristics. Before purchasing a skimmer for use in a particular area, the buyer should know what he requires and which characteristics best suit his needs.

Tests and experiences have indicated that present day skimmers do not operate efficiently in wave heights greater than 1.5 to 2.0 feet or in currents greater than 1.0 to 1.5 feet per second.

Prior to purchasing an oil skimmer, the buyer should determine physical restrictions, such as piers, shallow water, marshes, etc., estimate the quantities and types of oil to be handled, investigate the local weather conditions and become familiar with the body of water in which the proposed skimmer is to be used. Using this information the proper oil skimmer can then be selected.

Mechanical devices which physically remove oil from the water's surface contain three basic components: (1) the pickup head (2) the pump system and (3) the oil/water separator. These components may be constructed as one unit, separate units or any combination of the three. A brief discussion of each follows. The three most popular types of pickup heads in use today are: (1) weir type (2) floating suction type and (3) adsorbent surface type.

There are several parameters which will effect the selection of a particular pickup head.

These are (1) type of oil (2) sea state - waves and currents (3) debris - sorbents (4) physical restrictions - piers, etc. and (5) equipment available to separate oil/water mixtures. The weir type removes oil from the water's surface by allowing the oil to overflow a weir into a collecting device and holding the water back against the weir. The efficiency of weir pickup heads is highly dependent on calm water conditions and an adequate thickness of oil; however, viscosity of the oil is not too important.

Even so, a certain amount of water is drawn over the weir making it necessary to provide a means for separating the oil and water. Therefore, this type of pickup head would be used in areas where the water conditions are generally calm and the oil slick thickness can be maintained at greater than ¼ inch. The type of oil is not a major consideration.

The floating suction type operates on the same principal as the household vacuum cleaner. The unit is generally small in size and is connected to the oil/water separation section by either a suction or a pressure hose. The floating head may limit the amount of water either by a system of weirs or by the design of the intake openings.

Water surface conditions and oil viscosity greatly affect the efficiency of these devices. The floating suction heads are well suited for working in tight areas such as around piers and ships. Debris and sorbents tend to clog these units quite readily and therefore must be considered.

The adsorbent surface type operates on the principle that a hydrophobic and oleophilic surface will be preferentially wetted by oil and not water. As the adsorbent surface is drawn through the slick oil will cling to it. This oil is then removed either by rollers, wringers or wiper blades. The surfaces are constructed of aluminum or various oleophilic foams or fibers.

The problem of additional oil/water separation is eliminated with these devices. They have been tested successfully in wave heights up to 2 feet. Debris interferes with their efficiency and in some cases may cause severe damage. These units would be best suited for open harbor use. The type of oil to be collected determines the type of adsorbent surface required. For example, heavy oils tend to clog foams and fibers and light oils are not picked up efficiently on aluminum surfaces.

Pumps and other accessories used with skimmers need to have the following features incorporated in their designs:

1. Dependability - carburetors, magnetos, points and plugs need to be protected against ocean air and salt spray.
2. Portability - easy to transport. Equipment must have good balance and good handles for manual lifting, and sling attachments for mechanical lifting.
3. Quick and positive coupling and uncoupling of hoses and other connections is essential.
4. Low maintenance costs and easy cleaning and storage arrangements.
5. Ability to work in gangs to handle increased volumes.
6. Protection from clogging due to debris or sorbents.

The selection of the type of pump is very important. A centrifugal pump can handle large volumes of flow which may be required to remove very thin slicks. However, this type of pump imparts tremendous mixing energy and produces an oil/water emulsion which may not separate satisfactorily in the settling chamber.

On the other hand, positive displacement pumps, such as diaphragm or piston pumps, while they cannot handle as large a flow, do not mix or emulsify the oil/water mixture.

The efficiency of the settling chamber is thus increased. Positive displacement pumps also are very easy to repair, unclog and maintain and do not require priming. Here, the choice of a pump depends on the volume of water flow required, the size of the oil/water separator available and the expected amount of maintenance assistance available. Based on past experience, the chief cause for an unsuccessful oil skimming operation has been due to failure in the pumping system.

The oil/water separator is essentially a settling chamber where collected oil separates by gravity from collected water. A system of weirs and baffles may be employed to facilitate separation. The water may either be pumped out or forced out by gravity. The oil is normally pumped out when the tank has reached its oil storage capacity. The separator may be nothing more than a tank truck or it may be specially designed to meet particular needs.

Normally, the weir type and the floating suction head type pickup units require an oil/water separator. Likewise, if a centrifugal pump is used, a separator is usually required. Mechanical clean-up devices generally have not been capable of performing to the predicted level. At present, these devices are suitable only for calm water and well-defined slicks. Although proponents of particular systems have claimed rough water capabilities, none has been proved effective so far and the prospects are not promising.

WEIR DEVICES

This type of skimmer depends on gravity to drain the oil off the surface of the water. Once collected in a sump below the surface, the oil may be pumped to a storage area. A number of different kinds of weir skimmers are available, ranging in size from the very small individual unit to the large barge type. Simplicity and mobility are among the advantages of the weir type skimmer. Its efficiency rapidly decreases, however, in waves. In larger waves, small units may be swamped.

This skimmer as shown in Figure 131, removes oil from a mixture of oil and water by allowing the oil to overflow a weir into a collecting device and holding back the water against the weir. The height of the weir is usually adjustable by various means such as rate of pumping oil away, mechanical adjustment, etc., so that the thickness of the oil being skimmed from the water can be controlled. The weir can take on many configurations, smooth plate, saws teeth, circle, and many others.

FIGURE 131: SCHEMATIC OF WEIR-TYPE OIL SKIMMER

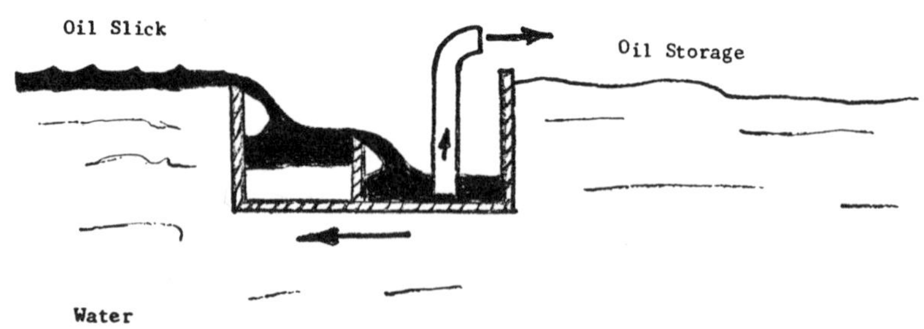

Source: EPA Report PB 218,504

The Standard Oil Company (New Jersey) reports in its publication, *Oil Spill Cleanup Manual,* prepared by Oil Spill Task Force, Standard Oil Spill Task Force, Standard Oil Company (New Jersey), Logistics Department, (1969) that:

> "The operating conditions required for skimming vessels which recover oil by drawing it over a weir are similar to those for suction skimmers. Their efficiency is highly dependent on calm water and an adequate thickness of oil so that water intake is minimized; however, viscosity is much less important. Even under ideal conditions, a certain amount of water is drawn over the weir, thus making it necessary to provide a means for separating the oil and water.
>
> Various weir arrangements are used to collect spilled oil. Straight-edged weirs are usually used as the primary collector at the forward end of the skimming vessels. The oil and water mixture collected over the straight-edged weir then flows by gravity to a sump where the oil and water separate. A secondary weir is sometimes installed in the sump to skim the oil which has separated from the water."

Cities Service Oil Company Device

A device developed by *J.W. Penton; U.S. Patent 3,762,556; October 2, 1973; assigned to Cities Service Oil Company* comprises a floating housing placed within a body of water, for example, a settling pond having an oil slick thereon. The housing includes multiple, interconnected hinged gates to provide weirs over which the oil slick can spill into an oil sump within the housing. Oil which accumulates within the housing is pumped out of the sump into a storage vessel.

The interconnected hinged gates are counterbalanced to hold them in a normally upright position but as oil is pumped from the housing the interconnected gates fall inwardly into the housing at a slight angle so that oil spills over the gates into the sump while also preventing excessive inflow of water. The housing has multiple compartments defined by the gates. Water is removed from the compartments through piping which communicates with the lower region of each compartment.

Figure 132 shows such an apparatus which contains a single gate for decanting oil into the housing for oil accumulation. Housing **101** has four sides and a bottom and is equipped with a gate **102** which forms a substantial part of one side thereof. The gate **102** is pivotally connected at its bottom to the housing **101** by hinge **104** such that the gate **102** is mounted flush with the adjacent walls **105** of the housing **101**.

Connected to the upper portion of gate **102** is an eye-hook **103**. A wire line **113** connects the eye-hook to a counterweight **115**, which balances the gate **102** in a normally upright position by passing wire line **113** over pulley **116**. Therefore, when the housing and exterior portion of the housing are submerged in a fluid such as water, the gate is normally positioned in an upright fashion.

Provided within the housing is a suction tube or siphon **107** having a quick release union **112** upon which may be attached pump hose **111** such that fluid contained within the housing **101** may be pumped through siphon **107**, union **112**, pump hose **111**, pump **108**, storage line **109** and into storage reservoir **110**. Axially positioned about siphon **107** may be a screen or mesh material **106** to prevent large entrained particles from contaminating the flow through siphon **107**.

In the typical operation of the apparatus depicted, the gate **102** is initially in an upright position. When the pump **108** is started, there is a suction produced within the housing which forces the body of water exterior to housing **101** against gate **102** pushes inwardly, and by control of the weight of the counterweight **115**, wire line **113**, passing over pulley **116** and attached to the gate **102** by eye-hook **103**, from about 1/32 to 1 inch of oil slick is allowed to decant into housing **101**.

FIGURE 132: CITIES SERVICE WEIR-TYPE OIL SKIMMER

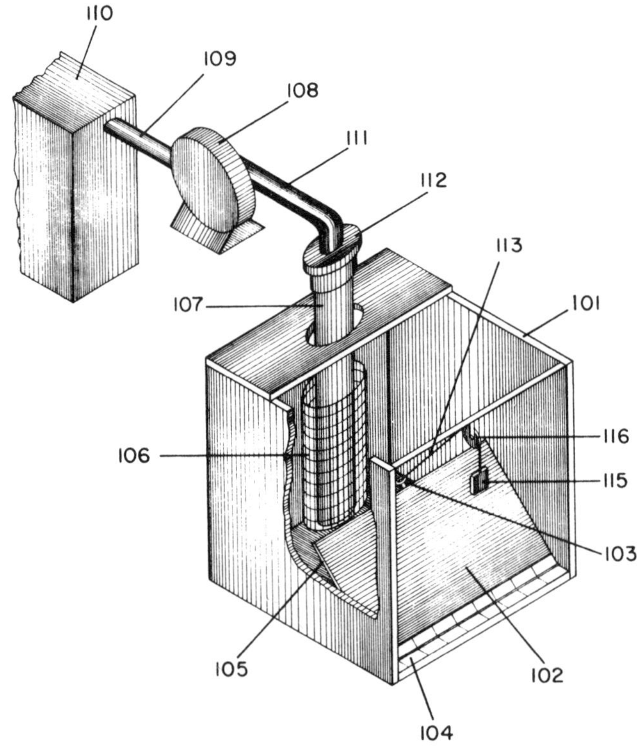

Source: J.W. Penton; U.S. Patent 3,762,556; October 2, 1973

This oil is picked up through siphon **107** and pump line **111** and conveyed through pump **108** and reservoir line **109** into storage reservoir **110**, thereby removing the oil slick from the body of water and accumulating it in an easily moveable storage container. The apparatus described can be permanently positioned in a stationary manner as by extending legs from the housing and sinking them to the floor of the body of water.

Alternatively, the apparatus can be constructed of a buoyant material such as wood so it can float and be moved freely about the body of water, thereby being positionable about the body of water in those locations where a heavy oil slick buildup has formed. A valve can be placed in the lower portion of the housing to allow the removal of water. In operation, this valve can be automatically manipulated by connecting it to a float which opens the valve when a predetermined water level is obtained.

Gulf Oil Corporation Device

A device developed by *J.L. Henning, Jr. and W.J. Robicheaux; U.S. Patent 3,702,134; November 7, 1972; assigned to Gulf Oil Corporation* involves an apparatus for skimming oil from the surface of water in which oil overflowing a weir drains into an overflow conduit joined to a counterbalancing duct. A float supports the juncture of the overflow conduit and counterbalancing duct at a substantially constant position relative to the liquid level. A drain line is connected to the counterbalancing arm whereby a difference in the rate at which liquid flows through the overflow conduit and the drain line causes a change in the liquid level in the counterbalancing duct. A change in volume of liquid in the counterbalancing arm rotates the overflow conduit to change the level of the weir and thereby adjust the flow through the overflow conduit to approximately that through the drain line. A float secured to the counterbalancing duct stabilizes operation of the apparatus.

Figure 133 shows one form of such a device in plan and sectional elevation in which the counterbalancing duct is a continuation of the overflow conduit, and the drain line is connected to the end of the counterbalancing duct remote from the skimmer.

FIGURE 133: GULF OIL CORPORATION WEIR-TYPE OIL SKIMMER

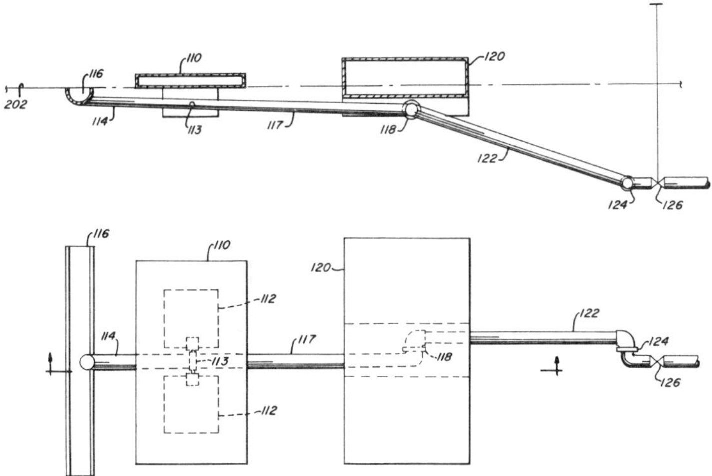

Source: J.L. Henning, Jr. and W.J. Robicheaux; U.S. Patent 3,702,134; November 7, 1972

In the embodiment illustrated, the oil skimmed from the surface of the water must flow through the counterbalancing duct before it enters the drain line. Float **110** has a pair of spaced, downwardly projecting, buoyancy chambers **112** between which an overflow conduit **114** is pivotally suspended by pivot **113**. At the left end of overflow conduit **114**, as shown, is a skimmer pipe **116**. Skimmer pipe **116** is a section of pipe, for example, 4-inch pipe, cut longitudinally and closed at its ends to form an open-topped channel through which liquid overflowing the edge of the skimmer pipe is directed into overflow conduit **114**.

Overflow conduit **114** is joined at pivot **113** to the end of a counterbalancing duct **117** which continues to a swivel joint **118** supported by a stabilizing float **120**. A drain line **122** extends from the swivel joint to a second swivel joint **124**. Counterbalancing duct **117** is approximately 1½ times as long as overflow conduit **114**.

In the operation of the skimming apparatus shown, liquid overflowing into the skimmer pipe **116** is delivered through overflow conduit **114** and counterbalancing duct **117** into drain line **122** and discharged from the system. Normally the valve **126** in the discharge line is set to maintain a liquid level in drain line **122** between swivel joint **118** and swivel **124**. If the valve is opened too wide, the level of liquid in drain pipe **122** falls which reduces the load carried by stabilizing float **120**. That float and the right-hand end of counterbalancing duct **117** rise and cause overflow conduit **114** to pivot about pivotal support **113** and lower the skimmer pipe **116**. The lowering of the skimmer pipe increases the rate at which liquid flows into the overflow conduit and thereby raises the level of liquid in drain line **122** and the overflow conduit **114**.

It will be noted that the entire counterbalancing duct **117** is filled before the liquid backs up into overflow conduit **114**. The greater length of counterbalancing duct **117** than overflow conduit **114** reduces the vertical movement of skimmer pipe **116** for a given vertical movement of the stabilizing float **120** to increase the accuracy with which the level of the overflow edge of the skimmer pipe is fixed.

This skimming apparatus is higly sensitive in that a small change in the volume of liquid in the apparatus will cause an adjustment of the level of the overflow weir. The high sensitivity contributes to the accuracy of positioning the level of the overflow weir by making insignificant any change in the level of the floats as a result of a change in the volume of liquid in the apparatus.

The accuracy of control of the inlet of the overflow weir is further enhanced by the greater distance of the stabilizing float than the inlet from the pivotal center. A relatively large movement of the stabilizing float will result in a reduced movement of the inlet of overflow weir and thereby contribute to the accuracy of control of flow of liquids into the ap-apparatus. The stabilizing float reduces hunting and thereby adds to the accuracy with which the level of the inlet can be controlled.

Gulf Research and Development Company Device

A device developed by *R.C. Amero and G.L. Karner; U.S. Patent 3,534,859; October 20, 1970; assigned to Gulf Research and Development Company* for removing and collecting oil floating on water contains a first inner member which serves as both a main flotation member and a notched weir, and an outer buoyancy member held above the flotation member and closely adjacent the surface of the oil. A flotsam screen is provided. An inflatable embodiment easily carried on vessels or other vehicles is also provided.

Figure 134 is a sectional perspective view of such a device. Referring to the drawing, **10** designates an oil recovery device which comprises an inner flotation member **12**, an outer stabilizing and buoyancy member **14**, and a plurality of rib members **16** interconnecting the members **12** and **14**. Suspended from inner flotation member **12** is a combined tank and funnel assembly **18** which comprises a tank member **20** within which is nested a funnel member **22**.

Suspended from outer buoyancy member **14** is a screen **24** which extends down from member **14** through the layer of oil **26** and into the water **28** below the level of the uppermost portion of inner member **12**, or lower. The screen will surround the inner member even where the outer member is discontinuous, or provided with a gap or gaps, as described below.

Means are provided to adjustably control the buoyancy of the overall device by controlling the amount of air and water within inner flotation member **12**, and to also control the horizontal position of the device so as to keep the top surface of the flotation member **12** level. The buoyancy control permits location of the top surface of the inner member at a predetermined level with respect to the oil/water interface. To this end, flotation member **12** is divided into a plurality of separate compartments by a number of transverse dividing members **30**.

Each compartment carries a first valve means **32** which may be a compressed air fitting to permit filling the compartment with compressed air; second valve means **34** which may comprise an air release valve; and a third valve means **36** which may comprise a combined water inlet and outlet valve. By manipulating the ratio of compressed air and water in each compartment of member **12**, the overall buoyancy of the entire device may be controlled. Thus, by selectively changing the air to water ratio in the various compartments, the device may be caused to float with the top surface of the flotation member **12** both level and at any predetermined level with respect to the oil/water interface.

Means are provided to present a notched or irregular surface or weir to the liquid which is to flow into and be salvaged by the device **10**. Such a surface provides advantages over a smooth surface in certain situations, which smooth surface is also operable particularly with thick layers of oil, because it is thought that a notched weir will improve buoyancy stability and improve the oil recovery efficiency of the device particularly while operating in thin oil films. The precise physics resulting in these advantages are not understood.

However, it is thought that the notched weir improves stability because the peaks serve as a part of the apparatus tending to be above the top of the water, and the valleys serve to promote cohesion of the droplets of oil making up a thin film into streamlets which flow more readily than the droplets to and then across the weir. To this end, flotation member **12** is torús-shaped and is provided with ridges **38** on its outside surface transverse to its circular axis.

FIGURE 134: GULF RESEARCH AND DEVELOPMENT COMPANY WEIR-TYPE OIL SKIMMER

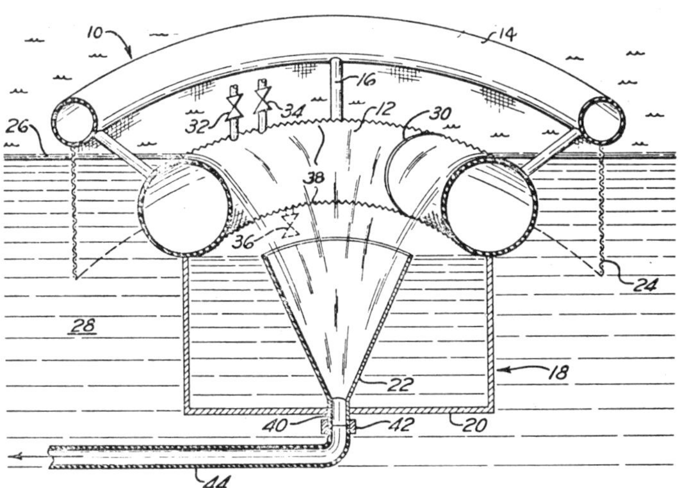

Source: R.C. Amero and G.L. Karner; U.S. Patent 3,534,859; October 20, 1970

It will be understood by those skilled in the art that device **10** could have any configuration such as square or rectangular, or the like, so long as it closes on itself, and the round shape shown is by way of example only. Similarly, the outer buoyancy member **14** is shown as a closed torus by way of example only, and it is anticipated that gaps in the outer member could be provided so that particularly when operating with thick oil layers, the member **14** would not block the flow of oil into the device.

Means are provided to transport the oil collected by the apparatus to other locations. To this end, funnel member **22** comprises a neck **40** which passes through a suitably formed opening in the bottom of tank **20**, and suitable connecting means **42** are provided to connect neck **40** to a hose or other liquid transmission member **44**, to carry away the collected hydrocarbon liquids. A suitable pump, not shown, will be provided to draw the collected liquids away, after which they may be reprocessed, which reprocessing basically comprises removing any water collected with the oil. It is noteworthy that both the necessary pump and a source of compressed air or other gas are already available in virtually all locations where the device would be used.

According to *B.J. Hoffman; U.S. Patent 3,753,497; August 21, 1973* this skimmer removes only oil and will not remove flotsam. Further the screen in this Amero et al device may become easily clogged and it is further claimed that the Amero et al device is not suitable for use in a moving body of water such as a stream.

Hikari Device

According to EPA, Edison, N.J. publication *Oil Skimming Devices,* Report PB 218,504, Springfield, Va., Nat. Tech. Information Service (May 1970) a weir type skimmer is manufactured by Hikari Industries Corporation, Japan and is installed on a catamaran 26 feet long and 13 feet wide. The weir is controllable. The equipment uses an open bottom settling tank to separate the oil and water mixture. As the oil reaches a certain level, it flows into a second of several successive settling tanks, each with bottom drains. The oil is concentrated in the last settling tank.

Hoffman Device

A device developed by *B.J. Hoffman; U.S. Patent 3,753,497; August 21, 1973* is a skimmer for removing supernatant matter such as oil from a liquid such as water. The device has a base portion containing a centrally located drain and a plurality of troughs extending outwardly laterally from the drain but somewhat skewed relative to radii from the drain to aid in the formation of a vortex within the drain.

The base portion is supported by a like plurality of laterally extending wings each having a density less than that of the liquid and each independently adjustably affixed to the base portion so that the elevation of the base portion relative to the liquid surface may be varied. A drain tube is connected to the lower end of the drain and after passing through a pump discharges the matter which is entering the drain into a surface material receptacle which may be a multiple outlet container for separating immiscible liquids.

A V-shaped weir partially surrounding the base and wings and mechanically connected thereto may be provided for use if the skimmer in flowing liquids and the skimmer, weir and pump may all be mechanically interconnected so as to float as a unit.

Figure 135 shows such a skimmer in plan and in cross section. The skimmer is seen in the cross-sectional view to consist of a plurality of laterally extending wings **11** or other means for supporting the skimmer in a liquid. These wings are, of course, less dense than the liquid and provide the floating support for the skimmer structure. The wings are attached to a base portion **13** by means of a pair of screws **15** and **17** in conjunction with a fulcrum **19** so that each individual wing **11** is independently adjustable to raise or lower the base portion in the liquid.

Thus, for example, to lower the wing **11** and thus raise the base **13** further out of the water in the area of the particular wing being adjusted screw **15** would be loosened somewhat and screw **17** tightened. Each of the wings may be adjusted in this manner to allow leveling for an all around flow or to compensate the skimmer attitude in a flowing stream environment. The base portion **13** of the skimmer is provided with a plurality of troughs having bottoms **21** and side walls **23**.

The piece forming the side walls **23** is configured similar to the corresponding overlying portion of the wing and thus these side walls are seen in the figure only near the center of the base. The trough side walls near the center of the base provide a point of attachment for a transporting ring **25** which is screwed to the base by means of screw **27**. The trough bottom **21** reaches its highest point at **29** and from thereon toward the center forms the beginnings of a funnel-like drain into which supernatant matter such as a surface liquid or other pollutant may flow. This drain is centrally disposed in the base portion of the skimmer and the lower portion of the drain **31** is, for example, threaded to allow the attachment of a drain tube. Such a drain tube **33** is illustrated.

FIGURE 135: HOFFMAN WEIR-TYPE OIL SKIMMER DESIGN

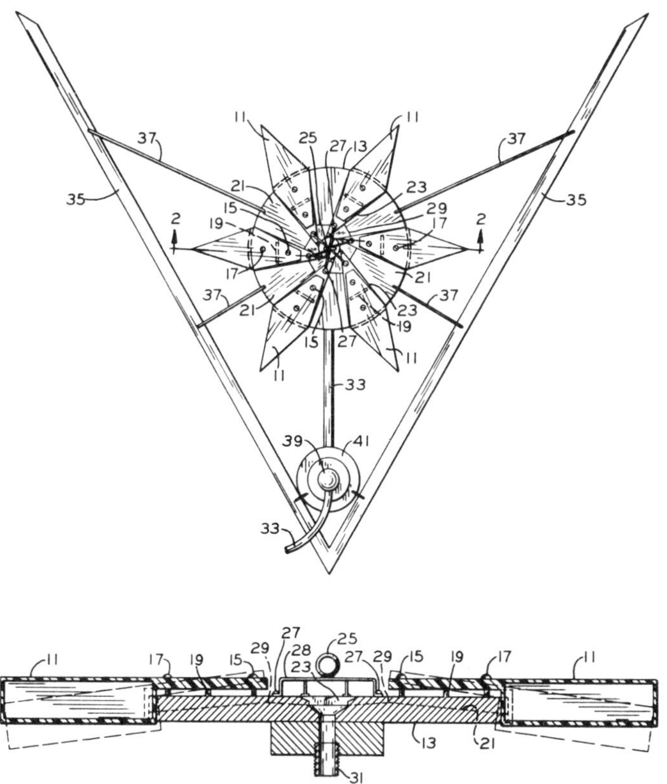

Source: B.J. Hoffman; U.S. Patent 3,753,497; August 21, 1973

The skimmer, as shown in the plan view, is mechanically interconnected with a V-shaped weir **35** which the skimmer may be used in conjunction with when it is desired to skim supernatant matter from a flowing body of water, the skimmer is attached to the weir by a series of ribs or lines **37**. A pump **39** is also mechanically interconnected with the V-shaped weir **35** and is, of course, in circuit with the drain tube **33**. In flowing water environments, the open end of the V-shaped weir or baffle **35** is directed up stream so as to collect the oncoming surface material.

The skimmer (with or without the weir) has the troughs as well as the inner portions of the laterally extending wings skewed somewhat relative to radii extending from the center of the drain. These troughs are skewed so as to induce a vortex within the drain to speed up the drainage portion of the skimming operation. As illustrated in the top view, the liquid entering the drain will tend to have counterclockwise vortices induced due to this skewing of the troughs. This counterclockwise vortex is, of course, best suited for use in the Southern Hemisphere and the direction of skewing of the troughs and wings might be reversed from that shown for a skimmer to be used in the Northern Hemisphere. While the drawing illustrates the pump as being supported near the vortex of the weir, the pump may be provided with a flotation collar **41** and float independently.

Hoyt Corporation Device

A device developed by *O.C. Schell; U.S. Patent 3,701,429; October 31, 1972; assigned to Hoyt Corporation* for skimming oil spills from water, comprises a floating annular body having gently downwardly centrally inclined upper surfaces that terminate radially inwardly adjacent an annular weir. Oil is drawn over the inclined surfaces and flows over the weir and collects in a central sump.

The weir is raised by an annular float disposed in the sump when the overflow in the sump is too great to be removed by a centrally positioned pump. A ballast chamber makes possible a rough adjustment of the draft of the skimmer. Peripherally spaced marginal floats disposed in upright casings that extend above and below the water line are vertically adjustable to trim the floating skimmer and to make possible a fine adjustment of the draft.

As shown in Figure 136, the present device takes the form of a buoyant body indicated generally at **1**, comprising a generally flat annular hollow float **3** that surrounds a central upwardly open well **5** that extends below annular float **3**. The material of float **3** and well **5**, as well as other parts of the device, is preferably inert to seawater and hydrocarbons and may, for example, be glass fiber impregnated with a thermosetting resin such as an epoxide resin, of any of the types well-known to persons having ordinary skill in this field.

FIGURE 136: HOYT WEIR-TYPE OIL SKIMMER DESIGN

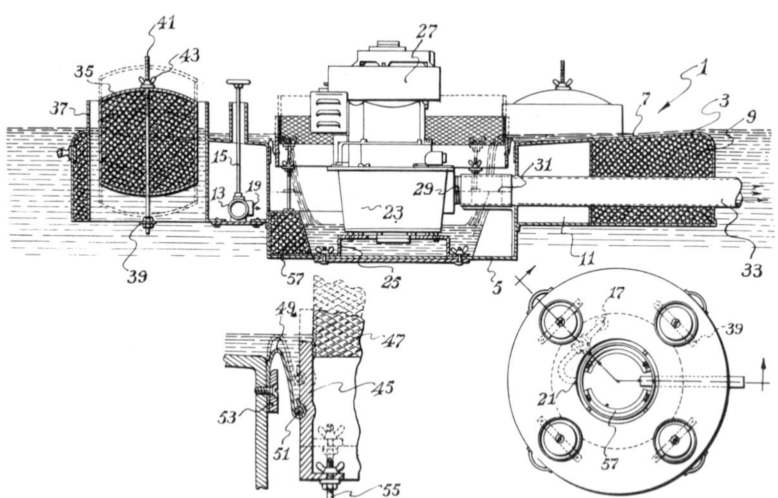

Source: O.C. Schell; U.S. Patent 3,701,429; October 31, 1972

The upper surface **7** of float **3** is conical and is pitched radially inwardly and downwardly at a small acute angle, preferably about 4° to the horizontal, so that the summit of float **3** is a circular annulus adjacent the radially outermost portion of float **3**. Surface **7** is smoothly rounded adjacent this summit so as to promote laminar flow thereover and to minimize turbulence. This summit provides the first weir over which the material to be skimmed from the surface of the water passes in a radially inward direction; and the height of this summit beneath the upper surface of the water greatly influences the flow rate. It is accordingly very important to regulate the buoyancy of the skimmer. Consequently, the float **3** is filled with plastic foam **9** of any conventional type, which may be foamed in situ but which is discontinuous at its radially inner portions so as to define a ballast chamber **11** for the reception of water thereby adjusting the draft of the skimmer.

To this end, a three-way valve **13** is supported in ballast chamber **11** and may be manually controlled by a valve operator **15** from the upper side of the skimmer. An intake **17** admits water from beneath float **3**; and valve **13**, in one position thereof, directs this water through an opening **19** into chamber **11**. In another position of valve **13**, water leaves chamber **11** through opening **19** and flows by gravity to well **5** through an outlet **21**.

Water from ballast chamber **11**, and oil that overflows the summit of float **3**, flows into well **5** and is discharged by a central intake centrifugal pump **23** with its axis vertically disposed, mounted on a support **25** and in turn supporting a superposed pump driving motor **27** which may, for example, be an internal combustion engine. Water and/or oil is discharged from pump **23** through its outlet **29** and passes through a flexible sleeve **31** to a discharge conduit **33** where the pumped liquid is conveyed to the shore or to a vessel for its reception. The use of a flexible sleeve **31** permits discharge conduit **33** to be rigid.

The draft of the skimmer is thus regulated in a general way by use of the ballast chamber **11**. However, fine regulation of the draft, and also the adjustment of the trim of the device, is made possible by a plurality of auxiliary floats **35** filled with rigid foamed plastic and marginally disposed in and equally peripherally spaced about annular float **3**. To accommodate floats **35**, float **3** has a corresponding plurality of upright cylindrical sleeves **37** therethrough that are open at both ends and at least at their upper ends extend above float **3** to prevent liquid from washing over the upper edges.

At their lower ends, sleeves **37** are bridged by horizontal straps **39** secured to float **3**, the straps **39** at their midpoints carrying upright rods **41** which extend through floats **35** and on which floats **35** are vertically reciprocable. Wing nuts **43** are in screw-threaded engagement with the screw-threaded upper ends of rods **41** thereby to permit selective individual adjustment of the height of each float **35** relative to the skimmer. Sleeves **37** reduce the tendency of the liquid within them to rise and fall with wave motion and hence lend a measure of stability to the skimmer.

Just as ballast chamber **11** permits coarse adjustment of the draft of the device while auxiliary floats **35** permit fine adjustment of the draft, so also the ballast chamber **11** permits coarse regulation of the quantity of skimmed liquid that reaches the well **5**. However, a more sensitive adjustment of that quantity of skimmed liquid is provided by an auxiliary weir **45** in the form of a cylindrical sleeve whose axis is vertical and is coaxial with the skimmer, weir **45** being spaced radially inwardly from the upper side walls of well **5** and extending vertically above and below the radially inner edge of conical upper surface **7** of float **3**. A cylindrical wire mesh screen **47** about the upper inner edge of weir **45** prevents floating objects from overflowing the upper edge of the weir and clogging pump **23**.

The space between weir **45** and the upper inner edge of well **5** is closed by a flexible annular apron **49** of waterproof and oilproof material such as synthetic rubber or plastic or the like. Apron **49** is secured about its inner periphery to the outer side of weir **45** by means of a wire **51** and is secured at its outer periphery to the inner upper edge of well **5** by a band **53**.

The weir is supported by upright brackets **55** on a sectional annular float **57** which may, for example, be of rigid foamed plastic or the like. If the rate of admission to well **5** of oil from the surface of the body of water and/or water from ballast chamber **11** is greater than the rate of discharge by pump **23**, then the buoyancy of float **57** with its supported weir **45** and brackets **55** will cause it to rise from the bottom of well **5** and raise the weir **45** so as to reduce or discontinue the flow of liquid over the weir. Float **57** is made sectional so that it can rise on either sie of the discharge conduit. In operation, valve **13** is actuated to admit water to ballast chamber **11** or to discharge water from ballast chamber **11** to well **5**, so as to make a coarse adjustment of the draft of the skimmer.

The auxiliary floats **35** are then adjusted vertically on their respective rods **41**, to trim the skimmer and to make a fine adjustment of the draft, so that oil overflows the summit of conical upper surface **7** in a radially inward direction at a flow rate limited by the thickness of the floating oil layer or the capacity of the pump, whichever is less. The oil flows down the inclined surface **7** until it joins the annular lake backed up behind weir **45**, and then overflows the weir **45** into well **5** where pump **23** discharges it through conduit **33**.

It is particularly to be noted that the placement of the summit of surface **7** at the radially outer portion of float **3**, in combination with the gentle conical incline or surface **7**, greatly increases the ability of the skimmer of this device to draw the surface layer, which in the illustrated embodiment is the oil layer, toward itself from regions relatively distant from the skimmer.

Thus the device has the result of skimming the surface layer at a high flow rate, while at the same time avoiding agitation of the surface layer and avoiding intermixing of the surface layer with the subjacent liquid. In other words, oil flows at a relatively rapid velocity over surface **7**, with substantially laminar flow and with a minimum of turbulence.

Ocean Pollution Control, Incorporated Device

A device developed by *R.M. Sorensen; U.S. Patent 3,662,892; May 16, 1972; assigned to Ocean Pollution Control, Incorporated* is an immiscible liquid separating apparatus having an adjustable weir which admits a predominant portion of the lighter of the two liquids and a minor portion of the heavier of the two liquids into a first chamber. Liquid in the first chamber may communicate with a second chamber over another adjustable weir to permit the lighter of the two liquids to flow over the top of the second weir into the second chamber.

The first chamber is provided with an aspirating slot disposed generally transversely across the bottom of the chamber to permit the heavier of the two liquids admitted to the first chamber to be withdrawn therefrom upon movement of the apparatus through the liquid body. The lighter of the two liquids collected in the second chamber may be removed by pump means disposed within the second chamber or through a line connected to the pump means positioned externally of the chamber.

Figure 137 is a longitudinal section through such an apparatus. The apparatus as shown has a top **14**, a rear wall **15** and a bottom member **16**. The body may be constructed of any suitable material which is resistant to water and oil, such as fiberglass. Disposed transversely between the side walls is a downwardly and rearwardly extending weir **17**. The weir comprises a first plate member **18** having an arcuate section which terminates in a substantially horizontal portion **20** near the bottom of the side walls. Parallel and proximate to plate **18** is a second plate **19**, which together with plate **18** defines a slot **21** within which is slidably carried a gate **22**. The gate may be vertically adjusted within slot **21** by rotation of crank **23**.

Wire **31** is affixed to crank **23** through top **14** and will be wound thereabout upon rotation of crank **23**. As illustrated, plates **18** and **19** define a U-shaped opening **33** within which the gate **22** may be adjusted. Thus, the amount of liquid admitted to the apparatus **10** may be varied by vertical adjustment of gate **22** which may be used to reduce the size of the opening defined between the lip of gate **22** and top **14** of the apparatus. Mounted on weir **17** across the opening is a large mesh screen which prevents large object from passing into the apparatus. Rearwardly of weir **17** is another weir **33** comprised of parallel plates **34** and **36**. Plates **34** and **36** define a slot **37** therebetween which slidably receives a gate **38**. Wire **46** is attached, through the top **14**, to clamps **48** which are, in turn, attached to the top lip **51** of gate **38**. Plates **34** and **36**, like plates **18** and **19**, define a U-shaped opening within which gate **38** may be elevated and lowered. Weir **33** defines with weir **17**, bottom member **16** and the side walls a first chamber **52**. Disposed transversely across chamber **52** are perforate swash plates **53** and **54**.

Weir **33** forms with back wall **15** and bottom member **16** and the side walls a second chamber **56**. Transversely disposed across chamber **56** is a perforate swash plate **57**. Also contained within chamber **56** is a fluid pump, generally indicated by reference numeral **58**, supported within a housing **59** which extends between the side walls. Housing **59** is provided with a plurality of depending legs **61** which support the intake **62** of pump **58** above bottom member **15**. Pump **58** discharges liquid passing therethrough through a conduit **60** which passes through top **14**.

FIGURE 137: OCEAN POLLUTION CONTROL, INCORPORATED WEIR-TYPE OIL SKIMMER DESIGN

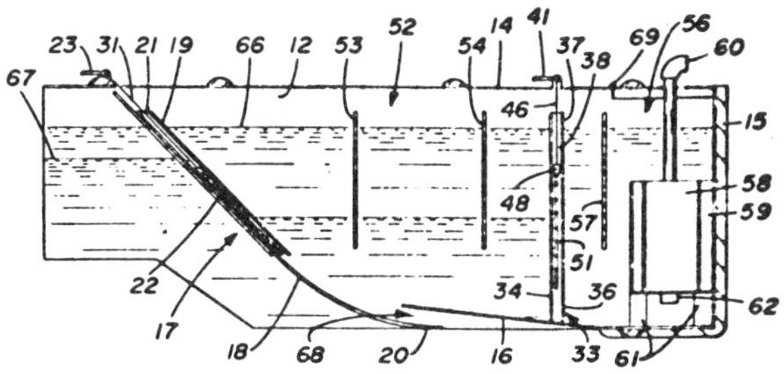

Source: R.M. Sorensen; U.S. Patent 3,662,892; May 16, 1972

In operation, the apparatus is affixed to a suitable oil concentrating apparatus and a flotation collar affixed so that the surface of the body of liquid within which the apparatus is suspended is in approximately the position indicated by the wavy line **66**. The gate **22** is then adjusted by manipulation of crank **23** to position the top lip of the gate slightly below the interface between the oil and the water. The surface of the water is generally indicated by the phantom line **67**. The apparatus is then towed through the body of liquid causing the oil and some water to be drawn over gate **22** into chamber **52**.

The water which is admitted to chamber **52** will settle to the bottom of chamber **52**. As the horizontal portion **20** of front plate **18** and bottom member **16** define a slot **68** therebetween, the forward motion of the apparatus, due to the shape of the slot **18** will create an aspirating effect causing the water to be withdrawn from the chamber **52**. The oil admitted to chamber **52** will pass over lip **51** of gate **38** into chamber **56** where it will pass under housing **59** and be drawn into the intake **62** of pump **58** and discharged through the exhaust line **60** of pump **58** to a suitable storage vessel located in proximity to the apparatus.

The position of gate **38** within the slot **37** may be adjusted depending upon the quantity of oil being admitted to chamber **52**, its viscosity and variables which the operator may observe and evaluate during operation by observing the quantity of water being discharged from the pump **58**. More specifically, if too much water is passing over the gate, it should be raised. Swash plates **53** and **54** facilitate the separation of oil and water in chamber **52** by minimizing turbulence within the chamber generated by wave action of the liquid within which the apparatus is suspended and swash plate **57**, also, minimizes turbulence in chamber **56**.

Reynolds Submarine Services Corporation Devices

A device developed by *R.F. Wirsching; U.S. Patent 3,722,688; March 27, 1973; assigned to Reynolds Submarine Services Corporation* involves a centrally disposed ballast chamber which is surrounded by a sump compartment into which liquids are deposited over the lips of a plurality of independently floating and an articulated weir member. The latter is provided with a substantially universal connection to the ballast chamber. A flexible skirt member forms an external wall portion of the sump compartment and is connected to the lips of the weir member.

In this manner a pressure differential across the skirt member is communicated to an individual segment of the weir member in order to vary the elevation of the weir member in a self-compensating manner. Means are provided to pump liquids out of a submerged exit port in the sump compartment and for imparting direction to these pumped liquids in order to provide controlled propulsion for the skimming apparatus.

Figure 138 is a simplified sectional view of such a device in use. A skimming apparatus is schematically shown at **10a**, and for purposes of reference floats at **40a** and a flexible skirt **44a** are shown to be provided with an open center or axial flow pump **90**. A flexible hose **92** provides a conduit for the oil and water pumped out of the sump compartment in skimmer **10a** to one or more collecting tanks **94** in a separate vessel **96**. The water which has settled out at the bottom of the oil/water mixture in tank **94** is then pumped overboard through pipe **98** with the aid of an auxiliary pump **99**.

FIGURE 138: REYNOLDS WEIR-TYPE OIL SKIMMER DEVICE

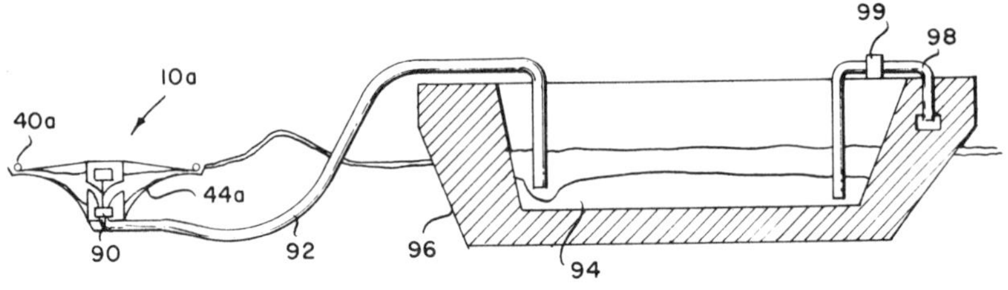

Source: R.F. Wirsching; U.S. Patent 3,722,688; March 27, 1973

A device developed by *A.L. Markel and R.F. Wirsching; U.S. Patent 3,722,689; March 27, 1973; assigned to Reynolds Submarine Services Corporation* gives details of various skimmer-separator devices of a related type.

Schuback Device

As noted by the EPA, Edison, N.J. publication *Oil Skimming Devices,* Report PB 218,504, Springfield, Va., Nat. Tech. Information Service (May 1970), a skimmer was developed by *G. Schuback; U.S. Patent 3,237,774; March 1, 1966.*

This is an apparatus for skimming oil layers from water surfaces and consists of a collecting tank connected to floats. It consists of two floats forming a channel in between, a collecting tank connected to the floats and suspended in the channel, and a bailer of shovel-like design provided on the collecting tank and directed toward one end of the channel. A guiding channel for the collecting tank may be formed by the floats at the skimming side of the apparatus with the guiding channel becoming narrower toward the bailer. A frame may be provided for mounting the collecting tank transversely to the direction of skimming, and the frame may be suspended between the floats on pins extending in the direction of skimming. The collecting tank may be provided adjacent to the bailer with a collecting receptacle which extends into the water. The preferred line-up is to include a suction pipe connected to a pump, which is secured directly to the floats and extends into the collecting receptacle of the tank.

Sea Broom Device

The Sunshine Chemical Sea Broom is a combination pump with a fiberglass skimmer attachment. The system includes a $3\frac{1}{2}$ inch pump which is powered by a 3 hp, 4-cycle gasoline engine. A fiberglass skimming attachment, filled with urethane foam for maximum flotation, converts the pump into a skimming assembly.

Adjustable inlet control covers allow for skimming $\frac{1}{2}$ to 3 inch depths on the surface of the water. The skimmed oil is pumped to tanks, trucks or pits. Further information on the Sea Broom can be obtained from the Sunshine Chemical Company, Jacksonville, Fla. according to EPA Edison, N.J. publication *Oil Skimming Devices,* Report PB 218,504, Springfield, Va., Nat. Tech. Information Service (May 1970).

Sea Sweep Device

This is an oil skimming system under development by Sea Sweep, Inc. and Core Laboratories, Inc. of Dallas, Texas. A single tugboat pulls a barge, and the skimming device is attached to the barge. The barge is equipped with pumps and storage facilities for the recovered oil. The skimming device is attached to the barge by means of a harness mechanism similar to that used in trawl fishing as shown in Figure 139.

FIGURE 139: SEA SWEEP COMBINATION V-BOOM AND NET

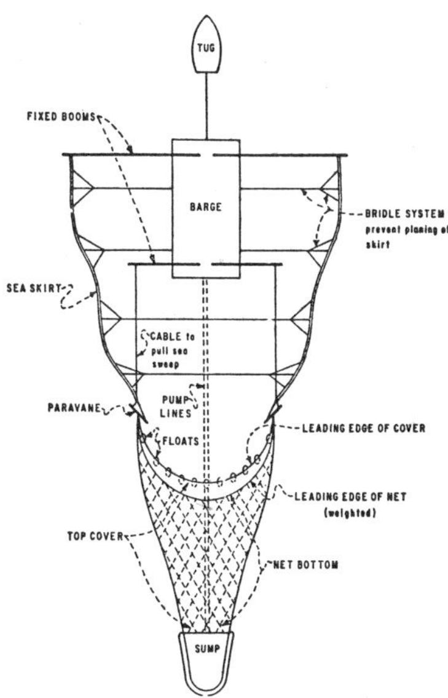

Source: EPA Report PB 218,504

The skimming device is a tapered, flattened funnel assembly of wide-spaced net material, having a cover or top surface of flexible sheet material such as plastic or canvas which is substantially impermeable to oil. The tapered shape of the collecting unit channels the oil toward the back and to a collecting sump at the apex of the funnel. The sump is provided with a bottom panel of wide-spaced netting to permit the water to move freely underneath and out of the sump. The sump is divided into partitioned sections to permit pumping off of the oil with a minimum amount of water. The oil is pumped through plastic lines to storage tanks in the barge.

Widths of the skimming device up to 200 feet are claimed to be feasible. If the assembly is towed at a speed of two knots per hour with a funnel opening of 200 feet and with an oil film of 0.1 inch, the volume of trapped oil will theoretically be 3,600 barrels per hour.

Shell Oil Company Device

A device developed by *P.E. Titus and J.R. Hanson; U.S. Patent 3,688,909; September 5, 1972; assigned to Shell Oil Company* is a skimming device in which a pivoted receptacle having a weir is buoyed to position the weir adjacent the interface of the liquids. Liquids accumulating in the receptacle are withdrawn for disposal. A stabilizing member extends around a substantial portion of the periphery of the skimmer to prevent submergence of the weir due to wave or current movement of the liquids.

The essential element of this device is shown in section in Figure 140 and as mounted in a bouyant platform in the right hand view of the figure. There is shown a skimmer **10** having as major components a buoyant platform **12**, a pivotable receptacle **14**, stabilizing means **16** and liquid removal means **18**. The skimmer **10** is positioned on a body of water **20** having a layer of immiscible liquid **22** thereon. The liquid **22** is a hydrocarbon, e.g., fuel oil, diesel oil, crude oil and the like.

FIGURE 140: SHELL OIL COMPANY WEIR-TYPE SKIMMER DESIGN

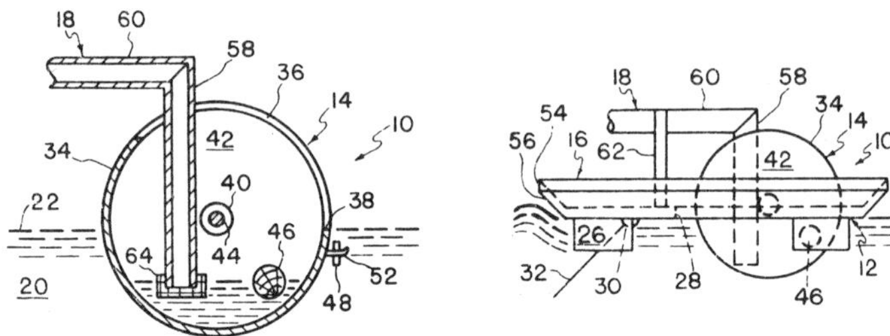

Source: P.E. Titus and J.R. Hanson; U.S. Patent 3,688,909; September 5, 1972

The platform comprises a deck made of any suitable material to which is secured a plurality of floats **26** which may also be of any suitable material. The deck **24** defines a recess **28** on one side of the platform in which the receptacle is located. A suitable connection **30** is provided on the platform for attachment to a line **32** extending to an anchor or deadman to tether the skimmer at a desired location. It will be apparent that the use of two or more divergent anchor lines may be preferred under some circumstances.

The receptacle comprises a cylindrical drum **34** from which a window **36** has been cut defining a weir **38** adjacent the open side of the recess. The window occupies the upper right quadrant and a part of the upper left quadrant of the drum. It will be seen that liquids are prevented from entering the drum opposite from the weir. A bushing **40** is secured, as by welding, to each end wall **42** of the drum along the axis thereof. A rod **44** loosely fits in the bushings and extends through the drum. The free ends of the rod are secured, as by welding, to the deck. It will be seen that there is provided means for pivoting the drum about the longitudinal axis thereof to position the weir adjacent the interface of the liquids. The drum is preferably of moderate size having a small moment of inertia about the longitudinal axis thereof. If the drum has a substantial moment of inertia, buoyant and gravitational forces applied thereto must be large in order to move the weir **38**. Furthermore, the reaction time between the application of such forces and movement of the weir may be so long as to preclude efficient skimming in a body of water having substantial wave movement. For these reasons, it is preferred that the drum be a 55 gallon drum, by which is meant that the drum **34** be approximately the size of a conventional 55 gallon drum.

A buoyant member **46** is located in a compartment defined by the drum below the weir. The buoyant member **46** is preferably solid to obviate splitting of a buoyant container caused by impact or the freezing of a liquid therein. The buoyant member may, for example, be wood and is preferably a log or timber secured to the end walls as by nailing. The buoyant member is secured between a vertical plane passing through the rod and a vertical plane passing through the weir. It will be seen that the weight of the buoyant member imparts a clockwise moment to the drum tending to depress the weir. It will also be seen that the buoyancy of the member imparts a counterclockwise moment to the drum tending to elevate the weir.

It is accordingly apparent that the gravitational moment exeeds the buoyant moment when insufficient liquid is accumulated in the drum thereby depressing the weir and allowing greater liquid flow into the drum. When an excessive amount of liquid is accumulated in the drum, the buoyant moment exceeds the gravitational moment thereby elevating the weir and decreasing the liquid flow into the drum. For a number of reasons it is desirable to have the capability to adjust the balance between gravitational and buoyancy induced moments.

For example, the layer **22** to be skimmed may be quite thick requiring that the weir be depressed to a greater extent. Similarly, the layer may be quite thin making it desirable to elevate the weir slightly and avoid processing substantial amounts of the underlying water. In addition, the density of the liquid **22** may vary from design parameters thereby affecting the buoyancy of the member. The use of a solid buoyant material such as wood necessarily means that the buoyancy of the member cannot be accurately controlled in a convenient manner.

There is accordingly provided a plurality of weights **48** removably attached to the drum in any suitable manner, as by the provision of a hook **52** attached to the drum on which the weights may be placed. As illustrated, the weights increase the gravitational moment imparted to the drum. It will be apparent that the weights may be placed on the opposite side of the rod **44** to increase the buoyant moment if the buoyant member is properly positioned.

The stabilizing means **16** comprises a member **54** extending beyond the periphery of the platform **12** along a substantial portion of the external periphery thereof. The stabilizing member presents an upwardly inclined downwardly facing surface **56** to the moving liquids **20, 22**. If the leading edge of the stabilizing member is depressed into the oncoming moving water, the surface acts to elevate the leading portion of the platform. This effect of the stabilizing member occurs because of the impact of the moving water thereagainst which may be in the form of current or waves.

The liquid removal means **18** comprises a vertical conduit section **58** extending through the window **36** into the compartment defined by the drum. The conduit section is connected to a horizontal conduit **60** which is supported by a bracket **62** affixed to the deck **24**. The horizontal conduit is connected in any suitable fashion to a pump (not shown) for removing liquids accumulated in the drum.

Positioned on the lowermost end of the conduit **58** is a foraminous member **64** which may be a perforated sleeve but which is preferably a screen. The screen prevents flotsam from entering the conduit thereby obviating damage to the pump. It would appear more advantageous to position a screen over the weir and thereby prevent flotsam from entering the drum. In practice, this approach has not proved practicable. Experience has shown that marine grasses, which are of substantial length and of minimal width, tend to orient in a horizontal fashion upon such a screen in a relatively brief period of time. An impermeable mat soon forms immediately above the weir which causes an insufficient amount of liquid to pass into the drum.

The balance between gravitational and buoyant moments in the drum is soon upset thereby depressing the weir exposing a fresh part of such a screen which in turn becomes clogged with marine grasses. Soon the skimmer **10** must be deactivated and the grasses removed from adjacent the weir. It has been found that the period of uninterrupted operation of the skimmer **10** may be increased substantially by allowing the marine grasses and other flotsam to collect in the drum. Since the grasses are of substantial length, they are unable to completely wrap around the screen and consequently do not mat so readily.

When the skimmer is initially placed in operation, the drum is empty so that the gravitational moment of the member **46** substantially exceeds the buoyant moment thereof. It will be apparent that contact between the buoyant member and the conduit **58** or contact between the conduit and the window **36** maintains the buoyant member on the right side of the vertical plane passing through the rod **40.**

The weir is consequently substantially depressed so that a substantial amount of water passes thereover into the compartment adjacent the screen **64.** Although liquids are being removed through the conduit, substantially more liquids enter the drum than depart. Accordingly, the liquid level in the drum rises thereby increasing the buoyant moment of the member until it exceeds the gravitational moment.

The drum then commences rotation in a counterclockwise direction to elevate the weir and decrease the amount of liquids drawn into the drum. In the event the weir is excessively elevated, the amount of liquid being withdrawn from the drum exceeds the amount passing over the weir such that the liquid level in the drum begins to lower. The buoyant moment acting on the member declines and the weir subsides. After a short period of time, the weir becomes fairly stationary.

If too much water is being withdrawn through the conduit, one or more of the weights **48** may be removed from the hooks **52** to elevate the weir. Similarly, if a substantial amount of the liquid **22** remains on the surface of the liquid **20**, one or more weights may be added to the hooks to depress the weir. It will be seen that the removable weights enable the skimmer to be used to skim thin or thick layers of the liquid **22** from the body of water **20.** Tests conducted in calm water with the skimmer showed the following results.

Oil Thickness Inches	Total Liquid Removal GPM	Oil Removal GPM
0.3[1]	68	5
0.3[1]	92	7
0.4[1]	58	8
1/4 - 1/2[2]	48	35-44
1/3 - 1/2[2]	44	25-37

[1] No. 2 Diesel Oil
[2] No. 6 Fuel Oil

Tests conducted in a wave tank revealed that the skimmer functioned adequately in waves up to two feet high, spaced eight feet apart.

Stebbins-Becker Device

A device developed by *G.B. Stebbins, J.F. Stebbins and J.G. Becker; U.S. Patent 3,722,687; March 27, 1973* is a floating platform, vertically adjustable relative to the surface of the water and an oil slick floating thereon, having a weir member at the outer periphery and a sump and centrifugal pump adjacent the center thereof, and an oil/water separator for receiving an aerated oil and water mixture from the pump and cascading it over vertically spaced pan-like separators for causing the air and oil in the mixture to float to the surface of the separator, and water to flow to the bottom.

Figure 141 is a side elevational view, partly in cross section of such an oil skimming apparatus and associated oil/water separator as well as a vertical section of the skimming apparatus proper (bottom view in the figure).

FIGURE 141: STEBBINS-BECKER WEIR-TYPE SKIMMER DEVICE

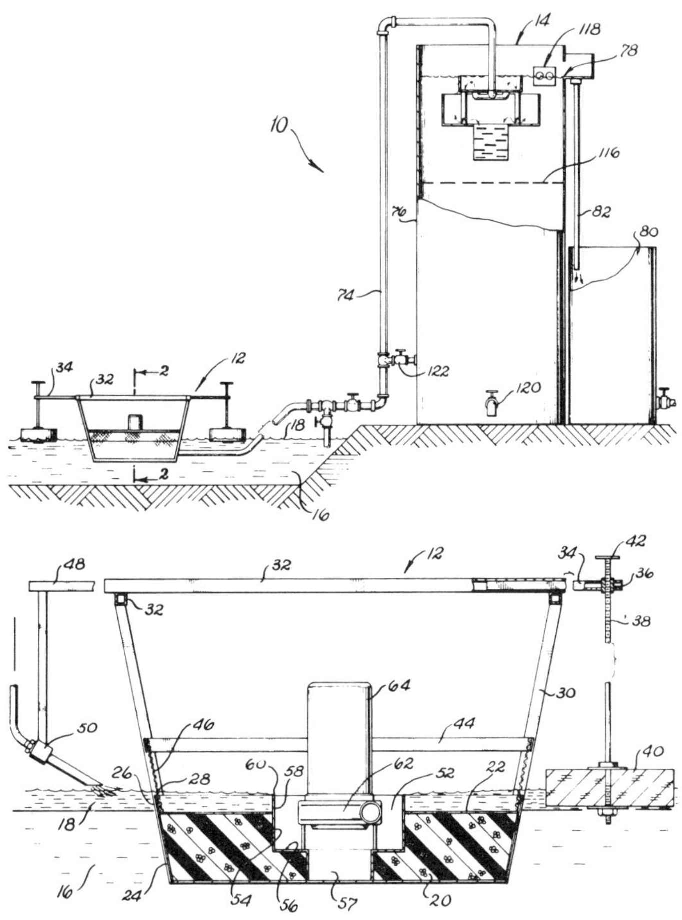

Source: G.B. Stebbins, J.F. Stebbins and J.G. Becker; U.S. Patent 3,722,687; March 27, 1973

The number **10** indicates generally the combination of the floating oil skimming apparatus and oil and water separator including an oil skimming apparatus **12** and an oil/water separator **14**, the skimming apparatus being shown floating on a body of water **16** having a layer of water and oil mixture **18** floating on the upper surface thereof, the oil and water mixture being referred to hereinafter as an oil slick. The skimming apparatus comprises a floating platform **20** of solidified polyurethane foam or similar material which provides a slight negative buoyancy. The platform could be made of a nonfloating material, but this would increase the size of the float members referred to below. The platform includes a flat upper surface **22**, and for convenience of construction is preferably of square configuration with four outer side walls **24** which extend downwardly and inwardly. In engagement with the side walls are four outer weir panels **26** which have upper edges **28** projecting a predetermined distance above the upper surface **22** of the platform.

The framework of the skimming apparatus includes an angle-iron upright **30** fastened to the weir panels **36** at each corner of the platform, and horizontally extending tubular bars **32** fastened to the upper ends of the uprights, each bar having a cantilever portion **34** which extends laterally beyond the uprights.

Adjacent the outer end of each cantilever portion is an internally threaded sleeve **36** which receives the upper end of a threaded adjusting rod **38**, the lower end of each rod being provided with a float member **40**. A handle **42** is provided at the upper end of each adjusting rod, for varying the vertical position of the floats relative to the upper surface of the platform, whereby the upper edges of the outer weir panels **26** can be adjusted relative to the bottom surface of the oil slick **18**, for a purpose to appear.

Fastened adjacent the centers of the uprights and extending transversely therebetween are supporting straps **44**, each of which supports the upper edge of a section of screening **46**, the bottom edge of the screening being fastened to the inside surface of the upper portion of each outer weir panel. The screens prevent trash and debris (not shown) in the oil slick from floating over the outer weir panels and onto the upper surface of the floating platform. If desired, additional outriggers **48** can be supported on the framework, such outriggers supporting nozzles **50** for directing a stream of water under pressure inwardly toward the floating platform to move the oil slick toward the platform and over the outer weir panels.

Adjacent the center of the floating platform is a sump **52** with an annular side wall **54** and a bottom wall **56**, and a recessed vortex chamber **57**. An annular inner weir panel **58** with an upper edge **60** is in engagement with the sidewall, the upper edge being substantially in horizontal alignment with the upper edges **28** of the outer weir panels **26**.

A centrifugal pump **62** which is driven by motor means **64** is supported by legs which bear upon the bottom wall, the pump being positioned directly above the vortex chamber **56**. The unique arrangement of the centrifugal pump with the bottom inlet positioned above the vortex chamber, cavitates and causes air to be drawn into the sump and to become intimately mixed with the oil/water mixture in the slick. This aerated oil/water mixture is then conveyed through a pipe **74** to the top of the oil separator **14**.

The oil and water separator includes a tank **76** with an oil overflow outlet **78** adjacent the top thereof, the top of the outlet is open to the atmosphere for the escape of the entrapped air and the bottom thereof is in communication with an oil storage container **80** through a drain pipe **82**. Although the oil and water separator assembly is relatively efficient, some oil is carried with the water through the discharge member to provide a water and water/oil mixture interface **116** below the lower end of the discharge member.

Located adjacent the top of the tank is a float control **118**, which is used for controlling the level of the aforementioned interface, the control being adjusted to maintain the interface a predetermined distance below the bottom end of the discharge member so as to provide an oil/water pad to receive the aerated oil and water mixture from the pipe **74**.

As shown, a first drain valve **120** is provided adjacent the bottom of the tank for drawing the water therefrom, and a second drain valve **122** interconnects the bottom portion of the tank with a portion of the pipe **74** so that when the pump **62** is not in operation, the water from the tank can be drained directly into the pond **16** or other body of water from which the oil slick is being removed. Thus, it is apparent that there has been provided a floating oil skimming apparatus with an oil and water separator which fulfills all of the objects and advantages sought. The centrifugal pump causes the oil slick to be drawn inwardly over the outer and inner weirs **26** and **58**, respectively, whereby the oil slick is skimmed from the upper surface of the body of water. The unique arrangement of the inlet of the centrifugal pump, the open impeller and the vortex chamber causes air to be intimately mixed with the oil and water mixture, whereby the separation of the oil from the water is facilitated in the separator where the mixture is caused to flow downwardly through circuitous passageways, whereby the air and oil float to the top of the container and the water flows to the bottom.

Thune Device

As developed by *T. Thune; U.S. Patent 3,219,190; November 23, 1965,* this T-T Skimmer from Norway skims oil spillage from the water surface and delivers it into a separator vessel which is maneuvered in the spillage area to sweep the water surface. The device consists of a seagoing vessel for separating liquids and materials of different densities from the surface of a body of water. It has a skimming board of spade form moved against the oil spillage so that the mixture, due to the relative movement, is guided aft into a chamber of the vessel. The bottom of this chamber is open and in contact with the water in which the vessel is in use to allow the water to escape and in the upper part of the chamber, oil is collected in an oil collector and later removed therefrom.

By means of this arrangement, the oil is concentrated in the chamber to such a degree that separation of the oil from the water may be obtained and reportedly only a small amount of water is in the collected oil when it is removed by pumping. The vessel is designed as a barge with a flat bottom and square bow open along a portion thereof with the inclined skimming board disposed at the bow. The lower end of the skimming board and a front inclined portion of the bottom define a horizontal leading edge at a suitable depth under the water surface at the open bow portion.

According to a further development of this device, it is possible to make the operation independent of the movement of the vessel. Accordingly, there is arranged over the inclined skimming board, means to urge the flow of water with lesser density contaminants over the skimming board and paddle wheel. The paddles of the wheel reach down in the water and will, when rotated, move the oil and water mixture over the read edge (weir crest) of the skimming board independently of the movement of the vessel.

The barge is powered by an outboard motor, and paravans can be used to assist in directing the oil to the paddle wheel. This equipment can be purchased from the East Coast Services, Inc., Braintree, Mass. The commercial unit has a capacity of 1,390 gallons per hour of oil recovered. It weighs 3,500 pounds, according to EPA, Edison, N.J. publication, *Oil Skimming Devices,* Report PB 218,504, Springfield, Va., Nat. Tech. Information Service (May 1970).

U.S. Navy Device

Oil may be skimmed by special apparatus from a box-shaped barge known as the Norfolk Skimmer. This weir-type skimming device was used by the U.S. Navy at Norfolk, Va. The skimmer barge is 12 feet wide, 25 feet long, 8 feet 2 inches deep, and has large holes in the bottom for the free passage of water. The top part of the barge consists of air or flotation cells 2 feet 10 inches deep which support the skimmer at approximately 18 inches freeboard. The tank section of the barge extends 5 feet 4 inches below the air cells and has a capacity of 10,000 gallons.

Under the air-cell section of this skimmer, there is a small diffusion box with numerous holes into which the surface oil and water are pumped and from which they flow with little turbulence into the barge. Here, the oil and water are separated by gravity. The water passes out through the bottom and the oil remains in the barge. A metal sump, triangular in shape, is fitted to one end of the barge at water level. At the open end of the sump is a dam, the top of which is seven inches above the bottom of the sump. A trash screen rests on the dam. Since the sump can be raised or lowered on the barge, the top of the dam can be adjusted to a point a little lower than the water level. In the bottom of the sump is a six inch connection pipe that leads to a centrifugal pump on the deck of the barge. The pump draws the oil and water over the dam into the sump and down through the diffusion chamber into the main compartment of the barge.

The pump operates at 65,000 gallons per hour and it can draw the oil and water continuously at this rate without changing the water line of the barge because of the large holes in the bottom. In this way, the skimming operation can continue until the barge is completely filled with oil. For example, in one test, the oil slick measured 1,525 gallons. This particular oil slick was removed in 2½ hours, or at the rate of 600 gallons an hour.

The large 65,000 gallons per hour flow of liquid over the dam creates surface movement of the oil and water. Thus, the concerted action results in withdrawing the slick from tight spaces. To induce movement from greater distances, a fire hose or work-boat propellers are used, according to EPA, Edison, N.J. publication *Oil Skimming Devices,* Report PB 218,504, Springfield, Va., Nat. Tech. Information Service (May 1970).

FLOATING SUCTION DEVICES

Similar in many ways to the weir skimmer, this type also can easily be put into action by placing it on the water and adjusting it to float at the oil/water interface. Because of its compactness and shallow draft, the floating suction skimmer is especially useful in shallow water and in confined areas under docks. As with the weir, self-priming pumps give the best results. This type of skimmer also works best in calm water. As the waves grow choppy, it rapidly loses its effectiveness because of air entering the suction hose. And it, too, can become clogged by debris if no screen is used.

Acme Products Inc. Device

The floating saucer skimming device is simply a modified pumping system. This pump handles up to 42,000 gallons per hour and floats unattended. It pumps clean water, sludge, gravel or any solids encountered which will pass through the coarse screen and impeller, without damage. It is rugged, durable and light in weight. It is powered with a heavy duty 4-cycle gasoline engine, electric motor or air motor. The drivers can be interchanged easily in the field.

There is a fiberglass skimmer attachment which easily attaches to the pump body. By removal of the pump intake assembly the pump is converted into a skimming assembly for removal of oil or any liquid floating substance from the surface of the water. The adjustable inlet control covers can be set to allow skimming of a half-inch to three-inch depths on the surface of the water. This device is made by Acme Products Inc. of Oklahoma City, Oklahoma according to EPA, Edison, N.J. publication *Oil Skimming Devices,* Report PB 218,504, Springfield, Va., Nat. Tech. Information Service (May 1970).

This floating surface skimmer as described by *H.E. Stanfield, G.W. Stanfield and G.F. Camp; U.S. Patent 3,693,800; September 26, 1972; assigned to Acme Products Inc.* is a device having a hollow bowl and a disc-shaped float member for supporting the bowl slightly below the surface of a liquid. The bowl has a substantially closed bottom and a substantially continuous vertical side portion, the upper edge of which forms a substantially continuous horizontal side edge constituting a weir.

Adjustment means are provided to adjust the position of the bowl vertically with respect to the float member which, in turn, creates the adjustment of the weir relative to the surface of the liquid. The skimmer is provided with an exhaust means which communicates, at one end, with the interior of the bowl and at its other end with a source of suction.

Cross Device

A device developed by *R.H. Cross III; U.S. Patent 3,706,382; December 19, 1972,* is a buoyant oil-removal device for use with a floating oil-confining barrier on a water surface with oil slicks of substantial thickness floating thereon. The device has oil inflow ports (of lesser vertical height than the oil slick thickness) with the ports normally floated at a level intermediate the thickness of the oil slick to provide inflow of the water-floated oil slick and to minimize either water or air inflow through the ports whether the water is calm or disturbed by waves. The ports are designed to collect oil from the water surface at a rate which is maximized subject to the requirement that intake of water and air is minimized.

The device has a longest overall dimension of not more than one-fourth of the wavelength of the shortest wave of significant amplitude, and includes a generally H-shaped skimmer structure; each arm portion of the structure is of generally rectangular or circular cross section, providing a plurality of inflow ports, each of vertical dimension of the order from 0.04 to 0.2 of the expected slick thickness. Inflow ports preferably are positioned immediately adjacent the top surface of the structure, and are horizontally elongated. The device is provided with a flexible, buoyant suctioning hose communicating with the inflow ports for the removal of oil.

Such a device is shown in Figure 142. The oil-removal device **10** includes a generally H-shpaed skimmer portion **12**, of generally rectangular cross section. The choice of an H-shape for the skimmer manifold is not critical, but provides a compact structure with relatively short arms, thereby ensuring a relatively small pressure drop over the length of the arm. The cross-sectional area of the arms **14** and **16** generally increases from the free ends toward cross-piece **18**.

FIGURE 142: CROSS FLOATING SUCTION SKIMMER DEVICE

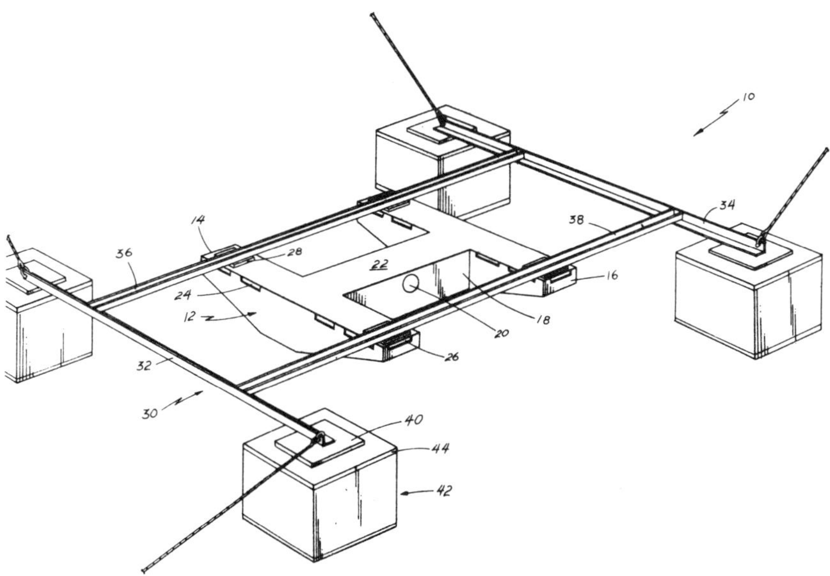

Source: R.H. Cross III; U.S. Patent 3,706,382; December 19, 1972

A hose fitting **20** through which oil is removed is provided in cross-piece **18**, positioned immediately adjacent top surface **22** of skimmer **12**. Inflow ports **24** in the form of slots are provided along the inner and outer sides of arms **14** and **16**. Further slots **26** are provided in the ends of arms **14** and **16** but none are provided on the cross-piece, for reasons that will appear more fully in what follows. The skimmer is preferably constructed of ⅛ inch thick aluminum alloy 5086, intended for marine use; seams may be welded or riveted, and the top or bottom plates may be bolted to the rest of the skimmer to simplify cleaning after use.

The purpose of the top plate on the device is to control the pressure gradient in the flow of oil within the skimmer. This pressure gradient is caused by accelerating the oil from the rest to the exit velocity from the skimmer body into the flexible hose. If the top plate were absent, free surface waves would form on the air-oil interface inside the skimmer, and the pressure gradient (and hence flow-rate) through a given orifice would be uncontrolled. Moreover, additional air would be entrained at the entrance to the hose. The bottom plate performs a similar function with respect to the oil-water interface.

Four frame-mounting plates **28** are provided for attaching the float frame. Float frame **30** is constructed of aluminum alloy structural sections. Two angle sections **32** and **34** are welded to two channel sections **36** and **38**. Skimmer **12** is bolted to channel sections **36** and **38** through frame-mounting plates **28**, using long studs to allow adjustment of the vertical position of the skimmer. Float mounting plates **40** are fastened to the free ends of angle sections **32** and **34**.

The floats **42** may be closed cell foam blocks, with plywood sections **44** bonded to the top and bottom of the foam for strength. Floats **42** are bolted to float mounting plates with stainless steel bolts or threaded rod. The choice of spaced float elements, rather than a single buoyant skimmer, results in a device that presents a small reflecting area to incident waves. Thus waves tend to move past the skimmer, carrying the oil slick with them past the inflow ports. This choice also provides superior wave-following ability.

An oil-removal device constructed as shown is conservatively designed to collect 5 tons of oil per hour, from an oil pool initially about 1 foot deep. For a slick 1 foot thick, with waves and a current, a ratio of oil to total fluid collected of 0.9 can be expected in 3 foot waves, and 0.80 or better in 5 foot waves, using properly selected pumps and hoses, and with proper operation of the entire system. A suitable suctioning means for use with the device may comprise a nonemulsifying, self-priming pump or pumps (such as a double Edson diaphragm pump) with variable speed control; using such equipment, the velocity in each inlet port is less than the maximum velocity at which oil can be drawn into an inlet port without entrainment of air or water, and no valves are required to control this velocity.

The oil-removal device is operable when there is a current in the water of as much as, say, 2 knots, and the efficiency of the device may even be increased, as the device may be oriented so that the current carries oil to the skimmer and concentrates it in a deep pool for collection. The device may be readily oriented, as desired, to take account of current, wave and wind conditions by tethering or anchor lines, adjusted to suit.

Donsbach Device

A device developed by *F.P. Donsbach; U.S. Patent 3,741,391; June 26, 1973* is one in which water contaminated with an oil slick is drawn into and confined within a large tub shaped vessel so that the lighter contaminating liquid can be drawn off at the top while the water is pumped away from a lower level.

Figure 143 shows sectional views of this device with the pumping systems switched on (at top) and switched off (at bottom). In the figures, the numeral **10** indicates a body of contaminated water which is confined by a generally cylindrically shaped vessel **2**, while numeral **1** indicates a buoyant ring. In the bottom of the vessel, a tubular sleeve **8** is arranged, to which is attached laterally the conveying system **9** for sucking off the water from the confined space **10**. The opening **15** at the upper end of the sleeve **8** lying in the space can be closed by a valve body **7** and the opening **16**, located in the open water, can be closed by valve **17**. The two valve bodies are connected with one another by way of the guide rod **6**. Numeral **5** indicates a float for the two valves. If the water level **18** in the confined space has dropped to a point where the valve body **7** rests in the opening **15**, then no water can be moved by means of the conveying system **9** from the space **10**.

The conveying system then sucks only outside water through the opening. In that way it will prevent liquid which is to be eliminated, from being sucked through the conveying system from the space and again fed to the outside water. The ring **1** may have an annular groove **19** in its undersurface, which sealingly engages with the upper marginal wall of the vessel **2**.

FIGURE 143: DONSBACH FLOATING SUCTION SKIMMER DESIGN

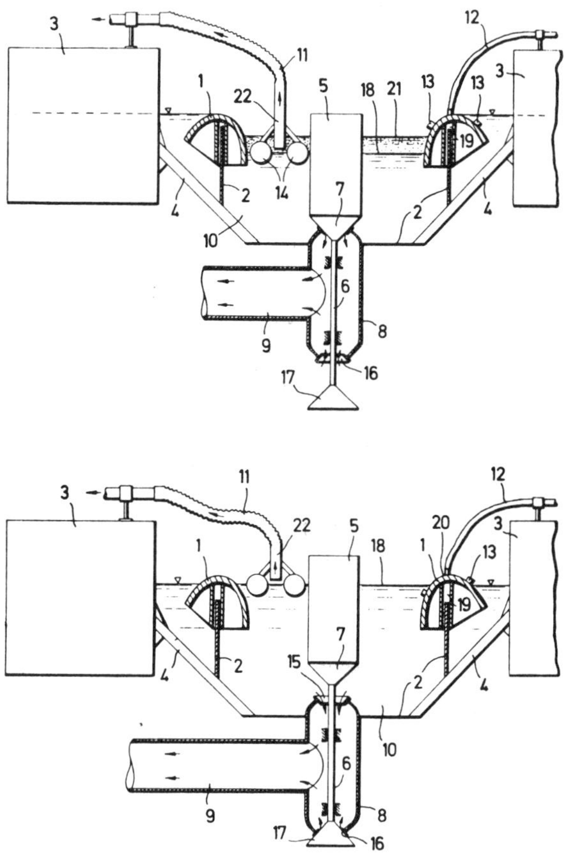

Source: F.P. Donsbach; U.S. Patent 3,741,391; June 26, 1973

Numeral **20** is the upper surface of the ring **1**, **21** is the surface of the liquid to be eliminated and **11** the conveying system for sucking off the liquid that is to be eliminated from the confined space **10**. Ring **1** is freely movable in a perpendicular direction.

In the figure there are furthermore floats **14** which carry the intake stack **22** for the liquid that is to be eliminated, **13** being a connection for a heating system, for example, steam heating system, while **12** is a gas connection, for example, an air connection, through which gas can be forced into the hollow ring **1**. The numeral **4** represents arms attached to the vessel **2**, the other ends of which are attached to floats **3**, which support the entire device.

When water is sucked from the confined space **10** by means of the pumping system **9** below the interface between the water and the liquid to be eliminated, as indicated by numeral **18**, the water level **18** in this space **10** drops. The floatable ring **1** is capable of limited vertical and angular movement with respect to the remainder of the vessel **2** so that it may assume a middle position between the two liquid levels inside and outside the space until liquid that is to be eliminated and some water will flow into the space across the upper edge **20**. As soon as the layer of liquid that is to be eliminated is so thick in the confined space that there is no longer any danger that water will be sucked from the confined space by means of the conveying system **9**, this pumping system is switched on.

Esso Device

As described by *D. Watt; British Patent 959,484; June 3, 1964; assigned to Esso Research and Engineering Company* this skimmer device consists of a suction hose having one end connected to a nozzle and adapted to a float as shown in Figure 144.

FIGURE 144: ESSO FLOATING SUCTION SKIMMER DESIGN

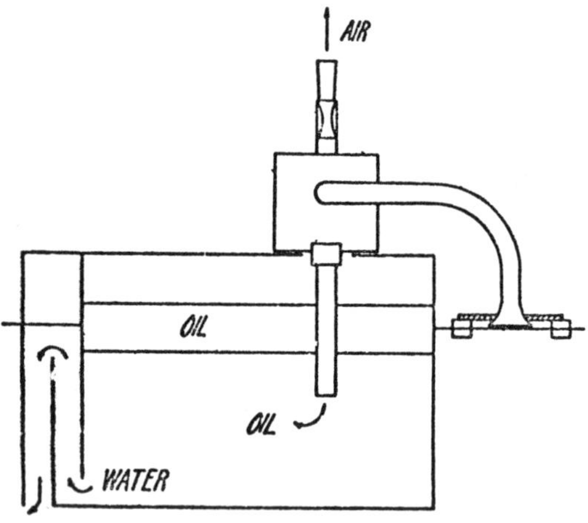

Source: EPA Report PB 218,504

The nozzle is above and directed downwardly toward the upper surface of the floating liquid. A cyclone separator is connected to the other end of the hose. In addition, a separator tank is connected to the cyclone separator as a means for extracting air from the cyclone separator. The end of the suction hose connected to the nozzle is adapted to float with the nozzle between 1½" and ¼" from the upper surface of the floating liquid.

Harrington Device

A device developed by *J.W. Harrington; U.S. Patent 3,534,858; October 20, 1970* consists of a flexible suction hose connected with a suitable vacuum source and a floatable skimmer capable of moving with varying wave motions in such manner that the suction apertures provided in the skimmer are maintained substantially at all times within the layer of pollutant. For sweeping operations to remove large bodies of oil or chemical pollutants on water surfaces, a bed comprising headers connected with a manifold to a common suction pump is utilized. A plurality of the skimmer apparatuses are connected to each header. The individual suction lines are then tied together in such manner as to allow freedom of movement by the individual units, but function as a sweeping unit to cover a large area.

Figure 145 shows this device mounted on a vessel ready for operational use (upper view) and in cross-sectional detail (lower view). This device is shown to comprise a frame **2** hingedly mounted on the side of a ship **4** and adapted to be lowered into operational position or raised when not in use by a conventional boom and pulley system.

FIGURE 145: HARRINGTON FLOATING SUCTION SKIMMER SYSTEM

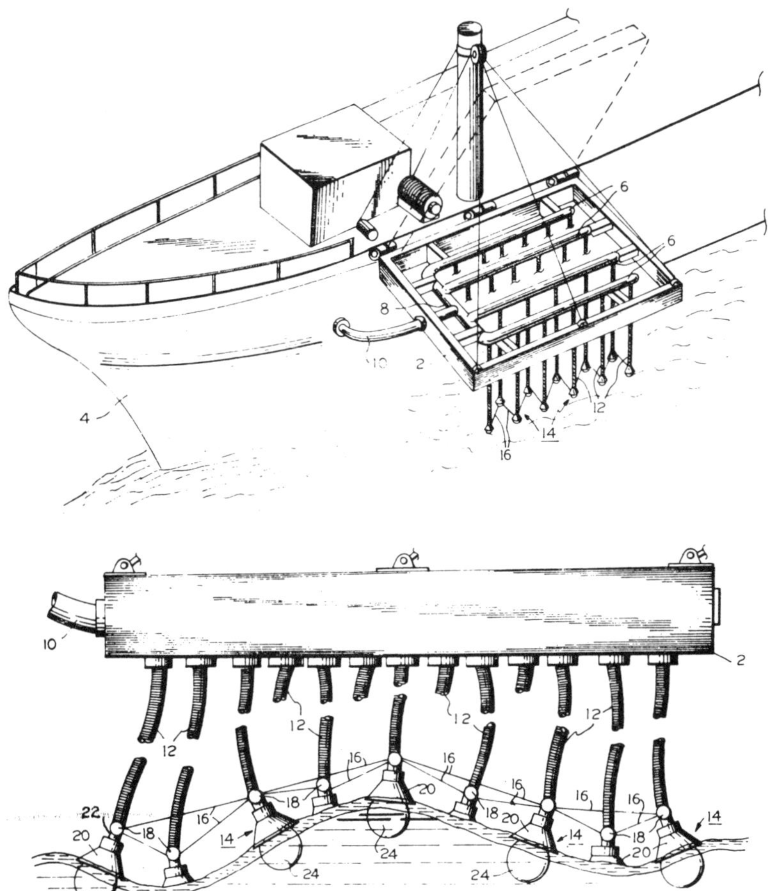

Source: J.W. Harrington; U.S. Patent 3,534,858; October 20, 1970

It will be appreciated that when the apparatus is used with ocean sweeping or waterways subject to heavier wave motion, equipment means must be provided to maintain the frame support **2** at the same height above the surface of the water irrespective of wave amplitude. Mounted within frame **2** are a plurality of suction headers **6** connected to a manifold **8**. A suction conduit **10** forms the connection between the manifold and a suction pump (not shown).

The suction pump may be conveniently mounted on the deck of the ship or within the hold of the ship. Any self-priming type vacuum pump of sufficient capacity may be satisfactorily used, for example, motor driven centrifugal high efficiency water pumps having hydraulically created vacuum systems that enable the pumps to continue operation when, on occasion, the skimmer aperture is prevented by wave motion from being completely immersed in the pollutant to be pumped. In other words, for most efficient operation, a pump is used which will not lose vacuum, when, for instance, because of sea turbulence, one or more of the vacuum hose inlets rises above the liquid surface and sucks air.

Connected with each of the suction headers are a plurality of flexible suction hoses **12**. Frame **2**, suction headers **6** and the manifold form so to speak, a bed horizontally disposed with respect to the liquid surface to be cleaned. The suction hoses hang downwardly from the bed for operational connection with a skimmer **14** mounted on the end of each suction hose. Adjacent to, but with allowance for freedom of movement of the skimmers, tie lines **16** are provided to allow the plurality of skimmers to sweep as a unit and additionally prevent blowing of the skimmer or skimmers out of the water in high winds.

As may be appreciated, the length of the suction hose is dependent upon many factors, e.g., state of movement of water to be swept, type of ship or barge used, or dock, speed of sweeping, ship, etc. In the preferred embodiment, and particularly where the device used to skim pollutants from the seas, each skimmer is joined with its respective suction hose **16** through a swivel joint **18** which allows a swinging movement of the skimmer throughout 360°.

As shown in the lower detail view in particular, nozzle **14** comprises a floatable hollow cone shaped member **20** slidably mounted on pipe section **22** and communicating with the interior of suction hose **12**. A substantially spherical shaped float **24** is connected with pipe section **22**.

Floatable hollow cone member **20** and float **24** are designed and constructed in such manner and of such material that the base edge of cone member **20** will float on the surface or just within the upper surface of the lighter liquid, e.g., oil, to be removed, while float **24** floats partially in the heavier liquid, to the end that a suction aperture is formed between the lower edge of cone member **20** and float **24**, which aperture lies wholly within the layer of pollutant. The cone member being slidably mounted on pipe section **22** will allow the intake aperture to automatically adjust itself to the thickness of the floating pollutant.

Muller Device

A device developed by *J. Muller; U.S. Patent 3,221,884; December 7, 1965* consists of a flexible pipe or pipes surrounded by a layer of buoyant material and an outer sheath, whereby the flexible pipes float on the surface of the body of liquid. It has a number of suction nozzles connected to the flexible pipe, a filter for separating the surface liquid from the liquid of the body and a method of discharging the separated water back into the main body of water.

In operation, the pipeline is laid on the surface of the sea so that it surrounds the patch of oil which is to be removed. The suction nozzles dip into the surface layer and the oil with a certain amount of seawater is sucked into the pipe and then carried into the ship where it is forced through a filtering apparatus to separate the oil from the water. The oil is stored in the tanks of the ship, so that it may subsequently be used for its original purpose and the water is returned by a pressure pump to the pipe for discharging to the sea. In addition, this device incorporates the provision of another pipe supported above and between the other pipes and a method (pump and piping) to supply liquid to this pipe in order to alter the buoyancy of the combination of pipes.

As shown in Figure 146 a body of liquid, in the case the sea, having a surface layer **1** of oil or other liquid floating on its surface which is to be removed. A pipeline **2** whose detailed construction will be described later, passes over the side of a ship **3** and is arranged around the patch of oil **1**. Both ends of the pipeline are in the ship. However, the two ends might be in different ships or the apparatus might be operated from the shore.

FIGURE 146: MULLER FLOATING SUCTION SKIMMER SYSTEM

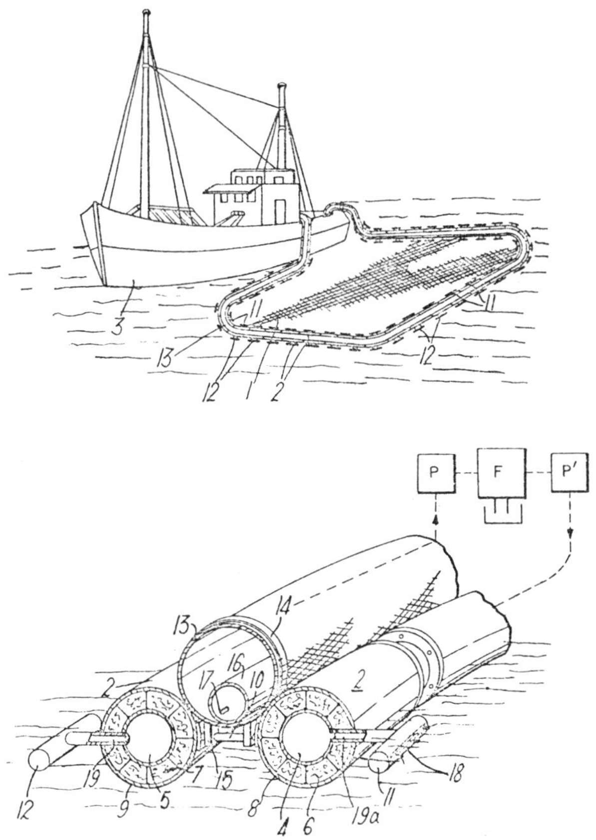

Source: J. Muller; U.S. Patent 3,221,884; December 7, 1965

The two pipes **2** are arranged parallel to each other. The pipe shown on the left in the detail at the lower part of the figure consists of an inner pipe **5** surrounded by a layer **7** of light buoyant material which, in turn, is covered by a waterproof sheath **9**. The right-hand pipe consists of an inner pipe **4** surrounded by a layer **6** of buoyant material and an outer waterproof sheath **8**. The two pipes may consist of rigid sections connected together, as shown in the case of the right-hand pipe, or they may be of any other convenient construction, provided that they are flexible. The two pipes **2** are separated from each other and held in substantially parallel relationship by spacer bars **10**. A third pipe **13**, which is also flexible, is supported on the spacer bars **10** and is held in place by bands **14** associated with each spacer bar, each band being provided with fixing lugs **15** by which it is attached to the associated spacer bar, so that the pipe **13** is carried by the pipes **2** above the center line between them.

Spaced at intervals along one side of the left-hand pipe **2** is a series of suction nozzles **12**, each attached to a pipe **19** which passes through the sheath **9** and the buoyant material **7** and communicates with the interior of the tube **5**. Projecting in a similar manner from the other side of the right-hand pipe **2** is a series of discharge nozzles **11**, having openings **18**, of similar form to the nozzles **12** and attached to pipes **19a** which pass through the sheath **8** and the buoyant material **6** communicates with the interior of the pipe **4**.

The pipe **13** has inside it a smaller pipe **16** which has openings **17** formed along its length and communicating with the interior of the pipe **13**. By pumping water, for example, into and out of the pipe **13** by means of the pipe **16** the degree of buoyancy of the three-pipe assembly may be altered to ensure that the nozzle openings are just at, or slightly below, the surface of the sea.

The ship **3** carries suction and discharge pumps **P, P′** and a filtering apparatus **F** of any known type capable of separating immiscible liquids of different densities. In operation the pipeline is laid on the surface of the sea so that it surrounds the patch of oil which is to be removed. The suction nozzles **12** dip into the surface layer and the oil, with a certain amount of seawater, is sucked into the pipe **5** and then carried into the ship, where it is forced through the filtering apparatus to separate the oil from the water. The oil is stored in tanks in the ship, so that it may subsequently be used for its original purpose, or for another useful purpose, while the water is returned by a pressure pump to the pipe **4** for discharging through the nozzle **11**.

Olsen Device

A device developed by *M.F. Olsen; U.S. Patent 3,745,115; July 10, 1973* is one in which one or more floats are provided for immersion in an oil-slick affected water area, the floats having a collecting compartment and a ballast compartment, and a limit valve for the ballast compartment, such that the floats will be partially submerged at the level of the collecting compartment so that the oil and water mixture may be collected.

Flexible tubes are also provided for the collecting compartment for transferring the collected oil and water mixture to a separation tank. The separation tank has two ball float control valves, one of which permits the clean water to drain back into the environmental water area and the other of which permits the collected oil to be drained off for further use or refinement.

Figure 147 shows such floats immersed in an oil-slick affected water area, and flexible tubing connected to the conical base of each float, and showing the floats accommodating to a moderate rolling sea, as well as an enlarged vertical sectional view of one of the floats. The device thus provides a plurality of floats **10** each having a hollow body portion **11** composed of an outer shell **12** and an inner shell **13**.

The latter, in turn, is connected to an inner tube **14**. The upper wall of the inner shell **13** constitutes the bottom of the oil and water collecting compartment **15** of the float **10**, such compartment being generally rectangular in plan view and having side walls **16**. Each side wall **16** is provided with a plurality of spaced semicircular cut outs **21** which approximate one-half the height of the side walls and extend downwardly from the upper edge of each such side wall. A cover plate **22** of rectangular shape in plan view, having crossed bracing ribs **23**, threaded eye bolts **24** at each corner and a lifting eye **25** is suitably mounted on top of the float to close the top of the oil and water collecting compartment, such cover plate being securely fastened to the side walls **16** and **17** by means of the eye bolts **24** which are threaded into the threaded holes **20**. With the cover plate **22** in place, the semicircular cut outs **21** constitute the only access of the oil and water mixture to the collecting compartment **15**.

FIGURE 147: OLSEN FLOATING SUCTION SKIMMER SYSTEM

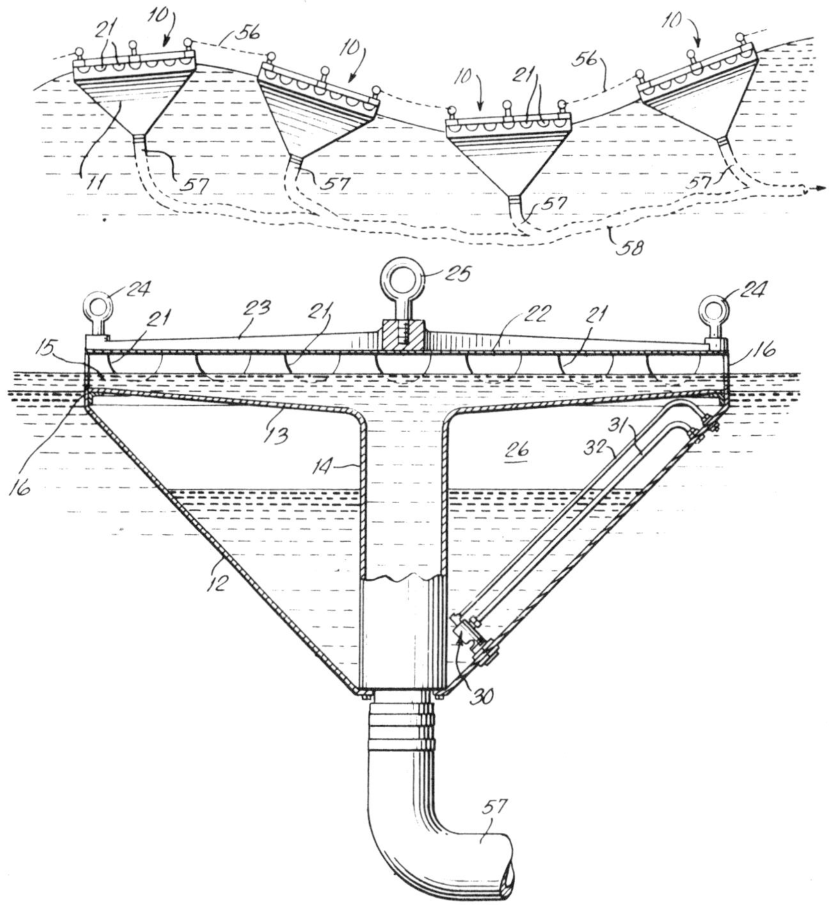

Source: M.F. Olsen; U.S. Patent 3,745,115; July 10, 1973

The float **10** is so designed and constructed that the cut outs in the side walls of the collecting compartment will always be maintained partially submerged. This is accomplished by providing a ballast compartment **26** between the outer shell **12** and the inner shell **13** for receiving an adequate amount of water to provide the necessary ballast to assure that the collecting compartment is always at the proper water surface level.

There is provided in the ballast compartment **26** a ballast limit valve **30**. Such ballast limit valve generally consists of a water intake conduit **31** and an air vent **32** both of which are disposed close to the inner shell **13** which forms the bottom of the collecting compartment **15**, with the air vent tube disposed nearer to such collecting compartment. A plurality of floats **10** may be connected in spaced relationship by means of flexible chains or other connecting members **56** which are attached to the eye bolts **24** located at each of the corners of the floats. With such flexible connections a series of floats **10** can follow the surface of a moderately rolling sea and still perform their function and purpose of collecting the oil-slick with a minimum amount of water.

It will also be noted that each float **10** is provided with a flexible hose **57** and that each such flexible hose is in turn connected to a common hose **58** which, in turn, is connected to a pump provided on a ship, barge, or other structure, also having on board a separation tank.

Rheinwerft Gesellschaft Device

As described by *Rheinwerft Gesellschaft; German Patent 1,062,407; March 22, 1967* one floating oil skimmer device is one in which flow to a bounded space is set up by means of a pump or pumps in the area of the water surface covered with the oil. This technique carries the oil floating at the top to the bounded space. The two flows are separated at the edge of the bounded space by a solid, separating member, having the form of a trough. The oil is removed by drawing off above the separating member.

Advantageously, the bounded space is formed by a body floating in the water. In order to prevent water from being also drawn off to any great extent from the bounded space, a feature of this process provides that the water level is first lowered inside the space bounded by the floating body by means of a water pump and the oil. It is then collected within the edge of the floating body in the bounded space and is removed by a second pump.

This system is now in production and is marketed by C.A. Bekhor Ltd., London, England. The largest available unit has a suction capacity of 10 tons per hour and a 40 ton holding capacity, according to EPA, Edison, N.J. publication, *Oil Skimming Devices,* Report PB 218,504, Springfield, Va., Nat. Tech. Information Service (May 1970).

Smith Device

A device developed by *M.F. Smith; U.S. Patent 3,556,301; January 19, 1971* is constructed of lightweight nonrigid materials and comprises two parallel-spaced sheets with flexible edges. The device floats on the surface of water and flexibly conforms to waves and swells on the water surface. Skimming is performed by exposing a negative pressure intake portal to a shallow skimming zone directly beneath the surface. The narrow elongated intake portal is defined between a flexible floating underflow edge of one sheet and a second flexible overflow edge of a second sheet spaced beneath the first sheet.

Figure 148 is a perspective view of this device and also shows a sectional view of construction and operation of the flexible skimmer head assembly. The floating flexible skimmer apparatus has two main components, one of which is the skimmer head assembly **10** and the other of which is a negative pressure source and delivery means **12**. The skimmer head assembly comprises first a section of aluminum conduit **14**.

The upper sheet **30** of the skimmer head is fabricated of closed-cell nitrile foam and is secured to the top face of the upper plate of the U-shaped plenum member. The sheet **30** is notched at **31** to accommodate the conduit **14**, which has already been welded to the plenum member. The sheet is anchored to the plenum member by sandwiching it between an upper plenum plate **22** and a correspondingly shaped plate **34**.

Positioned directly below the upper sheet **30** is a lower, less buoyant sheet **38** of nitrile rubber. It is also semicircular and is preferably shaped to correspond with the lower face of the upper sheet **30**. The facing areas of the two sheets **30** and **38** are held in spaced-apart relation by angularly separated spacer blocks **49** which may be either integral or cemented or heat fused to anchor them between the two sheets. The outermost semicircular ring of blocks are preferably spaced at 2½° intervals; the next three concentric rings of blocks are spaced at 5° intervals and the innermost ring of blocks are spaced apart at 10° intervals. The preferred material for these blocks, especially when used in an assembly with the removable ballast described above, is buoyant closed-cell nitrile foam.

The skimmer head assembly thus constructed is then connected by conduit **60** to a negative pressure source **61** preferably a diaphragm pump, and either disposal means, or an oil/water separation apparatus. The conduit may be fabricated in sections, all of which are supported on the surface of the water by a plurality of keg-shaped floats **62**. The several sections may be joined together by bellows like joints **64** which are extremely flexible.

FIGURE 148: SMITH FLOATING SUCTION SKIMMER SYSTEM

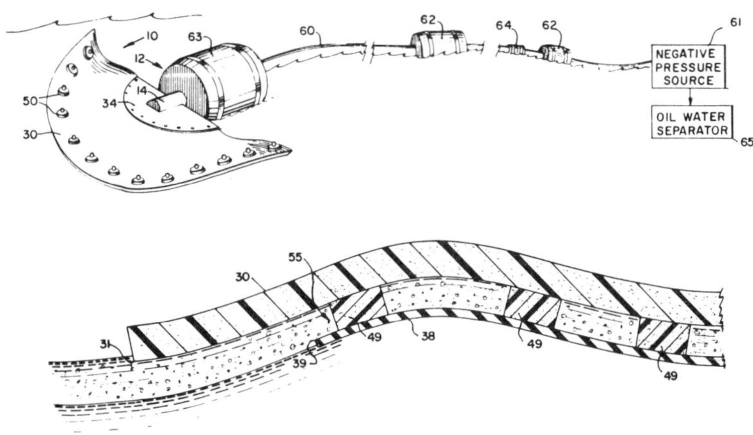

Source: M.F. Smith; U.S. Patent 3,556,301; January 19, 1971

The inside diameter of conduit **60** is telescoped over the mating outside diameter of conduit **14** extending from the skimming head assembly. Thus, the negative pressure source and conduit delivery means are connected to the skimming head assembly by slidingly engaging conduit **60** over conduit **14**. The other end of conduit **60** is sealably attached to the negative pressure source, thereby providing a closed vacuum delivery means connecting the negative pressure source to the skimmer head assembly. A large keg-shaped float **63** may be provided near the junction of conduits **60** and **14** to aid in buoyantly supporting the skimmer head assembly.

Because the intake portal **55** is defined between the two flexible, wave conforming sheets, the upper one of which floats on the surface, it is located immediately beneath the surface of the water regardless of the surface conditions. The elongated intake portal **55** is formed between the flexible underflow and overflow edges **31** and **39**, and is ideally positioned for the removal of waste from the surface of contaminated water.

Oil and most other common contaminants are less dense than pure water, and thus reside in a film on the surface. When the amount of pollutants is relatively great, as in a contained oil spill, the oil tends to form a layer which can be as much as an inch or more deep. Except for the wave damping effect of oil on troubled water, which is normally confined to a reduction of the cresting and breaking tendency of wind-driven whitecaps, films and layers of surface pollutants tend to conform with and behave in the same manner as the water. Thus the pollutants in effect form a conforming layer on the surface of the water, rising and falling with the crests and troughs of waves and undulations. The flexible buoyant skimmer apparatus mimics and joins in the motion of the waste on an undulating water surface, and thus the elongated flexible intake portal **55** always presents itself to the region in which the waste resides.

Henry Sykes Ltd. Device

As described in EPA, Edison, N.J. publication, *Oil Skimming Devices,* Report PB 218,504; Springfield, Va., Nat. Tech. Information Service (May 1970), Henry Sykes Ltd. of London, England has investigated the use of a Univac Pump Skimmer for oil pollution control. The Univac pump skimmer with horizontally extended suction pipes out beyond the side of a tanker. These pipes terminate in a right angle bend and a length of flexible pipe.

At the end of the flexible pipe is fitted a fan shaped suction entry of some three feet in length. The final design of the suction nozzle would be modified to meet a variety of conditions. Approximate calculation based on a vessel steaming at five knots and covering a 25 foot wide path, show that recovery rates could be in the order of 600 tons per hour of oil if the oil film were ½ inch thick.

The oil and water mixture would be pumped into the tanks of the coastal vessel and clear water would be extracted by arranging further pump suctions to draw water from a point near the base of the tank. This would allow the vessel to operate until the full cargo of oily sludge has been recovered.

The characteristics of Univac pumps are claimed to be particularly suitable for this type of work. They can operate with suction branches intermittently exposed to air and with suction lines up to 400 yards in length. They also have the ability to pass solid material up to four inches in diameter and thick sludges of high viscosity. They can handle appreciable quantities of sand and gravel. At present, they are in use handling oil sludges during tank cleaning at terminals and refineries.

Usher Device

A device developed by *D. Usher; U.S. Patent 3,578,171; May 11, 1971* utilizes a barge which contains means for encompassing a slick and confining it to prevent lateral spreading, skimming and withdrawing the pollutant under vacuum to a storage area on a barge, pumping into settling tanks and simultaneously siphoning off the water. The barge and its pollutant removal equipment may be transported by one from the group consisting of air, truck and rail, to any remote point for emergency use.

Figure 149 is a schematic plan and elevation showing a barge and associated slick bar floating upon a water body with the floating contaminants surrounded by the slick bar being skimmed off for storage upon the barge. Referring to the drawing, a flat bottomed barge generally indicated at **11** may be transported by air, by truck or by rail to remote points on emergency need for placement in a water body **W** containing a slick **S** or other body of floating pollutant.

This antipollutant self-propelled barge carries sections of the slick bar **12** or other pollutant confining device which is buoyant and which is made into long sections one to two hundred feet, for example, and which is drawn by a small tender or other boat to surround a large area of slick as at **S**, the respective ends of the slick bar being connected to confine an area of floating pollutant throughout 360°. Confinement may be between 180° and 360°.

The slick bar includes a cylindrically shaped buoyant body **13** made of a foam plastic material such as styrofoam or the like and depending therefrom it provided a flexible plastic skirt **14** along its length to provide retaining means for the floating pollutant to an appreciable depth. The barge includes a self-contained motor or other propelling means **16** and a suitable deck **17** mounting a swing boom **18.** As desired, the line **19** connected to the barge may be employed for pulling the secondary tandem barge **20** which includes a series of settling tanks **T** for excess storage capacity. Forwardly of the barge deck there is provided debris catcher **21** which is pivotally mounted as at **22**, and thus adapted to be swung into the water body and which includes screening on the bottom, ends and on one side to facilitate the picking up of floating debris such as logs or bottles, for illustration.

FIGURE 149: USHER FLOATING SUCTION SKIMMER SYSTEM

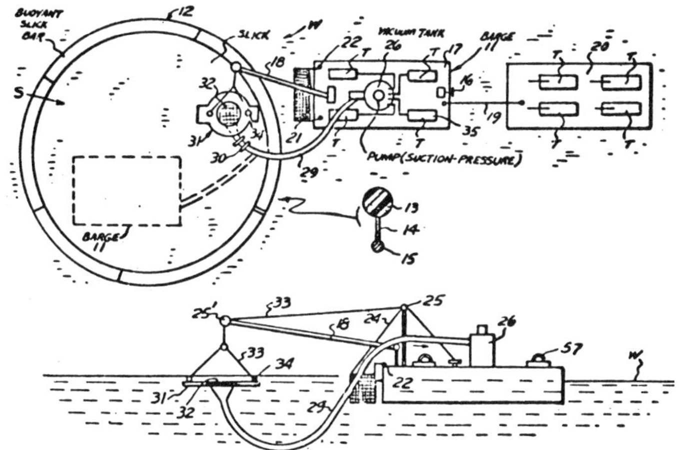

Source: D. Usher; U.S. Patent 3,578,171; May 11, 1971

The line **24** at one end is connected to the debris catcher and moves over the pulley **25** on a suitable mast to control raising and lowering of the debris catcher as desired. Mounted upon the deck is a suitable vacuum tank **26.** The skimmer hose **29** at its free end by the use of elbow fittings **30** is connected to a skimmer assembly **31** having a grilled or screened intake **32** communicating with the skimmer hose **29.** The cables **33**, connected at **34** to the frame of the skimmer extends over a suitable pulley **25'** on boom **18** by which the skimmer assembly including the intake **32** may be adjusted vertically so as to be sufficiently below the top surface of the water **W** and floating pollutant so that such pollutants may be skimmed under vacuum through the hose **29** into the vacuum storage tank. Provided within the barge, preferably below deck **17**, are a plurality of settling tanks **35.**

U.S. Navy Device

A device developed by *J.A. O'Brien; U.S. Patent 3,690,463; September 12, 1972; assigned to U.S. Secretary of the Navy* consists primarily of a floating suction head connected to a pump by a flexible hose. The oil/water mixture enter the head through a suction port, the latter being protected from debris by a series of screens. The device is shown in plan and sectional elevation in Figure 150.

There is indicated a floating suction head **10** which is preferably circular with a plurality of suction ports **12** located about the periphery. It is constructed preferably of a sturdy lightweight plastic material and is about 18 inches in diameter and about 7 inches in height. The head contains three tubes **14** into which weights **16** are inserted as needed to adjust the skimming depth. A central tube **18** interconnects with the ports **12** and a lightweight flexible and extendable hose **20** connects tube **18** to the floating suction hose **22** which leads to a pump not illustrated. The suction ports **12** are tapered and V-notched in order to help prevent mechanical emulsification of the oil/water mixture and thus aid in separation of the oil from the water. A debris screen **24** preferably of about ¾ inch mesh circumscribes the head **10** and is used to protect the ports **12** from clogging and other damage. The screen is supported independently of the head by the triangular shaped screen angles **26** which have a circular float **28** attached at each of three corners substantially as shown.

FIGURE 150: U.S. NAVY FLOATING SUCTION SKIMMER SYSTEM

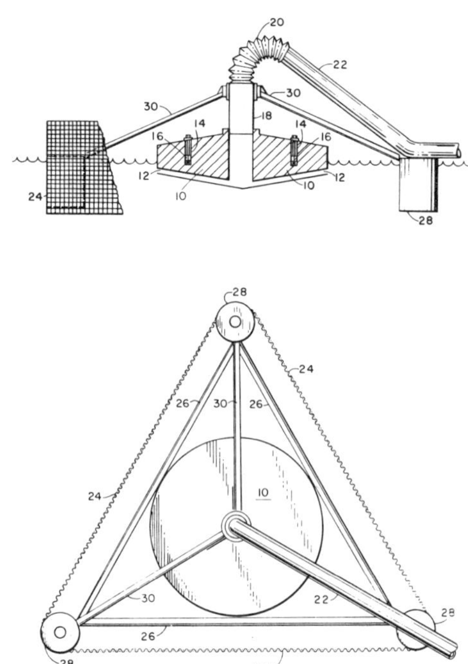

Source: J.A. O'Brien; U.S. Patent 3,690,463; September 12, 1972

The frame angles **30** are attached to each float **28** and to the tube **18** thereby providing additional strength to the framework formed by angles **26**. The angles **30** also support the flexible hose **20**. In operation the oil/water mixture enters the head through the suction intake ports and is sucked through tube **18** by suitable pump means into hoses **20** and **22** then into a storage area. Through a series of weights **16** which are added to head **10** as is required the skimming depth is maintained at between about ¼ to 1 inch. This ability to adjust the skimming depth enhances the oil-to-water ratio so that the volume requirement for an oil/water separation system is reduced.

Three suction head assemblies may preferably be used simultaneously with the same source to increase oil pickup efficiency. In case of substantial decreased output flow, the head is easily cleaned by backflushing. Routine cleaning with diesel fuel or strong detergent and water effectively removes the sticky oil and small particles of debris that may plug the suction head after severe use.

Verolme Device

The Rotterdam Netherlands Harbor is covered by a fleet of harbor skimmers constructed exclusively by the Verolme Shipyard Hensden, Ltd., of Rotterdam. They have operated this fleet for about ten years, have collected an average of 6,000 tons of oil per year, and can collect 50 tons per hour, per unit, according to EPA, Edison, N.J. publication, *Oil Skimming Devices,* Report PB 218,504, Springfield, Va., Nat. Tech. Information Service (May 1970).

The mixtures of oil and water are collected by a floating suction apparatus and go to a vacuum tank through hoses. The pump maintaining the vacuum tank will also forward the mixture so the first separating compartment wherein the separation of the oil mixture starts. For this purpose, the barge is divided into a number of closed hopper-shaped compartments.

In the first compartment, a good deal of the oil will be separated from the water; it then flows into the next compartment where the separation is continued. As soon as the level in the separating compartments is above the outside level, the water automatically syphons overboard from the last compartment. The oil, in the meantime, flows into the wing storage tanks. The separation process thus proceeds automatically, simply regulating the flow, and since no mechanical energy is necessary, it is performed in an economical manner.

The inlet of the oily water, after having passed the vacuum tank, goes through a coaming in the deck of the barge; the oily water flows through a grid into the first separating compartment. Because of the hopper-shape of the compartments, a great deal of the pollution is claimed to separate at once. The bottom of the hopper-shaped compartments form a rather narrow gutter; the baffles being fitted in the bulkheads at this point so that only the lower level of liquid flows from one compartment to the next.

The collected oil, which may be disposed of as a valuable commercial product, is temporarily stored in the side storage tanks which also act as buoyancy tanks. These tanks provide stability for the barge and, at the same time, produce a greater freeboard, thus assisting the automatic discharge from the separating compartments.

The barge may be constructed either as a dumb or a self-propelled unit. The pump maintaining the vacuum in the vacuum tank which draws its mixture from the water surface is so mounted that it can unload the oil from the storage tanks and also be used for draining the separating compartments.

Watermaster Industries Ltd. Device

A device developed by *A.J. Tudor and E.E. Tudor; U.S. Patent 3,762,557; October 2, 1973; assigned to Watermaster Industries Ltd., Canada* is a floating surface skimmer particularly adapted for use with a floating pump housed by the skimmer. The skimmer is circular in shape and has a plurality of peripheral radial openings communicating with a central opening in which the pump is housed whereby each of the peripheral openings acts as a weir. The height of the skimmer above a liquid surface can be adjusted to control the depth of liquid skimmed.

In Figure 151 is shown such a surface skimmer designated generally by the numeral **10** having an upper annular section **12** and a lower, truncated generally conical or cylindrical section **14**. The upper section has a central opening **16** adapted to receive side suction pump **18**, shown by ghost lines in the figure. A plurality of peripheral radial openings **20** are formed in the underside of upper section **12** and extend through section **12** in communication with the central opening, for reasons which will become apparent as the description proceeds. The lower section comprises conical or cylindrical portion **22** having rim **23** adapted to extend to the periphery of the upper section and adapted to be secured to the upper section by a plurality of connecting means such as bolts or rivets **24**. A receptacle or well **26** is formed in the central portion of lower section **14** by conical or cylindrical wall **28**, which is substantially coextensive with the inner wall **32** of the upper section, and a protruding lip defining annular flange **30**. Pump **18** is seated on flange **30**, effectively closing the bottom of the skimmer, and is secured centrally within the skimmer by overlapping clamp plates **34** held in position by eye bolts **35**. Eye bolts and handles **33** permit facile transport of the skimmer.

FIGURE 151: WATERMASTER INDUSTRIES LTD. FLOATING SUCTION SKIMMER SYSTEM

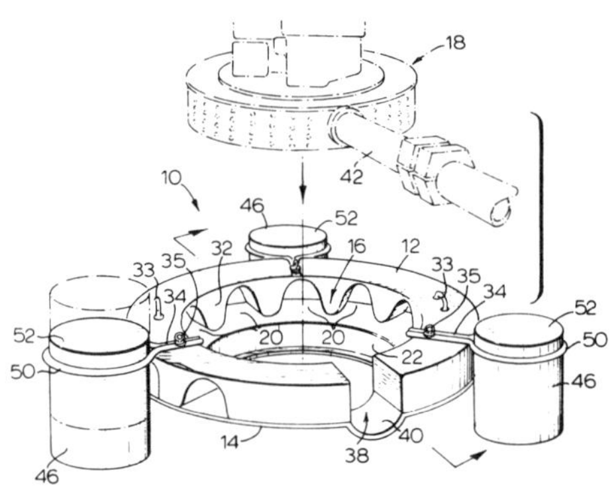

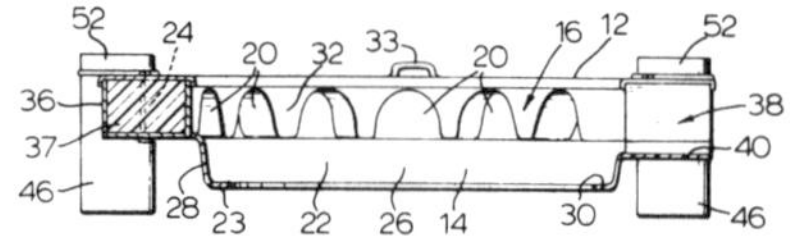

Source: A.J. Tudor and E.E. Tudor; U.S. Patent 3,762,557; October 2, 1973

Opening **38** is formed in upper section **12** and notch **40** in lower section **14** to accommodate the hose **42** of pump **18**. The upper section comprises an outer shell **36** of a rigid material such as fiberglass reinforced polyester resin and is filled with a foam plastic **37** having the flotation characteristics of urethane. Lower section **14** preferably is formed of a rigid material such as fiberglass reinforced polyester resin.

A plurality of height adjustment flotation devices **46**, preferably three in number, are secured to the top of upper section **12** by eye bolts **35** gripping connecting rods **50**. Each height adjustment device **46** comprises a rod **50** which defines a circular opening for receiving cylindrical, liquid-displacement flotation cylinders **52** in frictional engagement. The level of skimmer **10** can be adjusted by raising and lowering flotation cylinders **52**, as shown by ghost lines, within the openings of rods **15**.

This device provides a number of important advantages. The skimmer or skimmer with pump, can be readily transported by one or two men. The plurality of weirs permits the skimming of a large volume of liquid, in the order of 200 gallons per minute, and the depth of liquid skimmed and the level of the skimmer can be readily and effectively adjusted.

SORBENT SURFACE DEVICES

A third type of skimmer uses a surface to which oil can stick to collect the oil off the water surface. This type, employing a disc or drum or endless belt, is less vulnerable to wave action than are the weir and floating suction skimmers. Sorbent surface skimmers, however, are usually more expensive and less easily available. Too, their operators may need more training in order to operate them efficiently.

One type of sorbent surface device is the drum-type skimmer. These types of oil skimming devices use a rotating drum or cylinder covered with an oil absorbent material to absorb the oil from an oil and water mixture. These surfaces are generally hydrophobic and oleophilic (they are not wetted by water but are by oil). The drum surfaces used to date generally have been constructed either of aluminum or polyurethane. The oil is squeezed or scraped off the surface of the drum by a knife blade or by rollers and then is segregated from the water.

In the case where the drum has an aluminum surface, the best recovery rates have been achieved with higher viscosity oils. On the other hand, polyurethane coated drums have worked best for low viscosity oils which are nonweathered, since these oils can be absorbed more rapidly by the polyurethane.

The effectiveness of drum type skimmers has been reported to be reduced when the oil film thickness is less than ¼ of an inch, according to A.D. Little, Inc. in *Combatting Pollution Created by Oil Spills*, a report for the U.S. Coast Guard of the U.S. Department of Transportation. Because of this inefficiency, some method is required to concentrate the oil by some type of dynamic booming or some other means.

The Standard Oil Company of New Jersey, Logistics Department, has reported in their *Oil Spill Cleanup Manual* (1969) that:

> "The efficiency of recovery is dependent on the thickness of oil, viscosity of the oil, affinity of the drum's surface for oil, rate of cycling of the drum, and the velocity of the recovery vessel on which the cycling drum is installed. The velocity of the skimming vessel is important since a critical velocity exists above which a hydraulic jump will form in the oil along an arc ahead of the skimming vessel. This causes oil to escape around the end of the booms which are used to concentrate the oil as the vessel moves through the water. The angle between the booms is another important factor. If the angle is too large, oil tends to pile up at the boom and not move along the boom to the skimmer. The rate of cycling of the drum is important since water is picked up by the element if its speed of rotation is too great."

Under open sea conditions, or when waves are over two feet in height, the oil film layer comes in contact with the drum at varying rates. This action reduces pickup efficiencies (highest recovered oil/entrained water ratio) drastically. Therefore, to date, drum type oil skimmers have been limited to use in calm or mild sea conditions.

Another type of sorbent surface device is the belt-type skimmer. This type of oil skimmer utilizes an oil absorbent surface in a continuous belt to absorb the oil from an oil and water mixture. The belt carries the oil to the top of the belt mechanism where a blade or similar piece of equipment squeezes the oil from the belt before recycling. The belt is supported normally on two drums: one to submerge the belt in the oil and water mixture, and the other out of the mixture where the removal of the oil takes place.

A variety of materials has been used as the cycling belt on the endless belt type skimmers. These materials include specially treated fibers to make them hydrophobic: fine mesh wire screen, rubberized fabric, absorbent cloth and resilient foam. The belt type skimmers have many of the inherent problems previously discussed concerning the drum type skimmers.

Belt type skimming devices are generally limited to calm waters and where oil films are of considerable thicknesses.

Bennett Device

The Bennett International Company is the distributor of a device called the oil-evator. This device was developed by the Canadian government and was used in the Vancouver area. In operation, oil is recovered by a conveyor belt rated at 43,000 gallons per day. The belt will retrieve oily petroleum and leaves the sea water behind. The device weighs less than 12,000 pounds.

Oil can be recaptured from the device in numerous ways, such as in a plastic containment bag, a barge or pump from a self-contained reservoir onto a tanker. The oil-evator under testing recovered 98% of the crude oil during medium ocean swell conditions. Specific data as to type of belt material and removal mechanism of the oil from the belt is not readily available. Additional information on the oil-evator can be obtained from Bennett International Services, Inc.

Brill Device

A device developed by *E.L. Brill and B.M. Brill; U.S. Patent 3,640,394; February 8, 1972* is a device for skimming oil or the like floating on a pool of water including an endless substantially rigid loop of uniform cross section, generally circular. The loop is gripped at its upper edge by a pair of rolls rotating in opposite directions and drivingly engaging the loop at one zone in diagonally opposed quadrants, one above and one below the center of a section of the loop. The rolls rotate the loop in its own plane causing it to pass continuously into and out of the pool of water or hydrophilic liquid and to attract hydrophobic material, such as oil or the like or finely divided or colloidal material, which material is lifted by the coil and squeezed out upon passing through the rolls or separated by a scraper or by a blast of air. The loop may oscillate about an axis substantially tangential the loop at the driving zone. A modification utilizes a brushlike surface on the loop and on the driving rolls. Figure 152 is a simplified schematic view of this general type of apparatus.

FIGURE 152: BRILL ENDLESS ABSORBENT LOOP SKIMMER

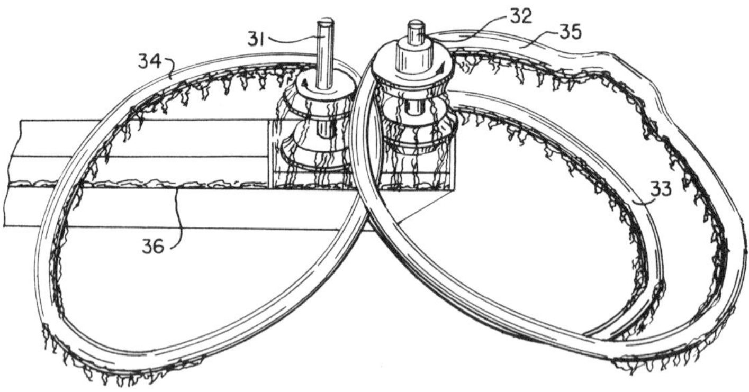

Source: E.L. Brill and B.M. Brill; U.S. Patent 3,640,394; February 8, 1972

The figure shows diagrammatically a pair of spaced parallel drive shafts **31** and **32**, generally vertical, on which are mounted a plurality of rolls to provide pairs of coacting arcuately concave annular surfaces, each pair of such surfaces engaging a different one of the loops **33**, **34** and **35**, to drive each loop for rotation in its own plane into the polluted water to carry the hydrophobic material upwardly out of the water where the hydrophobic material is squeezed out between the coacting drive roll surfaces to be diverted into the trough **36**.

This device is commercially available from Oil Skimmers, Inc. of Cleveland, Ohio. This skimmer includes a long tube as a belt over a sprocket and past a cleaning point where the oil is scraped off and flows to a container. The tubular belt is made of an oil absorbent material and is long enough to wind about the oil-water surface. The oil is absorbed as the belt leaves the oil-water surface. The tubular belt is small enough in diameter so that debris on the oil-water surface does not interfere with the operation. The sprocket handling the tubular belt is driven by an electric motor.

British Petroleum Company Devices

One belt type skimmer developed by British Petroleum Company is described in British Patent 1,026,201; April 14, 1966. As shown in Figure 153, the device comprises an endless belt of resilient foam material, several rollers between which the belt passes at its upper end, and means for collecting and removing liquid squeezed from the strip by the rollers. The resilient foam material is comprised of a number of interconnected pores and is compressible so as to enable a liquid contained in the pores to be removed. A suitable material is a plastic foam such as polyurethane foam. The compression may be in the form of one or more pairs of rollers between which the resilient foam material is arranged to pass. The rollers may comprise the means for driving the endless band of resilient foam material.

FIGURE 153: BRITISH PETROLEUM COMPANY ABSORBENT BELT SKIMMER

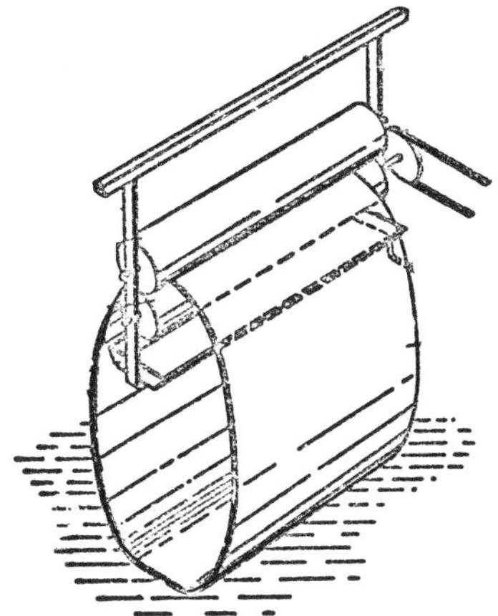

Source: Report PB 218,504

Another belt type skimmer developed by British Petroleum Company is described by *E.J. Lane; Canadian Patent 712,817; July 3, 1963; assigned to British Petroleum Company, Ltd.* This skimmer includes a device for oil pickup and transfer, and a receptacle for the transferred oil, comprising a craft, the base of the receptacle being at least partially open to the water. The oil pickup and transfer member is comprised of an endless belt. The belt being supported by two cylinders arranged so that when the craft is afloat, one cylinder will be at least partially immersed in the water, and the other cylinder can be mounted aboard the craft. The receptacle preferably comprises the craft itself and the base of the craft may be entirely open to the water so that, in effect, the craft becomes a floating boom. Since the craft is open at the base, it will, of course, be necessary to maintain the craft afloat; for example, by buoyancy tanks. The liquid level inside the craft is maintained at a similar level to the water level outside. As oil is transferred via the belt into the craft, water is displaced through the base of the craft.

The displacement of water in this manner allows a greater quantity of water to be accumulated in the device than in the normal type of collecting barge which is not open to the water. The actual material of the belt may be chosen from a variety of materials, one example being a woven cotton fabric coated with neoprene. It has been found that better oil transference is obtained by driving the belt so that oil is carried downwardly through the water than by driving the belt in the other direction. The belt is also provided with a suitable device to remove any oil from the belt and to direct this oil into the craft for storage. This device is a scraper or a roller at the top of the craft.

Preferably, the belt will be mounted in two cylinders. For example, a belt driving cylinder and a belt idler cylinder. The driving cylinder may be mounted on a shaft, supported by brackets, fixed in the bow of the craft and the shaft may be driven by any suitable means, for example, a diesel engine.

Centri-Spray Device

A belt-type device has been developed by Centri-Spray Corporation of Livonia, Michigan, according to EPA, Edison, N.J. publication *Oil Skimming Devices*, Report PB 218,504, Springfield, Virginia, National Technical Information Service (May, 1970). This unit is a belt skimmer, as shown in Figure 154, driven by a one-horsepower motor linked through a speed reducer which drives an upper pulley which rotates the endless belt.

FIGURE 154: CENTRI-SPRAY OIL RECOVERY UNIT

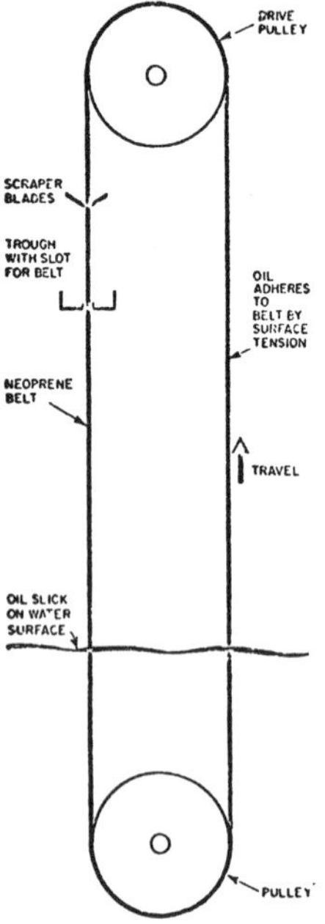

As belt rotates, surface oil adheres. The oil is scraped off by wipers below top pulley into trough where it is screened and dumped into steam heated tank.

Source: Report PB 218,504

The belt is dangled in the water and held tight by a lower pulley. As the belt is rotated through the water, the surface oils adhere to the belt and are drawn up over the top pulley. The oil is scraped off the belt by a set of wipers just below the top pulley and directed into a trough. The belt width is normally 24 inches and the scrapers are of plastic to reduce belt wear. There is some carryup of floating debris such as straw.

The oil from the belts usually contains less than 5% water. These units are available in capacities from 30 gallons per hour to 120 gallons per hour with belt widths ranging from 12 inches to 24 inches. The units are designed for sustained performance under continuous operating conditions. These units may be installed in rivers, lagoons and need only connection to electrical power for immediate operation.

These oil recovery units can be installed in weatherproof housings equipped with infrared or steam heating devices, to permit year around operation at below freezing temperatures. Centri-Spray Corporation has also recently developed multidisc oil recovery device. It is capable of removing 350 gallons per hour of 500 SSU oil at 70°F. Surface floating oil adheres to the continuously rotating multidisc assembly.

Scrapers remove the oil which is deposited in a self-contained oil storage tank. A sump pump transfers the oil from the skimmer tank to a permanently located storage tank. The unit generally contains 12, 24-inch diameter discs, 6 on each end. Other sizes are available, ranging from a single disc stationary mounted unit through any number of multidisc units.

Esso Device

A device developed by *J.H. McClintock; U.S. Patent 3,146,192; August 25, 1964; assigned to Esso Research and Engineering Company* operates by passing a polymeric film over the surface of the water and thus removing the oil from the water onto the film by means of the surface attraction between the film and the oil. Polymeric films which are suitable for use in this device are polyethylene or polypropylene sheets, or mixtures thereof.

It has been found that a strong surface attraction exists between polyethylene and polypropylene films and normally, liquid hydrocarbon compounds. That is, by passing a film over the surface of water on which there are liquid hydrocarbon compounds, the hydrocarbon compounds are selectively attracted to the sheet or film in preference to the water and are thus removed to the surface of the water.

FMC Corporation Device

A device developed by *T.J. Tillett and B. Straus; U.S. Patent 3,693,805; September 26, 1972; assigned to FMC Corporation* is a drum-type skimmer for separating a surface layer of material from a body of liquid. It comprises a rotating drum disposed for partial submersion within the liquid and a shroud with a depending baffle positioned radially from and adjacent to the lower portion of the drum to improve the skimming action between the drum and the material to be skimmed.

Figure 155 is a fragmentary, diagrammatic side elevation of such a skimmer apparatus. The lower detail view shows the shroud-baffle assembly referred to above. The apparatus **10** is adapted for use in a waste disposal facility for removing hydrophobic substances, such as oil **11**, from water **12** or other liquid material having polar characteristics differing from that of the oil. The apparatus **10** is supported within a large vat **13** containing waste and comprises a drum **14** rotatable about a central axis **15**. The drum **14** is suitably driven by a drive means that includes a driven pulley **16** fixed to a drive shaft **18**. An endless belt **20** is trained around the pulley **16** and is also trained around a pulley **21**. A drive assembly including a drive motor **22** is coupled to a speed reducer **24** through a coupler **26**, which speed reducer **24** rotates a shaft **27** to impart rotation to the driver pulley **21**. The motor **22** and speed reducer are supported upon a support assembly **28** which also supports the shaft **18** for rotation at a bearing block **30**.

FIGURE 155: FMC DRUM-TYPE OIL SKIMMER

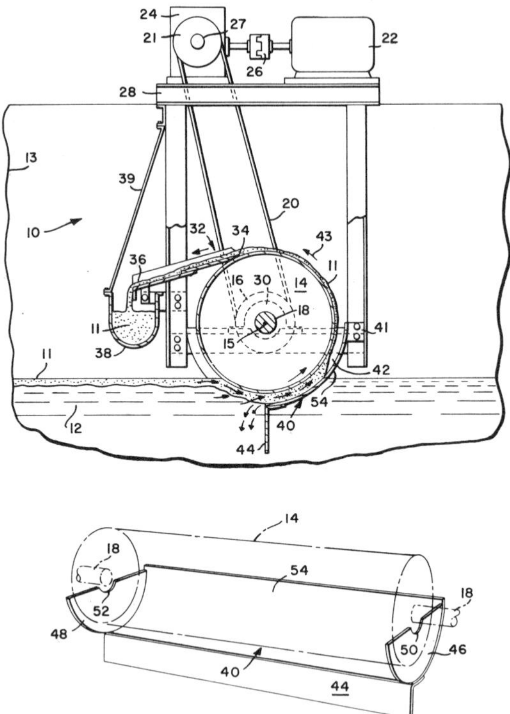

Source: T.J. Tillett and B. Straus; U.S. Patent 3,693,805; September 26, 1972

Adjacent the upper portion of the drum **14** is an oil scraper assembly **32** for removing oil from the surface of the exposed surface of the drum before the exposed surface of the drum is again submerged in the liquid. The scraper assembly **32** has an edge **34** for abutting the outer peripheral surface of the drum. The scraper forms a trough **36** extending from the edge to an oil conduit **38** or other oil container means. The oil conduit is secured in place to the support assembly **28** by an angle arm **39** and is adapted to provide a channel for the removed oil to an exterior collector, such as oil drums or the like.

The drum **14** may be viewed as comprising four quadrants. Adjacent to and spaced from the lower portion of the outer periphery of the drum within the fourth quadrant is an arcuate shroud **40** having a longitudinal center line coincident with the axis of the shaft **18** and the axis **15**. The shroud is fixed by a clamp **41** to the support structure **28** and is positioned from the surface of the drum within the fourth quadrant to establish uniform spacing **42** intermediate the outer surface of the drum and the inner surface of the shroud adjacent to the drum on the upward path of the drum beneath the liquid level. The drum rotates in the direction of the arrow **43** and defines an arcuate surface of 90° in the exemplary embodiment. The spacing extends from the lowest point of the drum within the vat **13** to a point above the surface level of the sewage. At the submerged end of the shroud is a baffle **44** extending laterally from the substantially normal to the end of the shroud or from the drum. The baffle extends the full longitudinal length of the shroud and serves as a header decreasing the velocity of the liquid relative to that of the drum.

The shroud **40** comprises a pair of semicylindrical laterally extending end plates or discs which are disposed to have a center line coincident with the axis of the shaft **18**. The end discs **46** and **48** are formed with end journaling members **50** and **52**, respectively, adapted to receive the ends of the shaft. Intermediate the end plates is an arcuate member **54** of substantially one-fourth of a cylinder centrally aligned with the end plates and extending over approximately one-half of the outer edge of the end plates.

In operation, it has been found that the shroud **40** tends to create an effective layer buildup of the oil **11** against the drum **14** peripheral surface. The oil tends to adhere to the drum surface on the upswing path of the drum, whereas with prior art structures the oil has been found to lose contact on the upswing portion of travel. It appears that the shroud precludes the drum from being wetted by the water once the drum passes below the oil strata level in the tank and an oil film is maintained at all times until removed by the scraper blade **34**. Consequently, more oil is delivered to the scraper assembly **32** and the overall oil pumping efficiency is improved.

Grabbe Device

A device developed by *F. Grabbe; U.S. Patent 3,314,545; April 18, 1967* involves cleaning a water surface by removing oil and other similar floating impurities via an endless conveyor belt, which is made of materials like polyvinyl chloride to which oil sticks. The bottom rim of the belt extends downwards at an inclination to and through the water surface, and in the direction of the flow. The belt is driven so that it moves down along the bottom run and picks up oil as it passes through the oil film surface. The belt then passes around a roller which is immersed within the water and up along an upper run to a cleaning station. The cleaner is formed of comb-like scrapers which scrape oil from bands adjacent to the upper roller and discharge the oil into a collecting tank.

It has been found in practical tests that the oil impurities picked up by the apparatus can be removed from the surface of the water in a single working step and that only small proportions of water are picked up together with the oil. Reportedly, another advantage is that the apparatus is light in weight and can be produced at relatively low cost.

Hale Device

A device developed by *F.E. Hale, Jr. and F.E. Hale, Sr.; U.S. Patent 3,670,896; June 20, 1972* includes a collection surface which is moved into and out of the water, the collection surface being made of a material including a high molecular weight solid hydrocarbon which is wettable with oil so that when the surface emerges from the water the oil collected by adsorption is wiped from the surface and collected.

Hercules Incorporated Device

A device developed by *N. E. Downs; U.S. Patent 3,669,275; June 13, 1972; assigned to Hercules Incorporated* utilizes a fibrous polyolefin body enclosing one or more perforated pipes which are attached to a pump on board an accompanying boat or barge employed to drag the fibrous body through the contaminated area. The polyolefin exhibits great affinity for oil and is thus able to absorb a substantial quantity of oil which is pumped off through the perforated pipe, or pipes, to an appropriate storage area on board the boat or barge.

The method just described derives its efficacy from the affinity of the olefin polymer, in fibrous form, for the oil. When such a fibrous body is thoroughly wet with oil, it is substantially completely repellent to water. Thus, in an ideal situation where the fibrous body remains completely wet with oil throughout the entire cleanup operation, substantially no water is collected with the oil. Such an ideal condition, however, will usually only exist in very still water. When the water is turbulent, there will frequently be water washing over the fibrous body or otherwise splashing into locations from which it can flow into the perforated pipe and be pumped to storage with the oil.

It is the purpose of this device to provide apparatus by which water is prevented from entering into the pipes employed for pumping off the oil gathered from the surface of the water, thus insuring the collection of substantially water-free oil even in very turbulent water. It has been determined that a flexible, water impervious shell, installed over the fibrous body in the area of the perforated pipe, or pipes, substantially completely eliminates water from the recovered oil. Figure 156 shows (in the upper view) a sectional detail of the device and (in the lower view) a plan view of the device in use.

FIGURE 156: HERCULES INCORPORATED ABSORBENT FOAM SKIMMER DEVICE

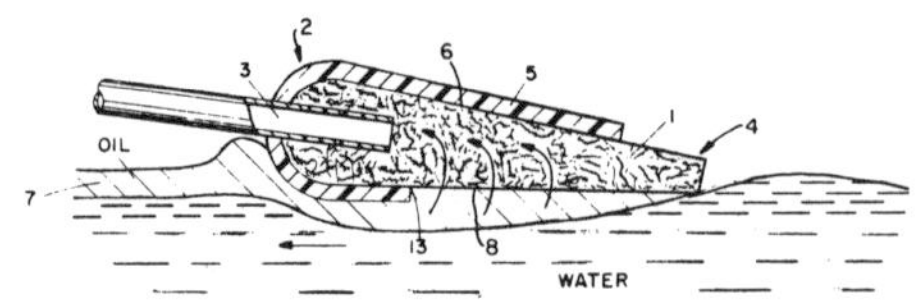

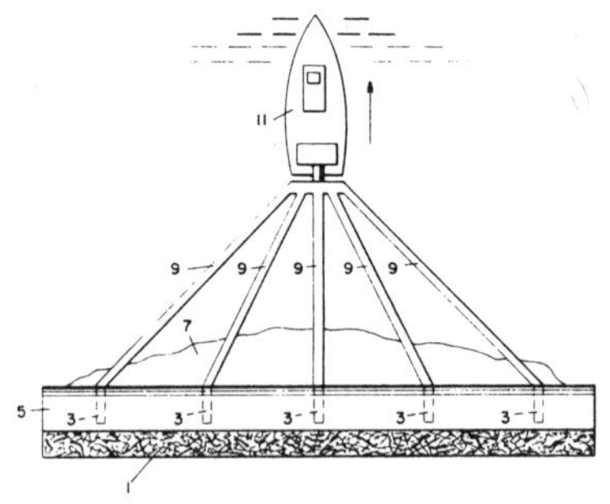

Source: N.E. Downs; U.S. Patent 3,669,275; June 13, 1972

The apparatus comprises a fibrous polyolefin body having a forward section **2**, an aft section **4**, an upper surface **6**, and a lower surface **8**. One or a plurality of perforated pipes **3** are embedded in the forward section of the fibrous body. The fibrous body has the periphery of its forward section enclosed within a water-impervious flexible shell **5**, adapted to enclose the area of the body housing the perforated pipe or pipes. A substantial portion of the fibrous body, especially including the lower surface of the aft section, is outside the shell, and in position to contact the oil on the surface of the water.

In removing oil from the surface of a body of water, the shell-encased fibrous body **1** rests on the water in the oil slick **7**. The pump-off pipes or fittings **3** are connected to pumping lines **9** which in turn are connected to a pump or a vacuum source on boat **11**. The fibrous body is drawn through the water in the direction of the arrow. The impervious flexible shell **5** completely enclosed the forward section of the fibrous body so that the nearest point at which liquid (oil and/or water) contacts the fibrous body is the point **13** which is spaced apart from the pump-off fittings in the aft section of the body. When the body is in contact with a layer of oil on the water surface as shown, the oleophilic fibers attract the oil and the oil is propelled by capillary action and by the pumping force on the pump-off pipes through the body to the pipes and thence to the boat. As depicted, the trailing end of the apparatus rests in the water, depicting a complete removal of the oil from the water. In such a case, the water which contacts the trailing edge of the body cannot travel to the forward section so long as the remainder of the body is wet with oil. In the normal course of operation, there will be occasions when the fibrous body will be in contact with water along its entire length or substantially along its entire length. This can happen particularly in turbulent water where waves cause the surface of the water to break, and water momentarily displaces all of the oil in spots throughout the area being contacted. It can also happen in places where the oil layer is relatively thin and all available oil is removed quickly, e.g., as might happen near the edge of the contaminated area. It can also happen, in rough water, that water will wash over the fibrous body and douse it. In any of these cases, water can be entrained with the oil if provision is not made to prevent this from happening.

When using the apparatus of this process, the point where oil is taken off from the fibrous body is shielded from contact with water by the flexible shell **5**. Due to the affinity of the oleophilic fiber for the oil, the oil is absorbed and can flow by capillary action to the pump-off fittings even though this involves a reversal of the flow direction as shown by the curved directional arrows. No such affinity exists for water and thus capillary flow of the water is substantially nonexistent. Moreover, if water does get into the fibrous body it can be expelled therefrom and displaced by oil while it still is well away from the pump-off point and not under the influence of the pumping force.

The oleophilic fiber is preferably a polyolefin such as polyethylene, polypropylene or any of the fiber-forming olefin polymers known to the textile art for use as fibers and filaments. The most commonly used and most preferred are those based on propylene, referred to generically as polypropylene.

The fiber can be used in the form on a nonwoven batt of either staple fiber or tow which can be needle punched or sewn to afford the necessary structural integrity. Alternatively, the batt can be prepared from a plurality of woven or tufted layers laid up and secured together into an integral structure suited for skimming the water surface. The individual fibers can be of any size up to about 70 denier per filament, preferably about 10 denier or less including microfibers of less than 1 denier.

It is found that the volume of oil picked up and the magnitude of the capillary force acting on the oil are inversely related to the denier of the fiber, i.e., the smaller the fiber denier, the greater the absorption capacity. Fibers of 1 to 70 denier can pick up as much as 1,800 percent of their own weight of oil. Fibers of less than 1 denier have been found to pick up as much as 4,000 percent of their weight. The fibers can be crimped or smooth, although crimped fibers are preferred due to their greater structural coherence.

Industrial Filter & Pump Mfg. Co. Device

A device developed by *H. Schmidt, Jr., R.W. Crain and J.F. Zievers; U.S. Patent 3,695,451; October 3, 1972; assigned to Industrial Filter & Pump Mfg. Co.* is a portable skimmer which utilizes a continuous flexible belt which passes through an interface between two liquids to adsorb one of the liquids and then passes between a pair of spring-loaded drive rollers which remove the liquid from the surface of the belt. These rollers are mounted in a collecting trough which directs the removed liquid to a desired location.

Maksim Device

A device developed by *J. Maksim, Jr.; U.S. Patent 3,702,297; November 7, 1972* for removing crude oil from a body of water includes a sponge-coated collection roll. The ends of the collection roll are supported from floats so that the lower portion of the roll is immersed in the oil, and a drive mechanism rotates the collection roll to continuously absorb the oil. A downwardly sloping wringer is held tightly against the lateral surface of the collection roll to deform the collection roll surface and squeeze the oil from the roll. A doctor blade or a roller disposed against the longitudinal surface of the wringer provides an inclined weir for the oil squeezed from the collection roll. The oil flows downwardly into a collection tank located adjacent to the collection roll. An elongated baffle plate below the collection roll limits the amount of water taken up by the collection roll. A worm gear removes grease accumulations trapped by the doctor blade or roller. Figure 157 is a schematic side elevation showing the operation of this skimming device.

FIGURE 157: MAKSIM DRUM-TYPE OIL SKIMMER

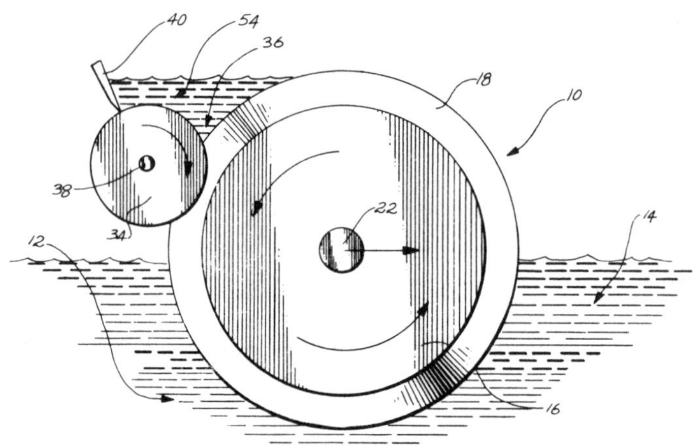

Source: J. Maksim, Jr.; U.S. Patent 3,702,297; November 7, 1972

Referring to the drawing this oil skimming device includes an elongated, tapered oil collection roll **10** rotatably mounted transversely to a body of water **12** having a top layer of crude oil **14**. The collection roll has a core **16** preferably in the form of a truncated cone with an absorbent outer covering **18** of a material having an affinity for oil. Preferably, the core is constructed of a high-strength, oil-resistant, relatively inexpensive material such as wood, or rubber. The outer covering of the collection roll preferably comprises any of a variety of oil-wet foam sponge materials. Such materials are well-known. The collection roll rotates about an elongated, centrally-disposed shaft **22**.

An elongated cylindrical discharge roll or wringer **34** is disposed tightly against the outer surface of collection roll **10** above the top of oil layer **14** and below upper longitudinal surface of the collection roll. The longitudinally extending wringer and collection roll define a downward sloping collection channel **36**. The wringer rotates about a centrally disposed shaft **38** supported at its ends. An elongated doctor blade **40** is disposed adjacent to the longitudinal surface of the wringer.

The wringer applies pressure to the longitudinal surface of the collection roll to deform the sponge coating and squeeze out the absorbed oil and water. The removed oil/water mixture **54** fills collection channel **36** and flows downwardly along the channel into a collection chamber. Doctor blade **40** is preferably used to provide an elongated transverse weir for the oil and water removed from the collection roll. Since the longitudinal surface of wringer **34** is held firmly against collection roll **10**, counterclockwise rotation of the collection roll causes the wringer to rotate in a clockwise direction, as shown by the arrows.

McRae Oil Corporation Device

A device developed by *B.G. Cornelius; U.S. Patent 3,536,199; October 27, 1970; assigned to McRae Oil Corporation* is a fire-extinguishing oil slick separator. More specifically, the device is an oil slick separator of the type comprising a rotated drum floated on oil slick water so that the drum picks up the oil and carries it into a chamber to be scraped off and to stand within the chamber which is thus not well calculated to support combustion, the oil from the lower part of the chamber being picked up by a conveyor to be carried on for further disposition.

Minnesota Mining and Manufacturing Company Device

A device developed by *J.F. Dyrud; U.S. Patent 3,627,677; December 14, 1971; assigned to Minnesota Mining and Manufacturing Company* comprises a flexible web and a frame used to support the web. The frame has retaining means attached to hold at least a portion of the edge of the web allowing the remaining portion of the web to form a receptacle for containing the separated oil.

The process provides for the separation and removal of water-immiscible oil from a mixture of oil and an aqueous medium. The process is comprised of first, forming a receptacle with a fibrous web, and then placing the web in contact with an oil and water mixture. The web and the oil and water mixture define a separation system in which the work of adhesion for the system as a whole is greater than one-half to one times the value of the work of cohesion for the oil-water portion of the system and the contact angle formed by the oil in an oil-water mixture with a smooth surface of a fiber-forming polymer is less than 90°. The oil is preferentially absorbed into the web until the web is substantially saturated. The receptacle formed by the web then becomes filled with oil as a result of the hydrostatic pressure imparted to the substantially saturated web.

Murphy Pacific Marine Salvage Company Device

As described in EPA, Edison, N.J. publication, *Oil Skimming Devices*, Report PB 218, 504, Springfield, Virginia, National Technical Information Service (May, 1970). On September 10, 1969, the Murphy Pacific Marine Salvage Company demonstrated a mechanical method of retrieving oil from a water surface at the Navy Base Port Service Dock area. The process reportedly developed by Shell Oil Company consists of an endless oil absorbent flexible floating vessel carrying a wringer unit which squeezes the belt to remove, collect and separate an oil and water mixture. This skimmer utilizes long, narrow belts with widths approximately 4 to 6 inches. The belts used are made of polypropylene. The device built to squeeze oil from the belt consists of two rubber covered rolls that both pull and compress the belt.

At a demonstration, the oil was dispensed on calm water and the moving belts were laid on the surface and threaded through the scrubbing device on the stern of an ICM boat. The demonstration showed that an oil scrubber process with a polypropylene belt has the capability to recover oil from a calm water surface. However, further development of the process under field conditions was recommended to be necessary before the feasibility can be firmly determined.

Oil Mop International, Inc. Device

A device developed by *H.M. Rhodes; U.S. Patent 3,668,118; June 6, 1972; assigned to Oil Mop International, Inc.* involves removing oil from the surface of a body of water with an oil mop made of thin gauge narrow strips of polypropylene or similar material passed through the oil on the surface of water and then through wringers and/or water or chemical sprays or both to remove the oil from the mop, depositing the oil in a receptacle and returning the nonoil ladened mop back into the oil covered water to pick up more surface oil.

Another patent by *H.M. Rhodes; U.S. Patent 3,744,638; July 10, 1973* describes constructional details of essentially a similar device which is shown in Figure 158. The figure shows plan and elevational views of the overall device and detail of a short length of oil mop pull line showing the buoyancy rods interlocked therewith.

FIGURE 158: CONTINUOUS POLYPROPYLENE MOP FOR OIL SPILL REMOVAL

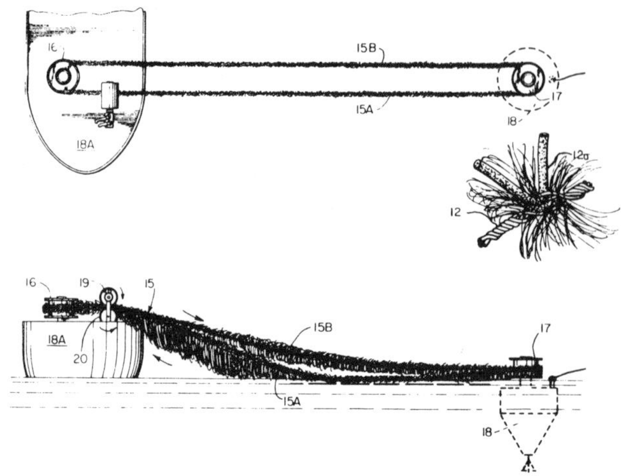

Source: H.M. Rhodes; U.S. Patent 3,744,638; July 10, 1973

The mop structure is made up of a plurality of thin gauge narrow strips of polypropylene, or similar material, having a specific gravity of less than 1.0, about 100 in number, the ends of which have been thermally fused to enhance the collection of oil. The groups are cut from a length of tow (2 to 5 feet) which contains more than 100 strips in cross section. The lengths are passed through or among the strands **12** of a multistrand polypropylene rope or line and attached to the line. Also passed through or among the strands of the multistrand polypropylene rope or line at frequent intervals of about 4 to 6 inches along the line are buoyancy units **12a**, such as closed cell foamed polystyrene in rod form **12a** approximately ⅜ inch in diameter weighing approximately 3.5 lb per cu ft. This construction constitutes the basic mop **15** which may be of any length, for example 100 or 1,000 feet. The line is spliced upon itself to make an endless mop.

The endless length of mop **15** is rooved about sheaves or pulleys **16** and **17** which are rotatably mounted on or near a barge or other support **18**. Pulley **17** may be rotatably mounted on a buoy **18** which could be anchored or tied to a tug which would keep the mop partially taut and directed in the area where the oil pick up is desired. Located between the pulleys **16** and **17** and being closer to pulley **16** on oil recovery barge or receptacle **18a** are a pair of pressurized power driven wringer rolls **19** and **20**.

The barge or recovery receptacle **18a** may be stationary, power driven, or towed to the polluted area and the mop **15** extended with the run **15a** being the pick up run and **15b** being the return run. The wringer rolls **19** and **20** when driven perform two functions, firstly, to move the mop from pulley **17** toward pulley **16** while secondly, squeezing the oil from the mop **15**. The oil recovered is drained into the receptacle beneath the wringer rolls. The gauge of the thin narrow strips of polypropylene is preferably 4 mils and a range of from 1 mil to 12 mils is operative. The width of the narrow strips averages ⅛ inch.

The cross section of the polypropylene strips may be circular, square, or rectangular and all have performed satisfactorily. The diameter of the circular sizes tested is 1 mil to 12 mils. The widths of the rectangular cross sections tested has ranged from 4 mils to ¼ inch with a thickness of 4 mils to 12 mils. The most successful operation is with a 4 mil thickness with a range of widths of a few mils to ¼ inch. The most successful run on the mop was 15 pounds of oil per pound of polypropylene per pass through the wringer.

Rex Chainbelt Device

A device commercially available from Rex Chainbelt, Inc. of Milwaukee, Wisconsin consists of an endless conveyor belt running over a lagged, crown faced 12 inch diameter head pulley assembly. This assembly includes a No. 14 galvanized housing with a hinged access section and an intermediate section as required to meet discharge height requirements. The housing is anchored to the floor.

The endless belt can be either 12 inches or 24 inches wide, three ply oil service Buna-N with vulcanized splice and nylon filled fabric. Two side guide idlers, directly mounted on the housing are included. The belt can be removed by unbolting and removing the upper housing only. The skimmer also includes two nonsparking wear resistant doctor blades and two collection troughs to convey the oil to a common collection point. The skimmer is driven by a motor not less than one-half horsepower.

Sewell Device

A device developed by *R.B.H. Sewell; Canadian Patent 735,254; May 31, 1966* involves the physical removal of a spilled liquid from a body of water by means of an absorbent belt, which on being brought into contact with the spilled liquid and the body of water, absorbs the spilled liquid in preference to water. It has been discovered that an absorbent material satisfactory for this purpose consists of one which contains a high percentage of cotton. Sewell has determined that dry cotton will absorb about 29 times its weight in bunker oil and cotton, which was soaked in water and subsequently wrung out. The cotton will soak up approximately 21 times its weight in oil on being dipped into the oil only once.

Various mechanical arrangements can be utilized for the purpose of passing the selectively absorbent material into the spill, drawing the soaked material out of the spill and recovering the oil by subjecting the material to pressure. The accompanying figure (Figure 159) shows schematically one such apparatus.

FIGURE 159: SEWELL BELT SKIMMER

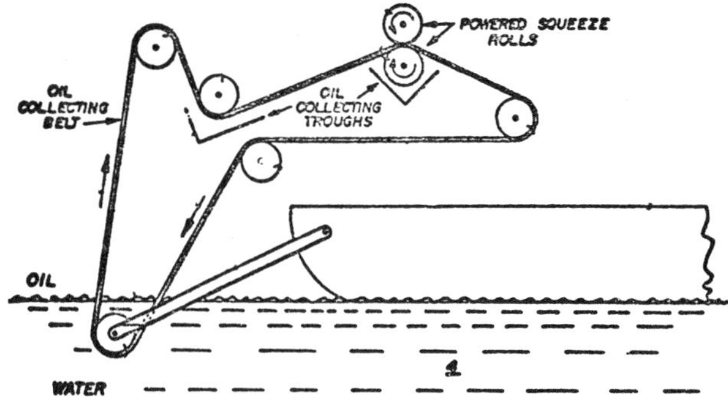

Source: Report PB 218,504

An endless oil collecting belt formed of suitable material such as loosely woven absorbent cotton is threaded about the idler rollers so as to pass between them and the trough and between the powered rollers. In operation, the powered rollers are rotated so as to cause movement of the belt and the roller support is lowered so as to bring the belt into contact with the oil spill. As the belt rotates, it selectively absorbs oil from the water surface, carries it over the various rollers and to the powered squeeze rollers where most of the oil is removed from the belt and it runs off via the trough. In this manner, continuous mechanical collection of the oil spill occurs and if the spill is large, the apparatus can be slowly moved through it by moving the barge.

Sharpton Device

A device developed by *D.E. Sharpton; U.S. Patent 3,643,804; February 22, 1972* is one in which an endless conveyor belt having an affinity for oil is mounted on the bow of a barge and projects downwardly and forwardly therefrom towards a water surface having oil floating thereon. Buoyancy means located at the lower end of the belt are adjustable such that the lower run of the belt can be selectively positioned slightly below the interface of the water and oil to assure maximum oil absorption with minimum water pickup. Figure 160 shows the device in plan and elevation.

FIGURE 160: SHARPTON BELT-TYPE SKIMMER

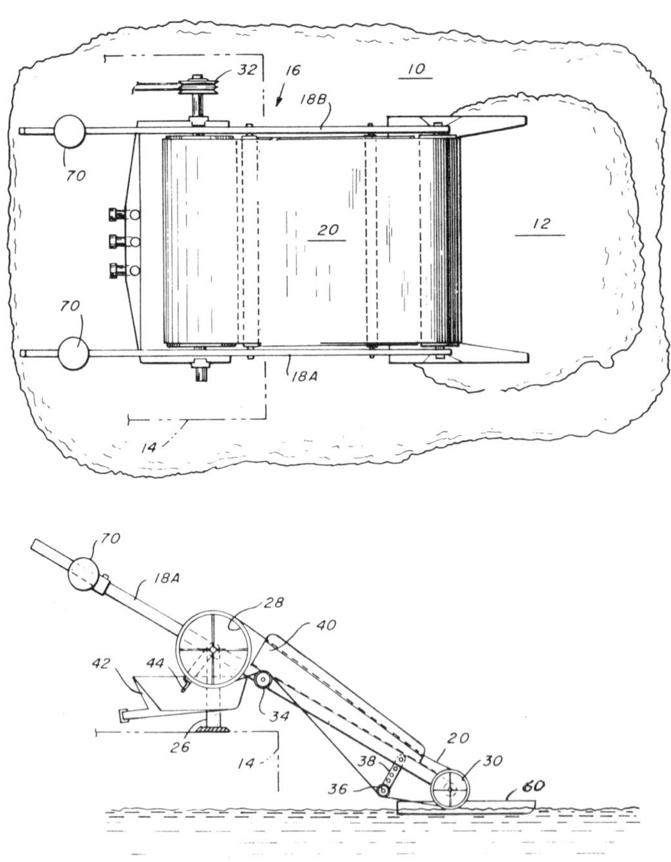

Source: D.E. Sharpton; U.S. Patent 3,643,804; February 22, 1972

There is shown a large expanse of water **10**, upon which a quantity of oil or other immiscible fluid **12** is floating. Projecting downwardly and forwardly from a floating barge **14** is the oil recovery apparatus **16**. A frame comprising two spaced-apart parallel elongated members **18A** and **18B** rotatably carries an endless belt **20**. The belt may be made of any material having an affinity for oil but not for water. A typical belt comprises an elongated neoprene core encapsulated by synthetic rubber.

Turning now to the side view, elongated frame member **18A** is pivotally connected to a support strut **26** which projects vertically upwardly from a point on the bow of barge **14**. Similarly frame member **18B** is also attached to the barge. Rotatably supported by and between elongated members **18A** and **18B** is an upper drum **28**; the axis of rotation thereof being transverse to the longitudinal axis of the elongated members. Rotatably supported by and between the elongated members at the lower end thereof is a lower drum **30**, the axis of rotation also being transverse the longitudinal axis of the elongated members. Belt **20** is wrapped around and between drums **28** and **30**. Preferably the diameter of upper drum **28** is twice that of the lower drum **30** such that for every revolution of the upper drum the lower drum makes two revolutions.

Power rotating means **32** attached to the shaft of upper drum **30** effects travel of belt **20**. A tension roller **34** is rotatably secured to and disposed between frame members **18A** and **18B** outside of the path of endless belt **20**. An adjustable pitch roller **36** is rotatably secured to the elongated frame members within the path of the belt via an arm **38**. An inner support pan **40** intermediate the upper and lower drums provides a planar surface upon which the upper run of the belt rides.

Carried by floating barge **14** directly under upper drum **28** is an oil pickup reservoir or sump **42** which has an outlet coupled to a suction pump for transferring oil therefrom to the main storage which may in some cases constitute the craft itself. Suitable wiper blades or squeegees **44** pressing against the width of the belt **20** wipe the oil from the belt and allow same to flow into the pickup reservoir.

Adjustable counterbalance weights **70** are slidably received on the upper ends of frame members **18A** and **18B** to cushion the rocking of pontoons **60**. In operation the barge with the oil recovery apparatus attached is transported to a point adjacent an oil slick. The pontoons are then moved relative to the frame members by manipulation of the crank handle until the lower run of belt **20** at the underside of drum **30** is slightly below the interface of the oil and water.

The belt **20** is then rotated in a counterclockwise direction to convey oil upwardly to upper drum **28** where wiper blade **44** squeezes the oil from the belt into sump **42** from whence the oil is conducted to the main reservoir within the craft. As before mentioned, the location of the lower part of the belt and the drum at the interface of the oil and water assures the maximum attraction of the oil to the belt while a minimum of water is conveyed upwardly onto the belt. As the reservoir in the craft begins to fill and the craft settles in the water, the crank is further manipulated to readjust the lower run of belt at drum **30** so that it remains slightly below the interface of the oil and water.

Thus at all times during the operation the lower drum is conveniently and easily positioned at the interface of the water and oil by an operator on the craft. In some applications oil recovery apparatus **16** may be conveyed through the oil slick by the barge's power plant. In other applications the apparatus may remain stationary and the flow of the water may be utilized to carry the oil to the belt. The dimensions of the components of the apparatus are determined by the particular requirements of each application. It may be feasible to mount the apparatus to a stationary support on shore rather than on a barge.

Shell Oil Company Device

A device developed by *C. Bezemer, H.J. Tadema and J.J.H.C. Houbolt; U.S. Patent 3,700,593; October 24, 1972; assigned to Shell Oil Company* is one wherein an elongated absorbent porous body is positioned along the water surface by base means which include squeezers for removing oil from the body, a container for collecting the oil and drive means for moving the body past the squeezers. Preferably the porous body has an average porosity of at least 80%; the porous body may comprise an organic polymer having a density between 0.85 and 1.50. Suitable polymers for this purpose are polypropylene or polyamides.

The polymer may consist of fibers with a fiber grade between 1.5 and 100 denier. The capillary properties of the porous body depend on the surface properties of the fibers, which define the oil wetting degree, on the packing density defining the porosity, and on the fiber grade, which is expressed in denier units (1 denier = 1 gram per 9,000 meters).

The finer the fiber and the higher the packing density, the stronger will be the capillary forces that cause imbibition of oil. The capillary rise of oil against air and the downward imbibition against water will set a limit to the thickness of the porous body. With relatively fine fibers, complete oil saturation can still be obtained at an immersion depth in water of about eight inches.

The contact time needed at the air-oil-water surface for nearly complete oil saturation is determined by the capillary properties counteracted by the flow resistance, which is influenced by the size and permeability of the porous body and the oil viscosity. An increase of the capillary forces by using a higher packing density and a finer fiber is associated with a simultaneous increase in flow resistance. On the basis of the above considerations, optimum results will be obtained when maintaining an average porosity of at least 80% and a fiber grade between 1.5 and 100 denier.

The density of the porous body is preferably kept below 1.00, but this is not necessary to keep the body floating, as special means may be provided, such as floats which prevent the body from sinking; however, preferably the density should not exceed 1.50, as otherwise it will become too bulky with respect to the porous body itself. As an alternative the organic polymer may consist of foamed plastic material. The porous body is preferably elongated and the body may be contained in a skin of permeable cloth.

A high rate of oil recovery may be obtained when the porous body is made endless. The body may be composed of sections which are interconnected which offers the advantage that an even distribution of the porous material can easily be obtained, and that, during operation plug formation will be avoided. The sections may suitably be interconnected in such a way that they are capable of rotating independently around a longitudinal axis. This may be achieved by rotatable coupling means known in the art. As an alternative, the porous body may comprise a flexible core provided with substantially radially arranged fibrous material fixed to the core, which results in a brush-type porous body.

Figure 161 shows two alternatives made of squeezing the oil saturated porous body. The porous body **20** is passed through a slit between pairs of cylinders or rollers **21, 22** and **23**, each pair being parallel and placed at such a distance from each other that the porous body can pass through the slit between each pair of cylinders. The slit between the cylinders **21** is of greater width than the slit between the last pair of cylinders **23**, the slits successively decreasing in width. A receiving trough **24** is placed below the set of cylinders for collecting the recovered oil and leading it to a receiving container **31**.

By placing the squeezer means for removing oil from the endless belt successively closer a more gradual removal of oil over a longer squeezing period is obtained as compared with the use of only one pair of rollers. This is of importance when handling viscous oil, with viscosity above 300 cs, since otherwise the pore pressure generated by the squeezing action would become excessively high, causing slippage and reduced life of the belt. With low viscous oils one pair of rollers generally appears adequate. Clearly the pore pressure generated also depends on the rate of circulation of the belt through the squeezer.

FIGURE 161: SHELL ABSORBENT BELT SKIMMER

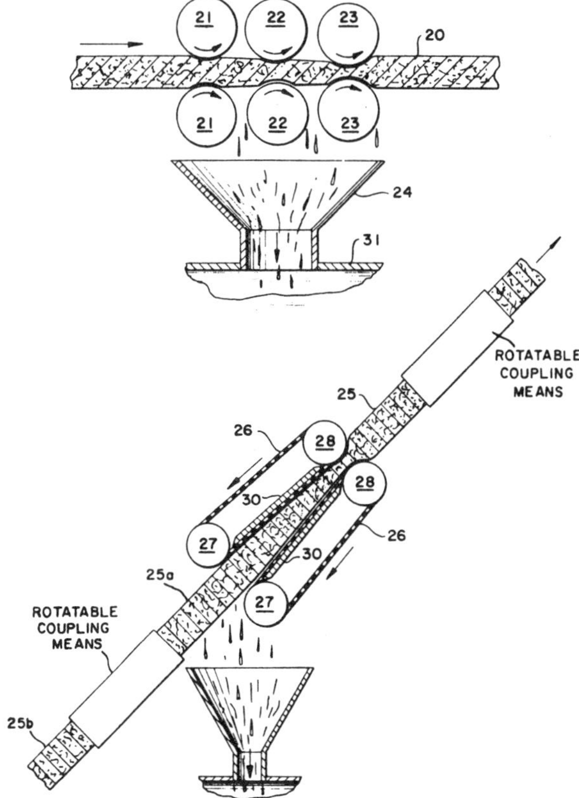

Source: C. Bezemer, H.J. Tadema and J.J.H.C. Houbolt; U.S. Patent 3,700,593; October 24, 1972

The lower view shows an alternative embodiment for squeezing the oil-saturated porous body. The porous body **25** is led through a slit which gradually narrows, this slit being formed between two endless belts **26** which are led over two pairs of cylinders **27** and **28**, the distance between the parallel cylinders **27** being greater than that between the cylinders **28**, thus forming the gradually decreasing width of the slit. A receiving trough **29** is placed at the lower end of this sloping arrangement and collected oil is led to a receiving container (not shown). Suitable supports **30** may be provided at the operating side of the endless belts. The distance between the belts may be made adjustable.

Standard Oil Company Devices

A device developed by *W.L. Bulkley, H.E. Ries, Jr. and R.G. Will; U.S. Patent 3,539,508; November 10, 1970; assigned to Standard Oil Company* is one in which at least one pair of spaced, revolving pickup members which dip into the liquid are used to recover the floating material. This material adheres to the members as they come into contact with the liquid, and means adjacent these members remove and collect the material adhering to them. The characterizing feature of this device is that the surface of one member is smooth and oleophilic, and the surface of the other member is porous and deformable. The member having a smooth, oleophilic surface is in advance of the member having the porous, deformable surface, so that the smooth surfaced member contacts the floating material before the porous surfaced member.

A device developed by *R.G. Will and W.F. Swiss, Jr.; U.S. Patent 3,546,112; December 8, 1970; assigned to Standard Oil Company* is a power driven apparatus having a rotation means with a closed supporting surface, absorber means for absorbing water and oil supported on the surface, removal means for sequentially removing water and oil from the absorber means, the removal means being a plurality of rollers exerting different pressures against the absorber means, and wiper means for effectuating the withdrawal of the oil. Figure 162 is an end elevation of such a device.

FIGURE 162: STANDARD OIL COMPANY ABSORBENT DRUM SKIMMER

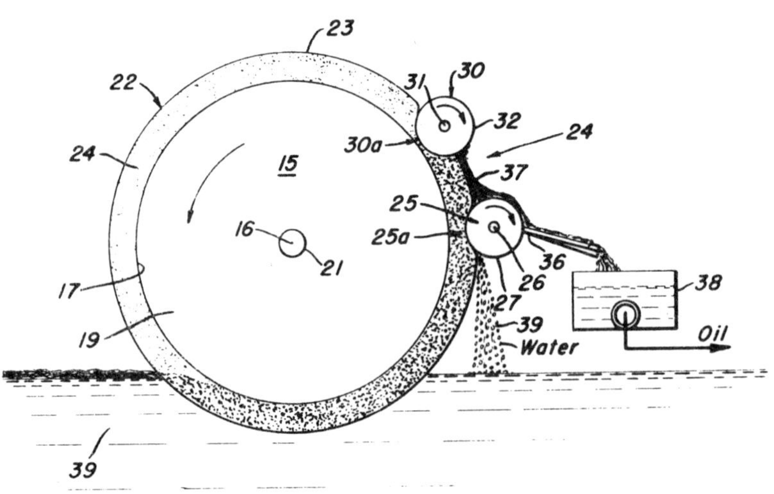

Source: R.G. Will and W.F. Swiss, Jr.; U.S. Patent 3,546,112; December 8, 1970

The rotating means is generally numerically designated **15** and includes the longitudinal axis **16** and supporting surface **17**. The rotating means also has closed ends **19** with extending shaft ends **21**. The supporting surface **17** of the rotating means **15** has absorber means **22** rigidly supported thereon for rotational motion about the longitudinal axis **16** of the rotating means **15** in unison with the rotating means. The absorber means **22** has an exterior surface **23** and a thickness numerically designated **24**.

The removal means **24** includes in combination, a first compression roller **25** having a longitudinal axis **26**, an essentially cylindrical surface **27** in contact with the surface **22** of the absorber. The shaft ends protrude through each end of the essentially cylindrical compression roller **25** for attachment to any desired type of operative drive means.

A second compression roller **30** of the removal means **24** has longitudinal axis **31** about which the roller is mounted for rotation. Surface **32** is in contact with the exterior surface of the absorber means **22** in a manner to provide a severe compression of the absorber means **22**. Like the first compression roller **25**, the second compression roller **30** rotates in a clockwise direction as the rotating means **15** rotates in a counterclockwise direction. The shaft ends of roller **30** accommodate any desirable operative type of drive means.

The wiper means includes a wiper blade **36** in contact with surface **27** of first compression roller **25**. Blade **36** extends at least the full length of roller **25** and is shaped to provide drainage of the oil **37** from the surface **27** of roller **25**. The oil **37** is depicted as flowing from the second compression zone **30a** to the surface **27** of roller **25** and into a retaining means **38** to segregate the oil **37** from the water body. The compression roller **25** forms compression zone **25a** in absorber means **22** and removes water **39** from the absorber means.

A device developed by *R.L. Yahnke; U.S. Patent 3,578,585; May 11, 1971; assigned to Standard Oil Company* is a rotating cylinder covered with a layer of porous polyurethane which absorbs oil flowing on a body of water, and a roller which squeezes absorbed oil from the layer into a collecting trough.

A device developed by *J.F. Grutsch and R.C. Mallatt; U.S. Patent 3,608,727; September 28, 1971; assigned to Standard Oil Company* is one in which oil, solid particles such as biological slime and floating debris, etc. are removed from water by an apparatus including endless chain means made up of a series of interconnected foraminous chambers holding a regenerable porous filter material such as polyurethane. Preferably the filter material has an outer large pore section and an inner small pore section. A plurality of buckets are attached to the chain means, and as the chain means moves through a closed loop path, these buckets catch debris and dump it into a holding bin. Simultaneously, the filter material absorbs surface and subsurface oil and the like from water traveling through the filter material. The filter material is regenerated by squeezing the filter material to release the oil.

Such a device is also described by *J.F. Grutsch and R.C. Mallatt; U.S. Patent 3,732,161; May 8, 1973; assigned to Standard Oil Company*. As shown in Figure 163, the apparatus **10** is mounted between the dual hulls of a catamaran. The chief component of the apparatus **10** is endless carrier means **12** including a plurality of foraminous chambers **16** which hold regenerable porous filter material **18**. Preferably, filter material **18** is reticulated hydrophobic foam such as polyurethane made up of an outer large pore section **18a** and inner small port section **18b**. Outer section **18a**, typically having 45 or less pores per linear inch, readily entraps viscous oils. Inner section **18b**, typically having 60 or more pores per linear inch, readily entraps light fluid oil. The pores of section **18a** are substantially larger than the pores of section **18b**. Such a multiporosity filter material is less likely to become clogged than a filter material having only a small pore structure. Alternately, the different chambers **16** may include foams having different size pore structures.

Each foraminous chamber **16** includes a bucket **14** which serves to collect debris, and each bucket has screen bottom **15** and a pair of end plates **21**. These buckets are connected between two series of endless chain links which make up carrier means **12**. These links include rollers which ride over tips **25** of the sprockets as these sprockets revolve. Sprocket **24**, the drive sprocket, is above the water's surface and its drive shaft **28** is connected to drive motor **30**. Sprocket **26**, the idler sprocket, is below the water's surface and offset relative to drive sprocket **24**. The offset arrangement of sprockets **24** and **26** inclines carrier means **12** relative to the horizontal, and the angle of inclination ϕ ranges between about 30° and about 90°, preferably about 60°.

The catamaran also includes debris storing bin **32** between the water's surface and sprocket **24**. Screen bottomed buckets **14** catch debris and then dump it into this bin **32**. Any liquid picked up by buckets drains through screen bottoms **15** of the buckets and through filter material **18** and is thus purified. To further facilitate debris collection, the catamaran is also equipped with spiral or paddle wheel skimmer **34** in advance of chain means **12**. Skimmer **34** directs in a positive manner surface debris and oil toward buckets **14**. Each of the chambers **16** includes rigid frame **36** secured to bottoms **15** of buckets **14**.

FIGURE 163: STANDARD OIL COMPANY BELT-TYPE SKIMMER

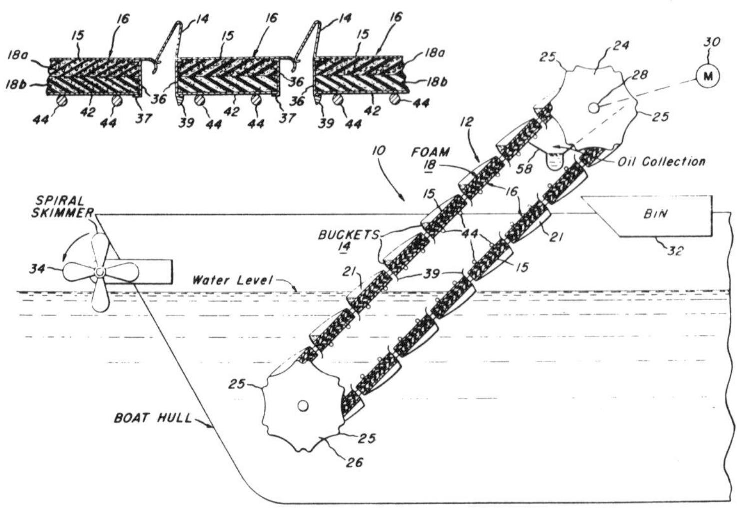

Source: J.F. Grutsch and R.C. Mallatt; U.S. Patent 3,732,161; May 8, 1973

Frame **36** includes lip **37** for retaining filter material **18** and a movable inner pressure screen **42**. Filter material is disposed between screen bottom **15** and movable pressure screen **42**, and it is regenerated by compressing it between the bottom screen and the movable screen to squeeze out absorbed oil. Edge **39** of the frame may be spout-like to aid in directing liberated oil into oil collecting means **58** mounted inside the looped path of carrier means **12**. Stiffeners **44** are attached to movable pressure screen **42** and these stiffeners act in conjunction with the pressurizing means to compress the filter material between screens **15** and **42**.

A device developed by *R.G. Will and J.F. Grutsch; U.S. Patent 3,617,552; November 2, 1971; assigned to Standard Oil Company* is one in which oil-contaminated water is purified using apparatus having a revolving polyurethane foam belt which is mounted on an incline relative to the horizontal. The contaminated water moves past and through the revolving belt or the belt is mounted on a boat which moves the belt through the water. In either case, as the belt and water move relative to each other, the oil-contaminated water filters through the belt and is purified. The belt is then squeezed twice. First gently to remove water, and then vigorously to remove oil. The belt may include inner and outer abutting sections which are reinforced by a network of threads. The outer section which first contacts the contaminated water has a larger pore structure than the inner section.

Figure 164 is a schematic cross-sectional view of the device; it also shows an enlarged fragmentary view of the construction of the foamed polyurethane belt employed. The apparatus **10** may be mounted in a stationary position in a stream of water so that oil-contaminated water flows past the apparatus, or apparatus **10** may be mounted on a boat so that it can be moved through a body of water such as a lake or even the open sea. In either case, the oil-contaminated water and the apparatus move relative to each other at speeds ranging between about one-half and about 3 miles per hour.

As illustrated, apparatus **10** has three basic components: (a) belt **12** made of a regenerable porous filter material which selectively absorbs oil, (a suitable filter material is foamed polyurethane), (b) regenerating means **14** which removes absorbed oil and water from belt **12**, and (c) collecting means **16** which collects oil removed from the belt. The belt is inclined, and the angle of inclination ϕ relative to the horizontal preferably ranges from about 30° to about 60°.

FIGURE 164: ALTERNATIVE STANDARD OIL COMPANY BELT TYPE SKIMMER

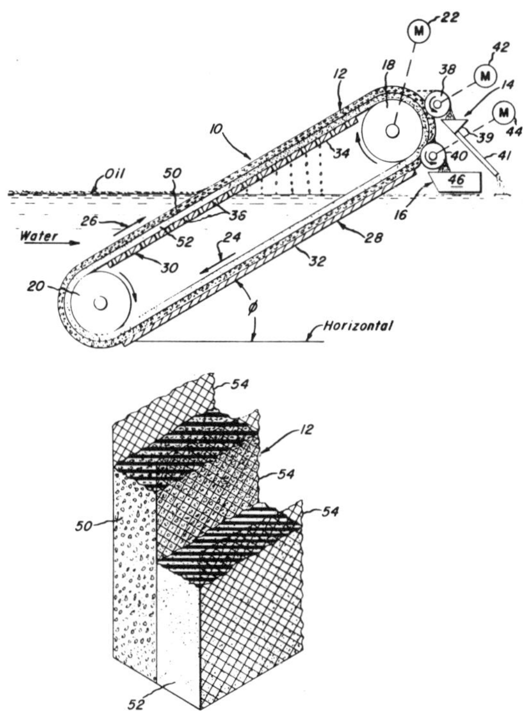

Source: R.G. Will and J.F. Grutsch; U.S. Patent 3,617,552; November 2, 1971

Belt **12** is trained about drive roller **18** and idler roller **20**. These rollers are vertically offset relative to each other, and the drive roller is above the water's surface and the idler roller is below the water's surface. Motor **22** turns drive roller **18** in a clockwise direction as viewed in the figure so that the belt moves through a closed looped patch with flight **24** of the belt dipping into the water and flight **26** leaving the water.

Guide means **28** and **30** are provided so that the belt will not stretch due to the pressure exerted on it by the water. Guide means **28** includes a solid plate member **32** which is adjacent the outside of the closed looped path of belt **12** and which extends transverse to flight **24**, and guide means **30** includes a screen or perforated plate **34** which is inside the looped path of the belt and which extends transverse to flight **26**. Plate **32** is only wide enough to support belt **12**, i.e., it is laterally coextensive with flight **24**. Braces may be used to attach plate **32** to a boat or sluice. Perforations **36** in plate **34** are substantially coextensive with the area covered by flight **26**; however, the ends (not shown) of plate **34** are solid and they extend beyond the periphery of flight **26**. These solid ends are attached to a boat or sluice as the case may be. Both plates **32** and **34** are preferably coated with a suitable material that reduces the coefficient of friction between the belt's surfaces and the plates. A suitable material would be a fluorocarbon polymer such as Teflon.

The regenerating means **14** preferably includes a pair of squeegee rolls **38** and **40** which press against belt **12**. Motors **42** and **44** are coupled to squeegee rolls **38** and **40**, respectively, and these motors turn these rolls in a counterclockwise direction as viewed in the figure. Roll **38** does not press against the belt with the same force as roll **40**. It only presses the belt gently so that substantially all the water is removed from the belt but none of the oil. Removed water flows over roll **38** into trough **39**, and then out pipe **41**, returning to the main body of water. Roll **40** presses against the belt with a greater force than roll **38**. This removes absorbed oil, which then flows into collecting means **16** such as trough **46**. Oil in trough **46** is then pumped into a suitable storage tank.

As illustrated, belt **12** is preferably made up of two or more sections **50** and **52**. Section **50** faces the outside of the closed looped path and section **52** faces the inside of the closed looped path. Thus oil-contaminated water first flows through section **50** and then through section **52**. Section **50** has a larger pore structure than section **52** so that most heavy, viscous oils are absorbed in this section **50** before they can penetrate into section **52**. Lighter, fluid oils are absorbed in section **52**. Preferably section **50** has 45 or less pores per linear inch. Section **52** preferably has 60 or more pores per linear inch. To make the belt more durable, networks **54** of nylon threads are glued to the exposed surfaces of the belt and between abutting surfaces of sections **50** and **52**. Threads **56** are substantially perpendicular to each other and they provide a ¼ to ⅜ inch mesh.

In operation, apparatus **10** is either moved through a body of water or disposed in a stream of flowing water. This will insure that oil-contaminated water impinges against flight **26**. Due to the relative motion of the oil-contaminated water and the apparatus, and with the aid of upward movement of flight **26**, contaminated water is lifted upwardly so that a differential in pressure is established across the belt. This differential in pressure causes the oil-contaminated water to flow through the belt and through perforations **36** in plate **34**. As the water flows through belt **12**, oil and some water are trapped in the pores of the belt. When the belt engages roll **38** virtually all the absorbed water is squeezed from the belt, but the oil is retained. When the belt engages roll **40** virtually all the absorbed oil is caught in trough **46** and pumped into a storage tank. The buckling and stretching tendency of the belt caused by the water pressure is counteracted by the reinforcement provided by networks **54** and guide means **28** and **30**.

A device developed by *I. Ginsburgh and R.G. Will; U.S. Patent 3,617,555; November 2, 1971; assigned to Standard Oil Company* is one in which oil and debris are removed from the surface of water using a revolving, partially submerged, endless brush belt. The brush belt has outwardly projecting bristles which ensnare the debris and pick up oil. Polypropylene bristles are preferred. The oil and debris are removed from the belt before the belt is reimmersed in the water. Alternatively, a brush-type drum could be used in place of the endless brush belt.

Surface Separator Systems Device

A drum-type skimmer has been developed by Surface Separator Systems, Inc. of Baltimore, Maryland. This unit is intended to be used in the recovery of persistent oil slicks in separators, lagoons and drainage basins where there is a continuous inflow of oily waste. This oil recovery device is a three cylinder hydraulic powered unit which recovers, separates and stores the oil in one continuous and combined operation. The oil film recovery mechanism is mounted on top of a 300 U.S. gallon oil storage tank. The buoyancy of both sections is such as to support the combined system at the proper operating draft. The oil recovery rate, when handling Bunker C, is about 600 gallons of 95% pure oil per hour. Wave heights up to 9 inches, wind, and rain are claimed to have practically no effect on the rate of retrieval. The recovered oil is delivered to the storage tank by gravity. To avoid significant changes in draft, this tank operates on a hydraulic displacement system, i.e., it will always be full of water, oil, or a combination thereof.

The prime mover for the power pack may be an electric motor, air motor, diesel engine or gasoline engine. In locations where very infrequent oil spills occur and where auxiliary marine equipment is available, this model may serve for emergency use; otherwise, the self-propelled unit should be used.

Texaco Devices

A device developed by *E.L. Cole and H.V. Hess; U.S. Patent 3,617,556; November 2, 1971; assigned to Texaco Inc.* permits separation and recovery of oil from oil slicks on the surface of water by continuously and selectively picking up the surface oil in a relatively thick blanket of a bulk fabric composed of highly oleophilic fibers expressing the oil from the blanket and leaving the fibers in an open condition highly receptive to additional oil. Figure 165 is a plan view of a small workboat carrying outboard two drums of the device (top view) and also has a detailed elevation through one of the oil pickup drums (lower view).

FIGURE 165: TEXACO DRUM-TYPE SKIMMER DESIGN

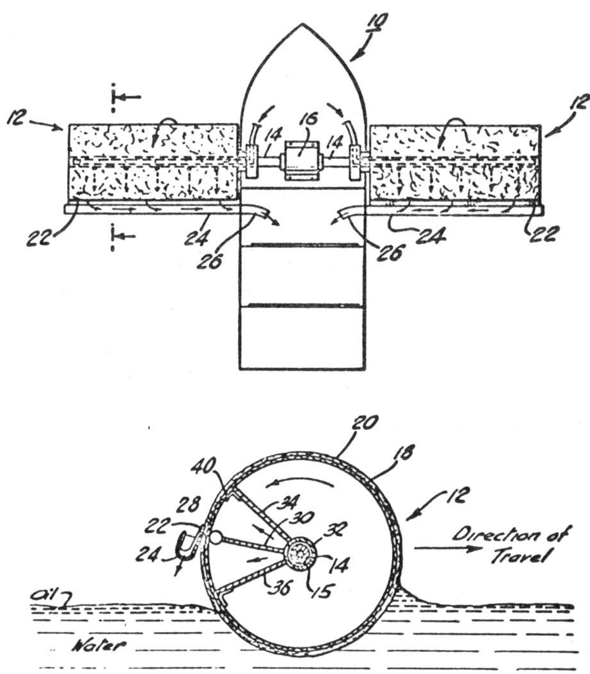

Source: E.L. Cole and H.V. Hess; U.S. Patent 3,617,556; November 2, 1971

As shown therein more or less diagrammatically, a vessel **10** supports two laterally extending drums **12** rotated on axle **14** by driving means **16**. The drum is, as indicated, so located that the lower extremity dips into the water a substantial distance.

The drum has the form of a cylinder which may comprise a screen **18** or a perforated sheet at least sufficiently rigid to maintain its cylindrical shape and support an outer layer of bulk fabric **20**, which completely envelops the cylindrical periphery of the drum. A doctor blade **22**, preferably of Teflon or similar flexible material is provided on the after side of the drum as shown, to remove the adsorbed oil which falls into trough **24**, from which it is recovered via outlet pipe **26**.

To support the gentle pressure of the doctor blade **22** against the drum, an internal fixed roller **28** is mounted within the interior of the drum, bearing lightly against its inner surface. Roller **28** is, in turn, rotationally mounted, by means not shown on spider or arm **30** which are mounted upon central sleeve **32**. It must be particularly observed that sleeve **32** is fixedly mounted against rotation, by means not shown in detail, on the main frame of the vessel, also sleeve **32** is spaced substantially from rotating axle **14** by annular space **15** to provide an internal duct for reasons which will hereinafter appear. Thus, while the drums rotate with the axle, the internal structures mounted on the sleeve remain angularly fixed.

Moreover, as is known in the art of lubricating oil dewaxing filters, the internal section of the drum in the region of the doctor blade is compartmentalized by fixed walls **34** and **36**, also mounted on the sleeve and which are impermeable and terminate preferably in wipers designated as **40**, which also bear lightly against the interior internal walls of the drum. Thus, the parts **30** to **36** are fixed as shown relative to the drum, which rotates about it.

The resultant compartmentalized segment of the drum is continuously provided with a supply of air under a pressure slightly above atmospheric, which performs the functions of supporting the bulk fabric in the area of the doctor blade while at the same time greatly facilitating the expulsion of oil from the spaces between the polymeric fibers to assist the wiping action of the doctor blade.

It is particularly important to note that the air blow is continued for a substantial angular distance after the point of contact by the doctor blade. This has the important function of counteracting the compressive effect of the doctor blade upon the relatively compressible fibrous fabric and expanding it again to a condition of substantial permeability so that the maximum volume of oil can thereafter be readily adsorbed.

As will be apparent from the foregoing, mechanical means is provided to continuously rotate the drum, preferably in the direction of the arrow, as shown, namely in a direction such that the lower surface of the drum moves continuously in the same direction as the direction of movement of the vessel itself. Therefore, as the vessel heads into an oil slick the surface of the drum tends to, at least momentarily, pile up the equivalent of a bow wave, which, since it is composed largely of oil, facilitates the absorption of oil into the previously prepared, highly receptive fibrous interior of the bulk fabric.

A device developed by *M.H. Van Stavern, W.T. Jones, H.F. Cossey and W.J. Clark; U.S. Patent 3,612,277; October 12, 1971; assigned to Texaco Inc.* is a rotatable drum type of oil skimmer which is continually rotated to pick up a film of oil and water on the surface of the drum, having a supplemental or transfer drum located substantially above the oil slick and well out of contact therewith, so as to come into contact with film on the pickup drum and receive a portion of the film, product oil being recovered from both drums.

Such a device is also described by *M.H. Van Stavern, H.D. Moorer, G.H. Miller and K.M. Gunn; U.S. Patent 3,685,653; August 22, 1972; assigned to Texaco, Inc.* Figure 166 is a perspective view of such a device. Referring now to the drawing, **10** represents the frame of the device which is carried by a pair of floats **12** which may take the form of any suitable marine vessel. Where it is desired to arrange the oil pickup device over a fixed channel, the frame can be simply mounted on fixed supports over the channel. A pickup roller, comprising cylinder **14**, is journaled as at **16** in frame **10**. Roll **14** is so arranged that its lower extremity dips into the water surface **18** as at **20**.

FIGURE 166: ALTERNATIVE TEXACO DRUM-TYPE SKIMMER DESIGN

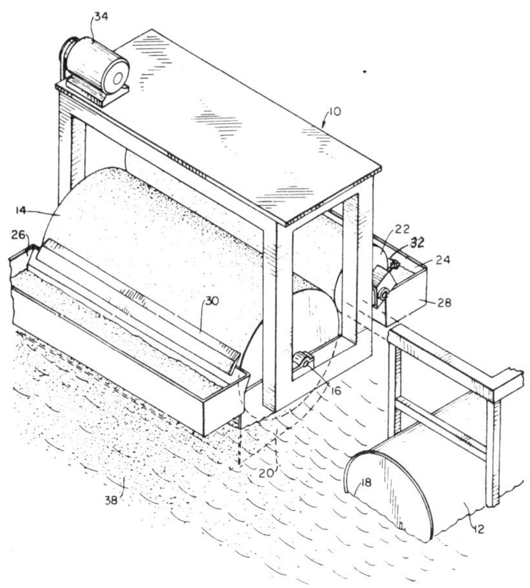

Source: M.H. Van Stavern, H.D. Moorer, G.H. Miller and K.M. Gunn; U.S. Patent 3,685,653; August 22, 1972

A supplemental or transfer roller comprises cylinder **22**, likewise journaled at **24** on the frame **10** and located in a position substantially above and out of contact with the oil slick **38**. Pickup troughs **26** and **28** are arranged respectively adjacent the pickup roll **14** and transfer roll **22** to receive film removed by scrapers or doctor blades **30** and **32**, which bear against the respective rollers. The rollers may be rotated in opposite rotational directions by drive motor **34** and suitable belt and gear drive so that the rollers move, at their point of close juxtaposition, at approximately the same speed and the same direction.

Trimble Device

A device developed by *L.E. Trimble; U.S. Patent 3,608,728; September 28, 1971* is a skimmer for removing oil from the surface of water which includes a cylindrical float mounted to a frame so that the float may rotate with its axis parallel to the surface of the water. A second float is connected to the other through the frame with paddles that cause the second float to rotate and actuate a bellcrank and connecting rod system to rotate the first cylindrical float. A scraper contacts the upper surface of the cylindrical float to remove oil collected on the surface of the float and allows it to drain down the trough in the scraper into a drainpipe that will flow the oil into a reservoir for storage. In operation, a push-bar frame is attached to the second float having paddle vanes and the entire apparatus would be pushed by a boat, thereby using the motion through the water to actuate the paddle wheel to rotate the oil skimmer float to pick up oil from the surface.

Welles Products Corporation Device

A skimmer device known as the Reclam-Ator produced by Welles Products Corporation of

Roscoe, Illinois was developed by American Oil Company's Research and Development at Whiting, Indiana. The skimmer is used to remove oil slicks from surface water in and around the company's refineries. A single polyurethane-covered drum rotates continuously in the water while absorbed oil and water are squeezed from the foam by metal rollers. A low pressure roller removes the water before the drum rotates under the oil removing roller where greater pressure is exerted.

The rotating drum is 12 inches in diameter and 4 feet long. It is mounted on a commercially available pontoon catamaran 24 feet long and 8 feet wide. These dimensions permit the unit to be transported by commercial air, rail or highway vehicles. Average grade Bunker C can be recovered at rates up to 50 barrels per hour at 21°C. Several more of these skimmers have been built by American and are kept in readiness at the company's major marine terminals.

VORTEX DEVICES

Bertin & Cie Design

A device developed by *J.-C. Mourlon and E.M.R. Dubois; U.S. Patent 3,635,342; January 18, 1972; assigned to Bertin & Cie, France* is shown in Figure 167. The oil slick **2a** on the free water surface **1a** is drawn past **2** into the vortex **3** which is produced by the vortex producing device which is positioned beneath the slick.

FIGURE 167: BERTIN & CIE VORTEX-TYPE SKIMMER DEVICE

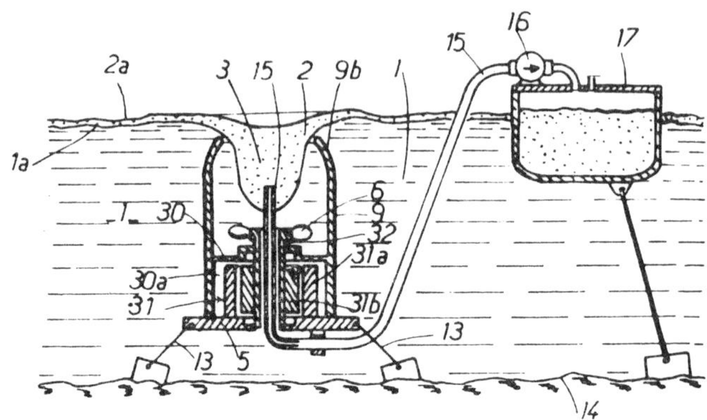

Source: J.-C. Mourlon and E.M.R. Dubois; U.S. Patent 3,635,342; January 18, 1972

The platform **5** is immersed in the liquid **1** and is connected to the bottom **14** by the anchoring device **13**. The fairing **9** is fastened above the platform, and the upper free edge **9b** of the fairing is located in the vicinity of the surface **1a**. A partition **30** makes it possible to form, along with the platform **5** and the fairing **9**, a fluidtight chamber **30a** inside the lower part of the fairing. A motor **31** placed inside the chamber includes a stator **31a** and a rotor **31b**. A hollow shaft **32**, rigid with the rotor, passes through the partition in a fluidtight manner, extending as far as the space located above the chamber. Its free end carries a screw propeller **6**. The pipe **15** for carrying away the light substance **2** is placed inside the shaft and opens the cavity **3**. It is linked with the tank **17** and with the pump **16**, which in this embodiment is carried on the tank.

Boyd Design

A device developed by *E.A. Boyd; U.S. Patent 3,753,496; August 21, 1973* is one in which a vortex generator in the form of a funnel-shaped casing is submerged with its upper edge or lip portion in close proximity to an oil slick. A plurality of vanes carried by the casing induce a rotary motion in fluids passing through the casing. A suction pump creates the flow by drawing a mixture of water and oil through the casing, the mixture then being carried by a conduit to a nearby tank or reservoir where it can be separated.

Reynolds Submarine Services Corporation Design

A device developed by *A.L. Markel; U.S. Patent 3,595,392; July 27, 1971; assigned to Reynolds Submarine Services Corporation* is an apparatus for separating fluids having different densities utilizing an axial flow pump which achieves vortex separation without emulsification of the fluids. A Pitot tube positioned downstream from the axial flow pump separates substantially all of a lighter density fluid from the discharge of the axial flow pump and delivers the fluid to a settling tank where gravity separation of the different density fluids is achieved very rapidly because of the absence of emulsification. A heavier density fluid which bypasses the Pitot tube is used to provide propulsion for a collecting unit. The fluids to be separated are contained by means of a floating boom attached to the collecting unit. The latter is provided with a weir or floating skimmer means which facilitates the delivery of the fluids to be separated to the axial flow pump.

Figure 168 shows an overall view of an oil collecting system and also a detail of the vortex separation device. There is illustrated a floating collecting unit indicated generally at **24** which may comprise an aluminum barge.

FIGURE 168: REYNOLDS SUBMARINE SERVICES CORPORATION SKIMMER AND VORTEX SEPARATOR

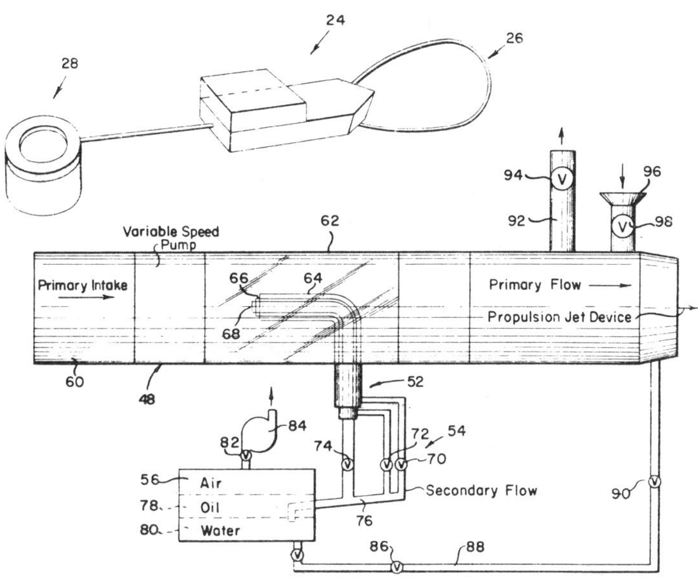

Source: A.L. Markel; U.S. Patent 3,595,392; July 27, 1971

At the front of the collecting unit **24** there is illustrated a floating boom means indicated generally at **26**. The purpose of the floating boom means is to provide local containment for the fluids to be separated at the surface of a body of water. For example, a boom may be placed in encircling relationship with respect to an oil slick or the like. A storage means **28** is positioned near at hand to the collecting unit and the storage means may take any suitable form. The barge **24** contains the vortex separation device shown in the figure.

Thus there is shown a primary intake **60** to the axial flow pump **48**. Downstream thereof a Pitot tube means **52** is positioned. In the preferred form of the device, the Pitot tube means is provided with means for changing its effective diameter such as by the provision of concentric tubes **64, 66** and **68** which are opened or closed by means of associated valves **70, 72** and **74**. These valves may be simply controlled by a single control lever or switch, if desired, whereby an operator may very quickly change from one effective diameter condition for the Pitot tube **52** to another in order to compensate for different flow conditions within the exhaust from axial flow pump **48**. The output of the Pitot tube under the control of valves **70, 72** and **74** is delivered to conduit **76** into one or more settling tanks **56** wherein in a typical application oil **78** and water **80** are separated by gravity.

Because of the fact that the fluid emerging from the axial flow pump has not been emulsified, the fluids of different density, such as oil and water, may be separated in separating tank **56** in a very short period of time such as in about one or two minutes. A valve **82** controls a variable speed air exhaust vacuum pump **84** for purposes of establishing a vacuum in the separating tank. The secondary flow in the Pitot tube is increased or decreased by the amount of vacuum in tank **56**.

It is desired to have means to remove the separated water from the settling or collecting tank **56**. This is accomplished by a valve **86** in conduit **88** and optionally may also include a variable speed pump **90** in series with the conduit in order to return the high-density fluid, such as water, which has been separated from the low density fluid, such as oil, to the primary flow.

If desired, and as an optional feature, a conduit **92** is taken from the primary flow downstream from the Pitot tube and may be led to an auxiliary jet manifold. Also as an optional feature, a chemical dispersant intake may be provided at **96** under control of valve **98** downstream from the separation of fluids into primary and secondary flows. Also, if desired, and as an optional feature, there may be straightening vanes downstream from the Pitot tube in order to assure laminar flow in the primary flow from axial flow pump **48**.

A section of the conduit immediately prior to the position of the Pitot tube but which also contains the Pitot tube is made into a sightglass or sight tube **62**. This greatly facilitates separation by sight of a darkened oil-rich concentrate or other low-density fluid from the exhaust of the axial flow pump by the Pitot tube which may be axially aligned therewith. For example, one suitable material for the sightglass or sight tube is methyl methacrylate (Plexiglas).

SOGREAH Design

A design developed by *B. Valibouse and J. Pichon; U.S. Patent 3,789,988; February 5, 1974; assigned to Societe Grenobloise d'Etudes et d'Applications Hydrauliques (SOGREAH), France* depends for its operability on the effect of the relative displacement speed of the heavier liquid and the overlying pollutant. Such relative speed is used to direct the removed layer of heavier liquid and pollutant, by tangential introduction, into at least one cyclone chamber wherein the induced rotation of the removed materials is caused as to create a whirlpool area in which the pollutant is concentrated. In the central portion of such concentration, the pollutant is extracted through a pipe fitting in the axis of and crossing the ceiling of the cyclone. The centrifuged water, free of pollutant, is discharged at the end of the cyclone.

Texaco Inc. Design

A design developed by *W.H. Pruitt; U.S. Patent 3,709,366; January 9, 1973; assigned to Texaco Inc.* acts to create plural vortices at the surface of a body of liquid which has a second liquid floating on top. The vortices are created by locating plural conduits with the open ends a short distance beneath the surface, and then applying a suction that is greater than the hydraulic head between the open end of the conduit and the surface of the liquid. The light surface liquid that is drawn off may be pumped into a storage tank for permitting separation.

United Aircraft Corporation Design

A design developed by *A.E. Mensing, J.W. Clark and R.C. Stoeffler; U.S. Patent 3,743,095; July 3, 1973; assigned to United Aircraft Corporation* is a vortex separator used to separate oil from an oil-water mixture which results, for example, from oil spills on a body of water. The separated oil may be stored or otherwise disposed of, while the separated water may be returned to the original body of water. The flow rate of the separated oil through the oil exit port of the vortex separator may be varied by using a movable tapered plug to change the area of the oil exit port. The area may be varied manually, or automatically in response to measurements of the oil content.

SCREW DEVICES

Brown Design

A device developed by *K.D. Brown; U.S. Patent 3,618,768; November 9, 1971* comprises an extended length of open screw conveyor constructed of material such that it will float on water. The conveyor may comprise a helical fin having a central bead or core about which stranded cables are wound, the stranded cables being of steel wire or plastic and when the conveyor is required to float on water, plastic materials are employed for all parts of the conveyor. The conveyor is connected to two spaced motor vessels one end being pivotally connected to one of the vessels for free rotation and the other end passing up into the other vessel through a surrounding conduit and being rotated by a motor mounted on the other vessel.

A reservoir is provided in the other vessel to collect the liquid. During operation the cable is rotated in a direction to draw the liquid toward the reservoir vessel and the two vessels are moved forward to sweep an area covered with oil or other lighter fluid. In another embodiment the flexible helical drive member is housed in a flexible tubing having a longitudinal opening for admitting the lighter liquid from the surface of the body of heavy liquid, the entire assembly thus formed floating on the body of liquid.

JET DEVICES

Stewart Design

A design developed by *J.K. Stewart; U.S. Patent 3,794,175; February 26, 1974* is one in which floating oil is picked up by water jets and carried over a vertical wall into a receiving chamber. The water jet forming nozzles and the receiving chamber are parts of a floating vessel. The water jet forming nozzles are positioned to discharge upwardly through the floating oil. Float controlled mechanisms automatically maintain the jet nozzles properly oriented with respect to the receiving chamber wall.

U.S. Navy Design

A design developed by *D.J. Graham; U.S. Patent 3,762,169; October 2, 1973; assigned to the U.S. Secretary of the Navy* is a device to direct floating oil into a recovery mechanism comprising a flexible rubber hose having attached floats positioned at spaced intervals. Water jets are operatively connected to the hose between the floats, and counterweights are added to uniformly balance the unit. A water pump provides water to the hose which is sprayed by the jets against the oil slick, forcing it to the recovery mechanism. Figure 169 shows the operation of this device in a boomed area containing an oil spill.

FIGURE 169: U.S NAVY JET-TYPE SKIMMER SYSTEM

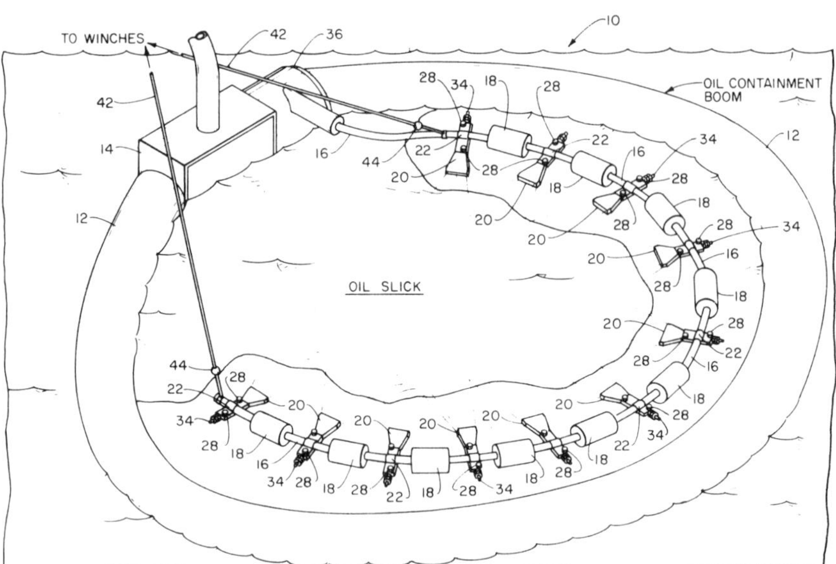

Source: D.J. Graham; U.S. Patent 3,762,169; October 2, 1973

As shown in the drawing, the device **10** is positioned where it will most likely be deployed, i.e., inside a boomed area containing an oil spill. The device is manually moved about inside the area enclosed by a boom **12** as necessary to direct the floating oil into the oil recovery mechanism **14**. Since the oil slick thickness is greatly increased in the region of the recovery mechanism, the effectiveness of the device is also greatly increased.

As prefaced by the above remarks, the device comprises a flexible hose **16**, preferably of rubber and having a 1½ inches inside diameter. The floats **18** of conventional fabrication are attached to the hose and are positioned at spaced intervals. The floats, while manufactured of any suitable material, are preferably made of closed-sell polyurethane. Water spray jets **20** are connected to the hose between the floats and are clamped in position by the clamps **22**, and by the nut and bolt assembly **28**.

Counterweights are threaded onto the rod **34**, mounted to the rear of jets **20**, the rod being an extension of clamps. The counterweights are added to uniformly balance the device. A water pump **36** is attached to any suitable floating unit, preferably the recovery mechanism **14**.

A torque stabilizer is mounted on the bottom of spray jets **20** in order to prevent excessive torque movement of the device during actual operation. Tension lines **42** which lead to hand winches located on the recovery mechanism **14** are attached to the flexible hose **16** by swivel joints **44** so as to maintain the position of the jets **20** when operating.

OTHER DEVICES

Daniel Devices

A device developed by *W.H. Daniel; U.S. Patent 3,667,609; June 6, 1972* is one in which oil is collected from the surface of a body of water by immersing a tent-shaped collector just under the surface of the water. The wave action drives oil down below the tent, and the oil rising under the tent then collects in the peak of the tent whence it is conveyed upwardly by hydrostatic pressure through a conduit into a flexible collection receptacle. The apparatus may be dropped from an airplane by a parachute, in which case the collection receptacle itself can be the parachute. The apparatus is buoyant yet weighted to maintain an upright attitude and to position the tent at a desired depth.

Another device developed by *W.H. Daniel; U.S. Patent 3,667,610; June 6, 1972* is one in which oil is collected from the surface of a body of water by immersing a tent-shaped collector from above the surface of the water to a depth such that the hydrostatic pressure of the oil in the collector will pump oil to an elevation above the surface of the water and into a collection receptacle. The collector slides vertically on a conduit and delivers the oil into the lower end of the conduit, the lower end of the conduit being positioned at a depth which determines the height to which the oil can be pumped above the surface of the water. In an illustration showing some of the essential components of the present device, Figure 170 shows a ship or barge **1** that is adapted to move across the surface of a body of water on which the oil spill occurs, to the site of the oil spill, and to carry the rest of the apparatus and to receive and retain the collected oil. Ship **1** carries a boom **3** vertically and horizontally swingable thereon and adjustably supported by guy lines **5**, from which the collecting apparatus is suspended.

FIGURE 170: DANIEL UMBRELLA-TYPE OIL SPILL SKIMMER DEVICE

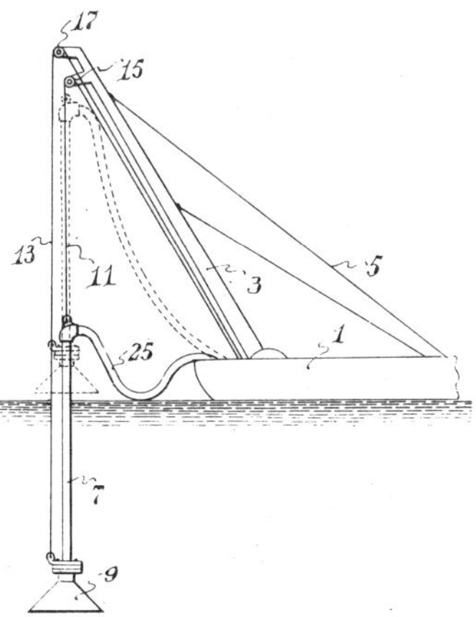

Source: W.H. Daniel; U.S. Patent 3,667,610; June 6, 1972

The collecting apparatus itself comprises principally a vertical cylindrical conduit of steel pipe or the like in the form of a hollow stem **7** with a tent in the form of a cone **9** encircling it and vertically slidable thereon. Cable **11** and **13** support stem **7** and cone **9**, respectively, for vertical movement conjointly with and relative to each other, the cable **11** being reeved over a sheave **15** carried by boom **3** and the cable **13** being reeved over a sheave **17** carried by boom **3**. Winches (not shown) mounted on ship **1** individually and/or jointly reel in or pay out cables **11** and **13**. Stem **7** is connected in fluid communication with a flexible tube **25** that extends from the upper end of stem **7** to the storage chambers (not shown) for collected oil on ship **1**. Cone **9** slides down on and relative to stem **7** and is retained at the lower end of stem **7** by stop **21**, whose conical upper surface matches the conical under surface of cone **9**. Cone **9** in turn extends upwardly in the form of a cylindrical sleeve **27** which slidably guides cone **9** on stem **7**. Sleeve **27** terminates upwardly in a radially outwardly extending horizontal flange **29** that provides one of the two seats for a packing **31** that is releasably compressed between flange **29** and a superposed ring **33** to which cable **13** is secured. When flange **29** and ring **33** are tightened together, packing **31** is in slidable sealing relation with stem **7** so as to permit the escape of air but substantially to prevent the leakage of oil upwardly between cone **9** and stem **7**.

In operation, ship **1** is moved to the site of an oil spill or other accumulation of oil on the surface of a body of water, of course with stem **7** elevated. When in the midst of the oil spill, the stem is lowered into the water and immerses by its own weight. Then, with boom **3** and cable **11** held stationary, cable **13** is paid out to immerse cone **9** further by its own weight. When the cone enters the water from above, it confines within its periphery a circular or annular layer of oil, which tends to consolidate laterally and increase in vertical height as the entrapped air leaks out past the packing. The cone continues downward to the position, where the oil can then flow into the opening at the lower end of stem **7**. Because oil is lighter than water, the oil does so and is forced by hydrostatic pressure upwardly in the stem, so that initially, a plug of oil forms adjacent the upper end of the interior of the stem.

This plug of oil will be lighter than water and hence its upper surface will be somewhat above the surface of the surrounding water level. Cone **9** is then raised again and lowered again, to introduce a second plug of oil into the lower end of stem **7**, which second plug rises and joins the first, the upper surface of the oil within the stem rising still higher. After a number of such operations, the level of oil within the stem has risen to the level of and filled the union, after which oil flows through tube **25** to storage as quantities of oil are successively introduced into the lower end of the stem. In effect, it is the buoyancy of the collected oil, immersed to a suitable depth, that pumps previously collected oil by hydrostatic pressure to any desired elevation, dependent on how deeply the device is immersed.

A variation developed by *W.H. Daniel; U.S. Patent 3,784,013; January 8, 1974* has the tent-shaped collector in the form of a plurality of superposed coaxial cones mounted on a sleeve that slides on a hollow multiperforate stem.

Texaco, Inc. Device

A design developed by *I.C. Pogonowski; U.S. Patent 3,693,801; September 26, 1972; assigned to Texaco, Inc.* is a skimmer device for use on the surface of a body of water. The skimmer is actuated by a vacuum source to promote a flow of a lighter than water coating material floating at the water's surface, through the skimmer by way of a submerged inlet. Valve means provided in the skimmer is so positioned to assure unidirectional flow through the skimmer to maintain the integrity of the vacuum source even though the skimmer be inadvertently raised from the water such that air should otherwise enter the vacuum system and cause fluids to drain.

Figure 171 shows an elevation, a sectional transverse elevation and a detail of the sectional transverse elevation of the device. In the figure, water skimmer **21** is shown floatably positioned at the water's surface to receive a flow of water and oil by way of the skimmer inlet **24**. It is appreciated that skimmer **21** at the conduit **18** lower end is representative of a number of such skimmer units. The latter are segmented and cooperatively disposed adjacent to each other as to sweep a predetermined front as a vessel is moved into an oil slick, or as the slick moves toward the skimmer assembly.

Skimmer unit **21** comprises an elongated, rigid casing **26**, normally formed of a metallic material such as sheet metal which defines a general arcuate lower surface **27**. The upper side of the casing is open and provided with a main float mechanism **28** here illustrated in a generally cylindrical cross section. Float **28** is fixedly positioned with respect to the skimmer and within casing **26** to define a liquid flow passage **29** at the float underside. The passage will be generally arcuate in configuration between the float walls, and the inner wall of the casing **26**. Casing **26** is further provided with an exhaust port **31** having a flange **32**. The latter is adapted to removably connect to conduit **18** by way of a similar flange **33** together with bolting means and sufficient sealing gaskets disposed between the respective flanges to provide an air tight connection.

FIGURE 171: TEXACO WATER SURFACE SKIMMER DESIGN

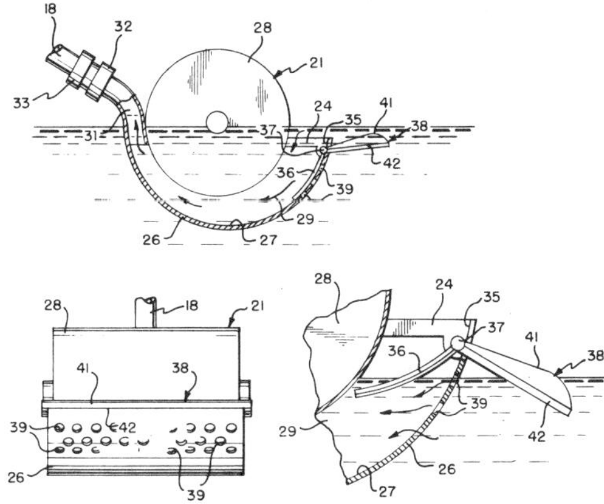

Source: I.C. Pogonowski; U.S. Patent 3,693,801; September 26, 1972

Inlet **24** to skimmer **21** defines a generally elongated opening lying in a horizontal plane and formed by the side wall of float **28**, which is spaced from the upper rim **35** of casing **26**. Said inlet **24** is rectangular in shape and receives a constant stream of water and oil. As shown, the normal disposition of skimmer **21** with respect to the water's surface is adjusted such that skimmer inlet **24** is slightly below the water's surface and preferably arranged to admit a minimum quantity of water in contrast to the amount of oil. Toward maintaining the fixed position of skimmer **21** with respect to the water's surface, the skimmer can be provided with supplementary support means as shown. The latter can consist for example of spring biased support cables or arms which depend from a vessel. In any event, the preferred position of skimmer **21** is such that inlet **24** is submerged immediately adjacent the water's surface and will consequently permit a continuous flow of water into said inlet. The liquid flow is thereafter directed through underpassage **29** and thence into the conduit **18** and up to the main collector tank **10**.

As above mentioned, occasion will arise during operation of the skimmer apparatus when, in accordance with the weather and other operating conditions, the skimmer will be forced from its preferred position. Thus, inlet **24** will be raised beyond the water's surface. Ordinarily under such circumstances, air rather than water entering inlet **24**, will cause a diminution of the degree of vacuum in the system thereby impairing or lessening the integrity of the latter. Toward overcoming any propensity of the skimmer to decrease the vacuum's effectiveness through such air leakage, skimmer **21** is provided with valve means which is automatically operable to maintain a flow of water and oil into liquid passage **29**, even though the skimmer itself is beyond the water's surface.

Skimmer **21** is thus provided with a check valve arrangement comprising a flapper plate **36** which is hingedly connected through an elongated hinge joint **37** with a generally horizontally projecting actuator **38** depending from the skimmer's forward side. Said valve member comprises the generally arcuate flapper plate **36**, which conforms substantially in contour to the arcuate configuration of the inner wall of casing **26**. The forward wall of the latter as shown, is provided with a plurality of discretely positioned supplementary openings **39**. Said openings **39** are formed in the lower wall of the casing wall, and normally beneath the water's surface. The openings thereby provide an alternate inlet for water normally contacting the casing **26** outer surface. However, an influx of water through supplementary openings **39** occurs only at such time as flapper plate **36** is displaced from its normal position contiguous with the casing **26** inner wall. In said normal position, the flapper **36** will substantially cover the respective supplementary openings **39** and preclude flow therethrough.

Toward promoting efficiency of liquid flow control through said openings **39**, the latter can be provided individually or cooperatively with gasketing means. The latter will comprise a resilient sealing member disposed about each opening, or a gasket carried at the periphery of the flapper plate **36** whereby to form a fluid tight engagement with the inner surface of casing **26** wall when the flapper plate **36** is urged into a closed position. While flapper plate **36** and the contiguous casing wall portion are shown in the instant figures as being generally arcuate in configuration the exact contour is of course of a relatively minor nature. The casing wall may for example be flat, or even of an appropriate curvature which would permit the desired supplementary flow into the fluid flow chamber at the proper time.

The actuator mechanism **38** functions such that flapper plate **36** can be either held in its closed position against the respective supplementary openings **39**, or displaced from the latter. Said flapper plate **36** is thus provided with an outwardly extending arm which in turn supportably carries float actuator **38**. The latter comprises essentially a buoyant element whereby to normally stabilize the position of flapper plate **36**. Since actuator **38** is at least partially buoyant, it is responsive to variations in the water level caused by waves, and the like. Thus, since in its normal operating position actuator **38** is submerged, it exerts an upward torque through hinge **37**, on the entire flapper plate **36**, the latter is urged into its closed position.

The actuator **38**, while being buoyant, is further provided with a curved upper surface defined by a top plate **41**. Said plate connects with lower plate **42** along a forward bead, and terminates to the rear adjacent hinge **37**. A preferred configuration of said top plate is such that the latter defines an aerodynamic surface possessing curvature characteristics. Thus, a rapid flow of water and oil passing across the actuator upper surface, and prior to entering the skimmer inlet **24**, will create a sufficient lifting force across said surface that actuator **38** will be urged into an upward disposition.

In contrast, as actuator **38** is raised to clear the water's surface due to the skimmer rising thereabove, passage of liquid across the actuator top plate **41** will terminate. The uplift force at said plate upper surface will thereby be reduced to zero. Concurrently, the buoyant effect of actuator **38** is reduced such that the latter will fall. Plate **36** will be thereby displaced to fully expose supplementary openings **39** along the casing **26** forward wall. In such a condition, the water will enter the supplementary openings to in effect maintain the flow of liquid to fluid passage **29**. This supplementary liquid introduction will preclude the entrance of an appreciable amount of air to the latter.

SKIMMING VEHICLES

Ainlay Design

A design developed by *J.A. Ainlay; U.S. Patent 3,623,609; November 30, 1971* is one in which a floating, seagoing skimmer has a separating chamber communicating at its lower end with the body of water into which it is placed, a weir at the front of the chamber having a forwardly extending curved surface which teminates in a skimming edge and a rotating vane impeller whose path of movement is disposed relatively closely to the curved surface of the weir so as to carry the combined floating liquid and water over the weir and into the separating chamber. The top of the weir is substantially above the water surface so as to create a hydrostatic head for causing waterflow out of the bottom of the chamber and the number of vanes in the impeller, its speed of rotation and its depth of penetration are such as to move the combined floating liquid and water with a minimum of turbulence. The skimming edge ahead of the weir is adapted to be held somewhat below the interface of the floating liquid and water surface.

Anderson Design

A design developed by *J.A. Anderson; U.S. Patent 3,762,558; October 2, 1973* includes a conveyor arrangement adapted to be used onto a barge or similar boat in combination therewith and including two endless-belt conveyors arranged to cooperatively pick up a floating pollutant, such as oil, and to convey the same upwardly between adjacent runs of the respective conveyors which are driven at substantially the same linear speed. Transverse ribs are secured to the top endless-belt conveyor and extend therefrom into liquid-tight edgewise engagement with the top run of the bottom endless belt in cooperation with longitudinally extending flexible wings arranged to laterally confine the picked-up pollutant and water during upward displacement thereof intermediate the two adjacent runs. The conveyors are pivoted onto the bow and the barge is provided with water ballast tanks to adjust the dipping depth of the outer end of the conveyors.

Otherwise stated, the apparatus consists of two interengaged endless-belt conveyors wherein the bottom conveyor has a smooth outer surface in order to avoid disturbance and repulsion of the surrounding pollutant and the top conveyor includes pocket forming means fixed thereto for conveying the picked up pollutant upwardly intermediate both endless belts. As shown in the plan and elevation views in Figure 172, the overall device includes a barge **1**, of any appropriate type, having a conveyor arrangement **2** mounted on the bow thereof and arranged for dipping one end into a body of water onto which a pollutant, such as oil **3**, is floating. Obviously, various types of boats may be used to support the conveyer arrangement and the term barge is meant to extend to all those types of boat.

FIGURE 172: ANDERSON SKIMMER BOAT

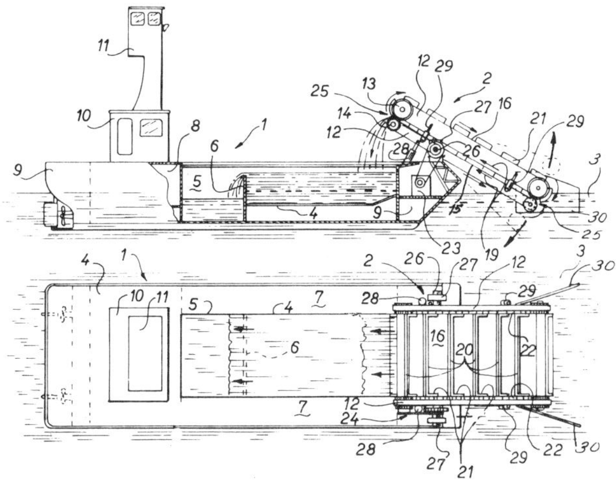

Source: J.A. Anderson; U.S. Patent 3,762,558; October 2, 1973

An open top tank **4** is provided into the hull of the barge **1** for the discharge of the conveyor arrangement **2** therein. The water also discharged by the conveyor arrangement **2** rests to the bottom of the tank **4** and is pumped out thereof.

An oil or the like pollutant collecting tank **5** is positioned adjacent the tank **4** and arranged to receive the surfacing layer of pollutant which flows over the edge **6** of the tank **4**. The oil, generally of the crude and heavy type, is pumped out of the collecting tank **5** with previous heating thereof, if required to increase its fluidity. The pumped oil is then transferred either directly to shore, if feasible, or into a standing by tanker, not shown. Buoyancy chambers **7** are provided laterally of the tank **4** to enhance flotation of the barge **1**. A machinery compartment **8** is provided at the rear of the tanks **4** and **5**. Water ballast tanks **9** are provided at the ends of the barge to control the relative depth of the barge and of the end of the conveyor arrangement dipping into the body of water **3**. Control cabins **10** and **11** are mounted onto the barge and arranged for the control of the latter therefrom.

The conveyor arrangement **2** includes a pair of endless-belt conveyors each having a pair of longitudinal beams **12** arranged in spaced-apart parallel relationship and arranged to rotatably support rollers **13** and **14** at opposite ends thereof. The bottom endless-belt conveyor has an imperforate flexible endless belt **15** wrapped around the supporting rollers **14** for rotation therewith. The top imperforate flexible endless-belt conveyor has an endless belt **16** wrapped around the supporting rollers **13** for rotation therewith. The top run of the endless belt **15** and the bottom run of the endless belt **16** are retained in parallel spaced-apart travelling relationship by sets of idler rollers.

The endless belt of the bottom conveyor is formed with a smooth outer surface, such that, upon travelling of the endless belt, minimum disturbance is made at the surface of the body of water **3** by the dipping end of the conveyor arrangement **2** and there results substantially no repulsion of the surrounding floating pollutant at the nip defined by the merging endless belts and travelling together in unison in the directions of the arrows **19**. The floating pollutant is then allowed to enter into the above nip to be picked up by the pockets or buckets, hereinafter described in detail. The belt **15** may preferably, though not necessarily, be covered with a layer of burlap or other oil-adhering material. The cooperating adjacent bottom and top runs of the top and bottom conveyors form substantially flat surfaces both longitudinally and transversely of said runs.

The endless belt **16** is formed with pocket-forming vanes of ribs **20** integrally formed therewith and extending transversely thereof. Longitudinal wing portions **21** are formed integral with the vanes and extend longitudinally of the endless belts at opposite ends of the vanes to cooperatively form said pockets therewith. It must be noted that the vanes and wing portions are all of the uniform depth and form a continuous straight outer edge into edgewise liquid-tight engagement with the endless belt when travelling along the above-mentioned bottom run of the top endless belt **16**. The latter, the vanes **20** and wing portions **21** are made of flexible material, such as rubber, and thence the wing portions **21** are allowed to flex around the end rollers **13**. The latter are selected larger than the rollers **14** precisely to facilitate such flexing. In order to effectively pick up the oil or similar pollutant, both endless belts **15** and **16** must travel at the same linear speed, whereby there is no relative displacement between the interengaging runs of the two belts.

Each endless belt **15** and **16** has a link chain **22** fixed thereto along each lateral edge thereof. Sprockets, not shown, are coaxially mounted at each end of each end roller **13** or **14** and are arranged for rotation therewith and to drive the link chains **22**. A motor **23** of any suitable type, is mounted into a hull compartment into the bow of the barge **1** and is connected by a sprocket and link chain mechanism **24** to the bottom conveyer to drive the latter. Gears **25**, of suitable tooth ratios, are connected to the top conveyor and sprockets thereof to positively drive the endless belt **16** at the same linear speed as the endless belt **15** for the above-mentioned purpose. A transverse pivot **26** extends through the longitudinal beams **12** of the bottom conveyor and is pivotally supported by upstanding brackets **27** to allow angular adjustment of the dipping of the conveyor arrangement. The link chain mechanism **24** has an articulation about the axis of the transverse pivot **26** to maintain driving connection for all angles of the conveyor arrangement about that axis. A pair of hydraulic cylinders **28** are connected to the longitudinal beams **12** of the bottom conveyor and arranged to vary the angle of the latter about the above axis and thence, the dipping height thereof.

Four hydraulic cylinders **29** are connected between the beams **12** of the bottom and top conveyors and arranged in pairs on opposite sides, to elevate the top conveyor relative to the bottom conveyor, to separate the bottom run of the former from engagement with the top run of the latter, when floating solid debris are encountered and desired to be conveyed upwardly to the tank **4**. In this case only the bottom conveyor is operative. Upright deflectors **30** are angularly attached to the dipping end of the conveyor arrangement **2** to converge the floating pollutant towards the nip defined by the merging of the endless belts at that dipping end.

Ayers Design

A design developed by *R.R. Ayers; U.S. Patent 3,684,095; August 15, 1972* involves a barge based skimming system for oil slicks wherein the oil phase is concentrated relative to the water phase prior to its introduction into a final separating compartment wherein the oil is reclaimed in a conventional manner. A barge mounted wave reflector and boom arrangement diverts oil and water into open bottom chutes positioned on either side of a barge. The wave action is damped at the upstream end of the chutes and a skimming operation is performed downstream thereof to divert floating oil upwardly onto a shelf-like structure while allowing the underlying water to continue its passage through the open bottom chute. The shelf structure then directs the oil to a relatively quiescent area defined by a separating container at the rear of the barge where a second separating operation is conducted and the oil reclaimed.

Brittingham Design

A design developed by *C.J. Brittingham; U.S. Patent 3,664,505; May 23, 1972* is a floating collection apparatus for skimming oil slicks from a body of water. The oil and water is subjected to pressure generated by the weight and movement of the apparatus to eliminate the water adjacent to and entrained in the oil film prior to sending the oil to a collection tank.

Brydoy-Sletsjoe Design

A design developed by *S. Brydoy and A. Sletsjoe; U.S. Patent 3,737,040; June 5, 1973* employs an oil slick removal vessel which has means for carrying out both a mechanical and a chemical removal of the oil. Thus, in the first part of the vessel there are means for taking in water and oil and for the separation thereof by a skimming device. In the rear part of the vessel which part communicates with the first part over an adjustable overflow means, are arranged means for spraying chemicals onto the overflow from the first part. An outlet wherein a propulsion means for the vessel is placed, is located in the rear part thereof.

Bucchioni-Forgione De Toffoli Design

A design developed by *D. Bucchioni and M. Forgione De Toffoli; U.S. Patent 3,517,812; June 30, 1970* involves a self propelled floating means which is provided along its entire length with a longitudinal canal provided with a bottom; at least one transverse floodgate being partially immersed in the superficial layers of the liquid current flowing in the canal; wherein the depth of the canal increases up to a maximum upstream of the first floodgate and then preferably decreases downstream of the last floodgate. The superficial layers of the water to be cleaned are preferably pushed into the canal by means of a system of one or more reels or belts provided with blades which are preferably flexible so as to prevent whirlpools from forming. The device is also characterized in that it comprises one or more hopper-shaped collecting means which may be arranged at an adjustable height for collecting the material gathered upstream of each of the floodgates. The hopper-shaped collecting means preferably has a lateral surface not diverging upwards, so as to avoid movement of the waste away from said collecting means upon lowering the latter; the collecting means being further connected, through straight conduits to the first of a group of decantation tanks provided with drainage of the decanted water from the bottom the tanks being arranged in cascade.

Chablaix Design

A design developed by *D. Chablaix; U.S. Patent 3,529,720; September 22, 1970* for the recovery of a liquid floating on the surface of water, for example petroleum, employs a vessel which contains a liquid separation chamber, pumping means, and a conduit one of whose extremities is connected to the pumping means and whose other extremity is immersed and supported near the water's surface.

There is shown in Figure 173 a floating support **1**, for example a specially constructed boat, comprising a liquid separation chamber **2**, a pump **3**, connected by conduit **4** to the separation chamber **2**, and to a decompression chamber **5** connected to an air hose **6** for removing air or other gases mixed with the pumped liquid.

FIGURE 173: CHABLAIX SKIMMING SHIP

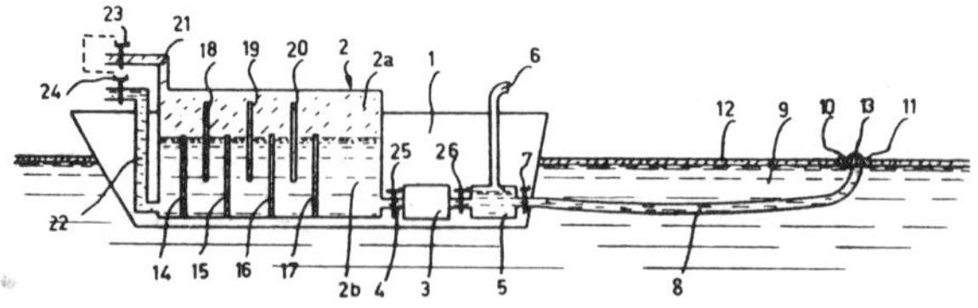

Source: D. Chablaix; U.S. Patent 3,529,720; September 22, 1970

The decompression chamber **5** is connected through at least one valve **7** to at least one flexible tube **8**, external of the boat **1**, immersed in water **9** and whose free extremity is supported by floats **10** and **11** near the water level, the distance separating the opening of tube **8** from the water surface being determined by the thickness of the layer of liquid **12** floating on the water surface. The opening of tube **8** is additionally surmounted by a grating **13** which is inverted so as to form a cupola whose grill work is intended to retain solid bodies floating on the water surface. The level of the extremity of tube **8** can advantageously be adjusted by means of floats having a variable flotation line realized by the introduction of a greater or lesser quantity of water in the bodies of the floats. In order to obtain greater efficiency, there will be used preferably several tubes **8**.

The separation chamber **2** has the shape of a rectangular enclave divided into several compartments by transverse walls **14, 15, 16, 17** rising from the bottom of the chamber to about half way up therein. These walls thus form a succession of basins successively filled with liquid arriving from pump **3** by flow of the liquid above the upper edge of the walls. Between walls **14** to **17** is positioned a second set of walls **18, 19** and **20** reaching from one lateral wall to the other in the chamber, of a height approximately equal to that of walls **14** to **17** but disposed on a higher level, the assembly of walls **14** to **20** constituting a labyrinth slowing down the flow of liquid pumped from one compartment to the other of the chamber and avoiding particularly the creation of a current at a level superior to the upper edge of walls **14** to **17**, such a current risking the rapid driving of oil towards the left-hand side of the separation chamber, the location in which separation of liquids must be finished. The separated liquids are recovered at the outlet of the chamber by two conduits **21** and **22** having valves **23** and **24** differentially coupled so that when one is open the other is closed. Conduit **22** is connected to the bottom of the separation chamber, while conduit **21** is connected to the upper part of the separation chamber. It is thus evident that conduit **22** serves to extract the purified water from the chamber, while conduit **21** serves for recovery of the floating liquid. It should be noted also that a valve **25** is provided between the pump and the separation chamber, and a valve **26** between the pump and the decompression chamber **5**.

The installation thus permits the pumping of the layer of floating liquid **12** by adjusting as best as possible the level of the conduit opening, or of the various valves. The floats **10** and **11** are positioned at a suitable place, and then valves **25, 26** and **7** are opened and pump **3** is started so as to pump the floating liquid and a minimal quantity of water and air. This mixture varies at first in the decompression chamber **5** in which air and possibly other gases can escape through air hose **6**. Owing to the position of tube **8** a small pressure in chamber **5** suffices in order that the liquid flows practically by itself through the tubes in this chamber. Pump **3** thus serves first of all to make liquid pass from chamber **5** into separation chamber **2**. During pumping, valve **24**, located on a level higher than the ceiling of chamber **2** is completely closed while valve **23** is open.

The liquid mixture pumped by pump **3** penetrates into the separation chamber in its lower part. It fills successively the compartments formed by the intermediate walls **14** to **17** to penetrate finally in conduit **22**. In each of the compartments takes place a separation of the water and of the floating liquid, the floating liquid ascending progressively to the surface. The labyrinth formed by intermediate walls **14** to **20** prevents the floating liquid from being carried by the internal current set up from the entrance **4** to the exit **22** of the separation chamber. This chamber filling progressively, the water level, occupying the lower part **2b** of the chamber, ascends and finally comes to compress the floating liquid into the upper part **2a**, under the effect of the pumping force exerted by pump **3**. The compressed liquid escapes finally through conduit **21** and valve **23** to be stored in suitable reservoirs or burnt on the spot in the case of a combustible liquid. When the level of water in the separation chamber reaches the ceiling thereof, valve **24** is opened, rendering possible the flow of this water which is thrown out to sea or in a lake.

According to the proportion of the two liquids and to obtain a continuous operation, it is advantageous to open more or less one of the valves **23** and **24** and to close more or less the other of these valves. It is possible to use advantageously two coupled valves in such a way that when one is open the other is closed, the coupling being adjustable, or yet still a differential valve. During the recovery of a layer of petroleum, the installation will be completed by a large diameter belt surrounding the floating gratings to prevent on the one hand a spread of the layer of the petroleum and on the other hand to reduce this layer by reducing the perimeter of this belt.

The separation chamber is naturally one example as other examples can be provided with a greater number of intermediate walls or a different labyrinth, for example the walls **18, 19** and **20** could contact the ceiling and have openings forming a baffle for the passage of the liquid. Other modifications can be made in which the tube is not immersed but held above water, only the extremity of the tube being emerged. Additionally, the separation chamber and the pump can be mounted on the ground.

Chastan-Bagnis Design

A design developed by *L. Chastan-Bagnis; U.S. Patent 3,695,441; October 3, 1972* is a self-propelled enclosure for treating polluted water surfaces. It comprises two longitudinal caissons, a bottom floor and transverse partitions to connect the longitudinal caissons and to define the enclosure. A door is pivotally mounted at the front end of the floating enclosure and is downwardly foldable between a closed position of the enclosure, intermediate positions of the door and a fully opened position thereof in a plane which substantially coincides with the plane of the bottom floor. In this manner, the upper edge of the door can be at a predetermined depth under the water surfaces to enable polluted water to enter the enclosure. A vault is formed in the bottom floor to define a channel underneath the bottom floor and a motor-operated helix is disposed at the rear end of the channel to suck water in the channel so that upon downwardly folding the door at a level lying immediately beneath a polluting layer, the self-propelled enclosure will be seen to advance over water to cause the polluting layer to slowly enter the enclosure until it fills the same. Viewed from above, as shown in Figure 174, the device sits inside a rectangle; its transverse section is an isosceles tetrahedron in which the shorter base is the bottom.

FIGURE 174: CHASTAN-BAGNIS SKIMMER VEHICLE

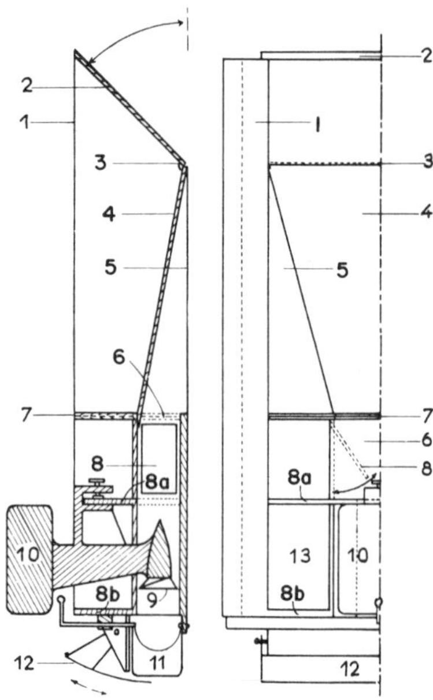

Source: L. Chastan-Bagnis; U.S. Patent 3,695,441; October 3, 1972

Two floaters **1** extend along the whole length of the lateral sides of the device, and these floaters **1** could be filled with water or emptied at will by means of pumps or compressed air. The front portion of the device comprises a door **2**, which is rotatable about the axis **3** in order to close the space between the bottom **5** of the device, and the floaters, **1** in the raised position of the door **2**. Eventually, when the door **2** is lowered down, it can reach the horizontal position. A prismatic type of inclined vault in the form of an open three-sided channel slowly rises from the axis of the door until it reaches the vicinity of the rear end of the device. The vault is identified by reference number **4**. The vault **4** projects upwardly from the bottom **5** and defines a channel until the latter reaches a duct **6** which has a circular cross section and in which a helix **9** is mounted. The bottom **5** also comprises shutters or flushing valves not illustrated in the drawings.

In front of its opening and perpendicularly thereto, the duct **6** is covered with a filter **7** which occupies the entire remaining inner transversal section of the device. Towards the rear, behind the filter **7**, the duct **6** comprises two vertically pivoting doors **8** which can be opened inwardly and used to flush the device after a filtration. Between the intermediate and upstanding transverse rear and partitions **8a** and **8b**, there is a motor **10** which is used to rotate the helix **9** inside the duct **6**. To steer the device, there is a rudder **11**, which is pivotally mounted at the rear end of duct **6**. A half-circular notch can be used to fold the rudder against the rear board. In this case, the motor must rotate the helix **9** in the opposite direction. It could eventually be possible to stabilize the device by means of a screen **12** which is perpendicular to the axis of the device and for which the immersion can be regulated. All the controls can be mounted in and carried out by anyone of the two small compartments **13**, provided on both sides of the motor **10**.

The operation of the device is as follows. The device is on top of the water, the door **2** is closed, and the device is then brought to a zone where water is polluted. The door **2** is lowered down in such a manner that the upper edge thereof lies immediately beneath the polluting water. The motor propels the device by sucking water in the channel and duct underneath the vault while causing the polluted liquid to enter the device and to slowly fill the compartment defined by the door **2**, floaters **1** and the intermediate partition **8a**. Once filling is over the door **2** is closed. The liquid which is now in the compartment defined by the floaters can be carried away, if the pollution results from the presence of hydrocarbons.

Cities Service Oil Co. Design

A design developed by *E.A. Bell; U.S. Patent 3,708,070; January 2, 1973; assigned to Cities Service Oil Co.* is one in which a floating oil skimmer barge is provided with a series of compartments, beginning at the prow of the barge, inflow to each compartment being effected over a respective floating baffle pivotally mounted at its bottom edge to swing into its compartment to a depth determined by the pressure differential across the baffle. Position of the baffle is controlled by pumping water at controlled rates from the bottom of the downstream end of each compartment to thereby cause an effective surface flow between compartments. Surface oil builds up in depth at the downstream end of the last compartment and is collected, substantially free of water, in a recovery chamber which is also provided with a floating baffle and from which oil is pumped at controllable rates

Craftmaster, Inc. Design

A design developed by *R.P. Smith, Jr.; U.S. Patent 3,785,496; January 15, 1974; assigned to Craftmaster, Inc.* for removing oil slicks from the surface of the water which includes two independently pivotable booms connected together to form a V. A paddle wheel is connected to the two booms across the open end of the V for skimming the oil off the surface of the water and forcing the oil towards the apex of the V. A sump is located at the apex of the V for collecting the oil forced to the apex of the V by the paddle wheel. A flexible stabilizer bar is connected between the two booms for stabilizing the booms and limiting the maximum amount that the booms may pivot with respect to each other. Such a device is shown in general perspective and in longitudinal section in Figure 175.

As shown there, a first boom **2** and a second boom **4** are coupled together by means of a shaft **6**. The booms are coupled to shaft **6** by pillow block bearings **8** in such a manner that each boom rotates independently about the axis of shaft **6**. Booms **2** and **4** are positioned with respect to each other such that they form a V, with the apex of the V at the point where the booms are coupled to the shaft **6**. A paddle wheel **10**, having a main shaft **12**, is mounted between the ends of the booms **2** and **4** at the open end of the V. The shaft **12**, which includes a telescoping portion **13**, is mounted on pillow blocks **14** on the ends of booms **2** and **4**. A motor **16**, mounted on boom **2**, is used to rotate the paddle wheel **10** and is coupled thereto by chain drive **18**.

Paddle wheel **10** includes spokes **20** and blades **22** mounted on the ends of some of the spokes. The blades **22** are positioned such that the extreme tip of the blade is at the depth of an oil slick on the surface, that is, the depth is to the oil-water surface. Therefore, when paddle wheel **10** rotates, the blades **22** push all of the oil on the surface of the water, but no water below the oil towards the apex of the V formed by booms **2** and **4**. The depth of blades **22** of paddle wheel **10** may be changed by changing the ballast in booms **2** or **4**. If the oil slick is very thick and it is desired to lower the depth of blades **22**, the ballast in booms **2** and **4** is merely increased, thus lowering the blades. The depth of blades **22** may also be varied by using variable height mounts for pillow blocks **14**. By adjusting the mounts the paddle wheel may be raised or lowered. The apparatus may therefore be quickly and easily adjusted to skim the maximum amount of oil from the surface of the water in accordance with existing conditions.

FIGURE 175: CRAFTMASTER, INC. OIL SLICK SKIMMING CRAFT

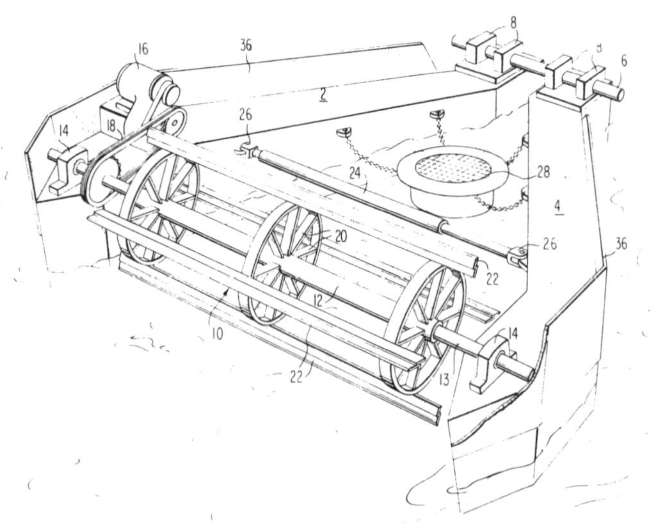

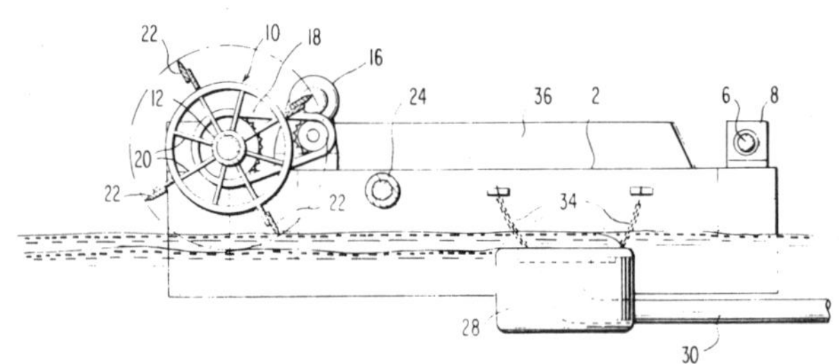

Source: R.P. Smith, Jr.; U.S. Patent 3,785,496; January 15, 1974

Pillow blocks **14**, which couple shaft **12** to booms **2** and **4**, are constructed such that the booms may rotate with respect to each other. If the water surface should swell at one boom and not at the other, the paddle wheel will then assume a position, which is parallel to the surface of the water, at an angle with respect to the horizontal. When the booms rotate telescoping portion **13** of shaft **12** telescopes out to provide the additional length to shaft **12**. Because of the independent rotation of the booms, the apparatus has great stability even in rough seas and even more important the paddle wheel always remains parallel to the surface of the water. A telescoping stabilizing bar **24** is connected between booms **2** and **4**. When one boom rotates with respect to the other, stabilizing bar **24** is lengthened through its telescoping action. However, the telescoping is limited such that the maximum angle that one boom may rotate with respect to the other is 30°. The stabilizing bar **24** is connected to booms **2** and **4** by means of eyelets **26** affixed to the booms. A sump or skimmer **28** of conventional construction is connected to booms **2** and **4** and at the apex of the V. The oil skimmed off of the surface of the water by paddle wheel **10** is forced back into the sump **28**. From the sump the oil is pumped through line **30** into a ship, a storage device mounted on the device, etc. (not shown). The pump may be mounted on booms **2** or **4** or may be located on the ship. Power to the sump and to motor **16** is received from a line which is connected to a power source also located on the ship. The sump **28** is connected to booms **2** and **4** by means of chains **34**.

Sump **28** has an adjustable ballast so that it may be raised or lowered with respect to the surface. It is positioned so that it will collect the maximum amount of oil, but no water. The device may be propelled through the water by a drive, mounted on the booms, a ship, or any other convenient method. One of the essential requirements of devices for skimming oil is that they be readily transportable from one location to another so that there is a quick reaction to oil leaks or spillage and oil may be removed from the surface of the water before it spreads over too large a surface area. The above described device may be easily separated into four major components, each of which is small with respect to the assembled device, in order to provide for easy transportation. To disassemble the device, shaft **12** is removed from pillow blocks **14**, thus separating the paddle wheel **10** from the booms. As an alternative, paddle **12** may be folded back on one of the booms by removing only one end from the pillow blocks.

Chains **34** are disconnected from booms **2** and **4**, thereby separating sump **28** from the booms. Finally, shaft **6** is withdrawn from pillow blocks **8** and the two booms may be separated. Once separated, the two booms may be folded together, using a hinge mechanism, or may merely be carried separately and aligned parallel with the paddle wheel **10**. It can readily be seen that in the disassembled state the apparatus is very small with respect to its assembled size. Although booms **2** and **4** are shown having a rectangular cross section, any cross section, such as circular, may also be used. The booms **2** and **4** may have flaps **36** mounted on the inside (not shown) or the outside as shown to prevent water from coming over the sides of the booms and then being collected by sump **28**.

Crisafulli Design

A design developed by *A.J. Crisafulli; U.S. Patent 3,756,414; September 4, 1973* consists of a module employed with a desired number of similar modules for connection with a floating barge or other vessel for collecting and skimming off oil when the barge or other vessel moves forward through an oil spill and discharging the collected oil or pollutant into storage tanks or the like incorporated into the barge or other vessel. Each oil skimmer module includes an open front receptacle having a horizontally disposed inclined front edge defining a weir that is capable of being raised or lowered for varying the depth of the weir in relation to the surface of the body of water. Each module also includes a pump for removing water and pollutants collected in the receptacle and discharging them into a suitable storage area such as settling tanks or the like on the barge or other vessel. Figure 176 shows a pan of side views of such a design, the upper view showing some details in broken away section.

The oil skimmer module is generally designated by the reference numeral **10** with it being pointed out that the number of modules employed may be varied depending upon the overall width of the area from which the oil is to be collected during each pass of the module or modules. The module **10** is adapted to be mounted in front of a floating barge or other vessel **12** for removing the surface layer **14** from a body of water **16** when the surface layer **14** includes oil or other pollutants. The oil skimmer module **10** includes a bottom wall or panel **18**, an upstanding rear wall **20**, and opposed substantially parallel and upstanding side walls **22**. This structure defines a generally rectangular receptacle having an open front. As illustrated, the rear wall **20** extends above the top edges of the side walls with the reinforcing gusset **24** being provided therebetween but the relative dimensions of the side walls and rear wall may vary as may the overall dimensions of the module. Attached to the front edge of the bottom wall **18** is an upwardly and forwardly inclined wall **26** hingedly attached to the bottom **18** by a hinge **28**. The two side edges of the wall or panel **26** are each provided with an upstanding side wall **30** which is generally triangular in construction and provided with a rear edge **32** generally perpendicular to the wall **26** so that it overlaps the side walls **22** of the receptacle when the wall **26** is upwardly inclined which is its normal position.

FIGURE 176: CRISAFULLI OIL SKIMMER VEHICLE

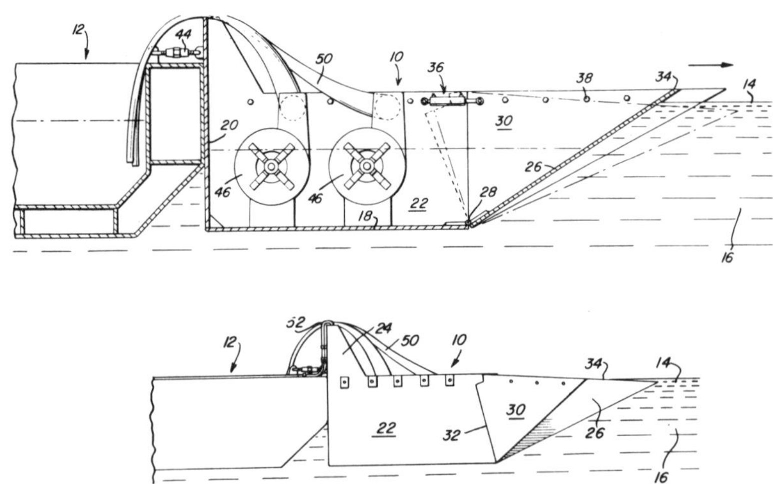

Source: A.J. Crisafulli; U.S. Patent 3,756,414; September 4, 1973

The forward edge of the wall **26** is defined by an inclined horizontal edge **34** which has its forwardmost portion at one side wall and its rearmost portion at the other side wall with the edge **34** being bevelled as illustrated to provide substantially a horizontally disposed cutting edge. The cutting edge **34** is swingable about the axis defined by the hinge **28** to orient the edge **34** at a desired elevation in relation to the surface of the body of the water. When in inoperative position, the edge **34** would be above the surface of the body of the water and while in operative position, it would be at a desired depth below the surface of the water for determining the thickness of the layer of oil and water removed from the surface of the body of water.

To vary the angular position of the panel **26** and the elevational position of the edge **34**, an adjusting mechanism generally designated by the numeral **36** is provided between the walls **22** and the walls **30**. Such adjustment mechanism may be in the form of a relatively small double acting hydraulically actuated ram connected with a suitable source of hydraulic pressure by suitable conduits, the details of which are not illustrated, so that the position of the cutting edge **34** may be varied from a remote point. Any suitable type of adjusting mechanism may be employed either hydraulically or manually actuated. The edge **34** defines a cutting edge and is submersible to define a cutting weir especially when the modules are oriented in pairs with the inclination of the cutting edge **34** forming a horizontally disposed V-shaped cutting edge which will remove the surface layer **14** with the least possible turbulence or disturbance of the surface of the water as compared with a straight across edge which would tend to build up a wave in front of the edge. The inclined construction of the edge **34** eliminates such a transverse wave and provides for substantially complete removal of a layer of oil or other pollutants of a predetermined thickness. When securing a plurality of modules together, suitable fastening bolts **38** are employed to secure the walls **30** together and also secure the walls **22** together with the walls **22** having suitable spacers thereon for providing a rigid interconnected structure.

The rear wall **20** of each module is connected to the bow of the barge **12** by any suitable means such as lugs being provided on the rear wall **20** and on the deck of the barge with an adjustable connecting device such as turnbuckle **44** adjustably interconnecting the lugs for securing the module securely but removably in relation to the barge **12**. Also, any suitable structure may be incorporated in the attachment between the module and the barge to provide for vertical adjustment therebetween within certain limits. For example, a plurality of lugs may be provided to enable the module to be elevated in relation to the barge as the barge is loaded and increases its draft. Of course, adjustment of the cutting edge or weir **34** also will compensate for variation in the draft of the barge so that a constant thickness of the layer skimmed off the water surface may be obtained.

Disposed on the upper surface of the bottom **18** and supported by suitable mounting brackets is a pair of pump assemblies **46** each of which is driven by a suitable hydraulic motor, such as a conventional orbit type hydraulic motor. Each pump assembly **46** is provided with a discharge pipe or conduit **50** which may be in the form of a flexible hose or the like extending from the discharge thereof to a storage tank or area within the barge which may be of any suitable configuration and capacity and may be in the form of a settling tank or the like so that water discharged into the tank on the barge may be removed from the bottom thereof and returned to the body of water, thus increasing the overall capacity of the storage area or tank in the barge. Also, fluid pressure conduits or hoses **52** are provided for powering the hydraulic drive motors for the pumps. The details of the pumps are not illustrated but it is preferred to use centrifugal pumps such as illustrated in U.S. Patent 3,371,614.

By governing the number of pumps as well as the size of the pumps in each module, the capacity of each module may be varied. One exemplary embodiment employs two pumps in each module having a diameter of 16 inches capable of pumping 160,000 gallons of water per minute when eight modules are used in the front of a barge. Any suitable structure may be employed for supplying hydraulic fluid to the hydraulic motors and ram. For example, a conventional hydraulic pump and fluid system powered by a suitable engine or motor may be employed and the discharge hoses or conduits may be any suitable conventional flexible material which enables flexibility of association of the barge and modules.

deAngelis Design

A design developed by *A.L. deAngelis; U.S. Patent 3,788,481; January 29, 1974* involves a boat furnished, in addition to the usual tanks, with one or more chambers fitted to be put in connection, through openings pierced into the plating and provided with interception organs, as gate valves or the like, with the surrounding water surface on which the boat is floating. The water with the floating oil layers enters through the openings into the chamber(s), being withdrawn from the action of the wind so that a quiet and horizontal sheet of liquid necessary to an efficacious separation of the oil is formed.

Within these chambers one or more known devices, formed for example by a funnel whose upper edge remains sumberged at an adjustable distance from the surface of the liquid, are placed. This funnel is linked at its upper end, for example by means of tie rods, with a float, while its lower end tapered to a pipe shape, is connected with a flexible pipe whose other end is connected to a pump. This pump sucks the liquid from the funnel and conveys it into a tank where it decants. The separated water is discharged, e.g., into the sea, while the oil so recovered can be conveyed to another tank.

Di Perna Design

A device developed by *J. Di Perna; U.S. Patent 3,651,943; March 28, 1972* utilizes a floating power operated vessel designed to collect by suction from the surface of a body of water, as it moves thereover, floating pollutant matter. Adjacent the prow of the vessel is a suction pump compartment and inlet pipes below the water level connected to a suction pump in the compartment. Adjacent the stern of the vessel is a clean water collecting compartment having a discharge pipe adjacent the top thereof and above the water level for discharging the same downwardly onto the surface of the body of water. Between the forward pump compartment and the rear clear water collecting compartment there is provided a series of transversely disposed and longitudinally spaced partitions forming communicating chambers designed to separate by degrees the mass of pollutant matter from the water.

At least one suction swing pipe is pivotally jointed at the prow of the vessel. Formed in the prow is a recess in which the swing pipe is received when raised to its inoperative position. A winch and derrick on the top deck of the vessel is cable connected to the upper free end of the swing pipe. A scoop structure which may be one of a variety of forms and sizes is attached to the upper free end of the swing pipe. The scoop structures are each provided with a forwardly extending blade arranged to lie flat substantially at the surface level of the body of water.

Galicia Design

A device developed by *F. Galicia; U.S. Patent 3,666,099; May 30, 1972* is an inverted V-shaped structure extending below the water surface and adapted when activated to move relatively thereto to carry floating contaminants to a subsurface collection point from which they may be removed for disposal relatively free of contained or occluded water. Figure 177 shows several views of such a device including a perspective view, a perspective detail and longitudinal and transverse sectional views. This device assumes the form roughly of a sea sled comprising a generally rectilinear deck **35** having depending side plates **36, 37,** a stern transom **38** and a composite keel plate **39** defining with the deck, side plates and stern transom an air-tight buoyancy chamber **40** which keeps the craft afloat.

Keel plate **39** comprises a plurality of relatively narrow strips **41** extending fore and aft along the sled at a small angle to the horizontal plane of deck **35** and defining inverted V-shaped troughs or channels **42**. Baffles **43** extending transversely of strips **41** angularly forward and downward restrict the eddying effects of the water while leaving the troughs **42** unobstructed for passage of less dense liquids rearwardly and downwardly along said troughs.

At the lower end of the latter a substantially pyramidal funnel **45** defines a collecting chamber **46** connecting with said throughs and carrying a baffle **47** which also assists in controlling eddying. A bottom plate **48** paralleling deck **35** defines with the forward edge of funnel **45** a horizontally elongated port **49** through which the troughs discharge into the funnel. At the stern edge of the bottom plate **48** a pivoted angle plate **50** is adjustably supported from rearward projections **51, 52** of the side plates to regulate the escape of water from the funnel through the stern discharge port **53** defined by stern transom **38**, the side plates **36, 37** and bottom plate **48** and insure coalescence and accumulation in the funnel of all buoyant liquids entering through port **49**.

At the forward end of the keel plate a splash guard comprising an angularly upwardly projecting strip or plate **55** confines surface liquids to the area of entrapment by the V-shaped troughs **42** while jet pipes **56** connected with a source of water under pressure, carried by the attending self propelled craft (not shown) provide for periodically or continuously flushing out the area of the troughs to prevent their obstruction by semi-fluid or solid debris which may be encountered.

It will be evident that when one or more of these craft are translated over the surface of a body of water, intercepting floating liquids during their passage, such liquids are accumulated in funnel **45** and are available for discharge by a pump or the like, (not shown) through a pipe connection **57** at the apex of the funnel which may be connected as by flexible hose, either to an accompanying floating tank or directly to the towering vessel.

FIGURE 177: GALICIA OIL SLICK SKIMMER VEHICLE

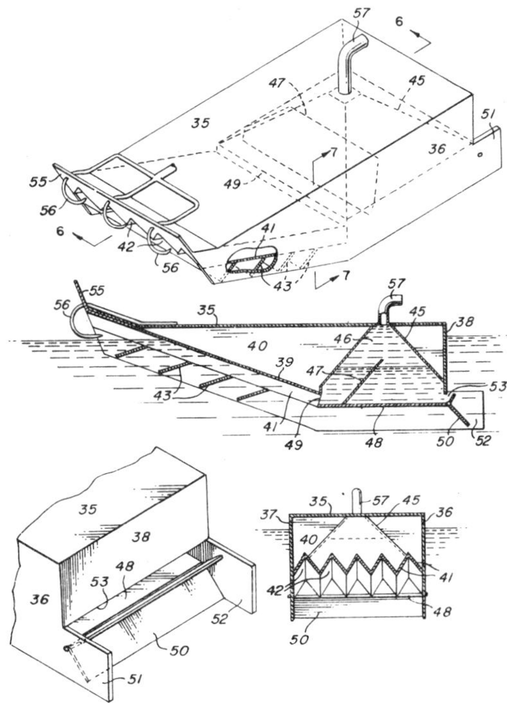

Source: F. Galicia; U.S. Patent 3,666,099; May 30, 1972

Harmstorf Corporation Design

A design commercially available from the Harmstorf Corporation of New York City consists of a hydraulic-powered steel open boat, fitted with a belt skimmer, slop collection tanks and transfer pump. It can be maneuvered through an oil spill at variable speed skimming and collecting the surface oil as it moves. The vessel is powered by a 22-horsepower diesel engine powering three hydraulic motors; one motor drives the variable speed propeller which is rotatable through 360° and hence, provides steering and direction. A second motor drives the roller of the belt skimmer. A third motor drives an eccentric worm pump which transfers oil from the collecting trough to any of the five slop tanks. Two hydraulic cylinders in the pump system control the height of the bow skimming trough. Four ballast tanks permit trimming of the vessel to suit conditions of oil load.

All controls are located centrally permitting one man operation. Depending upon the gravity and thickness of the surface oil layer, the vessel has an oil skimming capacity of up to 2,245 gallons per hour, according to EPA, Edison, New Jersey publication, *Oil Skimming Devices,* Report PB 218,504, Springfield, Virginia, National Technical Information Service (May 1970). The method of removal is by means of a number of endless belts driven around two rollers. One roller is immersed in the oil covered water and the other is mounted on the barge so that the floating oil is transferred from the water into the barge on the surface of the belts.

Headrick Design

A design developed by *E.E. Headrick, U.S. Patent 3,662,891; May 16, 1972,* is a boom arrangement for the collection of oil or other material floating on the surface of water, the boom being suited for towered or stationary positioning. The boom comprises flexible arms defining the sides of a converging channel and a harness located between the arms for providing the load bearing structure for supporting and maintaining the boom in proper configuration whether being drawn through the water or being used in a station keeping application. A plurality of wave attenuators may be located near the narrow end of the channel with a controllable gate linking this end of the boom and a collection device located on the side of the gate opposite the narrow end.

Heinicke Design

A design developed by *K. Heinicke; U.S. Patent 3,690,464; September 12, 1972* is one in which a novel oil recovery vessel for removing oil and other floating impurities from the water surface, is provided with means for collecting such impurities, means for conveying said collected liquid impurities through the vessel and means for the discharge of redundant water from the vessel. The conveying means used keeps the emulsification of the removed impurities to a minimum despite a rather high operational throughput. Such a vessel is shown in longitudinal elevation, transverse section and fragmentary plan views in Figure 178.

FIGURE 178: HEINICKE OIL SKIMMER AND SEPARATION VESSEL

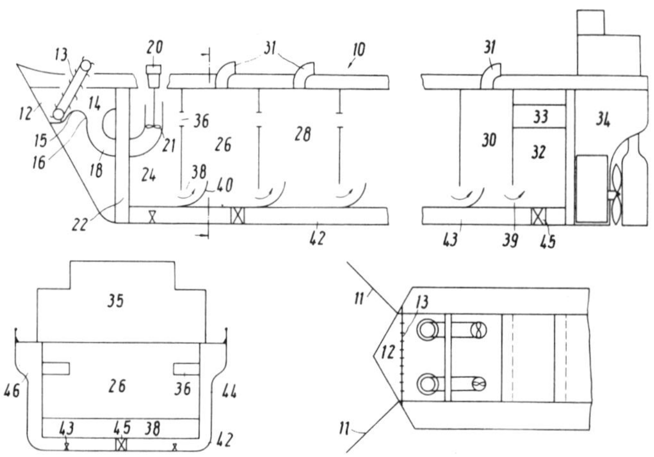

Source: K. Heinicke; U.S. Patent 3,690,464; September 12, 1972

An oil recovery vessel **10** is provided with a wide inlet at the forebody **12**. The draft of the oil recovery vessel **10** is adjustable in order to position the lower boundary of the opening **12**, in particular the upper edge **15** of a funnel **16**, immediately beneath the water level. On both sides of the opening **12** there are arranged retractable guide poles **11** at the bow of the vessel **10**. Further, a conveyor equipment **13** is arranged in the way of the opening **12**, which comprises mainly rotating, endless chains, with netting rigged between, which carries forked teeth pointing in outward direction. This conveyor equipment is intended to reject solid objects entering into the opening **12** and to prevent same from entering into the opening **14** of the funnel **16**, and to dispose of these solid objects on deck of the vessel **10**. From the funnel **16** a U-shaped channel **18** leads through a collision bulkhead **22** and terminates with its legs being open at the topside, in a first separation compartment **24**. In that leg of the channel **18** a propeller **21** is arranged, operated by a motor **20**. The propeller will be of such shape that its conveying effect is mainly concentrated to the area adjacent or near the channel walls **18**.

The compartment **24** is followed by further separation compartments **26, 28, . . ., 30**. These are connected with each other by through openings **36** and/or **38**, arranged in the upper and lower part of the compartment. Looking aft a baffle **40** is fitted behind each through opening **38** in the adjoining compartment. The last separation compartment **30** is followed by an outlet compartment **32**, with an opening **39** being provided only in the bottom part leading from the compartment **30** to the compartment **32**. The outlet compartment **32** is arranged with openings **33** in the upper part, same leading overboard.

Aft of the outlet compartment **33** the engine room **34** will be arranged, above which a wheelhouse **35** is located. The side walls and the bottom of the oil recovery vessel **10** are designed as floats **42, 44** and **46**. These floats are at least partly filled with air during the operation of the vessel. For initial adjustment of the draft of the vessel the floats may be flooded through valves **43**. Pumps not shown will discharge the water to such an extent that the edge **15** is slightly submerged. Additional valves are installed to flood also the compartments **24, 26, 28, 30, 32** should this be required.

The partial plan view shows in particular that with the arrangement of two funnels **16** with openings **14** arranged abreast in the forebody and with the guide poles **11** being extended, the oil recovery vessel will be able to operate with a comparatively large inlet area. With a suitable design of the oil recovery vessel **11** and the arrangement of a plurality of outfits **16, 18, 20, 21** placed abreast the operating capacity may be even increased.

During the opration of the proposed oil recovery vessel **10** the surface layer of the water in front of the vessel **10** will be drawn into the opening **12** and further into the opening **14** of the funnel **16** by the suction propeller, creating an eddy in the center containing the uppermost surface layer which is pulled through the channel **18**. Emulsifying with the water immediately below the uppermost layer is kept to a minimum, so that the separation of impurities or oil from the water will be completed in a comparatively short period, after the liquid has left the channel **18** and entered into the compartment **24**. The oil then separates in the upper part of the separation compartment **24**. Since liquid is continuously entering the impurities or the oil will move through the upper through openings **36** into the adjacent compartment **26, 28,** etc. The water settles in the lower part of the compartment **24** and will be forced by the liquid continuously entering through the lower openings **38** into the following compartment.

The baffles **40** will take care that the water when entering the next compartment has an upward component which promotes the separation of residual impurities. With the design illustrated, the last but one compartment **30** is arranged as a lock chamber which is only connected by a lower opening **38** with the preceding compartment. This compartment **30** is fitted with a bleeder valve **31** for removal of the separated oil, similar to the other compartments. In the compartment **30** residual impurities or oil not separated in the preceding compartments may separate from the water before the water enters into the outlet compartment **32** and is discharged through the upper outlet opening **33,** finally leaving the compartment.

Hercules, Inc. Design

A device developed by *R.H. Burroughs and P.R. Cox, Jr.; U.S. Patent 3,667,608; June 6, 1972; assigned to Hercules, Inc.* is one in which oil is removed from contaminated water by means of a fibrous structure of low denier polyolefin fibers attached to a pumping system. The polyolefin fiber structure can absorb may times its own weight of oil while absorbing little or no water. The oil is then easily removed from the fibrous structure by pumping. Figure 179 shows (center view) a plan view of a craft towing such a device, a perspective (upper view) of constructional detail of such a device and a section (bottom view) showing the optimum operating position of such a device on the water. The fibrous polyolefin structure comprises an elongated, coherent batt or mat **10** having a plurality of fittings **11** embedded therein to serve as pumping points. The fittings **11** are connected to appropriate pumping lines **12** which in turn are connected to a pump or to a vacuum source on barge of boat **13**. The polyolefin batt is dragged through the slick to pick up oil which in turn is pumped into appropriate storage facilities on boat **13**.

FIGURE 179: HERCULES OIL SLICK SKIMMER VEHICLE

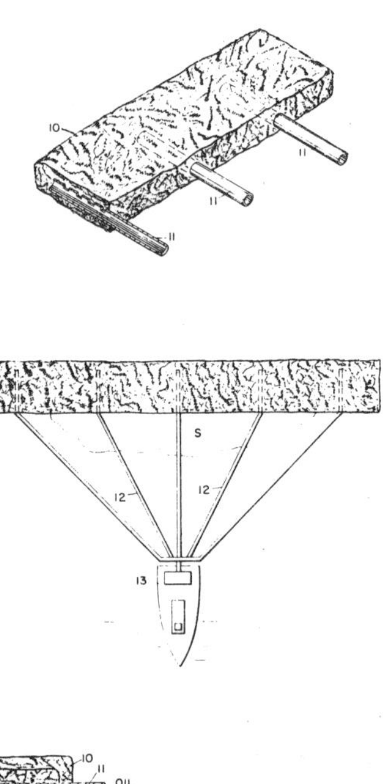

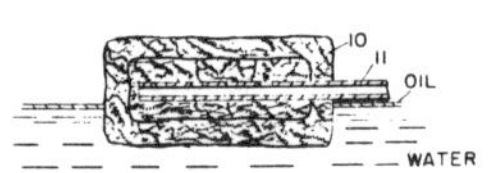

Source: R.H. Burroughs and P.R. Cox, Jr.; U.S. Patent 3,667,608; June 6, 1972

As shown in the bottom view in the figure, the fibrous batt **10** when in use, rests in the oil slick and on top of the water. The fittings **11** are located approximately at the oil-water interface. When oil is absorbed on the batt, and pumping force applied, oil is drawn into the fittings and removed from the batt. The preferred pumping force is vacuum. Capillary action causes the oil in the areas between the fittings to flow to the fittings for removal. The removed oil is immediately replaced by new oil picked up from the water surface by wicking action. The following is an example of the operation of the device.

Example: An oil slick about ½ inch thick was created by pouring kerosene on the surface of a body of water in a large, wide mouth vessel. A batt of polypropylene microfiber (0.03 denier per filament), 2' x 2" x 6" and with four ¼" metal fittings embedded therein along the long dimension to a depth of 4" was floated on the oil. The metal fittings were perforated throughout the area embedded in the fibrous batt and were attached by means of appropriate hoses to a vessel which in turn was attached to a vacuum pump so that the vessel served as a trap to collect the oil. Within a few minutes after contacting the oil slick, the fibrous batt appeared saturated with oil.. The vacuum pump was turned on to pull a vacuum through the metal fittings embedded in the polypropylene batt whereby oil was removed from the batt and collected in the trap. The oil was substantially free of water. When all of the oil had been removed, the pump drew air into the vessel, but still only a very small quantity of water was present in the oil.

Ivanoff Design

A design developed by *A. Ivanoff; U.S. Patent 3,715,034; February 6, 1973* is one in which oil floating on a body of water is collected by moving a shallow-draft water craft, such as a barge, having a sternwardly slanted bow section and below the water line an ingress opening in or near the bow section through an oil slick. The slant of the bow section forces oil in its path downwardly thereby causing the oil, possibly intermingled with water, to flow as a flat layer along the bottom of the barge. As the oil reaches the ingress opening it is propelled into a hold of the barge due to the pressure differential between the outside and the inside of the barge. Oil thus accumulating in a hold of the barge may be removed therefrom from time to time and clear water as may also enter the hold is returned to the body of water.

JBF Scientific Corporation Design

R.A. Bianchi; U.S. Patent 3,716,142; February 13, 1973; assigned to JBF Scientific Corporation describes a sweeping apparatus which consists of a pair of triangular shaped floats or pontoons which are secured to the bow of the skimmer. Typically, although not necessarily, these floats or pontoons, when viewed from above have the shape of a right triangle; the face of the pontoon whose edge forms the hypotenuse of the triangle is the sweeping face and is substantially planar. While the top surface of the pontoon or float is flat, the underbody curves upwardly from the lower edge of the sweeping face to the outer upper edge of the outside of the pontoon in the lateral direction. The underbody also curves upwardly to the point of the triangle in the forward portion.

Two triangular pontoons or floats, as described are secured to the bow of a skimming craft by hinges or the like which permit the floats to rotate about a horizontal axis at the bow. They are spaced so that the two sweep faces are opposed and so that they diverge outwardly from the bow of the skimmer. With this construction, efficient sweeping is obtained since the curved underbody of the pontoons causes water flowing under them to be substantially free of turbulence, thereby reducing power requirements, reducing the force on the sweep face and contributing to more efficient sweeping. Further, since the floats are buoyant, and articulated as described, they will follow wave action, lifting with the waves, and thus will contain any material floating on the wave surface.

Lasko Design

A design developed by *C.J. Lasko; U.S. Patent 3,700,109; October 24, 1972* consists of a pair of laterally spaced hulls supporting a basin between them and a skimming edge immediately ahead of the basin, the hulls including fore and aft chambers and means for admitting or expelling water independently into or out of each chamber to select the level and the fore and aft tilt of the hulls relative to the surface of the body of water so as to enable skimmed liquid to pass over the skimming edge and into the basin.

Lockheed Aircraft Corporation Design

R.L. Yates; U.S. Patent 3,576,257; April 27, 1971; assigned to Lockheed Aircraft Corporation describes a unique drum having a plurality of horizontal vanes for concentrating in a relatively quiet space a quantity of the fluid to be recovered while simultaneously utilizing the forces of surface tension and adhesion of such fluid to cause it to cling to a rotatable disc from which the fluid is scraped and allowed to flow into a central trough from whence it can be recovered, and further processed. Figure 180 shows a vessel equipped with such a drum separator as well as a detail showing a schematic view of the action of the drum.

FIGURE 180: LOCKHEED AIRCRAFT CORPORATION SKIMMER BOAT

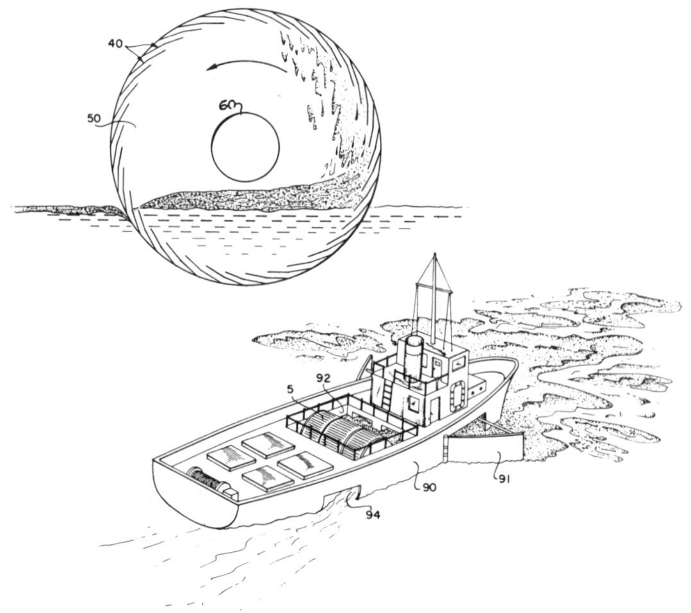

Source: R.L. Yates; U.S. Patent 3,576,257; April 27, 1971

The marine vessel **90** which may be provided with suitable inlet apertures and/or scoops **91** leading into a central well **92** within which is installed a separation drum **5**. Fluids to be separated from the surface of the water would be ingressed through the aperture and scooped into the forward part of the vessel, subjected to the action of separating drum **5** with the oil stored aboard the vessel and the water being exited through exit ports **94** near the stern of the vessel.

Operation of the device will now be discussed with oil and water as the fluids to be separated, these fluids being typical of the characteristics afforded by the separation/recovery unit. It is generally known that the surface tension of oil floating on water is such that if a portion of the oil is removed, the indentation or trough left in the oils is immediately filled by the action of oil drawn thereinto because of inherent surface tension characteristics. Such an indentation in oil or a break in the surface thereof on water is accomplished by means of the discs **50** which are suitably positioned on the surface of the water so that the bottom portion of the drum and discs extends substantially below the thickness of the oil film to be removed. As the drum is rotated either by applied power or by means of self-rotation due to the shape of the vanes as the drum is propelled across the water surface, the discs, due to the surface tension of the oil, pick up a portion of the oil and carry it with the rotation of the discs toward and against wiper blades where it is scraped off and allowed to fall by the action of gravity into the open trough in the hollow axle **60**.

It is especially important to note that as shown the vanes constituting the drum provide an essentially closed internal area within the drum in which the action of wave motion is substantially reduced from that outside the drum. Also, the effects of wind and other disturbances are minimized. Additionally, the vanes **40** as shown operate to thicken the level of oil within the drum over that floating on the water outside the drum and in this manner serve to concentrate the oil for pickup by the discs as they are rotated. Thus, the front descending vanes cause oil to be pushed toward the center of the drum. Simultaneously, the aft (ascending) vanes tend to push the oil toward the center, concentrating on thickening the oil to at least the overlap capability of the vanes, and preventing oil from escaping downstream from the rear of the vent. This is especially important in removing thin coatings from a fluid surface.

Marcocchio Design

A designed developed by *A.E. Marcocchio; U.S. Patent 3,688,506; September 5, 1972* consists of a barge-like float having a forward end shaped to be in skimming contact with the surface of the water. Two series of booms of special construction are attached to the forward end of the float, and the booms extend out from the barge in substantially an inverted V formation. The booms are intercoupled with one another in a particular manner, so that the booms may ride ocean swells without turning over, and without permitting the oil slick to seep under the booms. Figure 181 is a perspective view of such a skimming vessel and related apparatus.

FIGURE 181: MARCOCCHIO SKIMMER CRAFT

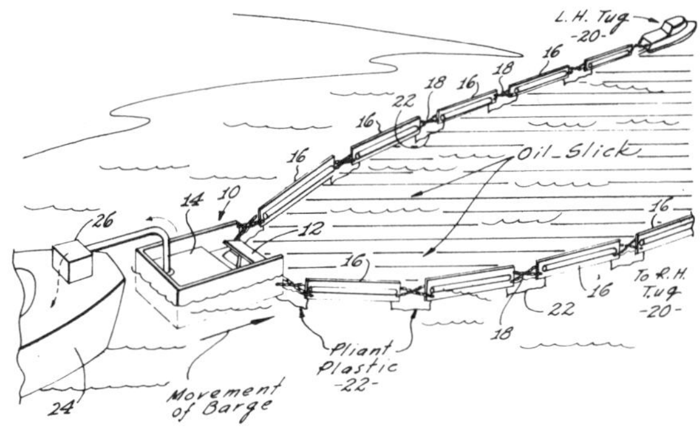

Source: A.E. Marcocchio; U.S. Patent 3,688,506; September 5, 1972

The float **10** in the illustrated embodiment is constructed to have a generally rectangular configuration, and to have upstanding side walls and an upstanding rear wall. The forward end of the float **10** is open, and a foreshortened wall member **12** extends across the open end of the float. The wall member **12** is inclined as a ramp, and it provides a weir effect, so that when the float **10** is towed through an oil slick, the oil flows up and over the weir and into the float with a minimum amount of water. The wall member **12** is preferably adjustable, so that it may be shifted down into the water to just below the level of the oil slick gathered at the front of the barge.

Therefore, when the float **10** is towed through the water, there is a tendency for the float to fill with oil derived from the oil slick on the surface of the water. An adjacent barge **24** may be provided, for example, equipped with a pump **26**. The pump **26** serves to pump the oil from the float **10** into the barge **24**. It will be appreciated that the intake to the pump **26** may be located at a position in the float **10**, for example, directly under a raised platform **14**, so that a maximum amount of oil and a minimum amount of water will be pumped from the float **10** into the barge **24**. The platform **14** serves to retain the oil in the barge **10** and to prevent the oil from splashing over the sides of the barge. An additional pump (not shown) may be provided on the float **10** having an intake extending to the bottom of the float, and which serves to pump sea water out of the float and return it to the body of water, preferably in front of the float to assure that any oil inadvertently pumped by the pump will not escape. As an alternative, the pump **26** may remove water and oil from the float **10**, and the mixture may then be subjected to a centrifugal, or other operation, within the barge **10**, so as to separate the oil from the water. In such an event, the oil will be recovered and stored in the barge **24**, for example, and the water returned.

Also included in the apparatus is a first series of booms **16** which are attached to one another by appropriate fasteners such as crossed anchor chains or cables **18**, and which are attached to one side of the open forward end of the float **10** in the same manner. The apparatus also includes a second series of booms **16** which, likewise, are attached to one another by crossed anchor chains or cables **18**, and which extend from the other side of the forward end of the float. The two series of booms extend in a general V configuration from the forward end, and may be attached to appropriate tug boats **20**.

The tug boats serve to draw the booms **16**, as well as the float **10** across the surface of the body of water on which the oil slick occurs, with the booms forming a general V configuration and surrounding the oil slick. As the apparatus is towed, the oil slick is directed into the forward end of the float **10**, and over the wall **12** into the interior of the float. It will be appreciated that as the oil slick is gathered together and directed towards the open end of the float **10**, it extends further and further down into the water, and care must be taken that the oil will not seep down under the booms **16** and escape.

The spaces between the individual booms **16** may be blocked, for example, by sheets of pliant plastic material, or by other appropriate means, designated **22**. Conversely the booms may be lashed together in an overlapping relationship so that there is no need to provide the sheets **22** between the successive booms. When these pliant sheets **22** are used, they are preferably attached to the ends of the corresponding booms. Also, the corresponding crossed chains **18** must be such that the booms are held sufficiently close together so that the pliant plastic material **22** is not taut. This is important, since when the apparatus is drawn over crests of waves it is important that the individual booms may rise and fall with the waves, and also so that adjacent pairs may form an apex with one another as the apparatus is drawn over the crest of the wave, without damaging the apparatus, and without interfering with the integrity of the boom and pliant plastic structure.

Ocean Pollution Control, Inc. Designs

A design developed by *H.J. Fitzgerald; U.S. Patent 3,523,611; August 11, 1970; assigned to Ocean Pollution Control, Inc.* for skimming an oil film from the surface of a large body of water includes a towed funnel assembly with a flexible cover and side skirts of impermeable sheet material with floats to keep the leading edge of the cover spaced above the surface of the water so that the oil film will pass beneath it, with the remaining portions of the cover supported on the floating oil, a bottom panel of netting to hold the side skirts in downwardly projecting position to confine the oil laterally, while permitting the water beneath it to escape freely, and a sump at the apex of the funnel to receive the oil for transfer to storage vessel.

Another device developed by *H.J. Fitzgerald; U.S. Patent 3,653,510; April 4, 1972; assigned to Ocean Pollution Control, Inc.* amplifies some of the constructional details of such a towed funnel device, as shown in Figure 182. The apparatus shown includes a funnel assembly generally designated **10** which is formed of a cover **12** of flexible sheet material such as closely woven canvas, nylon or dacron which is impervious and impermeable to oil, having a tapered shape with a concave leading edge **14** of catenary shape and tapered trailing edges **16** from which depend skirt members **18**, with the lower edges of the skirt members **18** connected to the opposite edges of a bottom panel **20** of open material, such as gill netting, which extends across and encloses the bottom of the funnel assembly. The leading edge **14** of the cover is reinforced by a heavy rope which is attached to and rests upon a series of floats **22** of such size as to keep the leading edge **14** spaced generally above the surface of the water an average distance of approximately 1 foot to insure that, as the funnel assembly is towed in a direction perpendicular to the leading edge **14**, the oil film will pass beneath the leading edge **14** and along the bottom of the cover **12**, with the trailing portions of the cover **12** being supported by the floating film of oil. The leading edge of the bottom netting panel is likewise reinforced by a heavy rope carrying spaced weights **23** to keep it submerged.

FIGURE 182: OCEAN POLLUTION CONTROL, INC. TOWED FUNNEL SKIMMER VEHICLE

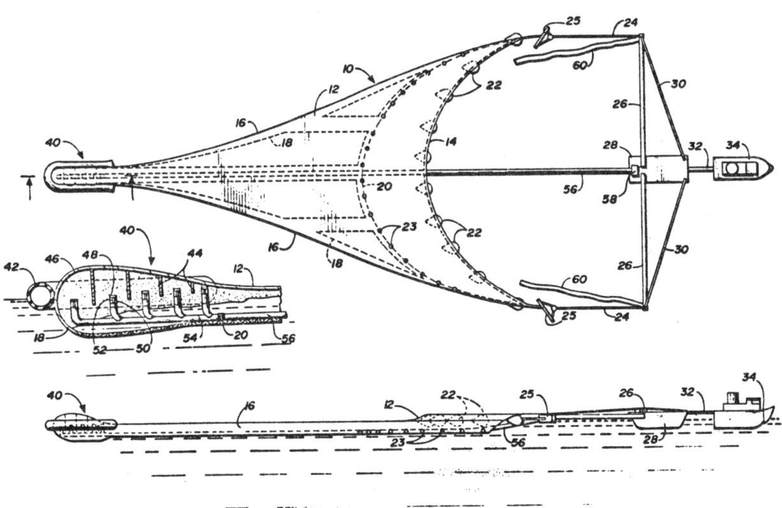

Source: H.J. Fitzgerald; U.S. Patent 3,653,510; April 4, 1972

As the funnel assembly is towed, the substantial pressure of the oil and water against the tapered sides **16** of the funnel and the lesser drag of the bottom netting panel **20** will maintain the somewhat flattened conical shape of the funnel, while its flexibility will permit it to conform readily and accurately to the changing surface of the water. The bottom netting panel **20** limits outward movement of the skirt portions **18** and insures their downward extension for a sufficient depth, for example 2 to 6 feet, to entrap all of the oil which enters the wide leading end of the funnel and prevent the escape of such oil with the water which is permitted to move freely through the bottom netting **20** and out of the funnel.

The two ends of the leading edge **14** of the funnel are connected to tow lines **24** which are provided with otter boards **25** which are provided with harnesses to keep them angled forwardly at their outer ends to convert the forward motion of the tow into outwardly directed lateral forces at the opposite ends of the leading edge **14** and thus keep the leading edge extended to maintain the shape of the funnel. The forward ends of the tow lines **24** are connected to the outer ends of outrigger booms **26** whose inner ends are pivoted on a storage barge **28** and which are braced by back stays **30** which extend between the forward end of the storage barge **28** and the outer ends of the outrigger booms **26**. The forward end of the storage barge **28** is connected by a tow line **32** to a towboat **34**.

At the apex of the funnel assembly is an oil sump generally designated **40** which is a narrow, elongated enclosure effectively constituting an upwardly enlarged rearward extension of the cover **12**, side skirts **18** and bottom netting **20** of the funnel. In the particular embodiment shown the sump is supported within an inflated U-shaped float **42** similar in construction to to a rubber life raft, with the sides of the float **42** supporting transversely extending vertical partitions **44** of a relatively stiff and impermeable sheet material which extend between the sides to divide the sump into a plurality of longitudinally spaced chambers. The upper edges of these partitions **44** adjoin the cover sheet **46** on top of the sump and extend downwardly into the sump to graduated depths increasing from the leading end to the trailing end of the sump. Thus, as the oil moves rearwardly into the sump, the lower edges of the successive partitions skim incremental layers off the top of the oil. An outlet pipe **48** extends vertically upwardly into each of the chambers, the upper ends of the pipes being held by clamps **50** which were supported on webs **52** extending transversely between the side walls of the float **42**. The positions of the upper ends of the pipes **48** are graduated in height, decreasing from the leading end of the trailing end of the sump. These pipes **48** all communicate with a manifold **54**, the forward end of which is connected to a flexible line **56** which extends forwardly to a pump **58** located on the storage barge **28** to move the oil from the sump **40** to the storage barge **28**.

Trailing from the outer ends of the outrigger booms **26** are a pair of hollow flexible tubular pipes **60**, closed at each end to keep out the water and maintain their buoyancy and insure that they will float along the surface of the water. These pipes are of a substantial diameter, for example 6 inches, and serve the purpose of guiding the film of oil inwardly toward the leading edge **14** of the funnel assembly **10** to prevent the oil from being spread beyond the ends of the funnel assembly by the wake of the tow boat **34** and storage barge **28**.

As will be appreciated from the foregoing description, as the apparatus is towed by a tow boat **34**, an area of the surface of the water having a width equal to the width of the input end of the funnel assembly **10** is swept of oil. This oil is funneled rearwardly and inwardly to the sump **40**. The oil film on the water may normally range in thickness from several molecules to several inches, depending upon the distance from the oil source. As the oil film is laterally concentrated toward the narrow end of the funnel assembly **10**, its thickness is of course increased, perhaps to as much as several feet. Due to the buoyancy of the oil, the upper portion of this oil mass may extend as much as a foot above the surface of the water and several feet below the surface.

The flexibility of the cover members **12** and **46** is sufficient to accomodate this upward bulging of the oil mass, while the skirts **18** extend downwardly beneath the surface a sufficient depth to trap all of the oil and channel it rearwardly into the sump **40** where it is removed through the pipes **48** and **56** as fast as it accumulates. This provides a continuous and rapid oil removal method of operation which can continue uninterruptedly until the storage barge is filled or even longer by providing means for pumping the oil through a flexible pipe from the storage barge to another vessel moving on a parallel course alongside the storage barge while the sweeping operation continues, in the same fashion as the refueling of ships by a tanker while underway.

While the arrangement of the partitions **44** and the inlet ends of the pipes **48** in the sump **40** will result in a high oil/water ratio in the liquid removed to the storage barge, under the conditions of extreme choppiness resulting in turbulent mixing of oil and water within the sump, a higher percentage of water may be received at the storage barge. Under such conditions it is desirable to provide an oil/water separator on the storage barge. Thus the water can be separated from the oil and discharged back into the main body of water so that only the oil is retained in the storage barge.

Although the funnel covers a large area, its light weight and flexibility and low water resistance limits the strain which is imposed upon it, and renders the apparatus practical and durable. Preferably the apparatus is towed at a relatively slow speed, on the order of 3 knots, to limit the strain upon it, as well as to insure capture of substantially all of the oil in the area swept. However, because of the considerable width of the funnel, it is possible to sweep a large area in a short time.

A device developed by *H.J. Fitzgerald and E.H. Koepf; U.S. Patent 3,557,960; January 26, 1971; assigned to Ocean Pollution Control, Inc.* involves a pair of generally similar funnel assemblies, one positioned behind and in the wake of the other with a harness for towing the same along their common central axis.

Another device developed by *H.J. Fitzgerald and E.H. Koepf; U.S. Patent 3,590,584; July 6, 1971; assigned to Ocean Pollution Control, Inc.* for collecting oil from the surface of a body of water consists of two V-shaped assemblies of flexible inflated floats, one arranged 5 to 25 ft leewardly of the other on the same central axis, whereby the wind and current drive the oil into the open end of the assemblies and cause it to be funneled rearwardly to their apices. Each V-shaped assembly is provided with a depending skirt of impermeable sheet material, the lower edges of the skirts at either side of the inner assembly being interconnected by shock cords and the lower edges of the skirts on the outer assembly being connected to the inner assembly by netting.

Still another device developed by *H.J. Fitzgerald and E.H. Koepf; U.S. Patent 3,612,280; October 12, 1971; assigned to Ocean Pollution Control, Inc.* involves a pair of wing assemblies attached to opposite sides of the hull of a marine vessel, each including a cover with tensioning means engaging its outer end to keep it extended outwardly, means to support its leading edge above the water to allow oily material at the surface to pass beneath it, an angled skirt portion at its trailing edge to funnel the oily material inwardly toward the vessel, conduits at the hull of the vessel to remove the oil, and lines engaging the outer ends of the wing members for hauling them inwardly to clear lateral obstructions.

Figure 183 shows such a skimming vehicle in plan view as well as a cross-sectional detail of the skimming wing. The apparatus includes a pair of wing assemblies **10** mounted at opposite sides of a marine vessel **12.** Each of the wing assemblies includes a cover **14** of highly flexible sheet material which is impermeable and impervious to oil, such as canvas, woven nylon, Dacron, or similar synthetic material, which may be impregnated with or laminated to a thin film of neoprene or the like.

Each cover is of a semidouble concave or sickle shape, with its wide inner end **14a** connected to a flange several feet above the waterline on the side of the hull of the vessel near its stern, so as to form a substantially watertight seal with the hull. The cover extends outwardly and forwardly and its tapered outer end **14b** is connected to a paravane device, such as an otter board **16,** which is also connected through a yoke **17** to a heavy towing line **18** which extends rearwardly and outwardly from a motor driven winch **20** at the bow of the vessel. The yoke keeps the otter board obliquely angled outwardly and forwardly so that the forward motion of the vessel will impose an outwardly directed tension on the leading edge of the cover to keep it extended laterally.

The leading edge **14c** of the cover is reinforced by a heavy rope, for example, of nylon or Dacron, which is supported on and connected to the top of a series of floats **22** which are spaced along the leading edge to support it a foot or more above the surface of the water. The extreme flexibility of the cover allows it to conform readily to the wave motion and chop of the water, as well as the bow wave of the vessel, and insure that all of the floating oily material will pass beneath it. The trailing portion of the cover is supported on the floating film of oily material.

At the trailing edge of the cover is a skirt portion **24** which projects downwardly for several feet below the surface of the water to trap the oily material. The inwardly and rearwardly angled direction of the skirt portion funnels the oily material inwardly toward the vessel **10** as the vessel moves forwardly through the water. Located at the side of the hull of the vessel **12** just beneath the cover and just forwardly of the skirt portion is a conduit **25** which extends to a pump **26** which transfers the oily material to a reservoir **28.** While the pump and reservoir are shown contained within the hull of the vessel, they may be remotely iocated, for example, in a barge towed behind the vessel, and connected to the wing assembly by a flexible pipe formed, for example, of fabric reinforced neoprene, which extends along the outside of the hull of the vessel.

FIGURE 183: WINGED SKIMMER VEHICLE

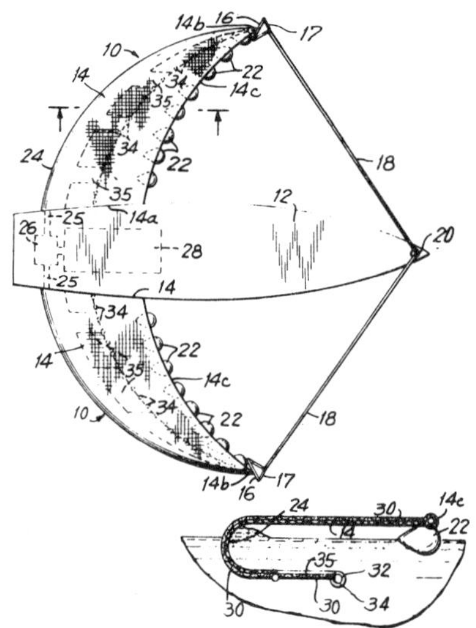

Source: H.J. Fitzgerald and E.H. Koepf; U.S. Patent 3,612,280; October 12, 1971

Screens are preferably placed over the input ends of the conduits **25** to catch floating debris and prevent clogging of the conduits or damage to the pump. Extending across the lower side of each of the wing assemblies is a bottom member **30** of material readily permeable to water, such as gill netting. As shown in the figure, the leading edge of the bottom member is reinforced by a heavy rope **32** which has attached at spaced positions along its length a series of weights **34** which keep the leading edge of the bottom member submerged.

The trailing portions of the bottom member are held downwardly by the water pressure against them as the boat moves through the water. The trailing edge of the bottom member is connected to the lower edge of the skirt portion to keep the skirt portion extending downwardly well below the surface and prevent its trailing rearwardly due to the water pressure against it and thereby allowing the oily material to escape beneath it. While the oily material at the surface is trapped by the skirt portion, the water beneath it escapes freely through the bottom member.

The flow resistance of the netting may be enhanced by a series of transversely spaced panels **35** of flexible sheet material, such as nylon, which is less readily permeable to water, extending fore and aft along the bottom member, thus insuring that the downward and forward tension imposed on the lower edge of the skirt portion by the bottom member will balance the rearward and upward pressure to which it is subjected by the forward movement of the vessel. The netting **30** may extend around the skirt portion and over the cover to provide additional reinforcement.

When the wing assemblies are fully extended, as shown, the apparatus will sweep a wide path, for example, up to about 150 ft. Thus, although the vessel preferably moves at a relatively low speed, for example, 3 knots, to minimize the strain on the apparatus, a large area can be swept free of oil in a relatively short time. The direct attachment of the wing assemblies to the hull of the vessel allows good maneuverability. When an obstruction such as a pier, a moored vessel, or buoy, etc., is encountered at either side of the path of the vessel, the wing assembly on that side of the vessel can be fully or partially retracted merely by hauling in the towline **18** on that side by means of the winch **20.** This allows the width of sweep to be reduced to clear a lateral obstruction without changing the course of the vessel. After the obstruction has been cleared, the wing assembly may again be extended merely by paying out the towline, and allowing the otter board to pull the wing assembly outwardly from the hull again.

Oil Recovery Systems Inc. Design

A design developed by *E.E. Lithen; U.S. Patent 3,752,317; August 14, 1973; assigned to Oil Recovery Systems, Inc.* provides a vessel for collection and salvage of oil spills having a vertically adjustable forward mounted scoop from which fluid collected under the action of gravity and the forward motion of the vessel is directed through conduits into submerged separation tanks under conditions of laminar flow. In the submerged tanks, the fluid is separated into oil, which is transferred to storage tanks, and water which is discharged.

Peterson-Cole Design

A design developed by *D.L. Peterson and C.M. Cole; U.S. Patent 3,661,263; May 9, 1972* is a V-shaped oil slick sweeping system including a log boom as one arm and an oil barge as the other arm, and wherein the barge is outfitted and compartmentalized to receive mixed water and oil, and wherein means is provided on the barge to separate the oil from the water, and to retain the former while discharging the latter.

Prewitt Design

A design developed by *C.H. Prewitt; U.S. Patent 3,730,346; May 1, 1973* includes an elongated sink-like or trough-like skimming unit adapted to be propelled through the water, with means for drawing fluid from the unit and conducting it to a separator for separating the floatable matter from the water. The elongated skimming unit extends laterally of the path along which the unit is propelled through the water and adjustable flotation means are included for controlling the vertical and angular position of the unit in the water.

The skimming unit has forward and aft edges lying in the same horizontal plane and maintained at substantially the same height in the water to prevent creation of a bow wave pushing the oil away from the unit and to permit the oil or other floatable fluid to be drawn into the trough across both the forward and aft edges. Flotation and propulsion adjustment means maintain the optimum position of the unit in the water during use.

Figure 184 is an isometric view illustrating the free floating skimmer unit self-propelled ahead of a mother craft, which has a boom to which the skimming unit is tethered for lifting the same out of the water. The skimming unit **10** is self-propelled and guided ahead of the mother craft **12** to which it is connected through a flexible conduit assembly **14**. It can be lifted from the water by a boom **16** extending from the front of the craft. This form of the skimming unit includes an elongated trough **18** having substantially parallel, horizontal fore and aft edges **17** and **19** and cowlings **20** and **22** at its opposite ends enclosing flotation tanks into which ballast fluid may be pumped to adjust the flotation level of the unit.

The boom may be lifted by the crane structure **23** mounted on the deck of the boat, by operation of the hydraulic cylinder **26**. The unit is connected to the boom by tethering lines **24**. The boom also carries compressed air lines **28** which are connected to the cowlings at opposite ends of the skimming unit and service flotation or ballast tanks and propulsion system by which the unit is moved and guided.

A flow through, self-priming submersible pump **30** is mounted in the flexible conduit coupled between the skimming unit and the boat, the pump being suspended from the boom by a tether line **31** so that its weight is not supported by the skimming unit. Power is delivered to the pump by a line (not shown) internal to the flexible conduit or attached to the tether line **31.** The flexible hose portion **34** between the pump and the skimming unit must be of the noncollapsible type because of the suction of the pump in this arrangement, whereas portion **32** between the pump and the mother craft may be of the collapsible type. As will be seen, the submersible pump may be mounted directly on the skimming unit to advantage, or it may be mounted aboard the deck of the mother craft, if the disadvantages of that arrangement do not render it unacceptable.

Aboard the craft is a control console **36** by which an operator may control operation of the skimming unit in cooperation with the pilot of the craft. It should be noted that the system may be operated from a dock on which the console and other equipment may be located. The control console is operable to control the compressor **38**, the separator **40,** valve control means **42** for adjusting the pump outlet valve **44**, flotation and propulsion valves, the hydraulic boom and suitable illumination means (not shown).

While the separator is diagrammatically representative of a centrifuge, the type of separator employed is not critical, the use of filters or gravity separators being acceptable under some conditions. The flotation tanks include fore and aft tanks **52** and **54** in the right hand end cowling **20** and fore and aft tanks **56** and **58** in the left hand cowling **22.** Also connected to the right and left hand cowlings are air supply lines which deliver air to right and left hand propulsion nozzles **94** and **96** protruding aft from the cowlings as shown.

As previously indicated, the flotation level of the skimming unit is adjustable by means of the pump outlet valve control **42** which helps to control the volume of fluid contained in the skimmer. However, the operator cannot always respond quickly enough to wave conditions and other factors causing the skimmer to become too full or not full enough and hence to float too high or too low in the water. Therefore, to further assist him in controlling the flotation level, a pair of automatic flotation controls are preferably built into the skimmer unit itself, namely the right and left hand float valves **106** and **108.** Each of these valves includes a flap valve door hinged to open and close and admit water to the trough **18.**

FIGURE 184: PREWITT SKIMMER VEHICLE DESIGN

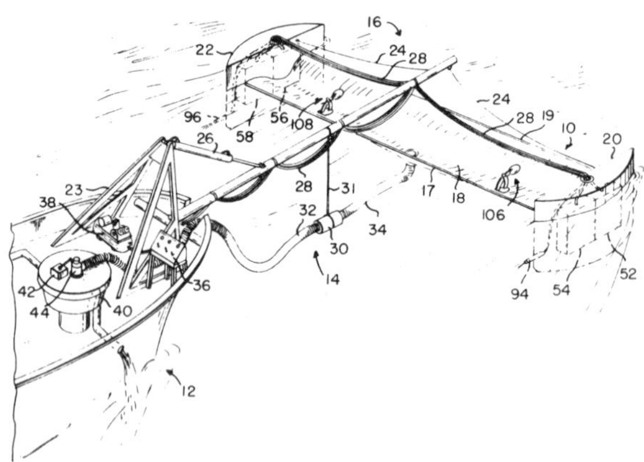

Source: C.H. Prewitt; U.S. Patent 3,730,346; May 1, 1973

Price Design

A design developed by *L.D. Price; U.S. Patent 3,630,376; December 28, 1971* for collecting and removing liquid and solid waste matter floating on the surface of the body of water comprises, in combination, first means independent of the movement of the vessel for selectively providing water under pressure substantially along a horizontal line beneath the surface of the water to form an upwardly directed wave in the body of water, means for permitting a selected portion of the wave thus formed to pass thereover and into the vessel, and means for withdrawing the liquid and solid waste matter from the vessel into suitable receptacles.

As shown in Figure 185, the vessel **11** is in the form of a barge having a relatively broad square bow portion **13,** and a relatively flat bottom or hull **44.** Suitable motors **16** may be affixed to the stern of the vessel for propulsion. In the example shown, the barge is self-propelled, but it may also be pushed by another boat.

A holding chamber **15** is positioned in the center portion of the vessel. This chamber has downwardly extending sides **17** and **18** and a downwardly extending back **19** all of which terminate at a point above the hull. More specifically, sides **17, 18** and back **19** extend downwardly approximately two-thirds the depth of the vessel, and can be extended downwardly even farther by provision of suitable flat plates which can be fastened to the sides **17, 18** and back **19,** as is well-known. A cover **20** which functions as the cover or deck for the vessel also provides a cover for chamber **15.**

The sides **17** and **18** of chamber **15** taper outwardly from the back **19** and abut against the sides **12** and **14** at a forward portion of vessel **11.** Note that sides **12** and **14** of the vessel extend forwardly of, or ahead of the bow **13,** and thus tend to force water into the vessel rather than splashing it aside as done by the usual type boat. Various vertically oriented pipes **21, 22, 23** and **24** extend into the hold **27** or inner part of the barge for purposes to be described.

A pipe **29** for carrying water is mounted in a horizontal position along the bow **13.** Pipe **29** includes a series of slits or holes **31** for permitting water to flow outwardly therefrom, as will be explained. One end, the left hand end, as oriented in the figure, of pipe **29** is affixed through pipe **33** to a pressure pump **35.** This pump pumps water from the hold **27** by pipe **21** and through **33** to pipe **29.** The free end of this pipe includes a valve **34** to permit more or less water to flow therethrough to control the amount and pressure of the water passing outwardly through slits **31.** The position of the slits, and hence the direction in which the water under pressure in pipe **29** is caused to flow, is adjustable by means of a suitable gearing assembly generally labeled **39.**

FIGURE 185: PRICE OIL SPILL SKIMMER BOAT

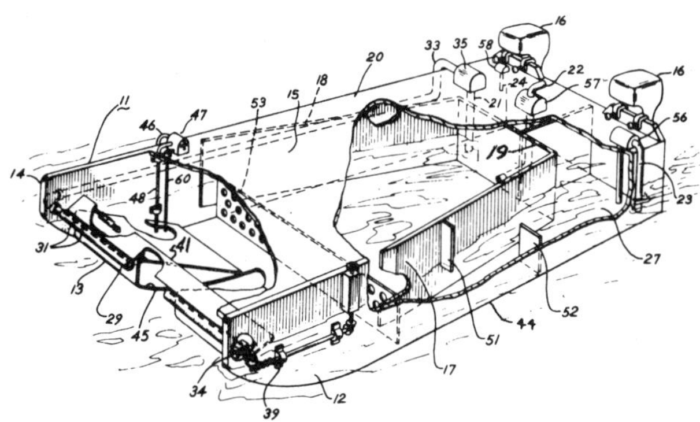

Source: L.D. Price; U.S. Patent 3,630,376; December 28, 1971

A baffle plate **41** is hingedly mounted in a horizontal position along bow **13** and is pivotably movable about a horizontal axis as by a power assembly. In operation, water is drawn into hold **27** of the vessel by means of one or more suitable pumps **56,** or by means of a sea cock, to raise the water level in the hold above the lower edge of the sides **17, 18** and back **19** of the holding chamber. The vessel is thus effectively lowered into the water by the foregoing flooding such that the pipe **29** is positioned below the surface of the water.

The pitch of the vessel, or the depth of the bow **13,** in the water is also controlled by pumping water in or out of a closed hold **45** positioned along the bow. When water is pumped into hold **45** through pipes **46** and **48,** suitable reversible pump **47**, displaced air will be allowed to escape through pipe **60** and the associated valve (not numbered), and the bow sinks deeper in the water, while if water is pumped out of hold **45** by pump **47,** the bow tends to rise, hence controlling the position of the pipe **29.**

Next, water from hold **27** and pipe **21** will be pumped by pump **35** under pressure through pipe **33** and to pipe **29.** The water under pressure will thus be forced out of the slits **31** to create a wave which is forced over the top of baffle plate **41.** The water pressure and the size of the wave depends on the pressure created by pump **35,** the size of the slits **31,** the setting of the valve **34** and the depth that pipe **29** is under the surface of the body of water. The water flowing out of the holes or slits is thus caused to form a ripple or wave sufficiently high such that a sheet of water carrying oil or debris goes over top of baffle plate **41.**

The vessel is then propelled forward, and pumps **35** and **56** continue to pump water. The baffle plate is adjusted to control the thickness of the sheet of water that goes over the top of the baffle plate into the holding chamber. As the sheet of water and oil moves over the baffle plate, the globules of oil are not broken up but rather the mass of oil on the water caused to move smoothly over the top of the baffle plate onto the holding chamber.

Since the water level in the holding chamber is above the lower edge of the sides **17, 18** and back **19** of the holding chamber, the water which is heavier than oil acts as a bottom or base surface for the oil and maintains the oil within the holding chamber. As more and more water and oil float into the holding chamber, the water in the barge is pumped out by pump **57** through pipe **22** to a suitable storage means on the deck **20** of the vessel or to an associated barge. Any excess water in the vessel can be pumped out through pipe **23** by pump **56** and the water level in the hold maintained at a suitable level.

The water pressure in the pipe **29,** the relative orientation of the holes or slits **31,** the angle of the baffle plate, and the pitch of the vessel may be continuously adjusted to provide good control of the sheet of water and oil moving over the baffle plate into the holding chamber. While the vessel may operate successfully without the holding chamber, the chamber is useful in collecting the oil and debris into a thickness which can be conveniently removed. Vertically positioned plates, such as **51, 52** and **53** can be placed in the hold and in the holding chamber to prevent excessive sloshing of the water in the vessel and to thus increase its stability. Pump **58** is used to pump air in and out of the hold as necessary, and in certain examples may comprise a vent means.

Reynolds International, Inc. Design

A design developed by *A.L. Markel; U.S. Patent 3,637,080; January 25, 1972; assigned to Reynolds International, Inc.* is one in which floating material, such as oil, is directed toward and into pockets on a continuously driven conveyor belt means which serves to submerge the floating material beneath the surface of the water. A transfer of the material is made to a suitable conduit means from which the floating material is pumped to a suitable reservoir or station where it is separated from the water. Figure 186 shows a perspective view of this design of skimming vehicle as well as details of the conveyor belt device. The vessel **10** has two parallel hulls **12** and **14** bridged in conventional manner by a platform or deck structure **16**. The bow sections of the hulls are each tapered at **17** to form a funnel-like opening for directing flotsam, such as oil, toward and into a skimmer apparatus **18.**

FIGURE 186: REYNOLDS OIL SPILL SKIMMER CRAFT

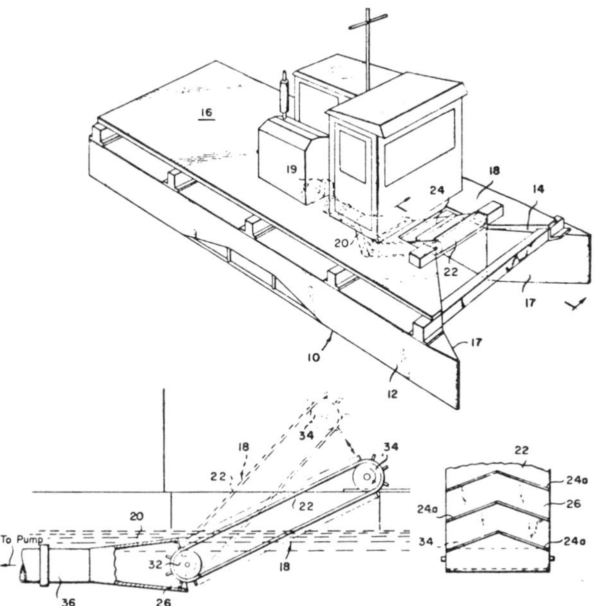

Source: A.L. Markel; U.S. Patent 3,637,080; January 25, 1972

After being carried beneath the surface of the water, as described below, the flotsam is transferred into a funnel **20** from where it is withdrawn through conduit **36** by the vacuum of a pump **19.** The skimmer apparatus comprises a driven endless belt conveyor **22** which is shown partially submerged below the surface of the water. With this conveyor partially submerged and driven to force the water toward the funnel, the flotsam is trapped under the surface of the water between the conveyor belt surface and hulls **12** and **14** where it can be readily collected, as will be explained in detail.

The conveyor is provided with transverse fins **24** which together with the oblique angle positioning of the conveyor itself cooperates to force the floating matter under the surface of the water. In order to accomplish this, pockets **26** are formed between adjacent fins **24** and the surface of the conveyor between the adjacent fins. The walls of hulls **12** and **14,** respectively, provide guide means or opposed ends for the pockets and prevent any substantial lateral escape of the flotsam entrapped in the pockets beneath the surface of the water. The fact that oil, debris, flowers, and other floating matter are lighter than water aids in the transport of the flotsam toward the funnel in that the buoyant force of the water serves to contain the floating matter in the pockets. Thus, an essentially closed chamber is formed for carrying flotsam to funnel **20.**

It is important that the funnel engulf the submerged end of the conveyor to ensure that no floating matter escapes the skimmer and returns to the surface of the water. In order to achieve this function, the funnel is provided with a mouth which is slightly larger than the submerged end of the conveyor so that the conveyor fins **16** passing the funnel mouth are close enough to prevent any flotsam from escaping the suction of the pump. The fins can be of any desired height to form a desired volume for the pocket depending upon the particular matter to be collected, with an appropriate adjustment being made in the mouth of the funnel. For example, if the flotsam is oil, relatively short fins may be utilized since a film of floating oil on water is normally quite thin. The fins may be V-shaped as shown at **24a**. The skimmer has a lower sprocket **32** mounted on a fixed rotating shaft, whereas an upper sprocket **34** is mounted on a shaft which is movable to increase or decrease the angle of inclination between the conveyor belt and the surface of the water. By having the upper end of the skimmer adjustable, varying amounts of flotsam may be conveyed by the skimmer, preventing overloading of the funnel or pump thereby making the skimming operation more efficient to run. The adjusting means may be either manual or motor operated.

Richards Design

A design developed by *F.A. Richards; U.S. Patent 3,700,108; October 24, 1972* for removing oil films, debris or other floating liquid and/or solid impurities from the surface of a body of water includes an impurity water conveyor, a perforated, separator-conveyor for separating solid impurity components from the conveyed impurities and water, and a holding tank for thereafter gravity separating water from the liquid impurity components.

The impurity water conveyor is pivotally supported at its upper outlet end adjacent a fore-aft midpoint of the float support and has its lower intake end independently float supported to rise and fall in conformity with the surface of the water adjacent the fore end of the float support. The impurity water conveyor includes a plurality of relatively flexible paddle elements, which cooperate to elevate separate charges of impurities water upwardly along a channel guide; each charge being formed and subsequently conveyed and separated with a minimum of emulsion inducing agitation, whereby to promote relatively rapid separation of the liquid impurity components from the water in the holding tank.

Ryan-Ledet-Colvin Design

A design developed by *D.J. Ryan, W.P. Ledet and J.R. Colvin; U.S. Patent 3,656,619; April 18, 1972* is one in which a group of liquid cyclone separator means is towed, propelled or otherwise moved through the body of water for directing the floating pollutants with a minimum of the water through the separating means. The pollutants may be rapidly removed from the body of water into the skimmer vehicle in this manner with substantially no mixing and emulsifying of the pollutant with the water.

Shell Oil Company Design

A design developed by *J. Cornelissen; U.S. Patent 3,348,690; October 24, 1967; assigned to Shell Oil Company* comprises a floating vessel having two elongated booms that are pivotally secured at one end thereof to opposite sides of the vessel hull. The booms float longitudinally on the water surface and are secured in a forwardly and outwardly divergent position with respect to the vessel axis of movement.

As the vessel moves, the booms are swept over the water surface, thereby funneling the oil film within the sweep of the booms axially therealong to the juncture between the booms and the hull. At the juncture, vortex producing inlets are provided in the vessel hull for the oil to flow into and then to be conducted to an oil-water separating tank within the vessel. The reader is referred to the discussion of the Waterwisser device (page 363).

Sky-Eagle Design

A design developed by *W.A. Sky-Eagle, Jr.; U.S. Patent 3,757,953; September 11, 1973* for skimming oil or other scum from the surface of water, has floats supporting a sump tank with a weir at its forward end, means for pivoting the sump around a transverse axis to adjust the depth of the weir below the surface of the water and pumping means for withdrawing liquid collected in the sump.

Spanner Design

A design developed by *W.F. Spanner; U.S. Patent 3,744,257; July 10, 1973* is a water surface cleansing ship having a hull provided with a channel extending longitudinally through the hull. At least surface water to be cleansed is caused to flow in operation of the ship, through the channel, by motion of the ship either self-propelled or towed and/or by flow causing means such as a paddle wheel in the channel. The ship has means for removing and recovering contaminants such as oil from water flowing through the channel. The contaminant removing and recovering means preferably include a mesh grill, one or more conveyor belt and tank assemblies, and detergent sprays.

As shown in Figure 187, the ship has a hull generally indicated at **1** which is roughly in the form of a rectangular box structure faired at the ends and of generally U-shaped cross-section. A double bottom **2** extends over the full width of the hull from station **S4** to station **S110.** A channel of generally rectangular cross-section extends longitudinally through the hull, preferably along the center line thereof from an opening **4** at the ship's bow to an opening **5** at the stern. Hinged flap means **6** and **7**, able to be secured watertight, are provided on the hull, with the flap means **6** at the bow opening and the flap means **7** at the stern opening.

The channel is bounded below by a rounded sill **8** at station **S8** conveniently 18 inches below the designed waterline **9** on the center line of the hull, and on each side by the main structure of the ship over a width of 35 ft 11¼ inches. The channel diminishes in width to 10 ft at station **S30** and thereafter is of constant width. The depth of the channel increases above 18 inches at the sill **8** at increasing distances aft so as to maintain an approximate area for the flow of water through the channel of about 60 square feet.

The channel has a depth of 6 ft 6 inches at station **S30,** this depth being maintained constant as far aft as station **S90,** whereafter the depth diminishes to 6 ft at station **S124.** The 18 inch depth of sill at the bow is maintained for 5 ft on both sides of the center line station **S8** and then increases uniformly on both sides to a maximum depth of 2 ft at the full width. It is of course to be understood that the foregoing dimensions are merely given by way of example and can be varied to suit particular design requirements. Water to be cleansed is caused to flow into the channel through the open bow flap means **6** and is treated whilst flowing through the channel to the open stern flap means, by means for removing and recovering contaminants. Basically the water may be caused to flow into the channel by towing the ship bodily through the water.

FIGURE 187: SPANNER OIL SKIMMING SHIP DESIGN

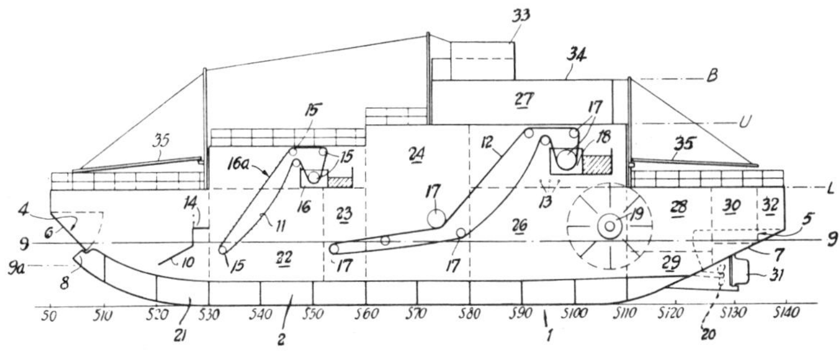

Source: W.F. Spanner; U.S. Patent 3,744,257; July 10, 1973

Additionally or alternatively the ship may be provided with means in the channel for causing water to flow therethrough. Indeed the ship may be self-propelled to cause water to flow through the channel by ship movement, with or without the provision of flow causing means in the channel. The ship illustrated is self-propelled and is provided with flow causing means in the channel.

The means for removing and recovering contaminants from water flowing through the channel includes a grill **10**, a first conveyor belt **11** and a second conveyor belt **12.** The grill is in the form of a wire coarse mesh screen, preferably with 3 inch mesh size, portably mounted at stations **S20** to **S27** in the channel near the bow opening and extending well below the operating waterline. A platform or walkway **14** extends transversely of the hull above the grill so that a seaman positioned thereon can remove large contaminants such as oil soaked seaweed from the grill. A bin (not shown) can be provided for receiving such seaweed removed from the grill.

Aft of the grill in the channel is arranged the first conveyor belt which is arranged to project at its forward end into water flowing through the channel. The belt is preferably in the form of netting made of synthetic plastic material such as nylon of ¾ inch mesh mounted on cross rods secured to roller chains which are preferably galvanized. The chains and belt pass over sprockets and rollers such as **15** through a first residue tank **16** containing solvent and/or washing medium such as hot water for removing tarry/oily lumps or masses over ¾ inch in diameter from the belt. The belt is provided with projections **16a** to facilitate the removal of tarry/oily lumps of between 3 and ¾ inch diameter from the water in the channel. The belt and tank **16** are portable for easy removal and cleansing, and drive means for the belt is provided conveniently in the form of a 7½ hp 230 v DC motor. A suitable operating speed for the belt is 1 fps.

Aft of the first conveyor belt in the channel is arranged a second conveyor belt **12** which also is arranged to project at its forward end into water in the channel. The belt **12** is in the form of preferably galvanized chains connected by cross supports carrying trays or rods which in turn carry sponges made of plastic material. The chains and belt **12** pass over sprockets and rollers such as **17** through a second residue tank **18** containing solvent and/or washing medium such as hot water for removing lighter/heavier fractions of crude petroleum picked up from the water by the sponges.

The oil initially sticks to the sponges whilst the water drains away and the oil is removed from the sponges in the tank **18.** The belt **12** and accompanying apparatus is portable for easy removal and cleaning and the belt is driven by electric drive means such as a 30 hp 230 v DC motor through couplings and gearing at a suitable speed of, for example, 4 fps. If desired, more than one such belt **12** may be provided. Aft of the belt or belts **12** are arranged detergent spray pipes **13** operative to spray detergent on the water surface in the channel to break up any residual oil slicks. These pipes may be accompanied by net containers of absorbent pads (not shown) which are suspended into the water to recover oil traces or slicks.

Finally, aft of the belt or belts **12** and pipes **13** in the channel is arranged water flow causing means in the form of a paddle wheel **19.** This paddle wheel is operative to cause or assist water to flow through the channel and to break up completely and disperse any residual oil slicks. The paddle wheel may also assist in the propulsion of the ship which in the main is effected by twin screws **20.**

Ballast and trimming tanks such as forward port and starboard trimming tanks **21** at stations **S20** to **S30** are provided for varying the draft of the ship so that in the water surface cleansing condition the ship floats at the operating waterline with flap means **6** and **7** open. In the ballast condition when proceeding to an area of operation, the flap means are secured watertight and the ship floats at a light waterline **9a**. In this latter condition the speed of the ship is appreciably greater than in the water surface cleansing condition. In operation of the ship for water surface cleansing, the ship is trimmed to a suitable draft to permit an adequate depth of water to be cleansed, including the water surface region, to enter the channel through the open flap means. The paddle wheel and/or screws **20** are operated to draw dirty water through the bow opening firstly through the grill which traps oil soaked masses of seaweed or other contaminants over 3 inches in diameter for removal from the grill by hand, then through the belt **11** which removes oily/tarry residues between 3 and ¾ inches in diameter from the water and carries them into the tank **16** where they are recovered and saved. Any remaining lighter fractions of oil flow through the netting belt **11** to the belt **12** where the sponges pick up light fractions of oil and carry them into the tank **18** where they are recovered.

Any slight traces of oil passing the belt **12** are dispersed by the detergent from the spray pipes **13** and the paddle wheel if the nets containing absorbent pads are not employed or do not absorb all remaining oil traces. The cleansed water then exits from the channel through the stern opening. Above the level of the double bottoms **2** and within the side compartments formed in the hull are accommodated, the two forward trimming tanks **21** at stations **S20** to **S30.** Over these tanks **21** are port and starboard chain lockers at stations **S20** to **S26,** and two stores compartments at stations **S26** to **S30.**

Also within the side compartments in the hull above the level of the double bottoms are a main generator room **22** at stations **S30** to **S52** on the starboard side and a standby generator room and workshop on the port side both extending to the level **L** of the lower deck, there being two pump rooms **23** (port and starboard) at stations **S52** to **S60.** The crews' accommodation extends port and starboard from station **S30** to station **S60** above the level of the lower deck, the two groups of cabins being connected by a passageway over the channel.

Two main oil/ballast/oily bilge/waste oil tanks **24** are constructed on port and one starboard with arrangements for separating oil from water and of capacity 170 tons SW each, 340 tons total port and starboard. Deck connections are to be provided one port and one starboard to receive oil waste from ships discharging contaminated bilge water. A coffer dam is provided for the full depth of the main oil tanks **24** port and starboard at stations **S78** to **S80.** Two main engine rooms **26** extend to the level of the lower deck port and starboard between stations **S80** and **S110.**

Above the level of the lower deck and extending to the level **U** of the upper deck are on the starboard side an auxiliary boiler room and laundry between stations **S80** to **S95** and a dredging pump room between stations **S99** to **S110.** On the port side is a main switchboard and auxiliary generator room between stations **S80** and **S110.** Above the level of the upper deck and across the full width of the ship is the galley, cook's, captain's and officers' accommodation **27** between stations **S72½** and **S107½.**

Below the level of the lower deck between stations **S110** and **S126** are a paddle wheel engine room **28** on the starboard side and the No. 2 dredging pump room on the port side. Beneath these two compartments are the stern tube compartments **29** port and starboard. Two steering engine compartments **30** lie between stations **S126** and **S135,** with stern frames and stiffening below to carry two rudders **31,** one port and one starboard. Tanks **32** are arranged, one port and one starboard between stations **S135** and **S140** for detergent.

Portable plates securely bolted are fitted as necessary to enable all machinery to be removed and serviced. Manholes are fitted to enable all closed compartments to be entered and watertight hatches and doors are fitted to enable stores and provisions to be shipped, and to provide access to compartments where officers and crew live and work, or which may require inspection during operations. Generally, the hull is constructed to conform with the requirements of Lloyd's Register of Shipping.

Three diesel engines are provided each of 300 brake horsepower at 1,500 rpm, to drive through suitable MWD reverse/reduction gearing in the case of the two screws **20,** and suitable MWD reduction gearing in the case of the paddle wheel. The diesel engines for the two screws, one port and one starboard are fitted in the main engine rooms **26** at stations **S80** to **S110** and are connected, through MWD reverse/reduction oil operated gear boxes to the screw propeller shafts by suitable flexible couplings. The propeller shafts are provided with thrust blocks and are mounted in plummer blocks and at the after end of the stern tubes Tufnol or similar lined bearings. The diesel engine for the paddle wheel is fitted in the paddle engine room **28** at stations **S110** to **S126** and connected, through a MWD oil operated reduction gear box, to the paddle wheel through a suitable flexible coupling and bevel reduction gear to give a suitable paddle wheel operating speed. Electrical generating machinery, switchboard, steering engines and pumps are installed in the designated compartments. All machinery is adequately protected and ventilated and generally conforms with the requirements of Lloyd's Register of Shipping.

The ship may be provided with ancilliary equipment to enable dredging operations to be carried out in rock, shingle, sand or other types of seabed and/or river bed, salvage operations, harbor tender duties, oil/gas drilling rig servicing and fire extinguishing and fire-fighting. Furthermore a well may be constructed in the channel and suitably positioned to enable divers, dredging equipment, etc. to be lowered from the interior of the ship.

Boats and life saving equipment are provided together with radio and navigation equipment. A chart house **33** and radio office are provided on the bridge deck **B** and a helicopter platform **34** is arranged on the bridge deck aft of the chart house. Derricks **35** are also provided. An example of dimensions, weights, etc. of such a ship is as follows.

Length overall	140 feet.
Breadth (moulded)	36 feet.
Breadth extreme	46 feet.
Depth to lower deck	22 feet.
Depth to upper deck	34 feet.
Draft in oil recovery/surface cleansing operating condition	12 feet.
Draft in ballast	8 feet.
Speed	8/10 knots.
Horsepower	600 total on two screws. 300 on stern paddle wheel when an oil recovery/surface cleansing operations
Electric Supply	1-150 K.W. 230 v.D.C. main Generator. 1-75 K.W. 230 v.D.C. Standby Generator. 1-30 K.W. 110 V.D.C. Auxiliary Generator.
Boiler	1-Auxiliary boiler for heating purposes
Capacity	500 tons for recovered petroleum products.
Lightship Tonnage	600 tons.

Surface Separator Systems, Inc. Design

A design developed by *M.M. Earle; U.S. Patent 3,259,245; July 5, 1966; assigned to Surface Separator Systems, Inc.* has been devised to take full advantage of the adhesion, cohesion, viscosity, wettability and surface tension characteristics of the various oils and their natural repulsion of water. The principle of operation is similar to that of the offset printing press wherein greasy portions of a cylinder may be utilized to repel water and attract ink.

It is through the unique series arrangement of multiple collecting and pickup surfaces, however, that one may concentrate, recover, separate and pump the contaminant oil film in one efficient, continuous, and combined operation. In this system, the components may be mounted for stationary recovery such as in a retention basin or for moving recovery such as from a barge or other floating craft.

Here, there is contemplated at least one collecting and concentrating surface, such as a cylinder, a portion of which is immersed in the contaminated water having an oil film, whereupon by motion of the surface, in this case the rotation of the cylinder, the thin oil film is transferred to an enclosed area which confines a comparatively thick oil film therein. To facilitate the action of the collecting and concentrating surfaces, mechanical wiping means are generally applied thereby positively depositing the collected oil from the cylinder into the enclosed area. Moving at a synchronous speed is a recovery and separation surface which has been adapted to immersion in the enclosed area for collecting therefrom the thick oil film and depositing it by a wiping means into a trough, tank, or other collecting facility or storage container.

Technocean Inc. Design

As described in EPA, Edison, N.J. publication, *Oil Skimming Devices,* Report PB 218,504, Springfield, Virginia, National Technical Information Service (May 1970), the Technocean Co. of Paris, France has proposed a pollution control ship, which is to be rather large (17,000 tons) and of a hybrid design. Forward, it will have a conventional single hull while the after half has a catamaran twin hull.

The ship will move backward and skim the surface of the ocean taking in the variable thickness oil layer. A suction valve, 32 ft wide and of adjustable height, will be located at the forward end of the slot between the twin hulls. This positioning of the suction inlet is intended to overcome problems with use in open waters. The twin hull creates a sort of artificial harbor where waves and swell are claimed to be dampened significantly.

According to specifications, the ship will be able to pump 2,640,000 gal/hr, separate the oil, store it, either in its own tanks or in floating inflatable tanks towed nearby, and pump clean water to the ocean to achieve good maneuverability when moving astern. The onboard storage capacity is 50,000 barrels.

Each catamaran hull will have a steerable shrouded propeller. This pollution control ship will have a main screw for normal forward motion immediately below the suction inlet. This screw will be useful in pollution control work as it will pump the mass of water pushed by the backward motion of the ship.

Tetradyne Corporation Design

A device developed by *F.N. Mueller; U.S. Patent 3,659,713; May 2, 1972; assigned to Tetradyne Corporation* is one in which impurities are removed from the surface of a liquid by impinging an elongated fluid stream, preferably a gaseous stream, along a length of the surface and moving the impinging stream relative to the surface causing the impurities to move and concentrate in advance of the impinging action of the stream, and thereafter isolating the concentrated mass of impurities.

An apparatus is also provided suitable for traversing a body of water and recovering impurities, such as oil and the like from its surface. A preferred apparatus includes a vessel having a pair of elongated booms which are adapted to extend forwardly and outwardly of the motion of the vessel. Each boom extends above the water and carries nozzles which are adapted to form a substantially continuous elongated gaseous stream for impinging upon the water between the booms to thereby concentrate floating impurities between the booms. In addition, the vessel is provided with impurity inlet means between the booms for collecting the concentrated impurities from the water.

Figure 188 shows a broken-away elevation of this skimmer vehicle in action as well as a reduced plan view showing the arrangement of the various components. Vessel **10** is equipped with a propulsion means diagrammatically indicated at **11** as a pair of propellers which have suitable actuation means, such as a diesel engine or the like (not shown). A rudder is also shown at **11a**. Alternatively vessel **10** can be moved through the water by a separate prime mover.

For example, the vessel can be pushed through the water by a barge or tugboat or the like. The vessel generally comprises a hollow body including a separation tank **21** having a perforated bottom and floating bands **12,** containing a suitable buoyant substance. For example, floating band **12** can contain a series of air compartments. Cabin **13** is positioned operatively on the vessel and includes suitable control means for the vessel.

FIGURE 188: TETRADYNE CORPORATION SKIMMER VESSEL DESIGN USING AIR SWEEPING OF SLICK

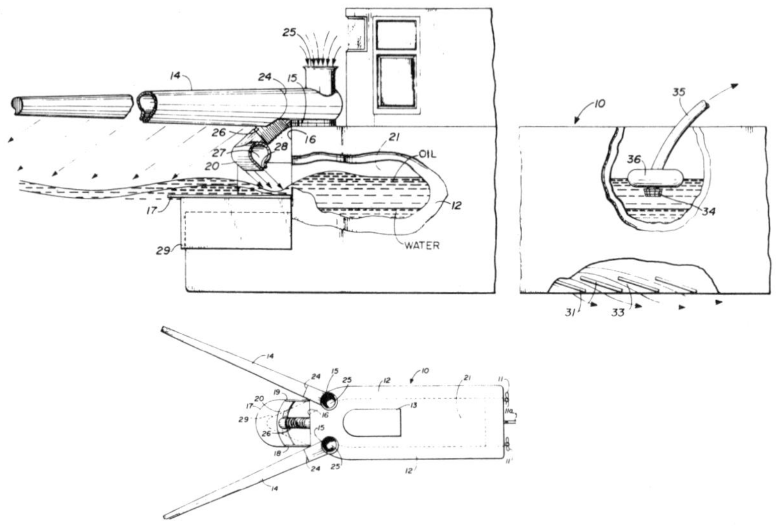

Source: F.N. Mueller; U.S. Patent 3,659,713; May 2, 1972

Booms **14** are mounted on the bow of the vessel, preferably in pivotal relationship at points **15** respectively, so that each boom can be rotated through about 90°. The booms are positioned to ride above the surface of the water, and are hollow and carry nozzle means for producing an elongated gaseous stream or curtain for impingement against the surface of the water. Suitable gas delivery means are operatively connected to the nozzles.

Impurity inlet **16** is operatively positioned through the bow of the vessel between the booms. Weir **17** also connects to the bow of the vessel and projects outwardly therefrom substantially parallel to the surface of the water and forms the bottom portion of the impurity inlet. The weir generally comprises a substantially flat member which preferably is movably mounted in a vertical plane as will be described below to control the opening of the impurity inlet. Baffle plates **18** and **19** extend from the sides of the weir, as shown in the figure. Curved nozzle bar **20** is suspended over the weir and carries back pressure nozzle means on the lower portion thereof to thereby exert pressure upon the impurity-carrying liquid on the weir in a manner described below.

As shown, each boom is pivotally mounted to the bow of the vessel so that it is suspended over the surface of the water but is free to rotate approximately 90° about pivot point **15.** Each boom generally comprises a hollow member. This configuration provides a generally widened gaseous flow channel which communicates with a nozzle slot along the length of the boom. Thus, the outlet of the nozzle slot is positioned to direct a gaseous stream to impinge against the surface of the water.

It is also preferred that each boom be rotatably mounted about its axis at point **24** in a manner so that the angle of nozzle slot in relation to the surface of the water can be varied as desired. It is noted that the nozzle slots can effectively operate at angles from about 0 to 90° relative to the surface of the water. However, it is preferred that the nozzles be positioned at an angle of from 45 to 75° relative to the surface of the water.

Suitable actuation means such as an electric motor, hydraulic or handcrank means or the like, can be operatively connected to booms **14** to provide for pivotal movement about points **15** and rotational movement about points **24.** A series of closely spaced apertures can be utilized on each boom instead of nozzle slots. The booms are operatively connected with a suitable gas supply means for supplying a constant flow of a suitable gas to flow channels **22**. In the figure the booms are operatively connected to the outlet of conventional air compressor means (not shown), and inlet air for the air compressor means is provided via inlet ducts **25.**

Nozzle bar **20** is suspended over the rear portion of the weir by a suitable gas supply duct **26**, which also communicates with the outlet of the compressor means which supplies air to flow channels **22** of the booms. The cross-sectional configuration of nozzle bar **20** is substantially the same as the cross-sectional configuration of the booms, and therefore provides for a flow channel **27** communicating with an elongated nozzle slot. If desired, the nozzle bar can be rotatably mounted over the weir to provide different angles of impingement of the gaseous stream emitted from the elongated nozzle slot upon the liquid on the weir according to the particular conditions. It is generally preferred that nozzle slot **28** be positioned at an angle within the range of from about 30° to 60° to the weir.

The weir is carried by frame **29** which includes a v-shaped member which fits in movable relationship over the lower portion of the bow of the vessel. The frame is movable in a vertical plane in relation to the bow of the vessel. Thus, the frame is operatively connected to a suitable actuation means such as an electric motor, hydraulic or handcrank means for adjustably positioning the weir a suitable distance below the surface of the water.

As previously described, the vessel comprises an inner separation tank **21** having a perforated bottom and partially encompassed by floating bands **12.** It is possible to eliminate the perforated bottom of the separation tank to yield a substantially open bottom. However, such a configuration is greatly preferred in that it will prevent turbulence and undue admixing between impurities such as oil, and water within the interior of the separation tank as the vessel moves across the surface of the water.

Thus, it has been found that the positioning of baffles **31** in the bottom of the separation tank, as illustrated, prevents water from flowing upwardly into the separation tank as the vessel passes thereover. The baffles should be angled in the manner illustrated in the figure, to deflect and ride over the water as the vessel passes thereover and allow water from within the separation tank to pass out between openings **33.** It is preferred that the baffles form an angle of from 1 to about 60° with the surface of the water.

In operation, the vessel will first move to the edge of a contaminated area on the water, for example, containing an oil film up to about $\frac{1}{16}$ inch in thickness. The booms can be adjusted to diverge at a desired angle, for example, about 45°. Next, the air compressor means is actuated to deliver a stream of air to flow channels **22** within the booms to thereby cause a substantially continuous flow of air from the nozzle slots along the length of each boom. The resulting flow of air will impinge upon the surface of the water at a desired angle, for example, about 45°, and will be maintained at a velocity of from about 20 to 100 fps. The action of the streams emitted from the nozzle slots will cause the oil floating on the surface of the water to move in advance of the impinging action thereof against the water and thereby concentrate in the area between the extended booms.

In the meantime, the vessel will be moving in the direction of the booms. This action will cause the concentrated oil entrapped between the two curtains of air emitted from the booms to pass over the weir, between baffle plates **18** and **19** and into impurity inlet **16**. During this time, a continuous flow of air is emitted from the elongated nozzle slot of nozzle bar **20**. This flow of air can be from about 20 to 100 fps and will be at an angle generally in the direction of the impurity inlet. This impinging action of the curtain of air emitted from the elongated nozzle slot against the floating impurities passing over the weir is shown in the figure. This action will keep the flow of the floating concentrated impurities over the weir and thereby prevent the backflow of any oil and/or water from the separation tank via the impurity inlet.

As more oil is collected within the separation tank of the vessel it will begin to displace water therefrom via openings **33**. In addition, as the vessel moves across the body of water, baffles **31** will deflect water from the body and thereby prevent its entrance into the interior of the separation tank. This action will prevent any turbulence from forming within the interior of the separation tank by uprising water and consequently prevent intermixing and possible emulsification of the oil therein. In addition, the action of the water deflecting against the baffles will enhance the flow of water from the interior of the separation tank via openings **33**.

Each baffle can be rotatably mounted about its axis, for example, so that it can rotate 90° from a substantially flat position with relation to the water to a substantially perpendicular position. In addition, the baffles can be made to interlock when they are in the substantially flat position. In this latter instance, they can be interlocked when the interior of the separation tank is substantially filled with oil and the vessel is returning to its port or dock.

If desired, the oil captured within the separation tank can be removed therefrom by suitable pumping means and stored in a barge, or if desired, deposited within large rubber bags, sealed, and left floating on the water for later recovery by a suitable vessel. In one form, the outlet **34** of hose **35** is operatively connected to float **36** so that it is constantly positioned on the surface of the oil within the separation tank. The hose is connected to a pumping means which functions to pump out the oil within the separation tank and deposit it in any suitable receptacle, such as those described above.

It is noted that it is possible to provide means for varying the height of the booms above the surface of the water. This can be done with various alternative methods including lifting means separated to each boom, for example, hydraulic adjusting means or by controlling ballast within the floating bands. Thus, pumping means can be provided for pumping water or any other fluid into the interior of the floating bands to thereby control the depth to which the vessel is carried within the water. This will provide means for adjusting the booms above the surface of the water, if they are in a vertically fixed position on the bow thereof.

The vessel is equipped with suitable means for compensating for various types and heights of waves on the water. As previously described, frame **29** is movably mounted in a vertical plane. This action will not only adjust the relative depth of the weir with relation to the surface of the water, but will also regulate the size and location of the impurity inlet relative thereto. Thus, frame **29** can be adjusted so that the weir rides below the trough of the largest waves on any particular day.

A device of somewhat similar design is described under the name of Air Boom by EPA personnel in EPA, Edison, N.J. publication *Oil Skimming Devices,* Report PB 218,504, Springfield, Virginia, National Technical Information Service (May 1970). That report states that the air broom takes advantage of the known, but previously unused physical fact that the movement of oil on a body of water is substantially influenced by the direction of the air current flowing above it. This device proposes to sweep oil using a high volume stream of air or gas in order to protect critical areas from the invasion of floating oil or to collect the oil into a less dispersed mass to facilitate collection and disposal thereof.

The source of the air or gas could be any of several high volume gas moving devices, such as blowers or fans. The exhaust gas discharge from a jet engine similar to those used for aircraft and other motive power application appears to be ideally suited for this application.

The air broom could be used in any of several applications where it may be desirable to move floating oil: (1) mounted on a vessel, the air broom could be shuttled back and forth along a side of an oil slick progressively sweeping the oil into a smaller area with each pass of the air broom; (2) mounted on a vessel, the air broom could be used in conjunction with a movable floating boom to move or convey the oil slick from a critical water area to a more suitable and less critical area for collection; and (3) the air broom could be mounted on shore and used to move oil carried by the current, such as oil floating on a river, to one side of the channel and into containment booms from which it could be collected and disposed.

TRW Inc. Design

A design developed by *P.G. Bhuta, R.L. Johnson and D.J. Graham; U.S. Patent 3,703,463; November 21, 1972; assigned to TRW Inc.* utilizes a hollow liquid surface tension separator having a surface tension screen wall. The separator is filled with the selected liquid and the outer side of its surface tension screen is placed in contact with the body of immiscible liquids to be separated, such that each screen pore exposed to the second liquid of the body contains a liquid-liquid interface whose interfacial surface tension resists passage of the second liquid through the pore. A pressure differential, less than the critical pressure differential necessary to overcome the interfacial surface tension force acting across the pore, is established across the screen to drive the selected liquid only through the screen into the separator.

Figure 189 shows a schematic view of the overall apparatus plus details of the construction and operation of the individual separator chambers **18.** As shown and described above, the apparatus is one for separating a selected liquid **12** from a liquid body **14** containing a second liquid **16** which is immersible. In this instance, the selected liquid is oil, the second liquid is water on the surface of which the oil floats.

As regards the details of operation of the individual separator chambers, each separator contains a chamber **20** bounded at one side, in this instance at its lower side, by a wall **22** comprising a surface tension screen. When the separator is conditioned for operation, its chamber is filled with the selected liquid to be recovered, i.e., oil, such that the oil wets the screen. The screen **22** is constructed as a material which is preferentially wetted by the selected liquid to be recovered. In this regard, it is significant to note that the liquid which is selected for recovery by filling the separator chamber is preferably the one which preferentially wets the separator screen. In the operation of the surface tension liquid separator, the outer side of its screen is placed in contact with the body of water, as shown in the lower left hand detail.

FIGURE 189: TRW SKIMMER SYSTEM FOR FLOATING VEHICLE

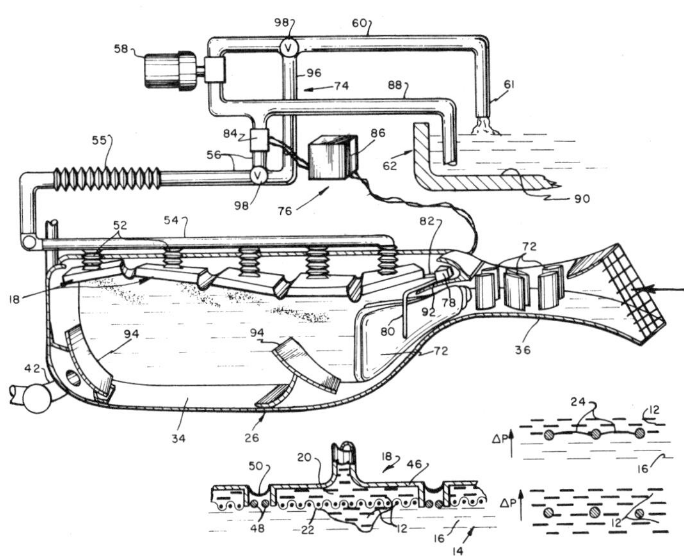

Source: P.G. Bhuta, R.L. Johnson and D.J. Graham; U.S. Patent 3,703,463; November 21, 1972

Under these conditions, each screen pore exposed externally to water contains a liquid or oil-water interface **24** (see upper right hand detail) which separates the oil on the inside of the screen from the water on the outside of the screen. Each interface possesses an interfacial surface tension which resists passage of the water through the pore into the separator chamber.

This interfacial surface tension blocks passage of the water through the pore under the action of a differential pressure ΔP (water pressure greater than oil pressure in the separator chamber) across the screen which is less than the critical pressure differential required to overcome the interfacial surface tension, i.e., required to drive the water through the pore against the interfacial surface tension of the corresponding oil-water interface **24.** This critical pressure differential is determined by the interfacial surface tension of the oil and water and by the characteristics, i.e., pore size, material, etc., of the surface tension screen. The critical pressure differential ΔP_c is represented by the equation:

$$\Delta P_c = p_w - p_o = K\frac{\sigma_{wo}}{r}$$

where p_w is the total water pressure outside of the screen; p_o is the oil pressure just inside the screen; σ_{wo} is the interfacial surface tension between water and oil; r is the radius of the oil-water interface, i.e., the radius of the screen pores; and K is a constant of proportionality having a value of approximately 2.0, depending on the type of screen.

In contrast, the screen pores which are exposed to oil on the outside of the surface tension screen contain no liquid interface, as may be observed in the lower right hand detail. Accordingly, passage of the oil through the screen into the separator chamber is not resisted by interfacial surface tension. The pressure differential maintained across the screen by the pressure regulating system thus drives the oil through the screen into the separator chamber.

From the foregoing discussion, it will be understood that if a pressure differential ΔP less than the critical pressure differential ΔP_c is established across the surface tension screen of the oil-water separator, only oil in contact with the outer surface of the screen will pass through the screen into the separator chamber. Passage of water through the screen is blocked by the interfacial surface tension of the oil-water interfaces.

The oil recovery apparatus comprises a floating oil recovery vessel having a relatively large stilling basin **26** supported by float means such that the basin floats at the surface of the water from which the surface oil is to be recovered. In this instance, the float means comprise pontoons or catamaran hulls which straddle the stilling basin and are attached to the basin by struts. Contained within the stilling basin is a relatively large stilling chamber **34.**

The forward end of the basin is necked down to form an inlet diffuser throat **36.** The diffuser passage within the throat opens rearwardly to the stilling chamber and forwardly to a surface skimmer inlet. This inlet projects a distance above and below the normal water line level of the stilling basin. A grill is placed across the inlet to block the entrance of sizable floating objects. A pair of surface skimming booms extend forwardly from the inlet in diverging relation to one another. At the rear end of the stilling basin is an outlet **42** connected to the inlet of a pump.

Within the stilling chamber is an array of oil-water separators **18.** Each separator has a flat rectangular housing **46** with an open bottom side across which extends the separator surface tension screen. Encircling each housing is a ring float **48** which provides the separator with positive buoyancy in water.

The separators are arranged side-by-side in rows extending lengthwise and crosswise of the stilling basin to form a rectangular separator array which encompasses virtually the entire water surface area in the stilling chamber. The adjacent separators are joined by hinge means **50,** in this instance fabric or plastic webs, which permit the separators to rise, fall, and rotate relative to one another under the action of wave motion in the stilling chamber.

Connected to the center of the upper wall of each separator housing is an axially and laterally flexible riser **52.** The risers of the several separators in each axial separator row connect to a header **54.** The rear ends of the headers connect to a common flexible outlet conduit **55**, connected to suction line **56** of a pump **58.** The discharge line **60** of the pump connects to a spout **61** which empties into an oil receiver **62.** The discharge of the basin suction pump empties overboard.

As the oil recovery vessel moves forwardly through the water, the skimmer booms skim the water surface and convey any oil on the surface toward and finally into the skimmer inlet. The surface water and oil entering the inlet pass through the inlet diffuser throat into the stilling chamber. The inlet throat passage flares outwardly as it enters the chamber to diffuse the entering, relatively high velocity water-oil stream to a relatively low velocity stream within the stilling chamber. Flow control and diffuser vanes **72** are mounted in the passage to aid this diffusing action and to damp wave motion. Water exits from the chamber through the chamber outlet **42** and the pump discharges the water overboard.

The oil-water separators extract oil from the water within the stilling chamber. In this regard, it will be understood that the separators are initially filled with oil to condition them for their earlier described surface tension oil-water separating or screening action. The separators are thus filled with oil through a filling system **74.** Actual oil extraction operation of the separators is initiated by operating the pump to reduce the pressure within the separator chambers and thereby produce across the separator screens the pressure differential required to induce oil passage through the screen into the separator by the surface tension oil-water separating screening action described previously.

In order to induce passage of oil only, and not water, into the separators, the pressure differential across the separator screens must be maintained less than the critical pressure differential ΔP_c. To this end, the oil recovery vessel is equipped with an automatic pressure control system **76** for continuously monitoring the pressure differential across the separator screens and regulating the separator pumping action to maintain the proper pressure differential across the screens. The particular control system illustrated comprises a differential pressure transducer **78** mounted within the stilling chamber.

This transducer is exposed, through a tube **80,** to the water pressure on the outside of the separator screens, and through a flexible tube **82,** to the oil pressure on the inside of the screen of an adjacent separator **18.** The lower end of pipe **80** is submerged to a depth sufficient to avoid uncovering of this end by wave motion. Electrically connected between the transducer and a solenoid proportioning valve **84** in the pump suction line **56** is a valve control unit **86.** Leading from this suction line, at a point between the valve **84** and the pump, is an intake line **88** which opens to the bottom of the oil collection chamber **90** in the barge **62.**

The transducer senses the water pressure on the outside of the separator screens and the oil pressure on the inside of the screen of the adjacent separator and generates an electrical signal representing the difference between these pressures. The valve control unit operates the proportioning valve in response to this electrical signal to proportion flow through the lines **56, 88** in such a way as to maintain the screen pressure differential at the proper level to accomplish the oil-water separating action. Thus, the control unit responds to a progressively increasing pressure differential across the screens by progressively closing the proportioning valve.

Under these conditions, oil flow from the separators is progressively reduced and the oil flow through that part of the oil recirculating loop comprising collection chamber **90,** and that part of the intake line before the point where line **56** joins **88**, is correspondingly increased. The internal pressure in the separators is thereby reduced. Conversely, the control unit **88** responds to a progressively diminishing pressure differential across the separator screens by progressively opening the proportioning valve to increase the oil flow from the separators and reduce the circulating oil flow through that part of the oil circulating loop comprised of **90** and that part of **88** before the point where line **56** joins **88.**

The internal pressure in the separators is thereby increased. The control system is adjusted to maintain the selected pressure differential across the separator screens. In this regard, it is significant to note that the control system maintains the proper pressure differential regardless of the presence or absence of oil in contact with the outer surface of the separator screens.

From this description, it will be understood that the oil recovery apparatus or vessel is effective to extract from the water in the stilling chamber any oil which enters into contact with the separator screens and to collect the oil in the barge collection chamber **90.** The only remaining requirement for efficient operation of the oil recovery vessel, therefore, involves contacting the incoming oil with the separator screens. This is accomplished by an inlet deflector **92** and bottom deflectors **94** within the stilling chamber. The inlet deflector deflects the entering water and oil downwardly to the bottom of the chamber, such that the oil then rises into contact with the underside of the separator screens.

This upward motion of the oil into contact with the screens is aided by the bottom deflectors and by the diffusing action of the diffuser inlet throat **36** and vanes **72** which slow the entering oil-water stream to increase the residence time of the oil in the chamber. As previously noted, water exits from the stilling chamber through the outlet **42** and the pump which discharge the water overboard.

As previously mentioned, the oil recovery vessel has means **74** for initially filling the separators with oil, thus to condition the same for their oil-water separating action. The particular filling means shown comprises a fill line **96** connecting the suction and discharge lines **56, 60** of the pump **58**, and two-way valves **98** which may be set in normal operating position and separator filling position. In their normal operating position, these valves block the ends of the fill line and connect the pump intake to both the separators and the intake line **88** and the pump outlet to the spout **61**. The apparatus is then conditioned for the described oil recovery operation. In filling position, valves **98** connect the pump intake to the intake line **88** only and the pump outlet to the separators only, through the fill line. Under these conditions, operation of the pump draws oil from the barge through the intake line and delivers the oil to the separator through the discharge line **60**, fill line, and separator headers **54**.

Underwater Storage Inc. Design

A design developed by *H.G. Quase; U.S. Patent 3,656,623; April 18, 1972; assigned to Underwater Storage, Inc.* is one in which a partially submergible platform has internal separation tanks, which are provided with screens for removing solid materials from water. Skimmers remove a relatively light immiscible liquid before returning the water to the body from which it is taken.

In one form of the device, contaminated water is pumped upward into separation tanks, and gravity drains the separated water from the tank and vessel. In another example, contaminated water flows into separation tanks, and purified water is pumped from the tanks, preferably rearwardly, for propelling the craft through the water. Gates form lower edges of intakes, so that the amount of water taken into the boat and the depth of the intake may be controlled.

Valdespino Design

A design developed by *J.M. Valdespino; U.S. Patent 3,615,017; October 26, 1971* is an oil slick entrapment and containment watercraft which has a pair of pontoons buoyantly supporting an open bottomed entrapment tower and a funnel shaped surface skimming shroud. The open bottom of the entrapment tower is below the water level, and, as the craft moves on the water having an oil slick, the oil passes into the skimmer and into the entrapment tower where the column of liquid is raised by vacuum applied to the top of the entrapment tower. Oil rises to the top of the water in the entrapment tower due to the difference in specific gravity and without emulsifying and is then pumped off the top.

A single pump carried by the craft is connected by suitable piping arrangements and provides multiple functions including: drawing vacuum in the entrapment tower by aspirating air from the entrapment tower through an aspirator, supplying air to the pontoons, and pumping the oil from the entrapment tower. The buoyancy of the pontoons is controlled by supplying air and water to the top of the pontoons allowing the water to escape out of a slot in the bottom of the pontoons and providing air purge lines in the pontoons.

Verdin Design

A design developed by *S.M. Verdin; U.S. Patent 3,754,653; August 28, 1973* is one in which an oil spill on the surface of the sea is collected by means of a funnel-like scoop which is moved horizontally through the water in a direction such that oil and seawater, including waves, flow into the open end of the scoop. An oil-rich mixture flows over a weir located at the apex end of the scoop and the water is returned to the sea after separation of the oil. The trim of the scoop is adjusted by ballast tanks.

Walton Design

A design developed by *J.F. Walton; U.S. Patent 3,656,624; April 18, 1972* provides a waste (oil and scum) collecting vessel including an impeller assembly comprising a cylindrical support and a plurality of flexible, circumferentially supported blades extending longitudinally of and radially extending from the support, the diameter of the support being greater than the height of each blade. Preferably, the blades are individually mounted and the vessel includes a lip member for folding the blades when the support is rotated in one direction, a skimmer for engaging the blades when the support is rotated in the other direction, and a system for removing materials from the bottom of a deep well of a waste collection tank and for transporting waste from within the tank to storage tanks within support pontoons.

Waren Design

A design developed by *F.A.O. Waren; U.S. Patent 3,682,316; August 8, 1972* is an apparatus for collecting buoyant foreign matter (particularly oil) floating on the surface of water comprising a tank adapted to float on water, an outlet at the bottom of the tank, an inlet track positioned to draw in water at or near the surface and means for pumping water together with the foreign matter through the inlet track into the tank. The foreign matter rises and is trapped in the tank, while the water passes out through the outlet. The bottom of the tank is preferably open. The pumping means may comprise one or more impellers located in the inlet track driven by a motor carried on the tank.

Waterwisser Design

A commercial design of an oil skimming vehicle known as the Waterwisser or water-wiper has been developed by Shell Oil Company engineers and is produced by NV Bonn en Montage Bedrijf Zwyndrecht, Netherlands and Holland Launch NV Zaandam, Netherlands, according to EPA Edison, N.J. publication *Oil Skimming Devices,* Report PB 218,504, Springfield, Virginia, National Technical Information Service (May 1970).

It has been used in the docks of Rotterdam. It is equipped with a 50 hp diesel engine, is 46.9 ft long, 13 ft broad, with a draft of 4.9 ft. When moving through the water at 1.8 miles/hour, the barge skims a strip of water 65 ft wide with rigid booms hinged at the rear end of the vessel. In front of the hinge on each side and reaching 15.7 inches below the surface, there is an inlet slot or weir through which the oil and water collected by the booms is sucked into the hold.

At this point where the boom is attached to the barge, the oil and water mixture enters the oil separation chamber via a vortex pipe. The submerged mouth of this pipe forms an open connection between the water inside and the water outside the barge. After entering, the liquid flows through the hold, and the oil separates to the surface. The clean water beneath this layer is then pumped overboard.

Total oil capacity of the hold is 20 tons. The clean water pump, which is also used to remove the collected oil, has a capacity of 100 tons/hour. Driftwood and other floating objects are arrested by a grating and then dumped at a suitable location. The barge is easy to navigate and when it moves backwards, the booms fold against the sides. They open again into a v-position when the vessel moves forward. Movement of the barge to the operating location is made with booms closed.

An operator can stand at the end of the booms and clean the embankments and harbor corners with a water jet and subsequently drive the oil between the booms. The embankments can usually be approached closely by the end of the booms since their draft is less than 10 inches. The reader is referred back to the discussion of the Shell Oil Company design (p.350) for the patent reference on such a device.

Worthington Corporation Design

A design developed by *A.R. Budris, F.J. McGowan, L.M. Evans, T.J. Wayne, E.E. Lithen and C.B. Darcy; U.S. Patent 3,646,901; March 7, 1972; assigned to Worthington Corporation* comprises a body member of the catamaran-type having spaced apart twin hulls between which is disposed an oil recovery system which can recover oil, for instance, in the form of an oil slick, which has been properly directed by the hulls of the craft.

The watercraft has a generally centrally located turning axis; first and second propulsor means secured to the body member for accepting water at an inlet opening thereof and discharging water at a higher velocity at discharge openings thereof; first thrust generating means in fluid communication with the discharge opening of the second propulsor means for generating thrust in a first direction about the turning axis; second thrust generating means in fluid communication with the discharge opening of the first propulsor means for generating thrust in the first direction about the turning axis; third thrust generating means in fluid communication with the discharge opening of the first propulsor means for generating thrust in a second direction about the turning axis; and fourth thrust generating means in fluid communication with the discharge opening of the second propulsor means for generating thrust in the second direction about the turning axis.

Further, there are maneuverability control means on the body member for selectively establishing and nonestablishing the fluid communication between the first and second propulsor means and the first, second, third, and fourth thrust generating means, respectively. Additionally, each of the propulsor means may be removably secured to the rear of the body member and includes a particularly advantageous construction comprising a multifunction one-piece casting; a propulsor housing secured to the undersurface of a rearwardly extending portion of the one-piece casting; a prime mover secured to the upper surface of the rearwardly extending portion of the one-piece casting; and a water gathering collection chamber secured to the propulsor housing for helping to establish the necessary pressure head used in the maneuverability system of the craft.

A device of this type known as the Mark I Worthington Corporation skimmer and also as the Mop-Cat Oil Skimmer, is described in EPA, Edison, N.J. publication *Oil Skimming Devices*, cited above. The Mop-Cat oil recovery system is rated at 20 to 40 bbl/hr depending on the viscosity of the oil and sea conditions. It is manufactured under a license from the Standard Oil Company of Indiana. The recovered oil is held in a 100 gal retention tank, then pumped into drums on the deck or to floating bags towed alongside.
The basic platform of the catamaran boat consists of two solid filled hulls, 25 ft long, which maintain buoyancy. The boat has been designed with sufficient structural strength to permit transporting by helicopter. The beam is 21 ft. At the bow, two bow wings, approximately 6 ft long, are mounted at 45° with the keel line to provide up to 18 ft wide oil slick recovery swath. The bow wings are manually retracted, but provision is made for hydraulic operation. The intake for the oil recovery system is protected from debris by screens. The prototype for this skimmer was tested in the New York area in the fall of 1969. See also the Welles Product Corporation device known as the Reclam-ator which was discussed on p. 311.

OIL REMOVAL FROM STREAMS, LAKES AND HARBORS

In recent times, a number of serious inland incidents have occurred. Compared to the major marine oil spillages, e.g., Torrey Canyon (30 million gallons), Ocean Eagle (1 million gallons) or Platform A in the Santa Barbara Channel (200,000 gallons), inland spills in general are smaller. However, this may not always be the case. A recent barge tow on the Mississippi and Ohio rivers amounted to about ⅓ of the Torrey Canyon load (i.e., 10 million gallons) and exceeded it in length. In harbors or near port facilities, vacuum or suction pumps can be used to remove oil spilled close to shore. Such pumps suck up a substantial amount of water along with the oil, making necessary the use of oil/water separation equipment. All things considered, however, the vacuum pump, when used with an appropriate skimmer, is one of the fastest ways to get oil off the water. In emergencies, vacuum pumping equipment of the type used by cesspool cleaning firms has been adapted. In some harbors and dock areas, skimming and separating equipment is permanently installed in spots where oil tends to collect due to wind, water current, or shore configuration.

An oil-slick collection system suitable for use on harbor craft is described by R.N. Cross of Alpine Geophysical Associates in Report EPA-R2-73-115, Washington, D.C., E.P.A. (February 1973). This system employs a pneumatic-powered vacuum cleaner to collect oil from the water surface by entrainment in a high-velocity air stream. The components are widely available commercial items. Tests show the system to be successful in picking up No. 4 fuel and lighter oils. The collection rate depends chiefly on the rate of oil supply to the skimmer. This system was designed to fill a perceived need for a system to pick up harbor oil spills that would be adaptable to fireboats, towboats and other available harbor craft.

The use of harbor craft for rapid initial response is attractive primarily because of their availability. Fireboats, for example, are on 24 hour call in many major ports, and can arrive within 10 to 15 minutes of a call (in New York Harbor). Coast Guard craft also have emergency response capabilities, and towboats are generally available in most ports. Moreover, these craft have other jobs to do; therefore, the cost of maintaining craft exclusively for oil spill control is avoided. This same fact poses complications, however. Equipment cannot be stored on board that would impair these crafts' primary functions, and both the equipment and method of operation should preclude the spillage of oil on personnel or decks. Fireboats, in particular, may have to respond to a fire from an oil spill, and oily decks would pose a substantial safety hazard.

The particular prototype system constructed was designed around the facilities available on the fireboats of the Fire Department of New York; many variations are possible to meet local conditions and take advantage of locally available components. The skimmer head in its simplest form consits of a short piece of 2 inch pipe on the end of the vacuum hose provided with flotation means. (See Figure 190.)

FIGURE 190: HORIZONTAL SKIMMER HEAD

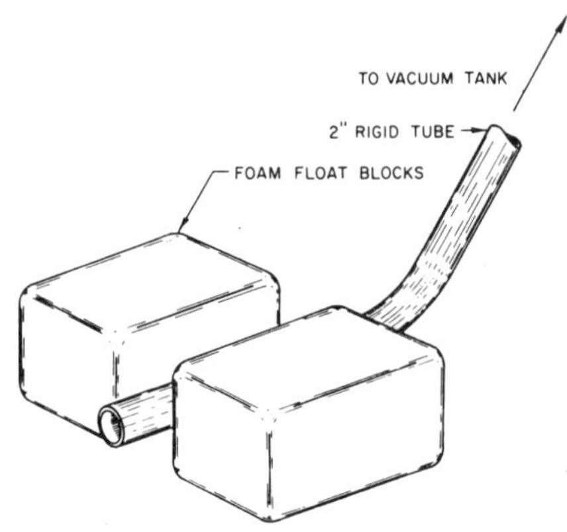

Source: Report EPA-R2-73-115

PROPRIETARY PROCESSES

Horne-Heyser-Neely Design

A device developed by *D.M. Horne, W.H. Heyser and H.M. Neely; U.S. Patent 3,727,766; April 17, 1973* is a boat-mounted vacuum system for removing floating liquid contaminants, such as oil spills, particularly from confined bodies of waters such as harbors, bays, and the like. The system employs one or more water jet eductors for creating a vacuum in a vacuum tank located between the suction heads and the pump supplying high pressure water to the eductors so that the pump does not lose its prime when the suction heads are lifted off the water surface from which it is collecting the contaminants. Figure 191 is a perspective view of a device of such a design in operation on an oil spill.

FIGURE 191: BOAT-MOUNTED SKIMMER SYSTEM FOR HARBOR AND BAY USE

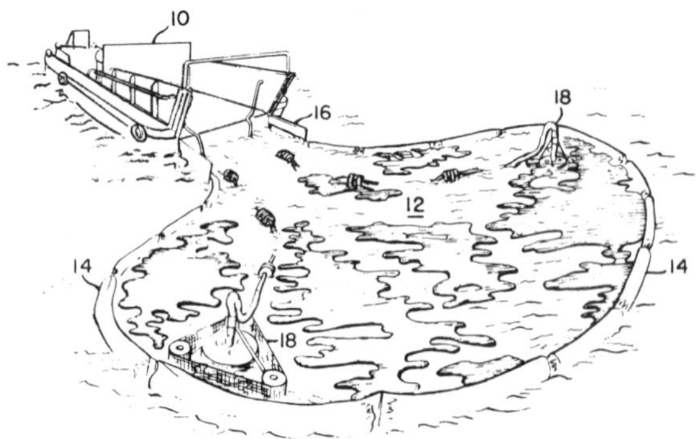

Source: D.M. Horne, W.H. Heyser and H.M. Neely; U.S. Patent 3,727,766; April 17, 1973

The system shown includes a barge or boat **10**, maneuvered to a position adjacent a contaminant slick **12**, such as one usually caused by petroleum spill. Where necessary, the oil slick may be confined by a floatable boom **14**, of conventional design, deployable from boat **10**. The boom is capable of being gradually hauled into the boat to gradually concentrate the slick toward boat ramp **16** which is shown in a lowered position. One or more floatable vacuum heads **18** and vacuum flexible hoses **20** are connected to the vacuum system. Inlet bow valves **21** are connected in vacuum hoses **20**. The oil/water mixture pumped aboard is sent to contaminant tanks on the deck of the boat and transferred as convenient to rafts alongside.

Pardee Design

A device developed by *R.L. Pardee; U.S. Patent 3,762,168; October 2, 1973* is a device for removing pollutants from water, particularly navigable streams, rivers and lakes. In one form, the device is disposed at an acute angle to the flow of water and is operable to direct the pollutants skimmed from the water to a collecting area such as a reservoir, from where it may be removed from time to time. Figure 192 shows a plan view of a waterway containing this device as well as a perspective detail of a portion of the skimming device. itself.

FIGURE 192: PARDEE DESIGN FOR OIL SKIMMING FROM RIVER

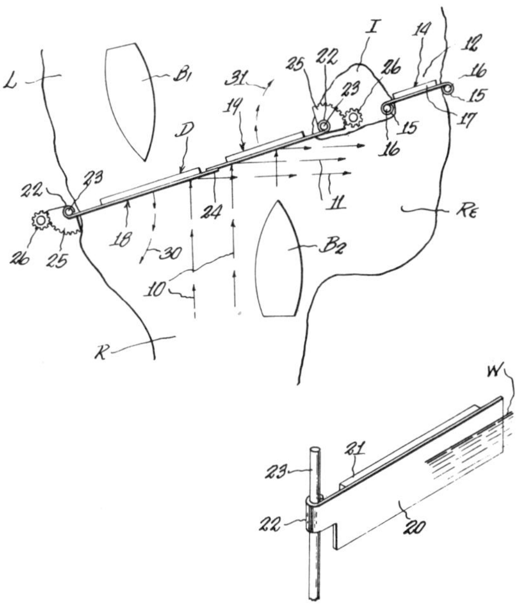

Source: R.L. Pardee; U.S. Patent 3,762,168; October 2, 1973

A river **R** is shown as flowing into a lake **L**, with improved skimming device **D** disposed crosswise of the river. A collecting area, such as a reservoir **Re** is in communication with the river, and an island **I**, either natural or man-made, is disposed between the reservoir and the lake. The flow of water in the river is indicated by the arrows **10** and it will be noted that the improved skimming device is disposed crosswise of such flow and at an acute angle thereto so that as the upper portion of the water strikes the barrier formed by the skimming device, it and any debris or pollutants floating thereon are directed toward the reservoir, as indicated by the arrows **11**. A relatively small water-way **12** is formed between the Island **I** and the adjacent shore, and a skimming barrier **14** is disposed thereacross. The barrier may be formed of rigid strip material, either metal or nonmetal, of sufficient width so that a lower portion is disposed below the surface of the water, but spaced from the bottom, to permit water to flow therebeneath from the reservoir to the lake. An upper portion of the barrier is disposed above the level of the water to block passage of floating debris and pollutants from the reservoir to the lake.

The barrier **14** is preferably mounted so that it will automatically rise and fall with the changing level of the water-way **12**, so that the barrier always maintains a definite relationship with the upper surface of the water and therefore is always operable to skim floating debris and pollutants. Periodically, the debris and pollutants accumulated with the reservoir **Re** may be removed. The rigid debris, such as logs and the like, may be removed by a power shovel, and the liquid pollutants, such as oil and the like, may be removed by a pumping operation.

In the form illustrated in the drawing, the barrier is formed with vertical loops or sleeves **15** at its opposite ends to receive upright posts **16** rigidly supported by the island and river bank, respectively. Suitable bearings may be interposed between the posts and loops to permit free vertical movement of the barrier. A floatation device **17** is attached to the barrier **14** to cause it to rise and fall with any change in level of the water. The skimming device **D** may be in the form of an unbroken barrier across the river mouth, in the event there is no navigation on the river. However, it is preferred to form the device **D** in a plurality of sections, such as the two indicated at **18** and **19**, and to support such sections for swinging movement to permit boats to pass to and from the river.

Each barrier section may be formed as shown in the detail view to comprise a rigid strip **20**, of metal or nonmetal, of sufficient width so that upper and lower parts, are respectively disposed above and below the level of the water **W**. The lower portion of the strip is spaced from the river bottom so that water may flow therebeneath, and the upper part extends above the water level a suitable distance to block flow of floating debris and pollutants from passing from the river to the lake. A floatation device **21** is secured to each section **18, 19,** to cause it to rise and fall with any change in level of the water.

The sections **18** and **19** are mounted for shifting movement to provide passage for boats and one manner in which this may be accomplished is to provide an end of each section with a vertical loop or sleeve **22** to receive a vertical post **23**. Suitable bearings may be disposed between respective posts and sleeves to permit free vertical movement of the sections. The respective posts **23** are shown as rigidly supported by the river bank and the island. The sections **18, 19** are of a length to overlap, as seen at **24**, to form a complete barrier across the river. The sections **18, 19** may swing in any suitable manner, and illustrative of the same, a gear segment **25** is rigidly secured to a sleeve **22** and is rotated by a motor driven gear **26**. The motors may be manually controlled, such as by a watchman, or may be controlled by a radio signal from an approaching boat, or by an electric eye device.

When a boat, such as illustrated at **B1**, approaches for entrance to the river, the section **18** is swung in the direction shown by the arrow **30** to provide for passage of the boat. This, of course, will interrupt the complete barrier across the river so that a small amount of debris and pollutants may flow into the lake, but since the section **18** is open only for a relatively short time, this is not of any serious consequence. When a boat **B2** approaches for entrance to the lake, the section **19** is swung in the direction of the arrow **31** to provide passage for the boat.

Steltner Design

A design developed by *H.R. Steltner; U.S. Patent 3,779,382; December 18, 1973* is an oil-guide-boom which is effective in the removal of surface oil and residue from flowing water.

The boom includes two support members and a sheet extending between the supports in a trough-like manner. The sheet comprises a wire mesh material and a nonwoven fabric. In operation the boom is placed in water toward the direction of flow so as to allow water to pass through the mesh portion of the sheet and oil to collect in the trough portion of the sheet. Figure 193 is a persepctive view of such a device in place as well as detail view of an alternate form of construction.

FIGURE 193: STELTNER DESIGN FOR OIL SKIMMING FROM RIVER

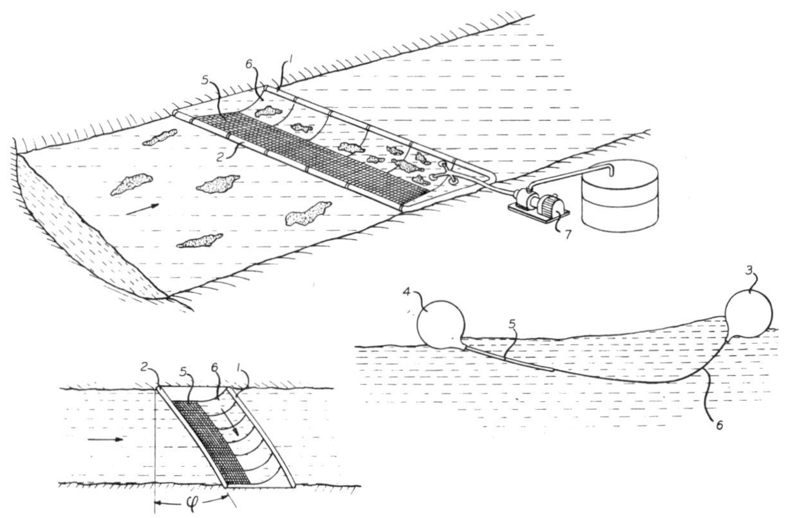

Source: H.R. Steltner; U.S. Patent 3,779,382; December 18, 1973

Two support members comprising two bars **1** and **2**, or floats **3** and **4**, are placed parallel at a certain distance on the water at an angle from the perpendicular to the direction of flow, as shown, which angle is optimal and relative to the velocity of the water stream. Both bars **1** and **2**, and floats **3** and **4**, are connected by means of two sheets of different materials, both of which have a specific gravity greater than 1.0 (voids not included). When bars are utilized instead of floats it is of course possible to use two bars, which are extending over the river and which are fastened on both sides of the river. If floats are used, they have to be anchored at the bottom of the river or the sea. In any case, the sheets of materials are arranged as follows: A net **5** with relatively wide meshes is fixed at float **4**.

Adjacently there is a nonwoven fabric **6** which may be reinforced by a fabric or a support member. The nonwoven fabric, in turn, is held by the second float **3**. As a result of the arrangement a trough forms naturally, in which by placing the whole arrangement at an angle ϕ to the flow as is shown, the diverted oil/water mixture does move to one side and towards a point of evacuation. The specific characteristic of the reinforced nonwoven fabric is that it contributes to a reduction of friction resistance. The selection of the most suitable nonwoven fabric providing for an optimum of oil retention at a maximum rate of water permeability proceeded via tests in the laboratory as well as in the field. The void content of the nonwoven fabrics needed for this device amounts to at least 80%, preferably 85 to 92%. If the void content is less, the water can no longer stream through the nonwoven fabrics. If the void content is too high, the oil may escape through the pores of such a nonwoven fabric. The efficiency of the above described nonwoven fabric in the present Oil-Guide-Boom is based on two factors. Once they act as filters, oil and solid particles which are penetrated into the nonwoven fabric, are retained therein. If the efficiency were solely based on a filter action, then the Oil-Guide-Boom would become ineffective after a relatively short time, because the pores of the nonwoven fabric would be clogged by oil and other residues. Since, however, the water is dammed up if it streams towards the nonwoven fabric, the penetration of oil and solids into the nonwoven fabric is prevented. The oil is concentrated in the backwater of the trough, from which it is continuously removed. The life time of the Oil-Guide-Boom is very long, because practically no oil and no solids deposit in the interior of the nonwoven fabric.

Thomas Design

A design developed by *F.B. Thomas; U.S. Patent 3,563,380; February 16, 1971* provides a method of removing oil, grease and other floating contaminants from a stream and it includes floating a barricade, which extends above the surface of the water, in a stream, and positioning this barricade at an acute angle to the direction of flow of the stream. The contaminants collect on the upstream face of the barricade and periodically are removed from the surface of the stream, and adjustable barricade means are present in the apparatus.

Tuttle Design

A design developed by *R.L. Tuttle, U.S. Patent 3,701,430; October 31, 1972* provides a structure to be used in conjunction with a barge for skimming oil spilled on the surface of harbor waters, lakes and the like. It comprises one or two outriggers which can be partially positioned just below the surface of the oil and which have harvesting booms that converge at the rear joining a collector sump from which the oil and water can be pumped into the barge. The system may also include a means for separating the oil from the water and also for separating out large particles of debris which are frequently found floating in these waters. For use in situations where the body of water has a light-swell condition, the outriggers include flexible leading ends which will follow the contour of the water surface.

U.S. Atomic Energy Commission Design

A design developed by *J.W. Strohecker; U.S. Patent 3,779,385; December 18, 1973; assigned to U.S. Atomic Energy Commission* is one in which floating contaminants such as oil and solid debris are removed from a moving body of water by employing a skimming system which utilizes the natural gravitational flow of the water. A boom diagonally positioned across the body of water diverts the floating contaminants over a floating weir and into a retention pond where an underflow weir is used to return contaminant-free water to the moving body of water. The floating weir is ballasted to maintain the contaminant-receiving opening therein slightly below the surface of the water during fluctuations in the water level for skimming the contaminants with minimal water removal.

OIL REMOVAL FROM UNDERGROUND WATERS

PROPRIETARY PROCESSES

Shell Oil Company Process

A process developed by *D.N. Dietz; U.S. Patent 3,628,607; December 21, 1971; assigned to Shell Oil Company* is one in which taste-spoiling components of oil spilled in a drinking water catchment area are removed from the aquifer by supplying liquid, such as water or tasteless oil to the spill area, and recovering the liquid together with the undesired components via an auxiliary well. Figure 194 shows, in sequence, from top to bottom: a cross section of a water catchment area traversed by a well used for the recovery of potable water in which area hydrocarbon liquids have been spilled on the earth; such a catchment area traversed by an auxiliary well; and the auxiliary well in operation according to this process.

Referring to the figure, an aquifer **1** overlying an impermeable layer **2**, such as a clay layer, is seen. Water is present in the pore space of the aquifer and recovered via a well **3** penetrating the aquifer. Suitable tubular equipment is arranged in the well for recovering the water. This equipment includes a pump **4** which transports the water from the bottom of the well to the earth's surface via a tubing **5**. The further equipment as used in the well is not shown in detail since it is known per se and is not of particular interest in this method. As can be seen, the surface **6** of the body of water present in the aquifer is curved towards the well, which withdraws water from the aquifer. The flow lines **7** indicate the flow of water from the various parts of the aquifer towards the well.

At some distance from the well an amount of oil **8** has been spilled on the earth's surface **9**. This amount of oil flows downwards through the subsoil **1** (see area **8a**) until it contacts the water level and spreads over the water surface until it is distributed in isolated or almost isolated clots and forms a pancake **10**, which is rendered immobile by capillary forces. To remove the objectionable components of the oil forming the pancake, a well **11** is drilled in the aquifer to a level below the water surface. This well is equipped with a pump (not shown) which, when in operation, can transport liquid from the bottom of the well to the surface of the earth. This liquid is guided away from the well by a conduit **12**. Withdrawal of water from the region near the bottom of the well causes a sink (not shown) to form in the water surface. At the same time, or thereafter, water not contaminated with oil is transported to the area around the auxiliary well and supplied to this area. This water may be supplied by the well via a conduit **14** provided with a valve **15**. The water is prevented from spreading over undesired parts of the surface **9** by a small dike **16**. By so choosing the area over which the water is supplied to the surface **9** that it is somewhat larger than the area of the pancake, the pancake will be washed by the water, which, as can be seen from the bottom view in Figure 194, flows downwards through the region **17** (as shown by arrows **18**), thereby removing the objectionable components such as mercaptans and phenols from the oil forming the pancake. At the same time, the objectionable components of the crude oil particles that have been retained in the area **8a** of the aquifer during the downward travel of the oil are washed from this oil. The water containing the objectionable components is removed from the subsoil **1** via the auxiliary well **11**. The flow lines **19** indicate the flow of this contaminated water.

FIGURE 194: SHELL OIL COMPANY SCHEME FOR OIL SPILL REMOVAL FROM UNDERGROUND AQUIFERS

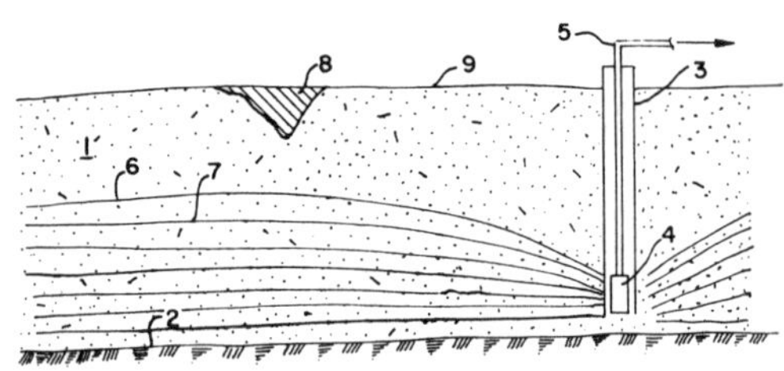

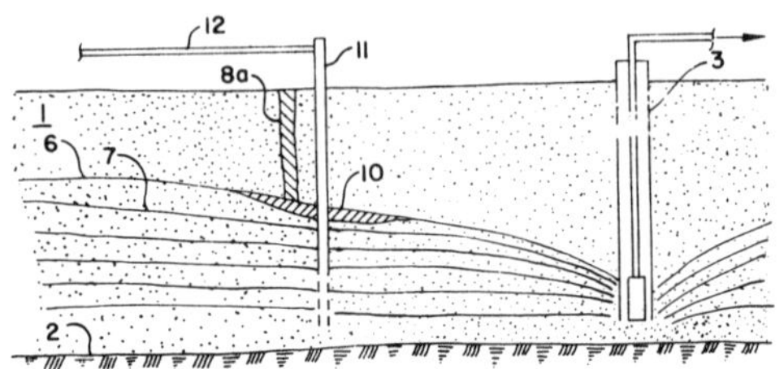

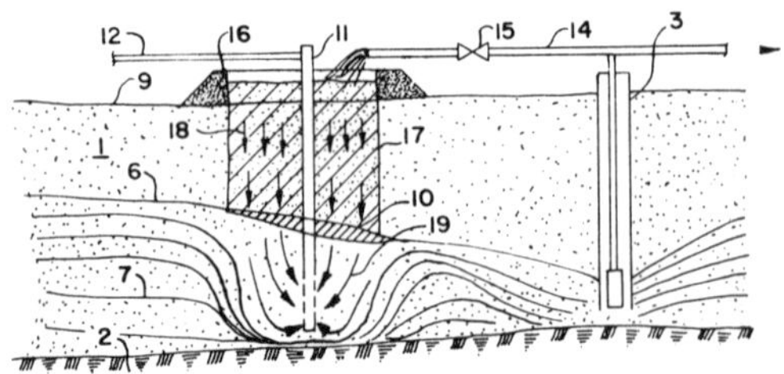

Source: D.N. Dietz; U.S. Patent 3,628,607; December 21, 1971

With the arrangement as shown in the drawing and the steps as previously described, the water as recovered via the well **3** will not be contaminated and will remain potable. The water contaminated with objectionable components from the oil is removed from the aquifer via the auxiliary well. The water as removed via this well is periodically tested for potability. If drinkable again, this means that objectionable components have been removed from the aquifer, and that the well can be put out of operation. It will be appreciated that at the same time the supply of water via the conduit **14** is stopped. By the above described method, the period over which the well should be kept operating is shortened to a considerable extent as compared with prior art methods.

Union Oil Company Process

A process developed by *W.E. Ridgeway; U.S. Patent 3,425,555; February 4, 1969; assigned to Union Oil Company* is one in which petroleum products which have escaped into the ground are removed by applying water to the area to form a mixture of water and petroleum. The mixture is percolated through the ground to the water table and collected by a number of collecting fingers, forwarded to a collecting basin and then to a phase separation tank.

OIL-WATER SEPARATOR DEVICES

However oil/water mixtures are present, whether they may themselves constitute the source of oil spills or whether oil/water separation is necessary after oil removal from water, oil/water separator devices may be important.

A report on oil separation from water discharged from ships has been made by W.C. McKay of the Office of Research and Development of the U.S. Coast Guard. The Coast Guard with support from the Navy and the Maritime Administration solicited proposals for a laboratory test program to develop and test unique concepts for separating oil from water discharged from ships. Eight concepts were funded, evaluated, and reported in Report CG-D-26-74, available as Report AD 770,346, Springfield, Virginia, Nat. Tech. Information Service (August 1973).

For comparison purposes test results on typical cartridge coalescers are included. The following are the systems evaluated: tubular membrane separator, centrifugal coalescer separator, laminar flow coalescing plate separator, ultrafiltration separator, electrochemical flotation separator, vacuum desorption air flotation separation, viscosity separation and vortex separation. A comparison of the systems is summarized in Table 16.

TABLE 16: OIL/WATER SEPARATION CONCEPTS

Concept & Contractor	Oils Tested	Mixing Method	Oil content analysis	PPM of Oil in effluent	Projection for 100 gpm Separator Lbs Wgt	Cubic Ft.	Dollars Cost	Recommended Use
Ultrafiltration, Abcor Inc.	No 6 FO Lube Oil Crude Oil Kerosene	Oil metered ahead of centrifugal pump	solvent extraction, galvimetric analysis	5 to 17	13,600	510	109,000	bilge oil
Centrifugal Coalescer, Foster-Miller Asso.	No 2 FO No 4 FO Crude Oil	batch mixing and oil injection ahead of pump	Particle count summation	61 av.	5,600	150	22,900	ballast with derated pump
Laminar Flow Plate Coalescer, General Electric	Navy SpF.O. Navy Distil. Crude oil Hydr&Lube	High speed propeller mixer gravity feed	solvent extraction colorimetric	26 av.	7,000	730	30,000	ballast with suction thru separator
Ultrafiltration Membrane, Gulf Environmental Sys.	Lube oil Diesel oil crude oil bilge mix	Centrifugal mixing pump loop on ressevoir	solvent extraction gravimetric and total carbon analysis	1 to 18	11,000	510	45,000	bilge water

(continued)

TABLE 16: (continued)

Concept & Contractor	Oils Tested	Mixing Method	Oil content analysis	PPM of Oil in effluent	Projection for 100 gpm Separator Lbs Wgt	Cubic Ft.	Dollars Cost	Recommended Use
Electrochemical flotation, Lockheed Air Service Co.	Diesel, Lube Hydrolic oil Grease, Navy SpFO Crude oil	Batch mixing with high speed propeller mixer	solvent extraction ultra-violet absorption analysis	0 to 11	Not Available	601	80,500	Not recommended
Vacuum Desorption, Mechanics Research	20 W Lube No 2 FO No 6 FO	Mixing pump loop on input stream	solvent extraction gravimetric analysis	126 av.	13,050	465	13,000	Not recommended
Viscosity Separation, Union Carbide Corp.	20 W Lube No 6 FO 40 W Lube 140 W Lube	Centrifugal pump mixer	visual and solvent extraction infra-red analysis	Not Available (This system was incapable of separating oil)	Not Available	Not Available	Not Available	Not recommended
Jet Vortex, United Aircraft Inc.	Crude Oil 40 W Lube No 6 FO Diesel Oil Navy Distill. 10-20-30 Lube	essentially no mixing		53 av.	2,350	120	10,000	Not recommended
Cartridge Coalescer	Navy Distill Turbine Lube	Static Vain Mixer inside pipe	Solvent extraction Infra-red analysis	52 av.	4,000	144	12,600	Bilge & ballast

Source: Report CG-D-26-74

PROPRIETARY DEVICES AND PROCESSES

American Cyanamid Company Process

A process developed by *E.A. Vitalis and R.J. Chamberlain; U.S. Patent 3,756,959; Sept. 4, 1973; assigned to American Cyanamid Company* is one in which a wide variety of mineral oil (petroleum) suspensions of the oil-in-water types, particularly those containing finely divided solid suspended matter are broken into an economically valuable mineral oil fraction low in suspended solids and water and an ecologically acceptable wastewater by using a high molecular weight water-soluble polyelectrolyte, such as a polyacrylamide, and at least one surface acting agent such as an alkyl sulfosuccinate.

FIGURE 195: SCHEME FOR BREAKING WATER-OIL EMULSIONS USING POLYELECTROLYTES

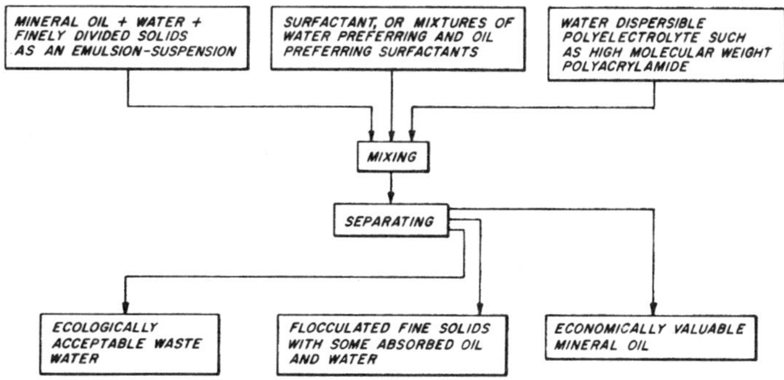

Source: E.A. Vitalis and R.J. Chamberlain; U.S. Patent 3,756,959; September 4, 1973

Other polyelectrolytes and other surfactants and mixtures thereof are effective with a wide variety of such emulsion-suspensions, and result in an ecologically effective separating procedure. The combination of (a) a preferentially water-soluble surfactant such as sodium isopropylnaphthalene sulfonate which is particularly effective in coalescing water particles, (b) a preferentially oil-soluble surfactant such as sodium di(2-ethylhexyl)sulfosuccinate which is particularly effective in coalescing oil particles, and (c) a high molecular weight polyacrylamide of up to about 20 million molecular weight which is particularly effective in flocculating finely divided solids, potentiate and synergistically aid each other and with heat to thin the oil layer, if it is viscous, gives particularly ecologically advantageous results. This process is shown schematically in the block flow diagram in Figure 195.

Beavon Process

A process developed by *D.K. Beavon; U.S. Patent 3,574,329; April 13, 1971* is one in which water containing oil and particulate solids, typically oil-wet solids, is filtered through a filter media, such as a sand, to retain particulate solids thereby yielding clear water or a mixture of solids-free oil and water, which will readily separate by gravity. The filter media is periodically regenerated by steam stripping to remove retained oil, then backwashed to remove oil-free particulate solids.

Bilhartz-Nellis Process

A process developed by *J.R. Bilhartz and A.G. Nellis, Jr.; U.S. Patent 3,594,314; July 20, 1971* for treating waste materials containing oil, water, oil and water emulsions, and oil coated solids, particularly slop oil from an integrated petroleum refinery, to separate material which can be used or disposed of without environmental harm, is one in which the waste material is subjected to ultrasonic treatment at subcavitation power levels and permitted to settle.

For example, such a mixture is treated at a power level of about 2 to 10 watts per barrel of oil with hourly alternate treatment for 5 to 30 min and settling for 30 to 55 min periods for an 8 hr treating cycle, followed by a 16 hr settling period, to provide a 24 hr total time cycle. A clarified oil phase is recovered as an upper phase; a lower phase is removed and subjected to ultrasonic treatment at cavitation power levels, for example, in a continuous flow operation and at a power level of about 1 to 10 kw hr/bbl of fluid treated; the cavitated product is then separated to recover an upper free oil phase and a lower water and solids phase; and the water and solids are then separated for use and/or disposal.

The lower phase of sonically treated products from either the subcavitation treatment or the cavitation treatment or both may be further separated to recover an intermediate emulsion phase and this emulsion phase may be recycled to the subcavitation treatment while the remaining water and solids phase is subjected to cavitation treatment. Alternatively, an intermediate emulsion phase from the first subcavitation treatment may be separated and subjected to a second subcavitation treatment prior to subjecting the bottoms product to the cavitation treatment. The water and solids from the process may be separated by filtering or centrifuging and, if desired, the solids material recovered may be washed with a solvent to remove any residual oil therefrom.

Brunswick Corporation Device

A device developed by *G.B. Peters; U.S. Patent 3,743,599; July 3, 1973; assigned to Brunswick Corporation* is a coalescing filter for an oil-water separator which comprises a winding of multifilament twisted nylon yarn completely enclosed by a winding of multifilament twisted polypropylene yarn. Both windings are impregnated with an oleophilic resin which, together with oil, forms a membrane that coalesces even emulsified oils.

Carlstedt Device

A device developed by *B.R. Carlstedt; U.S. Patent 3,574,096; April 6, 1971* for the separation of oil from an oil-water mixture involves introducing an oil-water mixture to a surface region of a layer of grainy oil-resistant material housed in a tank, flowing an aqueous phase substantially vertically down through the layer of the material, maintaining a ground water zero pressure level in the layer and finally collecting a substantially pure oil phase at a place in the layer situated at a distance from the region where the mixture has been introduced.

As shown in Figure 196 this device comprises an elongated tank **1** containing a bed which consists of an upper stratum of gravel, an intermediate stratum of sand, and a thin bottom stratum of gravel. The bottom stratum is drained by communication with a water-collecting basin or well **3** having an overflow **5**. At one of the longitudinal sides of the tank close to the tank wall an oil-collecting channel **2** is arranged in the upper gravel stratum. This channel has a wall in contact with the gravel layer and provided with vertical slots, and moreover has an overflow **6** at a level somewhat higher than the level of the water overflow **5**.

The apparatus is operated in the following way. The oil-water mixture to be separated is introduced into the surface of the bed through the conduit **4**, the flow of the mixture being maintained calm and continuous so as to prevent this oil-water mixture from spreading too much laterally. The bed surface, therefore, is somewhat inclined down towards the region where the mixture is introduced. In the capillary zone in the gravel stratum above the ground water level the oil phase then spreads laterally towards the oil-collecting channel, while the aqueous phase is flowing straight through the gravel-sand-gravel strata and is finally discharged via the overflow **5**.

The height of the ground-water level as defined by the overflow is determined mainly by the oil-water mixture added and by the flow resistance of the water in the gravel and sand strata. Oil that has now moved laterally towards the oil-collecting channel will enter the channel through the vertical slots and collect on the liquid surface in the channel, to thus form a layer increasing in thickness. When the top of the oil layer reaches the overflow **6**, oil will flow over into a separate collecting vessel (not shown).

FIGURE 196: CARLSTEDT OIL/WATER SEPARATOR DEVICE

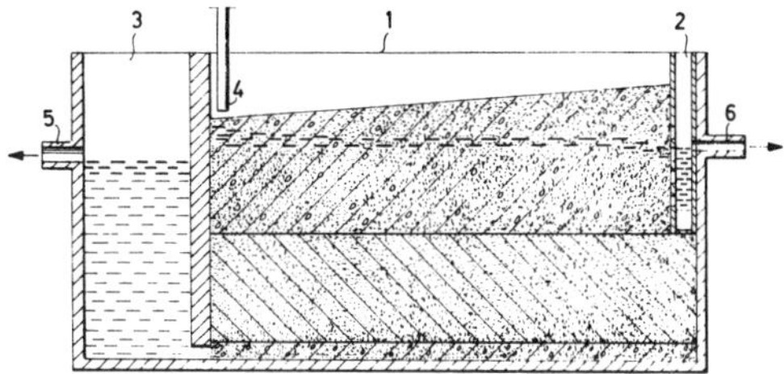

Source: B.R. Carlstedt; U.S. Patent 3,574,096; April 6, 1971

Laboratory tests for demonstrating the oil-separating effect have been performed with an apparatus of this type where the bed had a surface of 200 cm^2 (5 x 40 cm) and consisted of one 35 cm top stratum of gravel (grain size 3 to 8 mm), one 30 cm intermediate stratum of sand (grain size 0.5 to 1 mm) and one 5 cm bottom stratum of the same gravel as in the upper stratum. The difference in height between the oil overflow and water overflow was 2 to 4 cm.

The oil-water mixture that was tested had a content of 37.5 cc of colored fuel oil per liter of water. Samples of the discharged water which were taken after testing through the water storage basin of water volume corresponding to 3 to 5 times the volume of the basin showed a content of 20 mg of fuel oil in dissolved state per liter of water, i.e., a separation of 99.95% of the oil from the water had been achieved. In this connection it should also be pointed out that the fuel oil employed in these test runs had a maximum solubility in water amounting to 25 mg/l.

Cincotta Device

A device developed by *J. Cincotta; U.S. Patent 3,752,762; August 14, 1973* is one which can remove large quantities of oil from a body of oil contaminated water as well as be used for industrial purposes where it is necessary to remove oil from oil contaminated water. This device functions by passing the oil and water through a filtering material wherein the filtering material after becoming saturated with oil can be treated to have the oil removed therefrom and reused as required.

According to the process, pumice stones can be placed in a container having an open top, a removable bottom and porous opposed first and second side walls. The pumice stones are placed in the container through the top and oil contaminated water, such as from an oil spill etc., is directed through the first porous side wall, through the container and out the second side wall thereof. The pumice stones in the container absorb the oil from the water and when they become saturated the bottom of the container is removed allowing the pumice stones to drop on a conveyor.

The conveyor transports the pumice stones to a steam-cleaning area which removes the oil therefrom and the pumice stones are then recycled back to the container for further use. It is of course to be appreciated that the operation is continuous and that as soon as saturated pumice stones are removed from the container unsaturated stones are placed therein while the saturated stones are being cleaned.

Continental Oil Company Process

A process developed by *M.D. Gregory; U.S. Patent 3,779,908; December 18, 1973; assigned to Continental Oil Company* is one in which dispersions of an oleophilic liquid and water are coalesced by passing through a permeable, nonfibrous, consolidated bed of a polyurethane foam which is equilibrated with respect to sorption of the oleophilic liquid.

Dunlop Company Limited Device

A device developed by *J.E. Woolley; U.S. Patent 3,508,652; April 28, 1970; assigned to The Dunlop Company Limited, England* is a buoyant seaborne device in which an oil/water mixture is pumped continuously into one end of an elongated container in which the mixture separates into an upper oil layer and a lower water layer, and water is drained continuously from the lower water layer.

As shown in the cross-sectional view in Figure 197, the separating apparatus includes an elongated flexible container **1** which floats in the sea and is connected by an inlet pipe **7** to a pontoon **4**, the pipe being secured to the front end of the container. A pump **3** carried by the pontoon sucks in oil/water mixture from the surface of the sea through a pump inlet pipe **5** having a wide mouth **6** which can be maintained at any required depth below the surface of the water. The mixture drawn in along the pipe **5** by the pump is pumped through the pipe **7** into the front end of the container.

An outlet pipe **8** at the rear end of the container has its mouth **9** within the container near its rear end, and the pipe mouth is weighted by a weight **11** so that the pipe mouth always rests on the bottom of the container. The pipe **8** passes through the rear end of the container so that the pipe outlet **12** is outside the container. The upper part of the rear end of the container has a vent valve **13.** In use, the container and pontoon are towed through the water in the area from which it is desired to remove the oil by a tow rope (not shown) extending from a towing vessel to the pontoon, the pipe **7** also acting as a tow line between the pontoon and the container. The pump is then operated to continuously pump oil/water mixture from the surface of the water into the container.

FIGURE 197: DUNLOP FLOATING OIL/WATER SEPARATOR

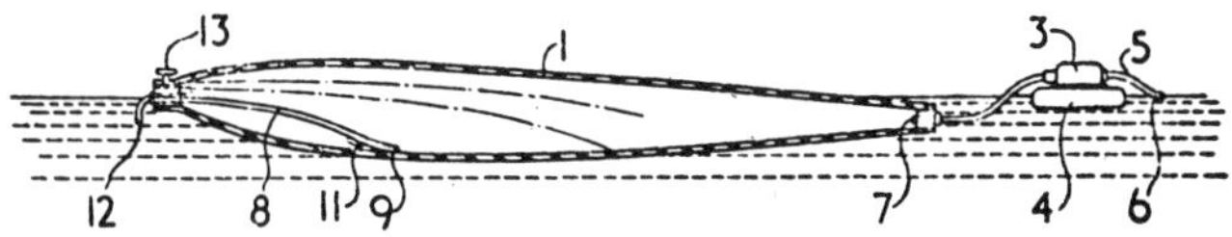

Source: J.E. Woolley; U.S. Patent 3,508,652; April 28, 1970

Initially, the container may be full of air so the vent **13** is opened until liquid is expelled from it, thus indicating that most of the air in the container has been removed. The vent is then closed. The mixture is pumped into the container at such a rate that it separates into an upper oil layer and a lower water layer before it reaches the mouth **9** of the pipe **8**. Thus, as mixture is pumped into the container, water will be expelled from the container through the pipe **8**.

The oil is therefore retained in the container and when the container is nearly full of oil, this will be indicated by the issue of oil instead of water from the pipe **12**. The pump is then stopped, the inlet and outlet pipes are closed, and the container towed back to shore where the container is emptied. The pump may be electrically operated by a power supply on the towing vessel, and in this case the tow rope between the towing vessel and the pontoon is preferably relatively short for convenience in feeding the electrical supply to the pump.

As described in the EPA, Edison, N.J. Report *Oil Skimming Devices,* Report 218,504, Springfield, Virginia, Nat. Tech. Information Service (May 1970), the Dunlop-Dracone Consortium has developed in their laboratories this automatic oil and water separation device. This system is based upon the British Dracone barge. The flexible towable container is available for bulk liquid transport by sea, and reportedly will allow the continuous clearance of oil patches on the sea and inland waterways.

A special forward fitting to the Dracone barge employs a suction device, operated by a pump (either mounted on the towing vessel or raft) which skims the top few inches of the water, the depth of pickup being adjustable. The oil and water mixture is pumped into the body of the barge where it rapidly separates. The length of the Dracone barge, up to 300 ft, assists almost complete separation of the oil and water.

A discharge nozzle, weighted to the base inside the barge, allows the separated water to escape. Tests have shown that even with as little as 5% oil, the final separation is up to 90% effective. A towing vessel with a suction device and pump installed on it could carry empty rolled up Dracone barges on deck. This system has been developed by the Dunlop-Dracone's Consortium, formed by Dunlop and the National Research and Development Corporation, and laboratory model tests have been successful.

General Motors Corporation Device

A device developed by *R.A. Willihnganz; U.S. Patent 3,617,548; November 2, 1971; assigned to General Motors Corporation* is one in which an immiscible mixture of oil in a liquid of higher specific gravity, such as water, is continuously passed through a knitted polyethylene, polypropylene or polyvinyl chloride packing which has an affinity for oil whereby the oil coalesces on the surface of the knitted polymer packing. The oil having a lower specific gravity than the water phase continuously rises along the surfaces of the knitted polymer packing and is released on the surface of the water phase as a layer which can be easily and continuously drawn off.

Kimberly-Clark Corporation Device

A device developed by *E.G. Greenman; U.S. Patent 3,494,863; February 10, 1970; assigned to Kimberly-Clark Corporation* for separating an oil contaminant from water utilizes a strengthened fiber body and a fluorocarbon which is oleophobic but readily wetted by water. In the process of oil removal the contaminated water is directed to the fiber body suitably in the form of a paper sheet and the water passes through while the oil is retained.

Nippon Oil Company Process

A process developed by *J. Yamamoto, K. Minakawa, H. Nishikado and S. Imon; U.S. Patent 3,770,628; November 6, 1973; assigned to Nippon Oil Company, Limited, Japan* is one in which oil-containing contaminated drainage can be effectively clarified by passing the same through a wax filled bed. In the preferred form, the treatment is carried out by the use of combination process selected from (a) a wax filled bed, an active carbon bed process; (b) a sand filter, a wax filled bed process; and (c) a sand filter, a wax filled bed, an active carbon bed process.

Ohta Device

A device developed by *M. Ohta; U.S. Patent 3,689,406; September 5, 1972* is one in which oil is removed from water, e.g., wastewater, by means of a fibrous filter composed essentially of a hydrophilic cellulose material such as regenerated cellulose, mercerized cellulose and acetylated cellulose.

Oil Mop, Inc. Devices

A device developed by *H.M. Rhodes; U.S. Patent 3,689,407; September 5, 1972; assigned to Oil Mop, Inc.* for separating oil from a mixture of oil and water involves pouring the mixture over two oil attracting media each having a different physical characteristic and placing the second oil attracting media between two floats having different specific gravities so that the second oil attracting media is wrung out due to its compression between the two floats to release the oil and to permit it to rise above the water level in the apparatus so that the oil may be drawn off separately from the water.

Such a device is shown schematically in Figure 198 where **A** designates a vessel or container which may be built in sections joined by bolted flanges to permit access to the vessel for replacing oil attracting media and floats as necessary incident to use. **A** is a pressure vessel of any convenient size or configuration to accommodate the pressure of the process at **X, Y** and **Z** (by example, 75 gal, 24 inch diameter).

B is a distribution pan to provide an even flow of the stream through the filter media, for example, 24 inch diameter, 1 inch thick and 24¼ inch diameter holes. **C** is the filter media consisting of an oil mop or other configuration of thin narrow strips or other configuration of polypropylene or similar velocity-viscosity sensitive material. The oil attracting media of $\mathbf{C_1}$ is coarse, for example, 12 mils, while the oil attracting media $\mathbf{C_2}$ is fine, for example, 6 mils, which enhances the operation of the filter.

D is a float of sealed polystyrene with perforations or any convenient nonabsorbent material with a low specific gravity and relatively high mechanical strength, for example, 24 inch diameter cast polystyrene, 2 inches thick with 24¼ inch diameter holes drilled symmetrically and having a specific gravity of 0.1.

E is a float similar to **D** except without perforations and with a higher specific gravity, for example, 0.9. **F** is a float chamber with two floats or a similar system for controlling liquid level at two set points. **Va, Vb, Vc** and **Vd** are valves of any convenient size, for example, ¾ inch nominal pipe size. Oil and water in any proportion enter the vessel **A** at **X** under pressure and are distributed evenly over the filter media **C** by distribution pan **B.** Separation of the oil and water takes place as the oil and water flow through the oil attracting filter media, the oil being retained by the filter media and the water falling freely to the bottom. As the water level builds up float **E** attempts to rise and is restrained by the filter media $\mathbf{C_2}$ which also has a specific gravity of 0.92 causing the filter media to be in slight compression. With valve **Vd** open, the water level rises until control is exercised by float **Fb** and valve **Vb.** Float **D** floats at the water level **B** and the flow stream and float valve **Fb** and **Vb** reaches equilibrium allowing the oil to accumulate on filter media $\mathbf{C_1}$ and the water to discharge at **Z.** As the oil accumulates on filter media $\mathbf{C_1}$ it forms drops large enough to overcome the attraction of the filter media and the drops rise to the surface of the water level **B.**

FIGURE 198: OIL MOP, INC. OIL/WATER SEPARATOR DEVICE

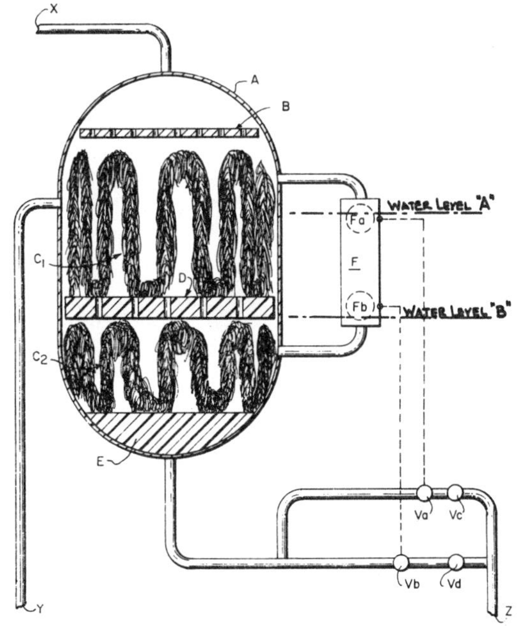

Source: H.M. Rhodes; U.S. Patent 3,689,407; September 5, 1972

The oil accumulates at water level **B.** If the oil is reentrained in the water flow stream, it is captured by filter media $\mathbf{C_2}$ until the drops become large enough to refloat to the surface by overcoming with buoyancy the attracting force of the filter media. If the oil has a specific gravity of more than one, or if the oil is of such type that it has a very strong attraction to the filter media, it will be retained by the filter media until the filter media is reused or is cleansed and replaced.

When a sufficient amount of oil is accumulated at water level **B,** valve **Vc** is opened and valve **Vb** may be closed by an automatic programming device or by manual operation allowing the water level set point at water level **B** to be temporarily transferred to the set point at water level **A** as controlled by float **Fa** and valve **Va.** When the set point is changed from water level **B** to water level **A** the water increases the buoyancy force of float **E** and float **D.** Float **E** compresses filter media $\mathbf{C_2}$ and float **D** compresses filter media $\mathbf{C_1}$ wringing the oil from the filter media which goes to the water surface and accumulates until it flows out at **Y.** After sufficient time for wringing is allowed the control is transferred from water level **A** set point back to water level **B** set point as exercised by float **Fb** and valve **Vb** by opening valve **Vd** and closing valve **Vc**. The filter is now ready to recycle.

Another device developed by *H.M. Rhodes; U.S. Patent 3,794,583; February 26, 1974; assigned to Oil Mop, Inc.* for separating oil from an oil and water mixture involves first subjecting the mixture to passage through a chamber containing an oil mop structure made from fibrillated strips of polypropylene secured to a polypropylene line so that the oil will be attracted by the polypropylene strips and the water will pass through the mop structure and will thereafter be subjected to passage through a fabric membrane having a 2% fluorocarbon solution impregnating the fabric which will permit passage of the water through the membrane rejecting the oil and permitting the water to pass on and be drawn off separately from the oil which may likewise be drawn separately from the chamber.

Still another device developed by *H.M. Rhodes; U.S. Patent 3,810,832; May 14, 1974; assigned to Oil Mop, Inc.* is an apparatus for accelerating the separation of oil from an oil/water mixture to bring the oil to the surface over as short a linear distance as possible in a linear flow of a mixture of oil and water. This is accomplished by directing the oil/water mixture through a barrier of filaments of polypropylene arranged across the path of mixture flow which barrier is anchored at its base at the bottom of the fluid confining means such as an API oil separator or a ditch or canal with the free ends of the strips of polypropylene directed upwardly forming an inclined plane up which the oil droplets amalgamate assisted by the buoyancy of the oil and the force flow vector of the mixture passing through the fluid confining means.

Passavant-Werke Device

A device developed by *W. Weiler; U.S. Patent 3,527,701; September 8, 1970; assigned to Passavant-Werke, Germany* involves spraying a liquid containing hydrocarbons, such as oil, under high pressure onto a filter made of particles containing an oleophilic compound, thoroughly mixing the liquid with the filter compound to remove the hydrocarbon compounds and then removing the cleansed liquid.

A filter packing formed of particles containing an oleophilic compound such as charcoal, bituminous coal, inflated mica material or vulcanic tufaceous limestone is placed in a supporting bed with an outlet beneath the packing. A plurality of high pressure liquid nozzles are movably mounted above and proximate to the filter packing surface. These nozzles project a jet of liquid, which may have previously been run through a skimming separator onto the filter packing with force sufficient to joggle the filter particles to cause a slow, steady shift and circulation of the filter as a whole which prevents pockets of stagnant liquid from forming, promotes the exposure of the surface areas of the entire filter to incoming liquid and provides a pressurized contact between the pollutants and filter media.

To facilitate contact between liquid and filter, the outlet can be dammed to maintain a desired level and degree of saturation. The dammed liquid also makes it easier to mix and circulate the filter material under the power of the nozzle jets. In addition, it assures uniform filtering conditions, even when the incoming liquid supply is unsteady. The filter packing material may be of a type that floats or sinks in water. In the latter case, a coarse grained, porous material such as gravel is used to support the filter and allow easy drainage.

Standard Oil Company Device

A device developed by *W.F. Johnston and R.G. Will; U.S. Patent 3,617,551; November 2, 1971; assigned to Standard Oil Company* is a cartridge type apparatus useful for purifying oil contaminated water. This apparatus includes a flow through chamber containing polyurethane foam and a piston-like member which responds to hydraulic pressure to squeeze the foam. As oil contaminated water flows through the chamber, oil is absorbed by the foam. Periodically, a hydraulic pressure head is established within the chamber which forces the piston-like member to compress the foam and squeeze the oil therefrom. Figure 199 shows the apparatus in oil absorption mode (at left) and in filter squeezing mode (at right). The apparatus includes vertical chamber **10** having at its opposite ends piping connections or port means **12** and **14.**

FIGURE 199: STANDARD OIL COMPANY OIL/WATER SEPARATOR DEVICE

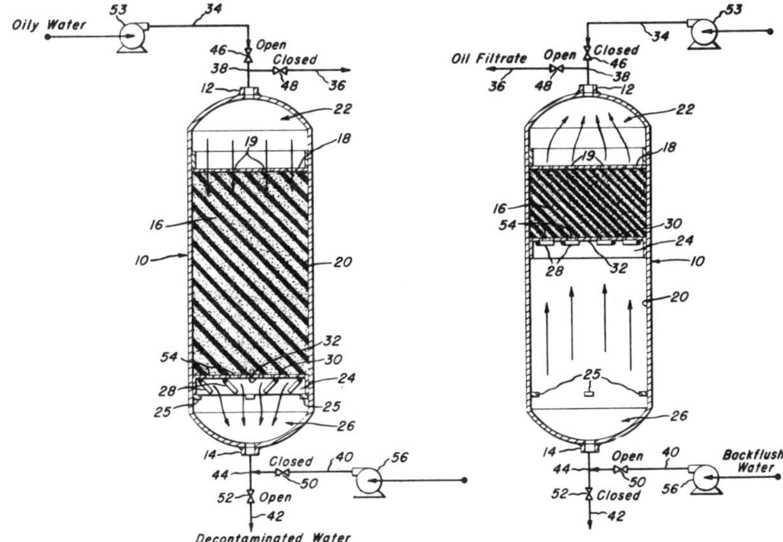

Source: W.F. Johnston and R.G. Will; U.S. Patent 3,617,551; November 2, 1971

The cross-section of the chamber is preferably circular, although other geometric shapes would be suitable. The chamber may be fitted with flanges, removable heads, or other means of access so that it can be filled with a cartridge of polyurethane foam **16.** Plate **18** attached to sidewall **20** of the chamber at upper end **22** abuts the foam. This plate may be a screen, wire cloth, perforated plate, etc., so that water can flow freely through the plate and into the foam. The openings **19** in the plate can be constructed so that large particles would be prevented from entering the foam.

Piston-like member **24**, at lower end **26** of the chamber between the foam and port means **14**, furnishes the means for squeezing absorbed oil from the foam. The member is supported by a plurality of stops **25**, and it preferably has one or more check valves **28.** These check valves are arranged so that they open when water presses against the side **30** of the member but are closed when water presses against the opposite side **32** of the member.

Port means **12** includes pipes **34** and **36** connected at T joint **38**, and port means **14** includes pipes **40** and **42** connected at T joint **44.** Valves **46** and **48** are disposed, respectively, in pipes **34** and **36**, and valves **50** and **52** are disposed, respectively, in pipes **40** and **42.** Three-way valves may, however, be used in place of two-way valves shown. Pump **53** is used to pump oil contaminated water through pipe **34** and open valve **46** into upper end **22** of the chamber. As indicated in the left hand view, this oil contaminated water flows through port means **12** downwardly through the plate **18,** foam **16,** and openings **54** in piston-like member **24.**

As the contaminated water flows through the foam, oil is absorbed in the pore structure of the foam. Purified water thus leaves port means **14** via pipe **42** and open valve **52.** Regeneration of the foam is depicted in the right hand view where water, either purified or unpurified, is pumped into the chamber by pump **56.** During regeneration, valves **48** and **50** are open and valves **46** and **52** are closed.

Water, under pressure, flows through pipe **40** and open valve **50**, through port means **14** into the chamber. This establishes within end **26** of the chamber hydraulic pressure which closes check valves **28**. Water pressure builds up and presses against side **32** of member **24**, pushing the member toward port means **12**. Foam **16** is thus squeezed between piston-like member **24** and plate **18**, and the oil entrapped therein is squeezed from the foam. This liberated oil flows through port means **12**, out pipe **36** and open valve **48** into a suitable storage tank. The liberated oil, as it is squeezed from the foam, also backwashes plate **18**, removing any contaminants trapped thereat.

The flow rate during filtration and the amount of hydraulic pressure needed to regenerate the foam will depend upon the size of the chamber. These parameters can be easily determined on the basis of maximum pressure drop across the polyurethane foam during filtration. Sampling of water periodically would also be a means for determining when regeneration was necessary. Upper end **22** of the chamber should be designed to hold the minimum amount of oil so that very little oil would be contained at this end after regeneration. After regeneration, valves **46** and **52** are opened and valves **48** and **50** are closed. With the valves in these positions, filtering operations are renewed. Flow of oil contaminated water through filter **16** causes the resilient, flexible polyurethane foam to readjust itself to its original volume as depicted in the left hand view of the figure.

Other forms could be devised. For example, instead of a single cartridge of polyurethane foam, several abutting layers of foam could be used with different layers having different pore sizes. In this instance, the larger pore sections would preferably be arranged so that oil contaminated water would contact these large pore sections before it contacts the smaller pore sections. Thus, viscous oil would be trapped first. This would minimize clogging of filter **16**. Moreover, filtration could be conducted in other positions, e.g., horizontally or upside down. Hence, flow does not necessarily have to be downward when filtering. Furthermore, piston member **24** could be loose-fitting, so that during regeneration, clean water would flow around it to some extent to assist in flushing out filter material **16** and upper end **22**. Oil contaminated water could be used to backflush the filter material, followed by clean water, or clean water could be used exclusively. In the latter case, auxiliary check valves could be provided so that water would backflush through the foam and upper end **22** of the chamber.

Alternate arrangements of member **24** are also possible. A more complicated alternate arrangement than check valve **28** could be used to move the member during regeneration. For example, the member can be connected to a shaft that passes through a packing box in the chamber and is connected to an external driving mechanism. In this alternate arrangement the member would contain valves or openings that were mechanically closed during regeneration and open during filtration.

A more simplified arrangement is also possible. This simplified arrangement would include all the parts depicted except member **24** and its check valves **28**. In place of the member, an element similar to plate **18** would be used, however, this element would not be attached to sidewall **20** of the chamber and it would be movable like member **24**. Compression of the foam during regeneration would be accomplished by using a backflush flow rate substantially higher than the filtering rate. The pressure drop across foam **16** would then become great enough to compress the foam and squeeze the oil out.

Wolf Device

A device developed by *N.H. Wolf; U.S. Patent 3,554,906; January 12, 1971* for removing oil particles from water comprises an absorption tower containing a diffusing medium, a solvent having a higher specific gravity than water and being insoluble in water, and a porous material which is permeable to water and impermeable to oil and solvent. A method for removing oil particles from water comprises the steps of passing water containing oil particles through a particulate diffusing medium such as sand, gravel or raschig rings, and high specific gravity solvent, the oil particles adhering to the solvent by a process of absorption, and the oil-free water discharging through a water-permeable porous mass. Electrically operated sensing means is located near the discharge of the oil-free water to verify that adequate separation is achieved and to guarantee efficient recovery of the solvent.

PHYSICAL-CHEMICAL TREATMENT OF OIL SPILLS

DISPERSION

Such chemical agents as dispersants and detergents cause oil to remain in suspension in water. Then, in time, nature takes care of it by bacterial action. Current U.S. government regulations, however, forbid the use of such chemicals except in unusual circumstances. When it is felt that a spill may endanger human life or waterfowl or is a substantial fire hazard, the federal On-Scene Coordinator of the cleanup operation has the authority to approve the use of these chemical agents. Chemical dispersants are usually applied by spraying, and agitation usually is necessary to insure proper mixing. Wind and wave action is sometimes sufficient. But in calm water, it may be necessary to use a boat's propellers or a towed board to create the desired churning effect.

Guidelines on the use and acceptability of oil spill dispersants have been published by the Environmental Emergency Branch of Environment Canada in Ottawa, Ontario as Report EPS 1-EE-73-1 (August 1973).

Dispersants theoretically serve to increase the surface area of an oil slick and disperse oil globules throughout the larger volume of water thereby aiding in accelerated degradation of oils by microbiological means. The chemical dispersants do not themselves destroy oil. They vary considerably in toxicity, effectiveness and ability to stabilize the oil after extended periods of time. Technology for proper application of dispersants over large oil slicks with necessary mixing is currently lacking. Use appears far more critical in harbor and estuary areas and in proximity to shore. Particular care must be exercised where water supply might be affected.

The desirability of employing dispersants in the open sea remains unresolved although their use here (the ocean) is potentially more promising pending additional field data. After widespread dispersant use during major incidents, reports led to the conclusion that dispersants or the dispersant-oil mixture caused more damage to aquatic life than the oil alone. For beaches, they actually compounded the problem by adding to the amount of pollutants present, by causing the oil to penetrate more deeply into the sand, and by disturbing the sand's compactness, so as to increase beach erosion through tidal and wave action.

As far back as the early 1930s water emulsifiable degreasers were used. These degreasers were developed to answer the need for effective methods of cleaning oily and greasy surfaces. They possessed the properties of dissolving or dispersing in the grease or oil, and making the resultant mixture dispersible in water so that it could be flushed away with water.

The early products were composed mostly of soaps and solvents. The demands of the petroleum shipping industry required products that could be used effectively aboard ship with seawater. This led to the use of materials other than soaps as emulsifying agents because soap breaks down in seawater. Sulfonated petroleum oils and later more sophisticated synthetic detergents made their appearance in these products.

These emulsifying degreasers were widely used aboard ship for engine room maintenance, as well as for the cleanout of petroleum cargo tanks prior to welding repairs and prior to upgrading of cargo. Because of their effectiveness in cleanout of oil residues it was natural that they should be tried for treating oil spills. In some cases, they were incorporated in oil slops prior to dumping overboard to minimize slick formation.

There are over 200 commercially available products which are claimed useful for dispersing oil from the surface of the water. These products have been referred to as soaps, detergents, degreasers, complexing agents, emulsifiers and finally dispersants. This latter term describes best what this type of product is intended to accomplish, the dispersion of oil from the surface into and throughout the body of water, and therefore is the preferred terminology. The primary components of dispersants are surfactants, solvents and stabilizers.

Surfactants or Surface Acting Agents (SAA): This is the major active component in dispersants. By their affinity for both oil and water, surfactants alter the interaction between oil and water so that the oil tends to spread and can be more easily dispersed into small globules, or into what is commonly called an emulsion. These agents are often defined according to their behavior in aqueous solutions. These solutions will usually wet surfaces readily, remove dirt, penetrate porous materials, disperse solid particles, emulsify oil and grease and produce foam when shaken. These properties are interrelated; that is, no surface active agent possesses only one of them to the exclusion of the rest.

SAA can be divided into two broad classes depending on the character of their colloidal solutions in water. The first class, ionic, form ions in solution, and like the soaps are typical colloidal electrolytes. The second class, the non-ionic, do not ionize, but owe their solubility to the combined effect of a number of weak solubilizing groups such as ether linkages or hydroxy groups in the molecule.

Commonly used SAA in oil spill dispersants include soaps, sulfonated organics, phosphated esters, carboxylic acid esters of polyhydroxy compounds, ethoxylated alkyl phenols and alcohols (APE, LAE), block polymers and alkanolamides.

Solvents: Since many of the surface active agents applicable to oil spill dispersant compounding are viscous or solid materials, some form of solvent is often necessary in order to reduce viscosity for ease of handling. In addition, the solvent may act to dilute the compound for economic reasons, to depress the freezing point for low temperature usage, to enable more rapid solubility in oil, and to achieve optimum concentration of surface active agent for performance reasons. The presence of a suitable solvent also serves to thin the oil to be dispersed, reducing viscosity and making it more easily emulsifiable. The solvent usually comprises the bulk of the dispersant product. The three general classes of solvents used in oil spill dispersants are petroleum hydrocarbons, alcohols or other hydroxy compounds, and water.

(1) Petroleum Hydrocarbons — Petroleum fractions with boiling points above 300°F are usually used, and these may produce finished dispersants with flash points as low as 110°F. The proportion of aromaticity is of concern, since this affects solubility and emulsification properties as well as toxicity. Wardley Smith, Warren Springs Lab, UK, in describing the Torrey Canyon incident, reported that the aromatic solvents used were 10 times as toxic to marine life as were the surface active agents. Some typical fractions of applicable petroleum solvents include mineral spirits, kerosene, #2 fuel oil, and heavy aromatic naphthas which contain significant quantities of higher alkylated benzenes.

(2) Alcoholic or Hydroxy Group of Solvents — Includes alcohols, glycols and glycol ethers. These solvents also lower the viscosity as well as the freezing points of finished dispersants. In addition, they furnish a cosolvent effect, often needed to mutually dissolve the various ingredients in a dispersant for stability of the compound in storage. This group of solvents may be used in conjunction with petroleum hydrocarbons as well as with aqueous solvent systems. Some of the more frequently encountered chemicals in this group include ethyl alcohol, isopropyl alcohol, ethylene glycol, propylene glycol, ethylene glycol monomethyl ether, ethylene glycol monobutyl ether, and diethylene glycol monomethyl ether. The more volatile members of the group are quite flammable.

(3) Water — Least toxic, least hazardous and most economical of the solvents. It lacks solubility or miscibility with oils. Where water is used as the solvent, special problems exist in the choice of surface active agents and other additives in order to provide the necessary miscibility with oils. Glycols and alcohols are used to aid in miscibility as well as freezing point depression when water is used.

Additives – Stabilizers: Third major component in dispersants, they may be used to adjust pH, inhibit corrosion, increase hard water stability, fix the emulsion once it is formed, and adjust color and appearance. Current information indicates that dispersants vary considerably in toxicity. The combination of oil and dispersant may increase the toxicity of either the oil, the dispersant chemical or both. The possibility of this synergistic action must be carefully examined before wholesale application of such a product is permitted. Dispersing of the oil, which has toxic components, may also compound the damage.

The toxicity of 40 dispersants has been reported by the Fisheries Laboratory, Burnham on Crouch, UK. It is important to point out that the dispersants used during the Torrey Canyon incident were mostly solvent based and highly toxic, killing marine organisms at concentrations of 10 mg/l. Chemicals available today are much less toxic.

A common method of dispersant application is by the water eductor method. A controlled amount of dispersant is educted into a water stream such as a fire hose. This water jet is an effective vehicle or carrier for the dispersant and provides good coverage in treating the slick.

This application procedure however, while compatible with a water base system, may be incompatible with a petroleum base system. This is because a dispersion of the petroleum solvent-in-water is formed as soon as the surfactant system is educted into the water stream. As illustrated in Figure 200, this accounts for the milky white appearance of the water after such applications. In this state, as graphically shown in Figure 200, it is difficult for the surfactant to transfer from its thermodynamically stable location at the petroleum solvent-water interface to the oil spill-seawater interface. Therefore, for a petroleum base system, neat application of the chemical directly onto the oil slick is a more effective application method.

The prompt application of mixing energy after the dispersant (solvent or water base) has been applied is particularly important. In essence, small oil droplets must be produced while the immediate water environment is surfactant-rich.

The cost of dispersants ranges from $2.00 to $5.00 per gallon. Using manufacturers' recommended doses, which are usually 1 gallon of dispersant to 10 gallons of oil, the cost of chemicals for dispersing a relatively small 500 barrel spill would be about $10,000. Based on EPA's actual field experience, which has indicated that frequently the necessary dosage is 1 to 1 or 1 to 2, the cost might run as high as $100,000 for this same size spill. Of course, the amount of chemical required is going to depend heavily on the type and age of oil to be treated, type of dispersant used, and the temperature of the water. Equally as important as toxicity, is the effectiveness of a dispersant.

FIGURE 200: DISPERSANT APPLICATION BY EDUCTOR METHOD

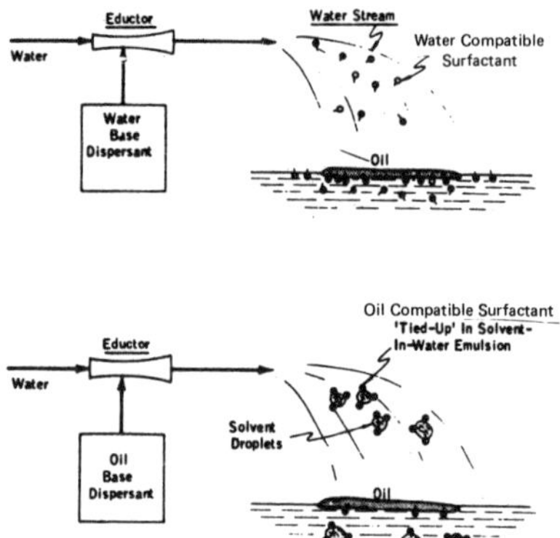

Source: Report PB 213,880

Evaluation of the effectiveness of dispersants during accidental spill incidents is difficult. Lack of adequate methods for measuring the amount of oil on water and the rate of natural dispersion make precise evaluation impossible. Their effectiveness during the Torrey Canyon is still being debated. Subsequent incidents which are claimed to have demonstrated their effectiveness have been at remote locations and without impartial, qualified observers. Application methods of dispersants and subsequent agitation, which are critical for effective performance, have not always been optimal.

FIGURE 201: OIL DISPERSANT EFFECTIVENESS

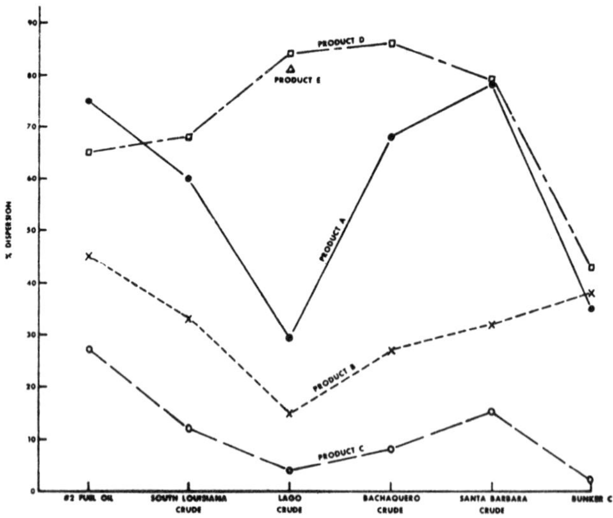

Source: Report PB 213,880

As an initial step in trying to solve this problem EPA developed a standard test for measuring emulsion efficiency. At this time, four private labs are testing the test to determine its applicability using a variety of oils and dispersants. Figure 201, which shows the results of testing done at the Edison Water Quality Laboratory adequately demonstrates the degree of variability which will occur when using different dispersants on the same type of oil and when using the same dispersant on different types of oil.

Continental Oil Company Process

A process developed by *D.R. Weimer and J.A. Wingrave; U.S. Patent 3,625,857; Dec. 7, 1971; assigned to Continental Oil Company* employs a dispersant for oil spills which comprises from 80 to 0 weight percent of an anionic surfactant or pine oil and from 20 to 100 weight percent of a nonionic compound which contains from 2 to 4 (CH_2CH_2O) units.

McNeely Process

A process developed by *W.H. McNeely; U.S. Patent 3,532,622; October 6, 1970* is one in which an oil slick on a body of water is dispersed by dividing the oil slick and concentrating the oil on the bow wave created by a boat propelled through the oil slick. At the same time a mixture of water and chemical dispersant is sprayed in high pressure jets which are swept across the bow wave in a cyclic oscillating motion substantially perpendicular to the length of the boat, thus producing a zig-zag spray pattern on the oil slick due to the forward motion to the boat. A near constant angular speed in the oscillatory motion of the jets automatically applies a greater concentration of dispersant adjacent the boat, where the oil is heaviest on the bow wave. The high dilution of the dispersant with environment water increases emulsification and turbulence for increased efficiency.

Figure 202 shows a suitable form of apparatus for the conduct of the process. The operation may be carried out by a boat **10**, which is preferably a wide beam type with a blunt bow to create a wide bow wave at low speed. When such a boat is driven through an oil slick, the oil slick is divided and the oil is concentrated in the bow wave, with the heaviest concentration close to the sides of the boat. It has been found that a boat with a 25 foot beam, moving at 3 to 6 miles an hour, will create a 20 to 25 foot wide surface disturbance of the oil slick area on each side of the boat. When the oil slick is treated according to this process there results a clear path at least 60 feet wide being made in the oil slick at one pass of the boat. This is merely an example to indicate the rapidity and extent of clearing possible with a single small boat.

A highly successful implementation of the method achieved the necessary lateral, high pressure spraying by a pair of hose nozzles **12** mounted on opposite sides of the boat. The basic method does not require mounting directly on the boat or any particular part of the boat, but a convenient installation is near the bow, either on the deck as shown or on superstructure of the boat if the arrangement is suitable. Since the flow rates and pressures are considerable, the nozzles are preferably mounted on supports **14** and are pivoted to swing in generally vertical planes substantially perpendicular to the length of the boat.

The angular range of swing is sufficient to enable a jet to be directed downwardly very close to the side of the boat and outward to the approximate width of the bow wave disturbance, as indicated in the full broken line positions in the lower view of the figure. For most efficient employment of the method the nozzles **12** are actuated by water powered monitor units **16**, of conventional type, which provide a constantly timed sweep over a preset angular range. However, manual operation of the nozzles is practical in some circumstances.

Supply hoses **18** extend from nozzles **12** to a branch coupling **20**, which is connected to the outlet side of a primary pump **22**, a pickup hose **24** leading from the inlet side of the pump into the water clear of the oil ordinarily at the stern of the boat. In this connection it should be noted that very high dilution of the dispersant by water from this environmental source produces a large volume in the dispersant spray. This large volume

FIGURE 202: PLAN AND ELEVATION OF VESSEL FOR DISPERSION OF OIL SLICKS

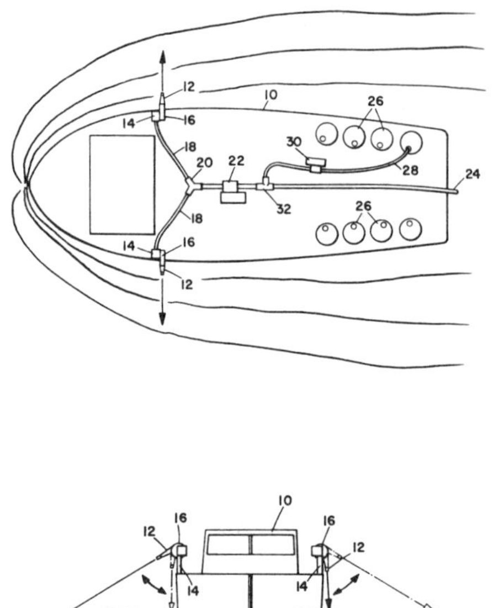

Source: W.H. McNeely; U.S. Patent 3,532,622; October 6, 1970

results in increased turbulence and emulsification. The boat carries a supply **26**, diagrammatically illustrated, of chemical dispersant, from which a pickup hose **28** leads to an injection pump **30**, the injection pump delivering dispersant at a controlled rate to an injection T-fitting **32** in hose **24** at the inlet side of the primary pump. On a boat not fitted with a suitable tank, the dispersant could be stored in barrels secured on the after deck.

A chemical dispersant particularly suited to the technique is sold by Ara Chem. Inc. (Gold Crew Dispersant) and is compatible with fresh or salt water. The water can thus be obtained directly from the main body of water on which the boat is operating and need not be stored on board, which is a limiting requirement with some other techniques. The efficient method of application permits the dispersant to be used in highly diluted form.

A ratio of one part of dispersant to 40 parts of water has been found effective on heavy oil and a ratio of one part dispersant to 80 parts water for light oil. The actual proportions are variable and can be quickly determined by tests. By comparison, sprinkler type sprays used on boats or aircraft normally use dispersant to water ratios on the order 1 to 4, and leave residual contamination in the water.

In a large oil slick the boat is driven at a constant speed through the oil, while the nozzles are operated to swing the jets with a cyclic oscillation across the bow wave. To avoid missing any spots the rate of oscillation must be fairly rapid, on the order of one or two complete in and out cycles a second, although this will vary with boat speed and jet divergence. It is in this mode of operation that powered monitor operation of the nozzles is necessary, since prolonged manual operation at such rates would be too tiring. For patchy oil, manually controlled nozzles could be used alone or in combination with the automatic nozzles.

Flow rates from the nozzles vary from 40 to 150 gal/min, depending on oil thickness, speed and other factors, the delivery pressure being on the order of 150 to 200 psi from pump **22.** The jets thus strike the oil with considerable impact and break up the oil into microscopic droplets or particles with an emulsifying action, and the particles are spread through the water by turbulence and the jet force, such that the film of oil from the surface is widely dissipated. The violent action of the high pressure jets in breaking up the oil makes it practical to use the chemical dispersant in highly diluted form. This results in a minimum of contamination of the water by the dispersion operation, well within the requirements of the Federal Water Pollution Control Administration.

It should be noted that there is a unique result from applying the jets of dispersant in the described manner across the bow wave. The oil will naturally be more concentrated close to the boat where the separation and disturbance are greatest, and the dispersant is distributed accordingly. From the view in the lower part of the figure, it can be seen that, with the nozzles swinging at a constant rate, considerably more dispersant will be applied to the oil close to the boat in the inner part of the swing motion where the nozzles are directed substantially downward. The distribution of dispersant and the force of the jet action is thus generally in proportion to the concentration of oil across the bow wave.

Once dispersed in the described manner the oil does not tend to collect again and the water remains clear. In one instance involving a spill of 80,000 gallons of black fuel oil, the oil was completely dispersed and there was no detectable damage to marine life in the area. The technique is practical in rough water and windy conditions, which would break up floating barriers and scatter light sprays. Operations have been successfully carried out in up to 10 foot ocean swells and gusty winds. One boat of the size described is capable of cutting a clear path 60 feet or more in width. With repeated passes, slightly overlapping to allow for drift, a large area can be cleared rapidly and the process can be safely used close to shoreline property.

Prial-Pigulski Process

A process developed by *G. Prial and C.J. Pigulski; U.S. Patent 3,681,264; August 1, 1972* employs an oil slick dispersant which is comprised of an aqueous solution of from 5 to 7% by weight of a mixture of alkylolamide alkyl aryl sulfonate and monobutylbiphenyl sodium monosulfonate and of from 1.5 to 3.0% by weight of tetrasodium pyrophosphate.

SORPTION

Oil is soaked up by an absorbent; it clings to the surface of the particles of an adsorbent. Nowadays, there are many such materials for cleaning up oil spills, including straw, foamed plastics, cotton waste, talc, and dried volcanic rock. Their properties for picking up oil have been carefully measured. In general, the lower the density of the sorbent, the more oil it can pick up per unit weight.

When it is first placed on water, a sorbent will usually float because of its low density. In time, however, many sorbents tend to become waterlogged. The air pockets of sorbents which originally provided buoyancy may become filled with water, sinking the water soaked mass. Only a few solid materials like polypropylene or polyethylene, which in foam form are lighter than water, float indefinitely. Polyurethane foam, because some of the cells in it are completely closed, is permanently buoyant.

Distributing straw, polyurethane foam, or other similar material on water can be readily accomplished with a mulcher of the type used by state highway departments for seeding highway rights-of-way. The wind, if strong, can be a deterrent to the operation.

Recovery of oil soaked sorbents may require a great deal of manual labor. Often, pitchforks are the only practical tools. In other instances, especially for large operations, modifications to conventional power equipment may serve. A propelled screen, for example, can be used to pick up sorbent materials, and moving mesh belts increase the rate of recovery. Worthy of consideration in this respect is modification of those vessels used to cut aquatic weeds in lakes or to harvest kelp along the coasts.

Sorbents are oil spill scavengers or cleanup agents which adsorb and/or absorb oil. Based on origin, sorbents may be divided into three general classes: (1) natural products, (2) modified or chemically-treated natural products, and (3) synthetic or man-made products. Sorbents may be further classified as to their physical characteristics: (1) powdery products, (2) granular products, (3) fibrous materials, and (4) preformed foam slabs or sheets. Types of floating sorbent materials available are:

Natural Origin: Types derived from vegetative sources comprise straw, hay, seaweed, kelp, ground bark, sawdust, reclaimed fibers from paper processing, and peat moss. Types derived from mineral sources may include the various clays, including montmorillonite, kaolin, fullers' earth, diatomaceous earth, etc.; vermiculite and the other micas, many forms of silicates, perlite, pumice, and asbestos. Sorbents of animal origin include chrome shavings from leather processing, wool wastes, feathers, and textile wastes.

Modified Natural Products: These materials comprise most of the sorbent types mentioned above but are chemically treated to produce a more desirable result. Some of the modified types are expanded perlite, charcoal, stearate-coated talc, asbestos treated with surfactant, and sawdust and vermiculite coated with silicones.

Synthetic Products: These sorbents include a vast array broadly categorized as plastics and rubber, but more specifically as the ethylenes, styrenes, resins, polymers and copolymers.

The desired characteristics for sorbent products have been listed by the Office of Water Programs of EPA in Report PB 213,880, Springfield, Va., National Technical Information Service (December 1971). The product should be:

(1) Oleophilic — have greater attraction for oil than water.
(2) Hydrophobic — should repel or reject water.
(3) Adsorptive — oil should adhere to the surface of the material.
(4) Absorptive — oil should be assimilated into the material.
(5) Possessed of High Oil Capacity — the rates of oil picked up to material applied (lb/lb) should be at least 5:1 but preferably 10:1 or even greater.
(6) Retentive — leaking of oil from the material should be minimal after harvesting.
(7) Low in Cost — on a nonreusable product. High initial costs are acceptable for reusable products.
(8) Buoyant — product should float under all conditions.

According to EPA, no existing products have all of these desirable qualities. Some advantages and disadvantages of the use of sorbent materials are presented as follows:

Advantages
(1) Aids in removing oil from water surface and alleviates or precludes undesirable after effects.
(2) Minimizes and decreases spread of oil.
(3) Generally, inexpensive and available in large quantities.
(4) Nontoxic.
(5) May increase both performance of booms and skimming techniques.
(6) Minimizes shore pollution when bleached.

Disadvantages
(1) No effective workable system at this time.
(2) High labor costs associated with acquisition, transportation, stockpiling, deployment, distribution on and working into an oil slick, retrieval and ultimate disposal.
(3) Manual retrieval only practical under calm conditions.

(4) Some products interfere with other forms of physical removal by clogging skimming and suction devices.
(5) Pollutional problems on disposal.
(6) Some sorbent products ultimately sink.
  (a) As their true density is greater than water, certain mineral products sink when air entrained in capillaries is replaced by oil and/or water or when products are wetted by water.
  (b) Some natural vegetative products such as straw, sawdust or waste pulp fibers become waterlogged upon prolonged exposure in water and may sink.

Sorbents may be used for many reasons:

(A) To agglomerate oil from a massive spill to minimize spread and potential damage. This could be applicable to spills on large open bodies of water and in areas where other control procedures are ineffective or unavailable;
(B) To polish a slick remaining after other control procedures such as booming and skimming have removed most of the oil;
(C) To sorb free oil from contaminated surfaces to facilitate cleanup;
(D) To deploy sorbents onto open waters and clean beaches in anticipation of uncontrolled slick; and
(E) To be used in conjunction with either fixed or towed booms.

To define sorbent effectiveness certain basic questions relating to the environmental use of sorbents under actual spill conditions must be considered. (A) Why were sorbents used? (B) What procedures will be used to harvest the oil-sorbent mass? In each of the use situations listed above, the effectiveness must be evaluated differently because criteria for evaluating effectiveness must be related to the original objective for each different use. For instance, the cost/application ratio (i.e., product costs/unit of oil sorbed) may be of primary importance in selecting a product for agglomerating a massive spill but would be of a lesser degree of importance for polishing action or for alleviating or minimizing potential damage.

The accumulation of oil by sorbent products is a complex phenomenon dependent upon several physical processes. For example, when straw is used the different processes most probably occur as follows: (1) Adsorption of oil to the straw surface; (2) Absorption of oil into the interstitial fibers and filtration into the hollow stem; and (3) Pickup of additional oil by oil-to-oil cohesion when saturation is reached.

Products have different adsorptive and absorptive capacities. For example, for free-flowing oils: (1) A highly pulverized mineral product manifests primarily adsorption. The pickup capacity is related to the surface area available. (2) A polymeric foam product manifests primarily absorption because of its high internal porosity. (3) The extraneous oil pickup of all products is related to the viscosity of the oil to be sorbed. The higher the viscosity, the greater the oil to oil cohesion and adhesion.

The criteria for the effective use of sorbents on unconfined oil slicks may be tabulated as follows:

(A) Products should be uniformly applied to the slick. Most unconfined oil slicks thin out very readily. Therefore, in order not to waste product, a thin uniform layer is preferred.
(B) The effectiveness of most products is increased by ultimately mixing the product into the oil slick. This can be accomplished:

  (1) By churning up the mass by running boats through at high speed after broadcasting of products.
  (2) By towing netting stretched between boats through the mass at slow speed.

(3) By existing environmental energy such as wind, wave and current action.

(4) If the above is not feasible, a time period of at least 3 to 6 hours should be allowed for the oil and product to mix. Most spills occur in tidal areas, and as such, the change of tide produces a minimum mixing energy. Additionally, the weight of most sorbent products will be sufficient to work the product into the oil if sufficient time is allowed.

(5) A harvesting procedure utilizing netting or booms which encircles a given area and concentrates the oil-sorbent mass by decreasing the area is also effective.

At this time, there are no standard tests for evaluating the effectiveness of sorbents. However, small scale laboratory testing of many products have been completed which at least give a measure of the relative oil capacity of many available products. Table 17 summarizes the results of bench scale testing of the oil sorbing capacity of selected products. The data was summarized from Edison Water Quality Laboratory tests, the oil pollution literature and manufacturers' brochures.

Table 17 reflects product costs varying from $20 to $20,000 per ton which breaks down to costs of approximately $0.02 to $1.00 per gallon of oil sorbed. These costs are product costs only, based on the reported oil capacity of the products. They do not reflect the equipment and labor costs associated with their use in the environment.

Compared with other oil spill cleanup techniques such as booming, skimming, dispersing and sinking, cleanup with sorbents is reportedly the most costly procedure. The costs vary from $0.50 to $5.00 per gallon of oil picked up, depending on the size of the spill and the equipment utilized. In the absence of an effective system for utilizing sorbents under actual spill conditions, the high costs are a reflection primarily of the labor costs of the multistep process.

TABLE 17: COSTS OF SORBENTS*

Type Material	Pickup Ratio Weight Oil Pickup: Weight Absorbent	Unit Cost Absorbent $ (Ton Absorbent)	$ Cost of Absorbent for Cleanup of 1,000 Gal Oil Spill
Ground pine bark, undried	0.9	6	27
Ground pine bark, dried	1.3	15	47
Ground pine bark	3	-	-
Sawdust, dried	1.2	15	50
Industrial sawdust	-	56	-
Reclaimed paper fibers, dried, surface treated	1.7	30	75
Fibrous sawdust	3	-	-
Porous peat moss	1.0	-	-
Ground corncobs	5	30	21
Straw	3 - 5	30	27
Chrome leather shavings	10	-	-
Asbestos, treated	4	500	440
Fibrous perlite	5	416	290
Perlite, treated	2.5	230	320
Talcs, treated	2	70 - 120	120 - 210
Vermiculite, dried	2	-	-
Fuller's earth	-	25	-
Polyester plastic shavings	3.5 - 5.5	100	80
Nylon-polypropylene rayon	6 - 15	-	-
Resin type	12	3,100	900
Polyurethane foam[1] **	70	20,000	1,000
Polyurethane foam[2]	15	4,500	1,050
Polyurethane foam[3]	70	-	-

(continued)

TABLE 17: (continued)

Type Material	Pickup Ratio Weight Oil Pickup: Weight Absorbent	Unit Cost Absorbent $ (Ton Absorbent)	$ Cost of Absorbent for Cleanup of 1,000 Gal Oil Spill
Polyurethane foam[4]	40	2,260	195
Polyurethane foam[5]	80	1,200	55

*Product costs per 1,000 gal spill. Does not include labor and equipment costs.
**Numbers refer to different types.

Source: Report PB 213,880

E.A. Milz of Shell Pipeline Corp., Houston, Texas has reported the performance of 15 floating sorbents tested on a relatively large scale. Summary of test data is presented in Table 18. Conclusions from tests were as follows:

(1) Tests showed that pound for pound polyurethane and urea-formaldehyde foams are the best oil sorbents. On a weight basis these materials removed about 10 times more oil than other sorbents tested.

(2) For long contact times the quantity of oil sorbed is primarily a function of oil viscosity and density.

(3) In most cases sorbents lose oil after pickup by drainage and evaporation. About eighty per cent of oil initially absorbed will be retained after draining for 24 hours. One exception to the above is Kraton rubber which dissolves oil into the body of the material and completely retains it. To facilitate sorption, samples were granulated to a particle size of four millimeters. Kraton rubber may also be foamed which should improve sorption.

(4) Power mulching machines can effectively spread up to 9 tons of straw or hay per hour with immediate mixing with oil. In contrast, when hay was simply dumped overboard, it remained in large dry clumps for over half an hour even in 4 to 5 foot waves.

(5) Power mulchers are satisfactory for shredding and spreading polyurethane foams.

(6) When sorbents are used with booms the same problems are encountered as by containment of oil alone by booms. Wave action will cause spill-over and currents above 1 fps will cause undersweep.

(7) Nets are more effective than booms for containing relatively small quantities of stringy material such as hay, bark and shredded foam. With 1 inch mesh nets, velocities of 2 to 3 fps are possible for small quantities of material without product loss. For large quantities of material, velocities of 1 to 2 fps are possible without failure (Figure 203).

(8) All collection or harvesting processes for oiled sorbents on open water involve screening processes. Therefore, the selection of mesh size for harvesting devices is critical and is related to the sorbent particle size.

TABLE 18: LARGE SCALE SORBENT TESTS

Test Conditions: Sorbent Applied to Oil Slick Confined to 20′ x 20′ Area, No Mixing

Sorbent Material	Quantity Used (lb)	Test Oil: Thickness (inches)	Test Oil: Quantity (lb)	Test Oil: Viscosity (cs)	Test Time (hr)	Oil Removed (lb)	Oil to Sorbent Ratio
Kraton 1101	25	¼	700	4	24	28	1.1
Kraton 1107	25	¼	700	4	24	ca 30	1.2
Urea-formaldehyde 2″ x 24″ x 60″	1.6	¼	700	4	24	41	26
Polyurethane[1]	20	¼	556	6	2	560	28

(continued)

TABLE 18: (continued)

Sorbent Material	Quantity Used (lb)	Test Oil Thickness (inches)	Test Oil Quantity (lb)	Test Oil Viscosity (cs)	Test Time (hr)	Oil Removed (lb)	Oil to Sorbent Ratio
Polyurethane[2]	20	¼	660	6	1	100	5
Polyurethane[3]	5	¼	300	6	5	230	46
Ekoperl	24	¼	300	6	24	120	5
Hay[4]	49	¼	700	6	24	210	4

[1] Fine ground polyurethane (ca ¼" diameter). For this test there was insufficient oil present to saturate the polyurethane. There was no trace of oil left after removing the polyurethane.

[2] Polyurethane in 9' long by 2' diameter bag. The bag picked up over 500 lb of water and only 102 lb of oil.

[3] Scraps of polyurethane of 1 and 2-inch thickness in various shapes and sizes up to 1' by 4'.

[4] Performance of straw and bagasse are similar to hay.

Source: Report PB 213,880

FIGURE 203: SORBENT BARRIER FAILURE MODES IN CURRENT

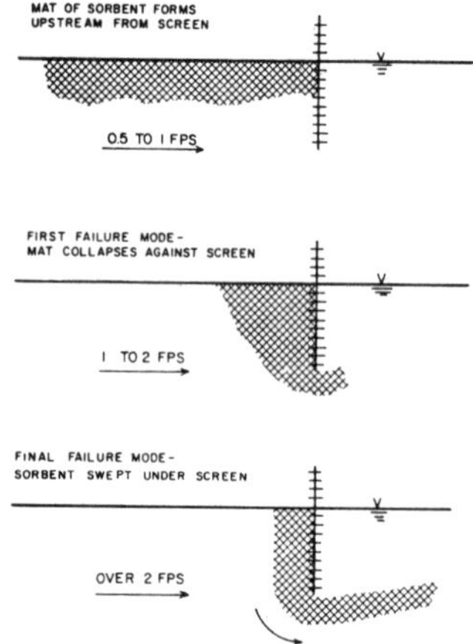

Source: Report PB 213,880

Because of the inherent limitations or regulatory restrictions for such techniques as booming, skimming, sinking, dispersing, burning or gelling, the advantages of sorbent systems for cleaning up oil spills show great promise. Reusable high-oil capacity sorbent products used in conjunction with self-contained mechanized systems will be developed which will have the following desirable features:

(1) Recovery of oil which decreases problem of disposal. The recovered oil can be taken to a separator and incorporated into refinery feeds rather than buried.

(2) Products will be reusable. Certain polyfoam products have an initially high cost which decreases geometrically for each subsequent reuse. Such products may be used dozens of times and potentially hundreds of times by reinforcement of product with plastic netting.

(3) On open waters under inclement conditions there is no expediency for harvesting. The products may be beached and subsequently the oil and product recovered on the beach or gathered later under calm conditions from the water surface.

Costs will decrease as manual labor is replaced by mechanical equipment incorporated into continuous self-contained systems. The EPA, U.S. Coast Guard and the API are evaluating proposed sorbent systems submitted for research funding.

The development and preliminary design of a sorbent-oil recovery system has been reported by E. Miller, L. Stephens and J. Ricklis of Hydronautics, Inc., Laurel, Md. in Report EPA-R2-73-156, Wash., D.C., EPA (Jan. 1973).

A development program has been completed and preliminary designs were prepared for 3,000 gallons per hour protected water and 10,000 gallons per hour unprotected water sorbent-oil recovery systems. The five phases in the development program were: (1) the characterization of the sorbent material, (2) the development of the sorbent broadcasting system, (3) the development of the harvesting conveyor and evaluation of overall recovery performance, (4) the development of the sorbent regeneration system, and (5) model tests of a ¼-scale model recovery platform.

The development program showed that a continuous sorbent-oil recovery system is feasible using 30 or 80 PPI polyurethane sorbent chips. In one pass about 90% of the oil in a 1.5 millimeter slick can be recovered. The water content of the recovered fluid is less than 10%. The preliminary designs are presented with detailed descriptions of the system components, operating procedures, and costs.

Depending on the options selected and the number produced, the initial cost of a 3,000 gph protected water recovery system will range between $40,000 and $80,000. For a 10,000 gph unprotected water recovery system the cost will range between $55,000 and $105,000. Operating costs for the 3,000 gph system will be between $100/hour and $200/hour and for the 10,000 gph system between $400/hour and $600/hour.

The sorbent material used in the recovery system was open cell reticulated polyurethane foam cut into chips approximately 3" x 3" x ¼". 80 PPI foam should be used for oils with viscosities less than 1,000 cp and 30 PPI foam should be used for viscosities greater than 1,000 cp. It was shown that the sorbent material could be regenerated in excess of 100 times without deterioration. Sorbent materials are both effective and relatively cheap. Even though they must be handled twice and then disposed of by burning or burial, they will probably continue to be widely used, since the use of dispersants, sinking agents, and biological agents is sharply restricted by federal, state and local regulations.

Anderson Cob Mills Incorporated Process

A process developed by *J. Vander Hooven and D.I.B. Vander Hooven; U.S. Patent 3,617,564; November 2, 1971; assigned to Anderson Cob Mills Incorporated* is one in which low-density corncob meal is placed on a polluting oil deposit in a body of water or on a land area. The corncob components absorb the oil and the contaminated corncob meal is then removed from the body of water or from the land area.

Anheuser-Busch Incorporated Process

A process developed by *J. Teng, J.M. Lucas and R.E. Pyler; U.S. Patent 3,788,984; January 29, 1974; assigned to Anheuser-Busch, Incorporated* involves removing spilled oil from water surfaces without further injuring the environment, by applying a carbohydrate fatty acid ester in powder, fibrous, or granule form to the oil covered surface. The preferred ingredient is cellulose acetate. The additive is nontoxic, biodegradable, water-insoluble, and is not degraded by the acids in petroleum fuels. The additive is sprinkled on the surface of the oil-coated water and after absorbing many times its weight in oil is easily removed. The picked up oil is easily removed from the cellulose acetate and up to 95% recovery of spilled oil is possible. The cellulose acetate also can be reused. Cottonseed hulls and sawdust as the starting cellulose material are desirable because of low cost and the enhanced oil absorbing character of the resulting acetate.

Anvar Process

A process developed by *A. Abadie, A. Gazost and H. Roques; U.S. Patent 3,729,410; April 24, 1973; assigned to Anvar, France* is a process for deoiling polluted waters which comprises contacting the oil-containing water with an oleophilic resin having a high surface area, for example, in the form of a powder or small spherules or granules, and removing the treated water from the resin, after several stages as necessary if a particular degree of deoiling is desired. The oleophilic resin can be an ion exchange resin; the oleophilic properties are imparted thereto by organic ions fixed thereon. Accordingly, an ion exchange operation can be conducted concurrently with the deoiling process, if desired.

BASF Wyandotte Corporation Process

A process developed by *J.V. Otrhalek; U.S. Patent 3,729,411; April 24, 1973; assigned to BASF Wyandotte Corporation* is one in which oil slicks and other oil films are removed from bodies of water by casting onto the water a silicone-treated expanded volcanic ash containing, optionally, a thickening agent, and thereafter, removing the oil absorbed composition from the water.

Baumann Process

A process developed by *H. Baumann; U.S. Patent 3,598,729; August 10, 1971* is based on the discovery that when a piece of fully cured urea-formaldehyde resin foam is dipped into an oil slick floating on water, only the oil slick is absorbed into the piece of foam. Entry of the water is blocked by the narrow capillaries in the cell walls which connect most of the otherwise sealed air cells in the foam to each other and to the atmosphere. They permit absorption of liquids of low surface tension, but not of liquids having a surface tension as high as that of ordinary water or of seawater which essentially consists of water. Urea-formaldehyde resin is much less costly than the corresponding resins prepared from melamine and phenol whose foams have a similar structure and are similarly effective.

Cabot Corporation Process

A process developed by *P.R. Tully, R.J. Lippe and W.J. Fletcher; U.S. Patent 3,562,153; February 9, 1971; assigned to Cabot Corporation* is based on the discovery that improved oil absorbent materials are provided when a liquid absorbing material is treated with certain colloidal hydrophobic metal or metalloid oxides. When contacted with oil contaminated water, the improved absorbents thus produced display superior qualities of buoyancy, water repellency and oil receptivity.

Absorbent materials suitable for treatment by the process are generally any inorganic or organic solid capable of imbibing liquids. Often, the starting absorbent material will be particulate, granular or fibrous in nature, and in the interests of facile handling and treatment, such materials will desirably be at least 50 microns in their smallest average dimension. Accordingly, crushed charcoal or coke, sand, kieselguhr, diatomaceous earth, peat, textile fibers, sawdust, chalk, mica, expanded mica, cork, felt, straw, wood chips, paper, nut shells, granulated corncobs and the like are all normally useful starting materials. Due to their generally close location with respect to water environments, diatomaceous earth, sand and dry marsh vegetation such as salt hay are normally advantageously employed.

The colloidal oxides useful in the practice of the process can generally be any metal or metalloid oxide having an average ultimate particle diameter of less than about 0.5 micron (preferably less than about 0.1 micron) and a BET-N_2 surface area of at least 50 m^2/g (preferably greater than about 100 m^2/gram). In order that the particulate metal or metalloid oxide be rendered substantially permanently hydrophobic by a chemisorption reaction thereof with a organosilicon compound it is of further importance that the starting oxide material bear on the surface thereof at least about 0.25 milliequivalent per gram and preferably above 1.0 milliequivalent per gram of hydroxyl groups. Specific examples of suitable available starting material oxides are: pyrogenic and precipitated silicas, titania, alumina, zirconia, vanadia, chromia, iron oxide, silica/alumina, etc.

Additionally, it is desirable that the oxide be relatively nonporous, i.e., that the preponderance of the total surface area thereof be external rather than internal (pore volume). The relative porosity of a given colloidal particulate solid can be determined by (a) calculating the surface area thereof predicated upon the average particle diameter (such as visually determined by electron micrographic analysis) and assuming no porosity; (b) experimentally determining the actual total surface area by the well-known BET-N_2 adsorption method. Accordingly, the porosity of the particulate solid is expressed as follows:

$$\%\ \text{Porosity} = \frac{\text{BET-N}_2\ \text{surface area} - \text{EM surface area}}{\text{BET-N}_2\ \text{surface area}} \times 100$$

For the purpose of the process, those particulate colloidal metal or metalloid oxides having a porosity of less than about 10% are to be considered relatively nonporous. Due principally to the above porosity consideration as well as their normally relatively high hydroxyl group populations surface areas and general availability, pyrogenic and precipitated silicas are starting materials of choice.

Ciba-Geigy Process

A process developed by *A. Renner; U.S. Patent 3,716,483; February 13, 1973; assigned to Ciba-Geigy, Switzerland* is a process for removing oil from water in which the contaminated water is brought into contact with a highly disperse, solid, water-insoluble organic polymer, for example melamine-formaldehyde resin or polyacrylonitrile of average molecular weight greater than 1,000 and a specific surface area greater than 5 m^2/g, and the polymer charged with the contamination is separated from the water.

Col-Mont Corporation Processes

A process developed by *C.O. Bunn; U.S. Patent 3,536,615; October 27, 1970; assigned to Col-Mont Corporation* is one in which oil leakage on the surface of bodies of water is treated by spreading over the surface of the water discrete particles coated with finely divided carbonaceous or the like material having a high affinity for oil. Oil is adsorbed on the surface of the particles, with the particles being thereafter collected for further treatment, for example, drying and agglomerating. The particles following such treatment constitute an economic fuel source immediately usable for fuel purposes.

Figure 204 is a flow diagram of the process of making the affinitive, coated particles. As shown in the figure, a feed conveyor **10** of conventional construction travels beneath a hopper **12** or like storage container for receiving the basic particles, designated **P**. The particles can, as above noted, be rather widely variable as long as the abovementioned characteristics are present. Commonly available materials which can be satisfactorily used are sawdust, shavings, wood chips, bast, bark, cork and various other vegetable fibers which have the requisite characteristics.

Although certainly not exhaustive, a list of such vegetable fibers includes cotton, flax, coir, abaca, hemp, henequen, jute, ramee, sisal, pina, kapok, alginate, lastex, rubber, paper, papier maché, palm, peanut and other hulls, etc. Animal fibers and materials, including wool, mohair, hair, casein, silk, animal dung, hide, bones, horn, and fat can also be used if necessary where availability permits consideration of such material. Foamed thermoplastic or thermosetting materials may also be used. Hereinafter, the particles will be referred to as wood chips, with the word chip being employed in its broadest sense to include shavings, sawdust, or wood particles of varying sizes and shapes.

FIGURE 204: PROCESS FOR PRODUCING OIL ABSORBENT PARTICLES

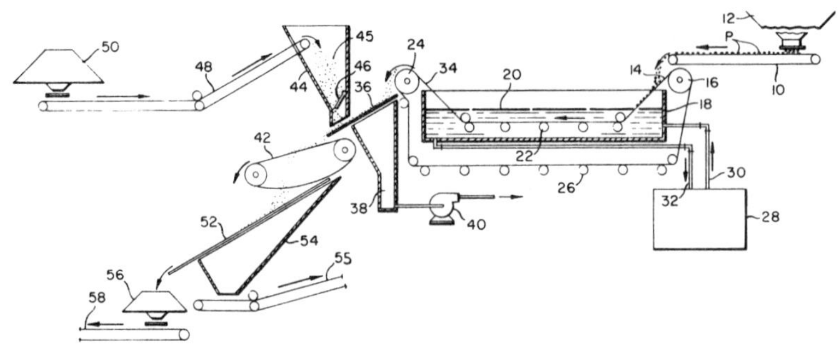

FIGURE 205: COL-MONT ABSORBENT SPREADER SHIP AND RECOVERY SHIP COMBINATION

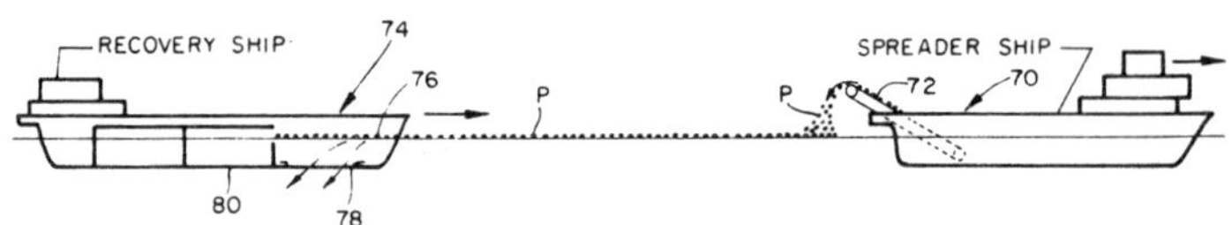

Source: C.O. Bunn; U.S. Patent 3,536,615; October 27, 1970

The wood chips **P** pass from the conveyor **10**, which is conventional in construction and need not be described in detail, onto a moving, endless conveyor belt **14**, one drive sprocket **16** for which is positioned generally below the discharge end of conveyor **10**. The conveyor belt **14** is trained downwardly through the confines of a tank **18** containing an adhesive and sealant material **20**. As shown diagrammatically, the level of the material **20** in the tank **18** is preferably maintained so that conveyor **14** is essentially submerged to agitate and coat the wood chips while passing through the tank in the direction indicated by the arrow. Guide rollers **22** are positioned in the tank **18** transversely to vertically guide the belt **14**, and sprocket **24** and bottom guide rolls **26** complete the closed conveyor loop system which agitates and coats the wood chips.

The fluid **20** is heated to the desired temperature to reduce viscosity by a heater **28** located exteriorly of the tank **18**, and fluid lines **30** and **32**, respectively, return the heated fluid to the tank **18** and deliver fluid from the tank to be heated. The fluid should possess two characteristics. One, it must function as the bonding agent or adhesive between the particle **P** and the final coating material which provides the necessary affinity for oil. Secondly, the fluid serves as a sealant to prevent the float particles from becoming water laden, which condition is a significant impediment both to the flotation properties of the particles and to the subsequent processing of the oil adsorbed particles for fuel purposes.

Although any fluid or fluids having the above noted characteristics may be used, the fluid is preferably selected from natural resins such as shellac or resins which have suitable adhesive properties. Synthetic resins are also acceptable, with polyester resins and furan resins being exemplary. These resins may be used alone or compounded with other materials, or can be mixed if desired with tar sands to reduce costs and improve the combustion characteristics of the end product. Under some circumstances, asphalt or manufactured tars are satisfactory. The adhesive can be applied to the particles in the manner illustrated and above described or with a suitable hardener directed to the particles just prior to the coating of the same with fine carbon.

It will be understood that the material or materials employed will to a significant degree depend upon the availability and local economic advantage of the noted materials in the area concerned, and the material from which particles **P** are comprised. It will also be understood that two separate fluids could be provided to obtain the necessary sealing and adhesive properties with suitable hardening and curing agents as required.

The particles after traversing submerged through the tank **18** are carried therefrom by the upper run **34** of the conveyor **14**, with the coated chips descending into a drain screen **36** of a suitable mesh size to pass the fluid but retain the coated particles. The fluid passing through the screen **36** collects in sump **38** and is returned by pump **40** to the source of fluid supply for return to the system. After traversing the screen **36**, the coated particles descend onto the down run of an agitated endless, fine screen conveyor **42**.

Disposed vertically immediately above the discharge end of the screen **36** is a hopper **44** which is adapted to contain finely divided carbon material which is adapted to be adsorbed on the surface of the coated particles passing down screen **36**. A suitable gate valve **46** or the like can be provided interiorly of the hopper **44** for regulating the quantity of material discharged therefrom. The carbon material is fed to the hopper by means of conveyor **48** on which the material is deposited from a storage bin shown diagrammatically at **50**. The quantity of carbon material discharged from the hopper is in excess of that needed to completely coat the particles passing therebelow from the screen **36**. The excess carbon particles pass downwardly through the screen **52** to a bin **54**. The latter discharges material from the bottom thereof onto a moving conveyor **55** for suitable disposition of the excess carbon dust.

The carbon material **45** can be selected from any material having the necessary characteristics. The material must have a greater affinity for oil than for water, must be adhesively attached to the coated surface of the particles so as to preferably completely coat the exterior thereof, and must possess when combined with the particles, a sufficiently high BTU rating to constitute an economically feasible fuel source. Crushed or ground coal are the preferred materials used, although carbon black, graphite, lignite, peat, charcoal, coal char, gilsonite and oil shale are satisfactory alternative material dependent upon local economic advantage. The carbon coated particles traversing the conveyor **42** descend onto the fine screen conveyor **52** and travel downwardly thereover to a coated particle storage bin **56**. A conveyor **58** travels below the bin **56** for receiving the coated particles for further handling.

Figure 205 shows the spreading of the coated particles on the surface of the water and the recovery of the oil absorbed particles from the water surface for further treatment. A spreader ship generally indicated at **70** provides the carbon coated particles **P** with a conveyor **72** communicating with the storage or manufacturing area of the ship to the exterior thereof. In this manner, the particles can be uniformly spread across an area generally coextensive with the width of the conveyor **72**.

A recovery ship generally indicated at **74** trails the spreader ship **70** and is provided with skimmer means (not shown) or the like for skimming in the particles from the surface of the water for further processing. The particles thus entering the recovery ship initially pass over a screen **76** through which the water passes to a discharge opening **78**, with the coated particles being carried by the conveyor **76** to a suitable storage area in the ship, indicated at **80**. Although the function of the recovery ship **74** may cease at this point, further processing equipment preferably is housed in the recovery ship for processing the individual coated particles to what is essentially a full product in consumable form.

The recovery ship is thus preferably provided with a suitable drying oven through which the coated particles are passed for reducing the water content to the lowest extent possible. The dried particles are then conveyed to a pressing facility where the individual particles are compacted into briquette shaped particles of predetermined size. If desired, a suitable binder material may be added just prior to the pressing operation to facilitate agglomeration of the particles. After briquetting, the particles are essentially in condition for immediate use as fuel and can be conveyed to a separate storage area in the recovery ship for discharge from the ship.

Conveying means are preferably provided for discharging the fuel briquettes from the recovery ship either in port or to a separate vessel employed to transport the briquettes to a shore installation. If the latter arrangement is employed, it will be apparent that the recovery ship can operate essentially continuously at site with the spreading ship to eliminate the oil leakage.

Another process developed by *C.O. Bunn; U.S. Patent 3,651,948; March 28, 1972; assigned to Col-Mont Corporation* involves mixing discrete core particles of wood material, polyethylene and finely ground coal to form particles having a high affinity for oil. The particles have a high degree of integrity, thereby permitting storage of the same at points relatively remote from the point of leakage and airlifting of the same to the oil spot. The oil-saturated particles are processed after collection for further use as a fuel source, with the BTU content being as great or greater than bituminous coal.

Figure 206 is a diagram of a suitable form of apparatus for producing the absorbent particles of the process. In the drawing, the inlets for the materials have been generally indicated at **10, 12** and **14**, respectively. The materials are fed through the inlets to a mixer generally indicated at **16**, which may comprise a series of longitudinally spaced and connected mixing devices of commercially available construction.

Various types of mixers are available for intimately mixing essentially any flowable solids, liquids or gases (Static Mixer, Kenics Corp.). This mixer is disclosed and claimed in U.S. Patent 3,286,992, and reference is specifically made thereto for a full understanding of the structure and function of the mixing device. Briefly, the device is provided on the interior thereof with a plurality of curved blades or elements, each of which affects the reversal of the flow from the immediately preceding element whereby the feed material is constantly being divided and intimately mixed with the other materials fed into the mixer.

FIGURE 206: ALTERNATIVE DEVICE FOR PRODUCING ABSORBENT PARTICLES

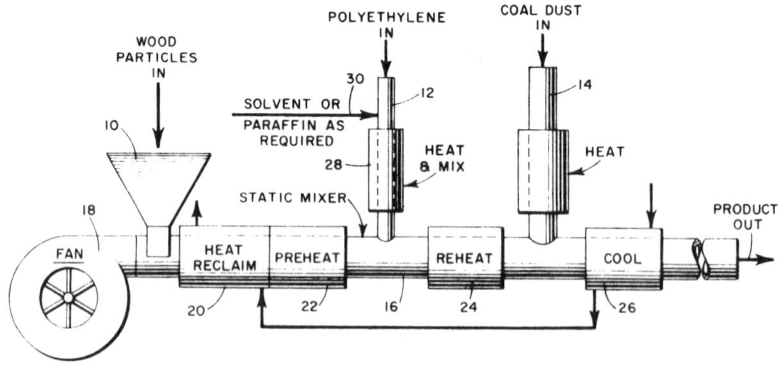

Source: C.O. Bunn; U.S. Patent 3,651,948; March 28, 1972

The material is forced through the mixer by means of a blower generally indicated at **18** mounted at the forward end of the apparatus adjacent the inlet **10**. Disposed circumferentially around the mixer **16** in longitudinally spaced relation are heat exchangers indicated at **20, 22, 24** and **26**. These exchangers are provided for the purpose of either heating or cooling the material as it longitudinally passes through the interior of the mixer. For example, heat exchanger **20** initially heats the wood particles as they enter the mixer, and preheat exchanger **22** further heats the material.

The heat exchanger **24** comprises a reheater which serves to elevate the temperature of the wood and polyethylene mixture, with the polyethylene being fed to the mixer immediately preceding the reheater **24**. The exchanger **26** comprises a cooler-heat exchanger which serves to cool the entire mixture after the coal has been introduced into the mixer through the inlet **14**. The product coming out of the mixer is preferably at a temperature sufficiently low to permit normal handling thereof. For purposes of efficient heat utilization, the heat resulting from the heat transfer at heat exchanger **26** may be used as the heat source for the heat exchanger **20**.

The wood particles entering the mixer through inlet **10** can be obtained from any suitable available source of material, and are preferably converted into chips approximately ¼" in size. From an economic standpoint, chips of this size can be manufactured completely from waste wood materials such as bark, edgings, rough trim, cull lumber, chippable residues, and the like. It will be understood that wood in the form of chips in excess of ¼" are also satisfactory, as well as wood in smaller size forms. In fact, sawdust can be satisfactorily employed. If desired, the wood material may be treated before introduction into the mixer through inlet **10**. For example, the wood may be coated with a sealant which may comprise materials such as, for example, shellacs, polyester and furan resins, natural resins, asphaltic materials, in addition to other well-known sealants such as silicones.

The polyethylene is fed into the mixer through inlet **12** after the wood has been introduced and preheated. The polyethylene is preferably of low grade, low density and low molecular weight, and can even comprise reground material for reasons of economy. The material when fed into the mixer is preferably in solid, ground form and can be supplied by air to the interior of the mixer. The polyethylene is preferably preheated prior to entry into the mixing chamber for the purpose of elevating the temperature thereof to approximately the temperature of the wood particles thereby reducing the heating requirements for the main chamber itself.

The viscosity of the polyethylene can be varied substantially by varying the feed temperature thereof or through the addition of solvents thereto through a solvent feedline **30**. Solvents suitable for this purpose include xylene, toluene, paraffin or similar known solvents and thinners. At 80°C, polyethylene is miscible in organic solvents such as toluene and xylene, and the fluid condition of the polyethylene and solvent carrier facilitates the mixing operation, as well as permitting reduction of the temperature at which the coal is mixed with the polyethylene coated wood. The solvent is boiled off at a predetermined elevated temperature.

In addition, although not shown, other materials can be blended with polyethylene to obtain the desired properties. For example, nylon or polymers could be blended with the polyethylene to provide permeability properties more compatible with a particular size or type wood feed material. Significantly, any component added to polyethylene should possess essentially the same characteristic of combustibility in order to provide a product which when collected serves as a useful source of fuel.

The polyethylene is intimately mixed in the interior of the mixer with the wood chips and carried toward the exit end of the mixer, under the influence of the blower **18**. The mixture of polyethylene and wood chips is heated by the heat exchanger **24** to approximately 160°C, at which elevated temperature the adhesive properties of the polyethylene are fully developed in the temperature regions below the char point of the wood, and at this temperature the properties of the coal are not adversely affected. At temperatures much below

160°C, the adhesive properties of the polyethylene are not fully developed, and at temperatures in excess of 160°C, the likelihood of combustion of the wood and coal is greatly increased. It will be understood that the optimum temperature will vary depending upon the type and moisture content of the wood, the type, size and moisture content of the coal, the presence of additives to the polyethylene, and the residence time in the heating chamber. The coal dust is admitted into the mixer downstream of the reheat exchanger **24**. The coal dust can be of substantially any mesh size although superior results are obtained where the coal dust is -200 mesh in size. The quality of the coal can also significantly vary, with sub-bituminous as well as bituminous coal being entirely satisfactory. The carbon coated particles are thereafter passed through cooling heat exchanger **26** for further processing.

The wood chips are delivered to the mixer and heated therein and thereafter coated with ground polyethylene admitted to the mixture through the inlet **12**. The mixing device greatly enhances the binding of the coal to the wood chips, and predetermined amounts of wood chips and polyethylene are delivered to the mixer for effectively conditioning the wood particles prior to passage of the same to the area in the mixer adjacent the coal dust inlet **14**. The water absorption rate of the particles can be controlled by the amount of polyethylene added, with relatively lesser amounts of polyethylene resulting in less than complete sealing of the wood chips, thereby conditioning the particles for absorption of water and/or oil.

As the polyethylene coated particles pass below the coal dust inlet **14**, coal is added to the wood chips in such predetermined amounts as to bind coal to essentially all the wood particles passing adjacent thereto. As previously stated, the wood chips are heated during the traversal thereof through the mixer and are preferably at a temperature at least as high as 160°C prior to the dusting of the same with the coal particles.

After the application of the coal dust to the particles, the particles pass through the cooler-heat exchanger **26** for cooling the same and setting the bond between the wood chips and the coal dust. The particles can be removed from the mixer in any suitable manner and with known apparatus. For example, the particles could be collected at the end of the mixer and conveyed therefrom to apparatus for further handling of the particles.

As above noted, one of the principal advantages of the process is the lack of criticality in the sequence of addition of the material ingredients which form the final, coated product. In lieu of the arrangement as shown in the drawing where polyethylene is applied to the surface of the wood chips and the resultant coated particles thereafter mixed with coal, an alternative method comprises the intimate mixing of wood and coal, with the resulting discrete particles being thereafter mixed with finely ground polyethylene. This process can be carried out in the same type mixing apparatus illustrated in the drawing.

In the alternative process, the sawdust and coal are conveyed to the mixer **16** illustrated in the drawing and intimately mixed therein. The coal is preferably -200 mesh and can be, as previously indicated, sub-bituminous in quality. The sawdust, as contrasted to the relatively larger wood chips, provides for increased permeability, reduced bulk density, and a greater coal/wood ratio. To enhance the initimate mixing of the wood and coal, the former is preferably dampened, for example, with water, so as to permit the sawdust to be thoroughly coated with the finely ground coal and polyethylene.

The coated sawdust is thereafter intimately mixed with finely ground polyethylene, with the latter preferably covering essentially the entire surface of the particles. The polyethylene is added in predetermined amounts to effect such thorough mixing, and highly satisfactory results have been achieved where the wood to polyethylene to coal ratio is 1:2:3, respectively.

The sawdust thus sequentially coated with coal and polyethylene is thereafter heated at approximately 160°C until the polyethylene becomes adhesive. The heating step can be effected in a manner similar to that above described and illustrated in which the coated particles are passed through a heat exchanger for a time sufficient to reach such temperature and effect the adhesion process. Any suitable source of heat may be employed. The fused particles or agglomerates are thereafter cooled providing discrete particles having a great affinity for oil. In lieu of the mixing device illustrated, other apparatus may be employed for mixing and heating the components in both of the described methods. For example, batch pans, conveyor belts, fluidized beds, twin pug mixers or rotary mixers can be employed to effect the mixing and heating. The important consideration is the ability of the apparatus employed to mix thoroughly and develop heat quickly so as to effect bonding of the coal to the wood material.

The integrity of the finally treated particles is an important characteristic, permitting the transportation and/or storage of the particles over long periods of time without significant deterioration of the product, provided excessive oxidation of the particles is prevented. In the actual employment of the particles in the oil recovering operation, the particles can be spread over the surface of the water in any suitable manner such as spreading the particles on the surface of the water by means of a spreader ship, with the oil adsorbed particles being collected by a recovery ship for further processing.

In view of the integrity of the oil affinitive particles, the same can be stored at relatively remote locations and transported by air to the location where needed. Relatively large amounts of the product could be airlifted and air scattered on the area of the water surface bearing the oil. A recovery ship would then be dispatched to the location and the oil adsorbed particles collected.

On the basis of experimental tests, conducted in accordance with the alternative method just described, approximately three tons of float material will collect approximately 2 tons of oil. In view of the ready availability and cheapness of the ingredients comprising the oil affinitive particles of the process, it will be seen that the oil can be recovered very economically. It should also be kept in mind that, unlike previous oil recovery methods and materials, the method makes use of the collected particles as a readily available source of fuel. In this regard, further treatment of the particles is normally desirable and may, for example comprise drying and/or compacting of the particles into briquet shaped sizes for more convenient handling and storing. It will be apparent that the oil can be recovered, as an alternative treatment, by flask distillation.

In regard to the fuel value of the particles both before and after the oil recovery process, testing has shown that the Btu content of the particles before the oil absorption is approximately 14,200 Btu/lb, approximately the value of bituminous coal, and the fuel value after adsorption is approximately 18,000 Btu/lb. All of the disclosed ingredients add to the fuel value of the product, and it will be understood that the fuel value will vary significantly with the separate heating values of the wood and coal employed in the formation of the oil affinitive particles.

Still another process developed by *C.O. Bunn; U.S. Patent 3,783,129; January 1, 1974; assigned to Col-Mont Corporation* is one in which a matrix material is provided for recovering oil from water comprised of finely divided coal particles bonded in spaced relation by polyethylene. The oil sorption capability of the matrix is exceptionally high and the matrix is highly selective to oil in the presence of oil and water. A closed system is provided for forming the matrix material and for separating the sorbed oil from the matrix material for reuse of the latter. The matrix material can be in the form of a fixed or movable bed through which the oil and water pass for selective sorption of the oil, or the material can be dispersed on the water surface and collected following oil sorption.

Figure 207 is a schematic plan and elevation of a vessel adapted to receive the oil-water film and to sorb the oil in the matrix material in accordance with this process, with the ship containing equipment for draining the oil from the matrix material and for recycling the latter for reuse. The recovery vessel is generally indicated at **10**. The basic construction and powering for the vessel **10** can be of conventional design, with the interior of the ship being specifically designed and equipped in accordance with this process.

FIGURE 207: RECOVERY VESSEL FOR OIL SPILLS USING ABSORBENT PARTICLES

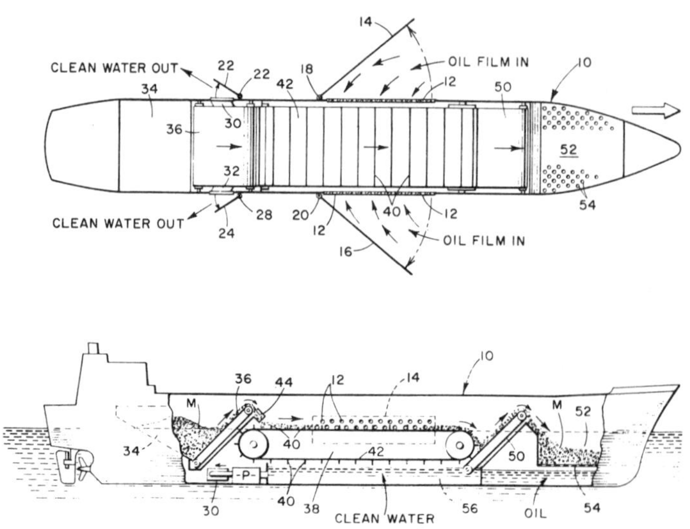

Source: C.O. Bunn; U.S. Patent 3,783,129; January 1, 1974

A plurality of openings, commonly designated at **12** are formed generally intermediate the vessel and are adapted to be closed when the vessel is inoperative by baffle members **14** and **16** mounted on the side of the vessel by pivotal connections **18** and **20**, respectively. The baffles **14** and **16** are shown in a fully open position and are adapted in such position to guide the oil and water into the interior of the vessel through openings **12** when the vessel is traveling in the direction indicated by the arrow. When the vessel **10** is inoperative for the purpose intended, the baffles are pivoted to a closed position contiguous the side of the vessel and retained in such position.

Outlet baffles **22** and **24** are pivotally connected on the vessel to the rear of baffles by pivotal connections **26** and **28**, respectively. Water entering the vessel with the oil flows outwardly of the vessel through openings **30** and **32** formed in the vessel, as indicated by the arrows, with the assistance of pump **P**, if necessary. Although not illustrated, it will be understood that the discharge of the clean water is accomplished without the entrance of water into the vessel through the openings **30** and **32**.

Located in the interior of the ship is a lean matrix storage area **34** for storing the matrix material indicated at **M**. A matrix feed conveyor **36** is provided at the forward, inclined end of the storage area **34** for conveying the lean matrix material **M** to a matrix bed conveyor **38** extending longitudinally of the ship in the intermediate portion thereof. The conveyor **38** includes spaced transverse plates or structurally adequate screen members commonly designated at **40** which cause the generally forward motion of the matrix counter to the generally backward flow of the oil-polluted water to insure integrity of the matrix bed and to maintain the counterflow action between the matrix and the oil and water. A pair of transverse distributor mechanisms or baffles **44** are provided at the discharge end of the feed conveyor **36** for both directing the material **M** onto the matrix conveyor and regulating the bed thickness. If desired, the plates **44** may vibrate or reciprocate to transversely level the matrix fed by conveyor **36** from the storage area.

A discharge conveyor **50** is located at the discharge end of the matrix conveyor **38** for carrying the oil-enriched matrix material into the rich matrix storage area **52**, with the latter being provided with a perforated plate **54** which retains the rich matrix material while permitting gravity draining of the oil therefrom into the bottom of the storage area as illustrated. In the operation of the recovery vessel shown, the vessel is directed through the area of oil spillage and the oil and water are directed into the vessel through the openings **12**, with the baffles **14** and **16** being open. The oil and water are directed over and through the bed of lean matrix material carried by the conveyor **38**, with the oil being sorbed by the matrix material and the water gravitating through the conveyor to the water sump **56** below the matrix conveyor **38**.

Lean matrix material is continually delivered to the matrix conveyor **38** by the feed conveyor **36**, and the enriched matrix material **M** is conveyed by the discharge conveyor **50** to the rich matrix storage area **52**. Depending upon the extent of the oil spillage, the rich matrix material can be stored in the storage area for continued draining of the oil from the matrix material or for subsequent processing of the material by other equipment. Alternatively, when the extent of the spillage makes reuse of the matrix material a desired objective, the separation of the oil from the matrix material can be accomplished by distillation equipment housed within the vessel by which substantially all of the distillable oil can be removed from the material and the latter conveyed to the lean matrix storage area **34** for subsequent conveying to the matrix conveyor. In either event, it will be understood that as the vessel passes diametrically through the oil spillage, the vessel will be turned around and the vessel again directed through the oil film.

As above indicated, the extent of the oil spillage may require as a logistical matter the complete separation of the oil from the rich matrix material on board the vessel thereby to permit recycling and reuse of the material.

Figure 208 shows a suitable form of recovery apparatus. It will of course be understood that the equipment illustrated could alternatively be land based, with the oil and water being pumped thereto or the matrix transported thereto in bulk or in bags or containers.

FIGURE 208: APPARATUS FOR OIL SEPARATION FROM SORBENT MATRIX

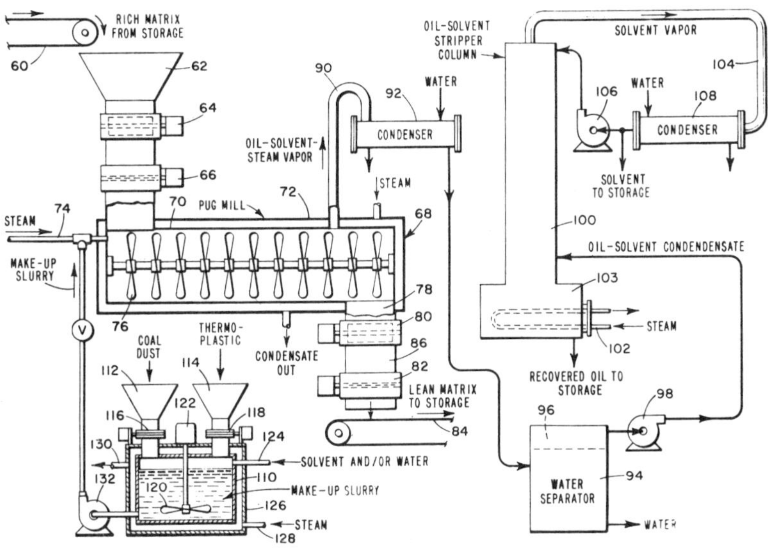

Source: C.O. Bunn; U.S. Patent 3,783,129; January 1, 1974

A rich matrix conveyor **60** communicates with the rich matrix storage area **52** and delivers the rich matrix material to a hopper **62**. Plate valves **64** and **66** control the flow of the matrix material into the inlet of a pug mill or twin pug mill generally indicated at **68**, being generally similar to the type widely used in the preparation of sheet asphalt for highway construction. Plate valves **64** and **66** can be of the type described and illustrated in U.S. Patent 3,449,013.

Exterior of the pug mill housing **70** is a steam jacket **72** to which heating steam is directed through line **74** to strip or boil off the oil in a well-known manner. The interior of the pug mill has a series of blades or beaters **76** for agitating the rich matrix delivered to the inlet thereof from the hopper **62** and advancing the same toward the outlet end of the pug mill. The outlet or discharge end **78** of the pug mill discharges the processed matrix material through plate valves **80** and **82** to a transfer conveyor **84** which transfers the lean matrix material to the lean matrix storage area **34** at the aft end of the oil sorbing bed. The plate valves **80** and **82** are similar to the plate valves **64** and **66** and function to alternately open and close the discharge pipe **86** for controlled discharge of the lean matrix material without affecting the pressure interiorly of the pug mill.

The pug mill functions to intimately agitate the matrix which causes steam to contact and vaporize the sorbed oil thus stripping the rich matrix material of any oil or solvents remaining therein and also functions to convert the therein added coal dust and powdered PE binder into makeup matrix material. The oil vapors, process steam, and such solvents or other minor treating reagents found desirable as the result of a unique characteristic of the particular coal or plastic used, are discharged from the interior of the pug mill through the gas discharge pipe **90** for delivery to a water condenser **92**. The matrix material after being stripped of its vaporizable components is discharged through pipe **78** as above described for recycling and reuse.

The condensed constituents emanating from the discharge end of the condenser **92** are conveyed to a water separator **94** from the bottom of which water is drawn for return to the ambient environment or to the ship's boilers. The relatively lighter liquid components comprising oil and solvent are pumped from the upper portion **96** of the separator by pump **98** to a stripper column **100**. A steam line **102** is located in the reboiler or kettle end **103** at the bottom of the stripper column through which steam is admitted for heating the now liquid oil and solvent condensate and for vaporizing the more volatile solvent for discharge from the stripper through top outlet line **104**.

The vaporized solvent is delivered to a conventional condenser **108** for condensing the solvent vapors. The solvent condensates which are not needed for reflux to cool the upper portion of the stripper are pumped to storage by means of pump **106**. The relatively higher boiling point oil is drawn off the bottom of the stripper column **100** and conveyed to an oil storage area which may be provided on the vessel or can be pumped to an adjoining vessel or other facility providing supplemental storage capacity.

The supply of lean matrix material provided by recycling the rich matrix material through the system may be augmented by the provision of means for providing makeup matrix material. A mixing tank **110** or other mixing means such as a static mixer is adapted to receive coal dust and thermoplastic material from hoppers **112** and **114**, respectively, through control gates **116** and **118**. An agitator blade **120** rotates within the mixer **110** and is driven by motor **122** thereby intimately mixing and forming a pumpable slurry of the materials fed to the mixing device. A water and/or solvent inlet line **124** is provided for forming the matrix slurry provided such solvent is needed for the particular plastic used as a coal dust binding agent.

An outer jacket **126** surrounds the mixing vessel **110** and steam is admitted thereto through line **128** and discharged from the jacket chamber through line **130**. The slurry is pumped from the chamber by pump **132** to the stripping steam inlet line **74** for entry into the pug mill **68**. By controlling the matrix slurry provided by the mixing chamber **110**, the quality of the lean matrix recycled can be closely regulated.

As above indicated, the mixing of the plastic binder and finely divided coal can be accomplished by commercially available equipment, and the system shown as noted employs state-of-the-art components. An alternative form of mixing and stripping equipment is shown in Figure 209, and reference is directed thereto. This system comprises a fluidized bed generally indicated at **150** formed with a porous plate **152** extending longitudinally in the bottom thereof through which the fluidizing medium passes. The fluidized bed pressure tank is preferably downwardly inclined to facilitate gravity discharge of the processed matrix material. The pressure tank is preferably steam jacketed (not shown) for exact process control in the same manner as the pug mill in the preceding figure.

FIGURE 209: FLUIDIZED BED SYSTEM FOR OIL SEPARATION FROM SORBENT MATRIX

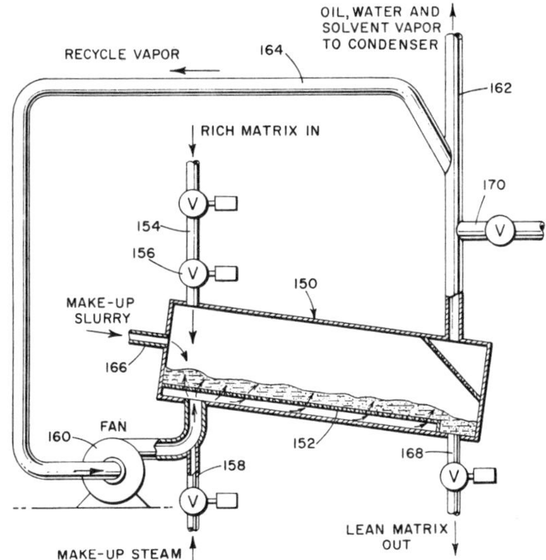

Source: C.O. Bunn; U.S. Patent 3,783,129; January 1, 1974

Rich matrix material is delivered to the tank **150** through inlet line **154** and feed valve **156**, which may be of the type previously described and illustrated at **64** and **66** in the preceding figure. The rich matrix material can be delivered to the inlet line **154** from the rich matrix storage area in the vessel, and it is of course contemplated that the system would be installed within the vessel for separating the oil from the rich matrix and replenishing the lean matrix material to the lean matrix storage area for subsequent reuse.

The fluidizing medium in this instance is preferably steam plus recycled oil/solvent vapors which are delivered to the tank through line **158** under the power of fan **160**. In a manner well-known to those in the art, the rich matrix material in the tank above the porous plate **152** is suspended in a fluidized state thereby permitting the steam to intimately contact the oil absorbed within the matrix material and to vaporize the same along with any vaporizable solvents in the material. The vapor constituents are drawn off through gas outlet line **152** to a condenser for component separation in the same manner as above described. In order to conserve heat energy, and to provide an oil enriched vapor to the condensing system (not shown), a portion of the vapor is recycled back to the tank **150** through line **164** which communicates with the inlet side of the pump **160**. Makeup coal and plastic binder in slurry form can be added to the tank through feed line **166**.

The rich matrix admitted to the fluidized pressure tank is stripped of the absorbed oil by the steam as the matrix material traverses the tank longitudinally toward the discharge outlet **168**. Substantially all the oil and other vaporizable constituents are separated from the matrix material by the time the discharge line **168** is reached, whereby the matrix material discharged through line **168** can be conveyed directly to the lean matrix storage area.

Oil cargo vessels or tankers pump sea water from their cargo tanks, and such water is polluted with oil. A lighter ship suitably arranged would be highly useful in purifying this type of polluted water before it is discharged, either after the tanker has docked or before it has reached port. In the latter case the lighter ship would meet the incoming tanker and the tanks emptied through a hose between the two vessels. This procedure assumes critical importance in the case of the very large ships, some of which are 312,000 dwt and sometimes load and unload in the open sea since few ports have sufficient depth to permit docking.

Much larger tankers are under consideration and there is no doubt that serious pollution problems are imminent which would be alleviated or prevented by suitable lighter ships equipped with this process. An arrangement which would be suitable for lighter ship service would combine the sorption function into an adaptation of the equipment previously described and illustrated.

Conwed Corporation Process

A process developed by *K.S. Peterson and G.R. Palkie; U.S. Patent 3,630,891; Dec. 28, 1971; assigned to Conwed Corporation* is one in which a felted fibrous sheet treated with a water repellent sizing material is used to remove oil floating upon the surface of water by absorbing the oil in preference to the water.

The preferred fiber is a wood fiber from aspen that has been defibrated in a known type of defibrating machine such as an Asplund Defibrator and which fiber is then sized with a suitable water repellent or size. The preferred sizing material is that disclosed in U.S. Patent 2,754,206. As disclosed in U.S. Patent 2,754,206, the size comprises crude paraffin wax, rosin, bentonite clay, aluminum sulfate and water which has been made into a substantive emulsion by first melting the solids content of wax and rosin, mixing in the clay and water, and passing the same through a colloid mill. This dispersion is mixed with the fibers whereupon the dispersion is exhausted by the deposition on the fibers of the solids of the dispersion.

Aspen fibers so treated with adequate size to provide 3 lb of the wax constituent per 100 lb of fiber were then tested in the laboratory as follows. 800 ml of water were placed in a beaker and 30 grams of crude oil added thereto. The crude oil rapidly floated to the surface of the water whereupon 1 gram of the treated fiber was applied to the surface of the floating oil. After 5 minutes, the fibers were removed with a spatula and the water remaining was visually inspected. It was found that no visual trace of the oil remained; however, the fibers were heavily soaked with the crude oil, thus making handling difficult.

While the above test indicates that so treated fibers will preferentially absorb oil to the extent of as much as thirty times their weight, it is preferred to use the fibers on a ratio of about 2 grams of fiber to 30 grams of oil in order to provide some clean fibers in the final oil saturated fiber in order to increase handleability. The above test was made both with the water in the beaker being still and with agitation and with both tap water and salted water with no discernible difference in the ability of the fibers to absorb the oil.

Refined sulfite pulp was similarly treated with the size of U.S. patent 2,754,206 in a quantity to provide 3 lb of wax to 100 lb of fiber and a pulp lap formed in known manner. Such a sulfite sheet was similarly tested on 30 grams of crude oil floating on 800 ml of water with 1 gram of the sulfite sheet. While the absorption time was somewhat longer for the sheet it was found that the chemically refined sulfite pulp was equally successful in removal of all of the visible oil from the surface of the water.

Kraft pulp was similarly treated with the same size to provide 3 lb of wax to 100 lb of fiber and formed into a sheet which was tested in the same fashion as the aspen fibers and the sulfite sheet. It was found that 1 gram of the kraft sheet similarly adequately absorbed all the visible oil (30 grams having been used) with little or no absorption of water.

Other tests of the same kind have been made in which the crude oil was replaced with crankcase oil, kerosene, gasoline, and soybean oil with similar results and complete removal of all visible oil. While reference has been made herein to the use of the particular sizing compound disclosed in U.S. Patent 2,754,206, other water repellent materials have been found to contribute this preferential absorptivity to wood fibers including abietic acid (rosin in water), polyvinyl alcohol, wax emulsions, and Mobilicer L and Mobilicer C.

Bonded felted fibrous sheets or blankets in which fibers of aspen, sulfite, kraft, or the like are felted and bound together with a suitable binder such as starch in which the fibers have been previously treated with sizing have also been tested as indicated above with similar results. Such felted blankets may be made by any one of several known ways including air deposition as disclosed in U.S. Patent 2,746,895. Such blankets of sized fibers also work well and have the added advantage of relatively easier handling of the fibers because of their blanket form.

After the oil soaked fibers, either loose or in blanket form, are removed from the liquid the fibers may then be treated to extract the oil therefrom by squeezing, centrifuging, or by other suitable means thus permitting recovery of the valuable oil constituent. It will be appreciated that because of the relatively low cost of treating the fibers and the relatively low cost of such fibers themselves, this method of removing oil from the surface of water is both efficient and extremely economical. When the fibers are used in blanket form the handleability is enhanced since the blankets may merely be rolled upon the surface of the water, left to absorb the oil, and then retrieved readily by picking up the blanket in any of a number of ways including rerolling the same.

It has also been found that when so treated fibers are applied to an oil slick in a quantity such that the oil is less than adequate to completely saturate the fibers, the fibers will merely continue to float upon the surface of the water awaiting more oil to absorb thus providing an opportunity to supply an excess of fibers when it is expected that the oil contamination will continue over a period of time as when the contamination source continues to discharge its oil onto the surface of the water.

Deutsche Perlite GmbH Process

A process developed by *H. Pape; U.S. Patent 3,382,170; May 7, 1968; assigned to Deutsche Perlite GmbH, Germany* involves removing an oil film from a body of water where mineral perlite, in an expanded state and coated with a silicone to produce an oleospecific adsorbent preferentially taking up oil from the water, is cast on the film.

Dow Chemical Company Process

A process developed by *D.H. Haigh; U.S. Patent 3,520,806; July 21, 1970; assigned to The Dow Chemical Company* involves separating oil from a water surface by contacting the mixture with a particulate cross-linked organic liquid-swellable, organic liquid-insoluble polymer, and separating the polymer with the oil imbibed therein.

Especially preferred for use in the practice of the process are cross-linked copolymers of alkylstyrenes and an alkyl ester derived from a C_1 to C_{18} alcohol and acrylic or methacrylic acid or mixture thereof. To insure buoyancy on water as well as the capability of imbibing or being swelled by a wide range of organic liquids, it is preferred that the copolymers, such as those of p-tert-butylstyrene and methylmethacrylate, contain at least 50 mol percent of the alkylstyrene.

Du Pont Process

A process developed by *A.N. Oemler; U.S. Patent 3,147,216; September 1, 1964; assigned to E.I. du Pont de Nemours and Company* is based on the discovery that the fluff of a crystalline solid polyolefin is capable of adsorbing more than its weight of liquid hydrocarbon and hydrocarbon-like materials. Crystalline polyolefin fluff is defined as finely divided, porous, crystalline polyolefin having a particle size of from 0 to 20% retained on a #20 mesh to 5 to 10% passing through #60 mesh. Polyolefin fluff differs from normally finely divided polyolefin powders in that it contains an extremely large number of capillary channels, whereas the normally finely divided polyolefin powder comprises solid particles.

Polyolefin fluff is obtained by several methods. One method comprises preparing a solution of the polyolefin in a volatile hydrocarbon solvent and then flashing off the solvent at temperatures at which the polyolefin solidifies. Another method comprises cooling a solution of the polyolefin in a hydrocarbon solvent to a temperature at which the polymer precipitates in the form of a gel and then flashing off the solvent. Other methods will occur to those skilled in the art.

The extraordinary ability of polyolefin fluff to adsorb hydrocarbon liquids is explained by a combination of the chemical hydrocarbon structure of the polyolefin and the physical structure of the fluff. The use of the fluff in the separation of hydrocarbon liquids from aqueous systems is made simple by the fact that both the hydrocarbon liquid and the fluff have a density below that of water. An additional benefit derived from the use of the polyolefin fluff is its chemical inertness to almost all inorganic compounds and many organic compounds. Little or no polar compounds are adsorbed by the polyolefin fluff; the adsorptivity of the polyolefin being extremely selective and substantially limited to hydrocarbon and hydrocarbon-like compounds immiscible with water.

Esso Research and Engineering Company Process

A process developed by *G.P. Canevari; U.S. Patent 3,487,928; January 6, 1970; assigned to Esso Research and Engineering Company* is one in which droplets of oil are separated from an aqueous phase using a mixture comprising a sodium montmorillonite clay and an agent selected from the class consisting of organic cationic agents and glycols. The organic cationic agent is preferably an amine.

Fahlvik Process

A process developed by *H.E. Fahlvik; U.S. Patent 3,617,565; November 2, 1971* is one in which the absorption of petroleum products is attained by bringing bark from trees belonging to the order Coniferae into contact with the petroleum products. The absorption facilitates the collection and removal of the products from a substrate such as a water surface or a solid base.

GAF Corporation Process

A process developed by *J.M.C. Whittington; J.E. Meyer and G.D. Tingle; U.S. Patent 3,728,208; April 17, 1973; assigned to GAF Corporation* is one in which a porous alkali metal silicate foam having oleophilic-hydrophobic properties is provided for use in oil spill control and removal. The silicate foam is preferably formed from a blend comprising solid and liquid alkali metal silicates and an oil absorption-water repellent agent. The blend is pelletized, heated in an oven to expand the material into foam particles, and then shredded, graded and retreated with an oleophilic-hydrophobic agent to coat the internal and external surfaces and thereby further enhance the oil-absorption characteristics.

Grantley Company Process

A process developed by *J. Weinberg; U.S. Patent 3,756,948; September 4, 1973; assigned to The Grantley Company* is one in which crumbs of a particular polystyrene foam are employed for the absorption of hydrocarbon oils, one use being for the cleaning up of oil spills on a body of water, another being the cleaning up of a sandy beach polluted by crude oil.

Although this process is useful in the handling of different hydrocarbon oils, it has particular application in the case of crude oil spills on the high seas and on other bodies of water which are occurring with considerable frequency. The ability of the oil absorbent to separate hydrocarbon oils on water could also lend itself to other uses, for example, the discharge from ship's bilges and the flushing of tankers and barges might be treated. The so-called blowing of bilges might be accomplished at dockside, using this process for the removal of oil and permitting the discharge of acceptably clean water into waterways.

There has been a wide search for materials for use in oil removal in the case of spills on the high seas and these were rated by good authority about a year and a half ago as to weight of oil removed from the spill relative to the absorbent, as follows: solid inorganics, 20 to 70% of the weight of the absorbent; porous inorganics, two to six times the weight of the absorbent; polymeric materials, five to ten times the weight of the absorbent; and natural organic materials (such as straw), five to forty times the weight of the absorbent depending upon conditions. The process utilizes a light and cheap material which absorbs about nineteen times its own weight of crude oil, which is effective within minutes where most of the other materials suggested take hours, which is easily picked up and clean to handle thereafter.

The material utilized in this process is a foamed polystyrene plastic which is of closed cell-type, made by inflating cells of polystyrene with air. This results in a nonpermeable, multicellular mass that contains about 97% air by volume. It is only $\frac{1}{42}$ as heavy as solid polystyrene and is 30% more buoyant than cork. The particular material utilized starts from an extruded foam which is produced by free expansion of a hot mixture of polystyrene, blowing agents and various additives through a slit orifice. This extrusion method allows variation of density and cell size so that foams can be tailor-made with specific properties.

The particular foam used in this process has a density between 1.4 and 2.0 lb/ft^3 and an average cell size between 1.0 and 3.0 mm, and absorbs about nineteen times its weight of crude oil, has a selective absorption for oil when subjected to a mix of oil plus water, and does not soil upon contact after absorbing a saturated load of crude oil out of a bath of water.

The absorbent foam of the process is prepared by shredding such an extruded polystyrene board. The shredding is necessary to open up the cells for better oil absorption. The material when shredded might be called crumbs, the acceptable size passes through ¾ inch mesh and is held on ¼ mesh screen. The specific gravity of this shredded dry foam in the form of such crumbs in 0.093. Before using this polystyrene crumb material, unless it has already been fireproofed during manufacture, it is preferably treated to make it more fire retardant, such as by immersion of the foam material into a 5% boric acid-water solution and then allowed to dry.

In utilizing this material in connection with an oil spill on the open seas or other bodies of water, the above described polystyrene crumbs are spread upon the oil layer, floating on water, by use of boats or airplanes. Preferably, the absorbent material is propelled in the nature of a jet stream into the oil layer from above which causes the same to penetrate the oil and to mix with it efficiently. The absorbent material becomes substantially saturated with about nineteen times its own weight in crude oil within a matter of a minute.

Sometimes it is desirable to soak the absorbent material in water before injecting it into an oil spill at sea. This raises its specific gravity to about 0.854 which causes the absorbent material, when forcefully propelled against the oil spill layer, to enter more quickly and more deeply into the oil layer. The absorbent material has a selective absorption whereby it shows a preference for the oil rather than for the water. Thus, even when previously soaked in water before spreading the same upon an oil spill, it will be found that this absorbent material is saturated with oil and rejects the water when spread upon a spill. In recovering the oil saturated polystyrene crumbs after absorption of oil from an oil spill, the crumbs of polystyrene are easily handled and do not deposit oil on the materials or equipment used to pick up and transport such absorbent material.

It may be desirable to aid in the agglomeration of the oil-saturated polystyrene crumbs so as to pick up the same more efficiently. To this end, a polyvinyl chloride solution may be spread upon the floating layer of saturated polystyrene crumbs, as by spraying. Such a solution may be prepared by adding to the polyvinyl chloride a straight chain ketone solvent, such as methyl ethyl ketone, which solvent is of such a character that it evaporates quickly into the air after leaving the spray gun. Another suitable agglomerate for this purpose is a low solids polyvinyl chloride solution which is a by-product from the cleaning of polyvinyl chloride reactors.

Hess-Cole Process

A process developed by *H.V. Hess and E.L. Cole; U.S. Patent 3,800,950; April 2, 1974* involves projecting open cellular particles of a highly oleophilic plastic foam such as polystyrene having a solvent affinity for petroleum substantially equivalent to that of polystyrene foam, to preferentially effect absorption of the oil and the agglomeration of the particles into lumps, mechanically recovering the lumps by screening and thereafter completely burning the oil saturated lumps to final disposal thereof.

The apparatus for carrying out the method includes a heated foaming vessel supplied with foamable plastic particles and means for projecting the particles, after they have been foamed or expanded, upon oil floating on a water surface. These means include a barrel to which the foamed particles are conveyed as well as propulsive means which disintegrate the foamed particles into relatively small particulates, which are then cast onto the oil. Preferably the oil is surrounded by the foamed particles. The apparatus can be mounted on a barge or other vessel.

International Minerals & Chemical Corporation Process

A process developed by *J.P. Harnett; U.S. Patent 3,771,653; November 13, 1973; assigned to International Minerals & Chemical Corporation* is one in which an oil film is removed from the surface of water by contacting the same with compost prepared by the bacterial digestion of organic waste material.

The compost which is used as an absorbent for recovering the oil is prepared by the aerobic digestion of organic putrescible material. The compost is prepared from the organic waste material by techniques well-known in the art utilizing bacteria. Composting plants utilizing such processes for converting refuse into compost are located in Altoona, Pennsylvania; Houston, Texas; St. Petersburg, Florida and Mobile, Alabama.

Typical of such composting processes is that used in the plant operated in Houston, Texas by the Lone Star Organics, Inc. division of Metropolitan Waste Conversion Corporation. In this process the municipal refuse delivered to the plant is initially sorted, and salvageable cardboard, paper, rags, glass and nonferrous metals are removed. Large pieces of iron and steel are also removed since they can damage the grinding mills that are subsequently used for processing the refuse. The salvage material is baled and shipped. Inert nondecompostables and noncombustibles such as rubber, heavy plastics, ceramics, and the like are selectively removed for clean landfill.

The material remaining after the salvage material is removed is comminuted, i.e., in hammermills, into uniformly small sized material which permits quick, controlled decomposition of organic wastes. Ferrous materials, e.g., tin cans, are mechanically removed after the grinding operation. The comminuted refuse is then conveyed into a digestion tank or bay where it usually remains for 4 to 6 days, and air is injected into the digesting mass to provide optimum oxygen and temperature control. The comminuted refuse may be initially mixed with dewatered sewage sludge to enrich the same with compostable organic matter. A temperature of about 135°F is maintained in the digestion tanks during the first 24 hours. This is raised to in excess of 145°F at the end of 48 hours and a final range of between 165° to 170°F. The material leaving the digestion tank is a dark brown tobacco-like textured compost, which is conveyed to a finishing area for final grinding, screening and drying.

The process utilized in the Houston plant of Lone Star Organics, Inc. for producing compost from refuse is described in further detail on pages 18 and 19 of the Spring, 1968 issue of *Compost Science,* published by Rodale Press Inc. of Emmaus, Pennsylvania, the brochure entitled, *A Modern Municipal Waste Disposal Process,* distributed by Metropolitan Waste Conversion Corporation, and U.S. Patents 3,385,687 and 3,533,775.

The compost materials employed are oleophilic and hydrophobic, so that they will preferentially absorb oil from water. The oils which are absorbed include all oils which are immiscible in water, whether they are of mineral, vegetable or animal origin. Mineral oils that are lighter than water, such as crude oil, gasoline, fuel oil and lubricating oils, are the most common oily contaminants of water and, therefore, the most likely to be recovered.

Compost produced from organic waste material by bacterial digestion utilizing processes such as employed in the Houston plant of Lone Star Organics, Inc. are in many respects superior as an oil absorbent to other materials that have been proposed or used for removing oil from water. The absorbent utilized in the process is relatively inexpensive since it is a product of a process for alleviating the refuse disposal problems of municipalities. Another decided advantage of the compost is that it can be readily compressed to relatively high density for shipment and storage without destroying any of its absorptive characteristics. The compost may then be easily dispersed on the water for use, or it may even be utilized in compressed form.

Because of its hydrophobic and oleophilic characteristics and the low bulk density (6 pounds per cubic foot) the compost floats on water and absorbs oil so that compost containing the absorbed oil can be easily scooped up for recovering the oil. Unless the compost is weighted these materials possess the advantage that they will even float when oil has been absorbed to the maximum degree. The compost is equally effective for absorbing oil from both seawater and fresh water.

The oil-containing compost can be readily removed from the surface of the water by any suitable means such as scooping from the surface through a wire screen. The water will drain from the compost and pass through the screen while the oil-containing compost is retained by the screen. The mesh of a wire screen used for removing the compost from water will necessarily depend on the form in which the compost is used. Specifically, a finer mesh screen is necessary where the compost is dispersed on the surface of the water than when it is used in a compressed and baled form.

The absorbed oil may be recovered from the compost by any suitable method, such as compressing the oil-containing compost to squeeze substantially all of the oil from the compost, which may then be used again for absorbing oil. If desired, the compost containing residual oil that is not removed by the compressing may be briquetted for use as a fuel. The absorbed oil may be recovered from the compost by a number of other methods, including centrifugation and the use of wringer-type devices.

The lighter-than-water compost may be weighted to such a degree that it becomes heavier than water as it absorbs oil. The weighting of the compost may be accomplished by a number of ways, such as contacting it with a relatively fine granular material, e.g., sand, and a material effective for binding the sand to the compost even in water, e.g., crude oil. For example, one part of compost mixed with two parts of sand wetted with 1% crude oil will initially float on water and absorb oil until it becomes heavier than water and sinks. The quantity of sand or other weighting material used should be somewhat greater than that marginally needed to weight the compost so that it is heavier than water at almost maximum oil absorption since, as the oil begins to decompose, a liberated gas that lends the oil-containing compost buoyancy forms and sticks to the surface of the sunken mass. The oil-containing compost is either embedded on the bottom of a quiescent body of water or carried away by the undercurrent. The weighted compost may be inoculated with oil-decomposing bacteria to hasten the decomposition of the oil in the water.

The compost used as an absorbent may be contacted with the oil by any suitable means. The contact may be effected by dispersing a layer of shredded compost upon the surface of the water in the area of an oil leak, preferably agitating the water by boats to provide turbulence that achieves contact between the oil film and the compost. The compost containing absorbed oil is then gathered by using a wire net or other suitable means to gather the same. The compost may also be used in a compacted form for controlling and collecting oil spills. In this method a plurality of linked masses of compost, each of which is preferably held together by netting or wire gauze, are employed to encircle an oil spill and contain the same within a definite area. The oil may then be recovered by dispersing loose fibers of the compost within the confined area to absorb the oil.

Still another method of employing the compost is to pass the oily water through a filter or bed of the compost so that it absorbs the oil from the water before it is discharged. This method is especially useful for removing oil from the ballast of oil tankers before it is discharged into the ocean. An alternate approach to ballast cleanup would be to charge compost to the compartment containing the oily ballast in an amount sufficient to absorb the oil. This is preferably done at the time the ship takes on the ballast water. The movement of the ship during the return voyage will affect contact between the compost and the oil residue in the ballast water, and the compost will thus absorb the oil.

The ballast is then removed in the usual manner by pumping, but a screen is used to catch the oil-laden compost. The screen may be advantageously located on the inlet pipe to the ballast pump, and the compost is left in the compartment to be mixed with the new cargo of oil. The compost would represent only a negligible impurity (typically about 0.01 to 0.02% by weight) in the new oil cargo. The following example will serve to further illustrate the method.

Example: A quantity of 33 degree API crude oil was added to water to form a film thereon. Compost from Lone Star Organics of Houston (Alive) was added to the water while it was being agitated to simulate wave action. The compost completely absorbed the oil, picking up about 5 times its weight in oil. The compost containing the absorbed oil was removed from the oil and finally squeezed at 1,000 psi to effect a 90% recovery of the absorbed oil. A briquette which may be used as fuel remained after the oil was removed by the compression.

Membrionics Corporation Process

A process developed by *J. Orban and J. Brooks; U.S. Patent 3,681,237; August 1, 1972; assigned to Membrionics Corp.* for controlling oil pollution of open seas or calm waters is provided where a stable open cell resilient foam material specially treated with a hydrophobic-oleophilic composition to enhance its oil sorbability and water repellency is employed to remove oil from oil-contaminated waters.

The foam material which is employed is an open cell cellular plastic or foam of rigid or semirigid construction, preferably strong enough to support its own weight, but sufficiently flexible and resilient so as to be capable of being compressed between mechanical rollers or other compression apparatus and upon release from compression returning to its substantially original shape. Thus, the foam material should have a modulus of elasticity at 23°C of from 50,000 to greater than 100,000 psi and a density within the range of from 0.01 to 0.20 g/cc.

Although it is preferred that the foam material be of substantially open cell construction, the foam may contain a small fraction of closed cells to increase buoyancy thereof. Thus, for example, the foam can have a ratio of open cells:closed cells within the range of from 10,000:1 to 5:1.

Examples of foam material which can be employed include polyurethane foam, foamed or expanded polyvinyl chloride, phenol-formaldehyde resin foams containing at least 80% open cells, foamed elastomers such as natural rubber foam, styrene-butadiene rubber foam, polychloroprene foam, chlorosulfonated polyethylene-ethylene-propylene terpolymer foam, butyl rubber foams and polyacrylate foams. The hydrophobic-oleophilic treating composition is comprised of a hydrophobic agent (A) and an oleophilic agent (B) in a weight ratio of A:B within the range of from 1:4 to 1.5:1 and preferably from 1:3 to 1:1.

Mo och Domsjo AB Process

A process developed by *K.E. Eriksen; U.S. Patent 3,591,524; July 6, 1971; assigned to Mo och Domsjo AB, Sweden* is one in which an oil absorbent is provided that is capable of preferentially absorbing oil in the presence of water, and that will float on water for a considerable period of time when loaded with oil. The absorbent carrier has upon it the hydrophobic residue of a heat-decomposed ammonium or amine salt of an aliphatic or cycloaliphatic carboxylic acid and an oil.

Ocean Design Engineering Corporation Process

A process developed by *R.E. Hunter; U.S. Patent 3,723,307; March 27, 1973; assigned to Ocean Design Engineering Corporation* utilizes a method for separating oil from a water surface by distributing many small buoyant bodies of flexible foam material upon such surface, continuously lifting such bodies from the surface, compressing the bodies to remove the absorbed oil, and again distributing the bodies upon the surface for reuse. The method preferably utilizes booms for gathering the distributed bodies of foam material toward a conveyor which lifts the bodies upwardly. The booms are articulated and include floats so that the booms rise and fall with any wave action of the water, such as would exist in the unprotected waters of the open sea. The method preferably includes the steps of compressing the oil from the bodies, and thereafter dropping the bodies onto the water surface for recycling.

The oil absorbing material used is illustrated by number **72** and is in the form of a great plurality of relatively small open-celled or porous cubes or chunks which have many fine capillary passages. The bodies have a density less than water, both before and after absorbing any oil. Consequently, they always float upon the water surface **14** which is extremely desirable so that absorption is confined to oil, rather than water. That is, any pronounced submersion of the bodies beneath the surface of the water would subject them to a hydrostatic head which would tend to force water into the pores of the bodies along with, and perhaps in place of oil. Thus, the buoyancy of the bodies results in no appreciable relative movement between the bodies and the water surface even during conditions of extreme wave action. As will be seen, the bodies are necessarily resiliently flexible so that any absorbed oil can be extracted from them by the compressors **22**.

The bodies are made of a material which is hydrophobic and oleophilic. A suitable material has been found to be flexible urethane or polyester foam having a density of approximately 2 lb/ft^3 with a cell count of 27½ cells per linear inch. Bodies of this foam material, when spread upon an oil-water mixture and thereafter removed and compressed, have yielded absorbed material which is 95% fuel oil versus 5% or less of the water. Apparently the hydrophobic character of the foam material allows oil to readily displace any water which may be in close proximity to the surface of the foam bodies. However, even when the foam material is completely saturated with oil, it will still float upon the water surface.

The method basically requires only a conveyor system, generally designated **74** which is effective to lift the oil saturated foam bodies from the water surface by means of the conveyors **20** passing over rollers **36** and **38**. It delivers the bodies to a foam-oil separator such as the compressors **22** and, finally, redistributes the relatively oil-free bodies back upon the water surface as by distributors **26**. Preferably some form of reservoir or storage container **76** is provided to compensate for any foam bodies not recovered by the conveyors. In addition, suitable pumping equipment **64** would be provided for transferring the oil collected in the holding tank or bunkers **65**.

FIGURE 210: OIL SPILL ACCUMULATOR USING PLASTIC FOAM SORBENT

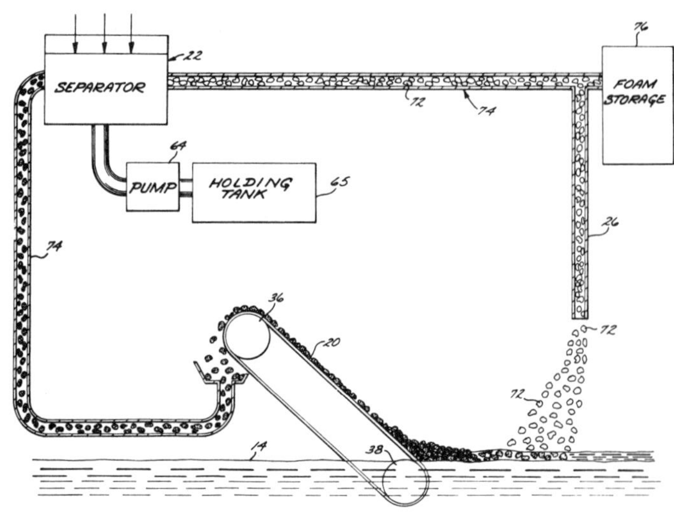

Source: R.E. Hunter; U.S. Patent 3,723,307; March 27, 1973

Osman Kogyo Process

A process developed by *K. Oshima, J. Kajiyama, S. Fukimoto and N. Nagao; U.S. Patent 3,617,566; November 2, 1971, assigned to Osman Kogyo, KK,* in which oil is separated from oil-containing water by contacting the oil-containing water with an adsorbing material consisting mainly of atactic, noncrystalline polypropylene having a molecular weight of 10,000 to 100,000. Straw, wood, wool, or natural fibers can be used as a carrier for the atactic-noncrystalline polypropylene.

Philip Morris Incorporated Process

A process developed by *N.B. Rainer; U.S. Patent 3,674,683; July 4, 1972; assigned to Philip Morris Incorporated* is a process for removal of oil from the surface of a body of water, where a material comprising a particulate microporous hydrophobic vinyl chloride polymer, such as polyvinyl chloride, is applied to the oil.

Phillips Petroleum Company Process

A process developed by *D.O. Hitzman; U.S. Patent 3,414,511; December 3, 1968; assigned to Phillips Petroleum Company* is one in which expanded vermiculite is floated on oil polluted water to absorb oil which is removed from the water by skimming the vermiculite from the water.

Polymer Research Corporation of America Process

A process developed by *C. Horowitz; U.S. Patent 3,494,862; February 10, 1970; assigned to Polymer Research Corporation of America* is one in which water bearing a layer of hydrocarbon liquid is contacted with porous polystyrene which completely absorbs the oil. The polystyrene containing the absorbed hydrocarbon liquid floats on top of the water and can be easily scooped up and recovered. The oil can then be recovered therefrom by heating until the polystyrene melts whereupon two liquid layers are formed, the oil layer on top and the melted polystyrene layer on bottom so that the oil can be easily recovered.

Shell Oil Company Process

A process developed by *A.C. Evans; U.S. Patent 3,518,183; June 30, 1970; assigned to Shell Oil Company* is one in which hydrocarbon oil films may be removed from the surface of water by applying a large surface area of a block copolymer to the oil, absorbing the oil thereinto and separating the oil impregnated block copolymer from the water.

A highly porous particulate form of the block copolymer is preferred. The useful block copolymers absorb large quantities of oil while still retaining a high degree of their original stress-strain properties and the oil absorption is within the interior portions of the copolymer rather than merely on the exterior surfaces. Their retention of a high degree of physical strength is especially noteworthy. Furthermore, the oil impregnated block copolymers are relatively dry and do not exhibit highly viscous, sticky, characteristics. The particles may be formed by any method known to those skilled in the art.

The oil removal process comprises the spreading of the finely divided block copolymer on the oil slick which normally would be situated at the surface of a body of water. This may be done by blowing the particles in the direction desired. Confined or shaped porous bodies of the block copolymer may also be used to surround or control an area of water where an oil slick is present or expected. For example, a collar surrounding an offshore drilling platform may be provided which comprises a tubular net filled with porous block polymer foam or shredded foam or other high surface areas such as particles of the block polymer. The collar forms a potential barrier for any oil which may be emitted from the offshore drilling zone. Similar arrangements may be used in the form of booms or the like to stop the invasion of oil slick into the mouths of harbors or along sections of coast line.

The amount of block copolymer used will depend in part upon the thickness of the oil film to be absorbed, the time needed for economic collection of the oil, the agitation of the body of water from which the oil is to be removed, and similar physical factors. When the oil slicks comprise a film from 0.001 to 0.5 inch in thickness, it is preferred that the finely divided block copolymer be applied in an amount between 3 and 7 lb/gal of oil. The time required for absorption of the oil into the high surface of the block copolymer will vary with the temperature, the degree of agitation, the ratio of oil to block polymer, the viscosity of the oil, and the surface area of the block polymer being utilized.

The time required for substantially complete absorption of the oil into the block copolymer will normally be between 2 and 8 hours. The oils particularly considered here comprise crude oil, fuel oil such as normally utilized in the marine engines and the like as well as other oil products including kerosene, gasoline, furnace oil, etc.

The process allows virtually complete absorption of the oil within the body of the block copolymer leaving the surface of the oil-impregnated block polymer essentially dry. This facilitates and enables the collection of the oil-impregnated block polymer by raking. The presence of the oil in the structure of the block copolymer enables the original particles to coalesce at the surface into a more or less firmly bound structure, again facilitating recovery of the oil-impregnated polymer from the surface of the water.

The process is especially effective and rapid when the proportions of oil and polymer are about equal. The oil-soaked polymer composition may be readily collected and economically utilized in a subsequent operation. It has been found that the block polymers as well as the oil can be dispersed in all proportions in asphalts as compared with the limited miscibility of other types of rubber. It is thus possible and practical to combine the collected oil-impregnated block polymer with asphalt and utilize the compositions so formed in the preparation of roads, etc. This is especially suitable when the oil being absorbed by the block copolymer is a relatively nonvolatile oil or fraction thereof.

Snam Progetti SpA Process

A process developed by *B. Ciuti and S. Del Ross; U.S. Patent 3,676,357; July 11, 1972; assigned to Snam Progetti SpA, Italy* is based on a composition useful in eliminating surface water pollution caused by crude petroleum oil or its fractions consisting of: (1) a carrier having a specific gravity sufficiently low that it floats (e.g., sawdust, silicates, diatomite, vegetable oil residues, vegetable flour, animal flour, silica gel, charcoal, polyurethane resin and polyvinyl chloride); (2) a surface active agent, viz, sodium stearate, sodium alkylsulfonate, potassium stearate, sodium dodecyl-benzene-sulfonate, potassium palmitate, nonylphenol ethylene oxide(1:7) or triethanolamine oleate; and (3) a wetting agent, viz, isoamyl alcohol, glycerol, ethylene glycol, butyric alcohol or amyl alcohol.

Sohnius Process

A process developed by *A. Sohnius; U.S. Patent 3,607,741; September 21, 1971* is one employing physical means for removing oil slicks from water and other surfaces utilizing chemically treated cellulosic bulk material contained in encasements of netting. The chemically treated material is hydrophobic but exhibits an affinity for oil, which can then be reclaimed.

Strickman Industries, Inc. Process

A process developed by *R.L. Strickman; U.S. Patent 3,657,125; April 18, 1972; assigned to Strickman Industries, Inc.* is one in which oils, particularly petroleum oils may be removed from water, from beaches and from wild life, by contacting the oils with a collector comprising granular polyurethane particles substantially devoid of cellular structure. Preferably the collector particles are of a jagged, spiny, cragged nature. When applied to an oil contaminant, the collector particles agglomerate the oil into a gel which can be skimmed or otherwise removed easily.

Wacker-Chemie GmbH Process

A process developed by *E. Pirson, M. Roth and S. Nitzsche; U.S. Patent 3,464,920; September 2, 1969; assigned to Wacker-Chemie GmbH, Germany* is one in which natural and synthetic oils floating on and contaminating the surface of bodies of water can be removed by absorption on a comminuted organic solid which has been rendered water repellent but remains oil absorptive after treatment with organosilanes.

Zerbe Process

A process developed by *J.J. Zerbe; U.S. Patent 3,734,294; May 22, 1973* involves removing a pollutant from the surface of water where the pollutant is confined to a recovery area and is directed to a recovery station by a plurality of linked booms. A particulate, floatable sorbent material is continuously distributed over the pollutant confined in the recovery area as it moves to the recovery station. The sorbent material is distributed in sufficient quantity to adsorb substantially all of the pollutant prior to its arrival at the recovery station. At the recovery station, the sorbent material and the adsorbed pollutant are removed from the water surface.

Figure 211 shows the essential elements of this system. There are a plurality of linked booms **11** for positioning downwind or downcurrent from a pollutant source **12** (or **21** or **22**) to confine the pollutant to a recovery area **13** and to direct the pollutant to a recovery station **14**. Distributing means **16** are provided for continuously distributing a particulate, floatable, sorbent material over the pollutant confined in the recovery area as it moves toward the recovery station. This material is distributed in sufficient quantity to adsorb substantially all of the pollutant prior to its arrival at the recovery station. Recovery means **17** are provided at the recovery station for continuously removing the sorbent material and the adsorbed pollutant from the water surface.

FIGURE 211: RECOVERY OF OIL SPILL ON PARTICULATE SORBENTS

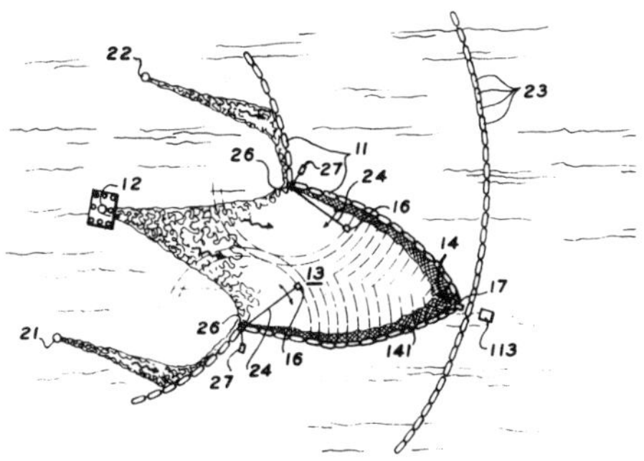

Source: J.J. Zerbe; U.S. Patent 3,734,294; May 22, 1973

The system is illustrated in connection with an offshore drilling structure **12** for an oil well or oil wells. The drilling structure constitutes the source of pollutant such as would occur in the event of a defective well or well blowout. Several other sources **21** and **22** of pollutant in the vicinity of the platform **12** are also shown. Overall containment of the oil spill or surface pollutant is provided by a surrounding barrier consisting of a plurality of linked booms **23** secured and anchored in any suitable manner. In the even of an oil well blowout or other leak in the vicinity of the drill structure, the surface of the ocean in such vicinity will become covered with oil. Depending upon wind, current, or both, the oil will move in a direction away from the source and gradually spread out.

The linked booms are arranged to contain the oil within the recovery area and to direct it to the recovery station which is accomplished by arranging the booms in a semicircular arrangement suitably anchored at the outermost ends. Because of the curvature of the booms, the oil is confined and directed toward the recovery area and merges with the oil from the platform. The semicircular portion of the boom arrangement is open towards the recovery area and the booms are redirected by suitable anchoring to funnel the oil in the recovery area towards the recovery station.

The amount of sorbent material distributed is such as to be greater than the amount at which saturation would take place. Thus, there is an overabundance of sorbent material present in the recovery area after seeding to insure that no oil or pollutant remains after the pickup operation described subsequently. The means **16** for distributing the sorbent material comprise seeder units deployed at the ends of tethers **24**. Two seeder units **16** and two tethers **24** are illustrated. A pivot platform **26** is provided at the corner between the funnel section of the booms **11** and the semicircular sections thereof. Each pivot platform is anchored on four sides for stable positioning and the tethers **24** are attached thereto. A supply or tender barge **27** is anchored near each of the pivot platforms for supplying sorbent material to the distributing means through flexible ducts.

A support tender **113** is equipped with suitable pumping capacity to effect continuous removal of the pollutant plus the sorbent material. To insure that the sorbent pollutant mat will not escape beneath the auxiliary boom, a small meshed fabric net **141** (½" mesh for example) is attached to the keel of the boom segments as shown. The bottom of the net consists of a tube in which a cable or nylon line is threaded with ends terminating at the two pivot platforms shown. By attaching the mid-portion of the tube at the bottom of the net to the recovery device **17**, the ends may be tensioned so that drawstring action shrinks the base of the net and prevents any sorbent material from escaping underneath the booms by undertow. The net is sufficiently porous as to allow full flow of current without restriction.

SINKING

It is also possible to dispose of floating oil by sinking it. Again, however, government regulations forbid the use of sinking agents, unless authorized by the Federal On-Scene Coordinator. In sinking operations, special powdered materials are spread over the oil. The oil adheres to the surfaces of these materials; the combination, being heavier than the water, sinks to the bottom. Common sinking agents include treated sand, brickdust, cement, silicone-coated materials, fly ash, and special types of clay. In general, the use of sinking agents should be limited to spills in offshore areas where no damage to organisms living near or on the bottom of shallower waters can occur.

Sinking agents are oil-attracting and water repelling sorbent materials designed to sink oil slicks out of sight rather than agglomerating oil on the water surface. The use of oil sinkants would seem advantageous in deeper waters outside heavy fishing zones, such as off the continental shelf and where adverse effects upon biological bottom life may be held at a minimum. Existing EPA policy on sinking agents restricts their use to waters exceeding 100 meters in depth.

Oily discharges into inland rivers and coastal waters do not remain floating forever since much of the oil will be naturally absorbed onto clay, silt and other particulate matter normally suspended in the water, thus causing eventual sinking of the oil. Sinking by natural means, for example, is believed to be one of the primary causes for the disappearance of oil slicks in the New York Harbor complex.

Various types of natural materials and commercial products are available which are claimed to be effective in sinking oil slicks. Typical agents include sand, brick dust, fly ash, china clay, volcanic ash, coal dust, cement, stucco, slaked lime, spent tannery lime, carbonized-siliconized-waxed sands, crushed stone, vermiculite, kaolin, fuller's earth, and calcium carbonate, including the Oyma type chalk used for the Torrey Canyon.

Sinking agents are believed most efficiently employed on thick, heavy and weathered oil slicks in contrast to relatively light and fresh oils. These agents are granular or fine particulate solids with a specific gravity generally between 2.4 and 3.0. These agents must be evenly distributed over the surface of a slick and suppled with proper mixing, agitation and time interaction. The particle-coated and agglomerated mass eventually becomes heavier than water and sinks to the bottom.

The major problem in sinking oils is that the bonding of the agent with the oil must be nearly permanent. Experiments in both the laboratory and field show that many agents will release entrapped and sunken oils back to the water environment. Increasing the application rate two or three times over prescribed amounts has served to minimize this release.

Studies conducted by various investigators indicated that:

(1) Sulfurized oils may show greater sinking abilities than desulfurized oils because of higher potentials for hydrogen bonding

(2) When sands are used for sinking it may be expected that the oily mass will be stable under water when the sands are closely packed and the interstices are filled with oil giving an oil content around 40% and a bulk density about 1.7. However, when the sands are loosely packed, the mass will become internally mobile so that oil drops will separate and escape to the water until equilibrium is reestablished.

(3) With carbonized sand, it was indicated that up to 3 lb of sand would be required to sink 1 lb of oil. Large-scale applications of amine-coated sands envision spraying the sand in a slurry form from a large vessel or hopper dredge.

In ports and harbors it has been mentioned that turbulence or agitation of the water body caused by storms or passing vessels may tend to release oily masses previously sunk. For siliconized sand, it has also been suggested that 1% bone flour or an inexpensive fertilizer may be added to promote bacterial growths and accelerated decomposition of the sunken mass. Table 19 summarizes the types, application rates and relative costs of various sinking agents.

TABLE 19: OIL SINKING AGENTS, APPLICATION RATES AND OTHER OPERATIONAL DATA

Type Material	Application Ratios*	Comments
Silica, untreated	-	Released oil easily after sinking
Carbonized sand	1:1 to 3:1	Higher proportion of sinking agent considered most appropriate; $27/ton carbonized sand (1948); particles SG, 2.7
Sands, amine-coated	3:2 to 2:1	Large-scale application costs estimated $5 to $10/ton of oil treated
Sands, 0.5% silicone-coated	-	More effective on heavier oils
Fly ash, untreated	-	Released oil easily after sinking
Fly ash, 0.5% silicone-coated	2:1 or higher	Permanent oil sinking reported; more effective on lighter oils
Spent tannery lime	-	Good oil retention properties after sinking
Kaolin clay	-	Released oil easily after sinking
Bentonite, montmorillinite clay, treated and untreated	-	-
Calcium carbonate	-	Released oil easily after sinking
Fuller's earth	1:1 to 9:1	1:1 to 3:1 for sulfurized oil; 3:1 to 9:1 for desulfurized oil
Calcium carbonate (Champagne chalk treated with 1% sodium stearate)	≈ 1:1	Used off the coast of France during the Torrey Canyon; $60 to $80/ton; particles SG, 2.7
Synthetic silicate		High efficiency in absorbing fuel and crude oils; some difficulty in sinking the mass
Synthetic plus filler material	4:3 or higher	Estimated $120/ton sinking agent

*Weight sinking agent:weight oil treated

Source: PB 213,880

The most prominent large-scale application of sinking agents was that undertaken by the French to sink large masses of Torrey Canyon oils in the Bay of Biscay. Some 3,000 tons of calcium carbonate (blackboard chalk from the Champagne area of France) coated with about 1% sodium stearate, was used to treat and sink about 20,000 tons of floating oils, although this amount was never precisely defined. The powdery chalk was sprayed or sprinkled over the thick, highly viscous and weathered oil patches and thereafter the water surface was mixed by vessels criss-crossing the area. The oil-chalk mixture reportedly sank in about 60 to 70 fathoms of water. The area of sinking was known to partially cover fishing grounds and thus there was fear of bottom inundation and oil resurfacing.

Numerous observations, since this incident, however, have reported no adverse effects upon fisheries and bottom life, and there have been no sightings of resurfacing oils upon the water or adjoining shorelines. Lack of knowledge on the precise fate of the oil shows need for further trials before this method may be recommended in other situations.

Opinion still remains divided on the use of oil sinkants as to their efficiency, cost, application, and possible adverse effects. With sinking agents, the same problems are encountered in the application and distribution as are encountered with the use of floating sorbents. However, with sinking agents, operational logistics become increasingly more difficult because of the much larger amounts of material required for treatment.

Although damaging effects have been ascribed to toxicity and smothering of bottom life by sunken oils, the sinking approach serves to localize oil pollution, to prevent its spread over the water surface, and theoretically submerge and anchor the oil near the source of pollution. Considering that the bottom of harbors and bays near many industrial ports are grossly polluted and nonproductive, emergency sinking of oils in these areas may not increase damage to fisheries. Dilution by flowing waters in certain areas may also be sufficient to adequately minimize toxicity to nearby shellfish grounds. The other important point of view is that sinking agents, even if used in harbor areas because of fire or explosion danger, can at best be of temporary benefit. The resurfacing oils, although less objectionable due to weathering, may require pickup. Furthermore, sinking may greatly extend the period over which aquatic fauna and flora are affected.

A discussion of the effect of oil on the sea floor with particular reference to the role of sinking agents in putting oil on the sea floor has been presented by the staff of the National Science Foundation and published as part of *Patterns and Perspectives in Environmental Science,* Washington, D.C., U.S. Government Printing Office (1972).

They point out that recent observations concerning the fate of oil in the ocean after spills and leaks such as those in the Santa Barbara channel and from the S.S. Torrey Canyon off the English coast have led some investigators to conclude that dispersal methods that involve removing the oil from the surface by overpowering its natural buoyancy (thus transferring it to the sea bottom) are potentially more harmful to the environment than methods that leave the oil dispersed but floating on the ocean surface.

Since sinking methods involve the use of extremely cheap agents (sand, ashes, and the like) and since they generally remove the oil before it can contact beaches, yacht hulls, and other recreational surfaces, there are strong economic and aesthetic arguments in favor of their continued use.

On the other hand, if it could be shown that the transferral of toxic petroleum constituents to the sea floor would result in damage to demersal fisheries, there are strong arguments for establishing an effective international regime to control both drilling for and seaborne transportation of petroleum, wherever the possibility exists that it may be deposited in quantity upon the sea surface, and to fix responsibility, assess damages, and compensate those economically injured in case such an event occurs.

Current scientific knowledge relevant to the problem of petroleum on the sea surface and sea floor is far from adequate with respect to reliable predictions of the possible harmful effects of removing petroleum or petroleum residues from the sea surface by sinking them to the sea floor.

It is known already from extensive investigations of the chemical composition of ocean sediments in may parts of the world, that detectable quantities of paraffins, aromatics, and asphalts; chemically indistinguishable from petroleum fractions; are present in ocean sediments. Ironically, these investigations have been carried out primarily to determine the sources of oil in sediments, not the fate of oil in the sea.

Emery summarized much of this work in 1960 in his book *The Sea Off Southern California.* He found the greatest rate of accumulation of hydrocarbons in marine sediments to be in certain stagnant basins, where they could amount to as much as 0.15% of the dry weight of sediment. Emery's calculations showed that about 880 tons of such material were deposited annually in the sediments over an area of 78,000 square kilometers, compared to an annual production of 135,000 tons of similar materials by the phytoplankton over the same area. Disregarded entirely in this computation is the possibility that any of the hydrocarbon material currently being deposited in the sediments is reworked from the numerous seeps in this region of the California coast.

Recent work by Horn, Teal, and Backus of Woods Hole Oceanographic Institution not only shows that floating lumps of petroleum residue are common on the sea surface but suggests two methods by which the constituents of such lumps can be transferred to the sea floor through natural processes as well as a natural method for disposing of the material at the sea surface.

There are natural sinking processes in operation. Goose-neck barnacles, which at certain seasons of the year attach themselves to any suitable firm substrate near the sea surface, were found adhering to floating lumps of petroleum. Since these creatures secrete a calcareous exoskeleton, they are significantly heavier than sea water; thus, as they grow they unquestionably transfer lumps of petroleum residue to the sea floor by adding weight. It is in all probability this effect and not slight toxicity that accounts for the observation that the largest barnacles attached to oil lumps were 8 mm long, whereas barnacles attached to pumice reached 11 mm.

The existence of floating pumice itself suggests another possibility in the transfer of floating oil to the sea bottom. Floating pieces of pumice on the sea surface are observed to decrease continually in size as the result of abrasion through wave action. The abraded particles in turn conceivably can be accumulated by (or accumulate) petroleum particles to the extent that the mixture is heavier than sea water and hence sinks to the bottom.

In shallow coastal water, supposing that oil is delivered to the sea surface at a rate greater than that at which it can be naturally oxidized, it seems likely that airborne dust and other solid residues will act as additional agents in increasing the density of floating oil and causing it to sink to the bottom. A layer of tarry residue will then exist on the bottom in such localities, its thickness increasing with time at a rate equal to the rate of delivery of oil minus the rate of oxidation in situ. Such layers can indeed be observed on the bottoms of industrial harbors.

Although present knowledge tells us that, at least in some cases, no harmful effects can be attributed to the presence of petroleum on the sea floor (the sea off southern California, for all its dozens of oil seeps, is one of the more productive fishery areas in the world) it would be a mistake to assume that we already have all the information required to settle the question of whether oil on the sea floor is preferable to oil at the sea surface. For one thing, crude petroleum varies widely in its chemical makeup. We need, therefore, to examine the relative toxicity of crudes from a variety of sources to marine plants and animals, pelagic and benthic. We need also to examine the rate of bacterial oxidation of various crudes and to establish the effect of temperature on these rates.

We need also to study bottom conditions in the vicinity of oil terminals and tidewater oil refineries as compared with control areas lacking such industrial activity to determine the extent to which areas of the ocean floor have already undergone the type of modification that has been observed in New York's East River (where there is a thick layer of blacktop in the vicinity of the Brooklyn Navy Yard) and the influence that incorporation of petroleum residues into bottom sediments has had on the benthic biota. And we need to map the various areas of the continental shelves and slopes of the world, down to the depth below which bottom conditions are without influence on fisheries, and to evaluate their productivity in terms of current fishing operations.

At depths greater than about 750 meters, the sinking method of oil dispersal can presumably be used without fear of harmful effects. Over lesser depths, where important demersal fisheries exist, only laboratory studies of the effect of sunken oil on the biota can provide pollution-control authorities with the information that will enable them to evaluate whether removal of floating oil through causing it to sink to the bottom is economically preferable to attempting to collect it on the surface, to speed its natural removal by spreading emulsifying agents, or letting it drift ashore.

Dutch Shell Laboratories, Holland, was one of the first groups to really evaluate sinking agents on a field scale basis. Their studies, still underway, involve using sand treated with an amine. One very important finding from their investigations was that there was a correlation between clay content and performance as measured by the percentage of oil sunk. The lower the clay content the more efficient the sinking agent. The Warren Spring Laboratory, United Kingdom, impressed by the results of these studies, conducted similar field investigations during 1969. Conclusions reached by the Warren Spring Laboratory were as follows:

(1) The sinking was not as effective as first hoped and it would appear that the thickness of the oil film is an important factor. Nevertheless, between 50% and 70% of the oil put on the sea was sunk.

(2) The skin divers reported that the oil sank to the bottom in small particles which were only slightly more dense than the sea. Consequently, the tidal current along the sea bed was sufficient to carry the oil over the ripples of sand on the bottom.

(3) Trawling was carried out and although no actual lumps of oil were collected, sufficient oil was rubbed onto the netting of the trawl to foul both the catch and the fishermen when they were pulling it back into the vessel.

(4) The oil which remained afloat was particularly difficult to dispose of using either more sand slurry or the normal solvent emulsifier mixture agitated by water hoses. Further investigations on this particular phenomenon are being carried out both in England and in Holland.

In the United States, the Army Corps of Engineers, under contract to U.S. Coast Guard, is evaluating various types of sinking agents, treated and untreated, as well as investigating the possibility of using their hopper dredges, normally used for dredging harbors, to handle and apply the sinking agents.

Clyne Process

A process developed by *R.W. Clyne; U.S. Patent 3,536,201; October 27, 1970* involves mechanically removing oils and floating nonadherent emulsions from waste water subsequent to the latter attaining a substantially quiescent state, the oils and emulsions thereof having been treated with a preferential absorbent material to form nonfloating masses. The apparatus includes an elongated horizontally disposed planar member mounted within a reservoir slightly beneath the level of the waste water accumulated within the reservoir.

The planar member is disposed between a water inlet and a water outlet for the reservoir and is adapted to have deposited thereon the nonfloating masses. A conveyor means having a plurality of elongated flights is also provided wherein the flights thereof engage and move across the planar member carrying therewith the deposited masses and discharging them at a predetermined discharge station.

Dresser Industries, Inc. Process

A process developed by *W.B. Patterson, Jr.; U.S. Patent 3,634,227; January 11, 1972; assigned to Dresser Industries, Inc.* involves rendering innocuous and/or eliminating an oil slick on a body of water by using an oil absorbent clay. An emulsifier can be used to allow the clay to sink in the body of water after absorbing the oil of the oil slick.

The oil bearing clay can be removed from the surface of the water by physically separating the clay from the water and hauling same away or by adding to the clay one or more emulsifiers in amounts which reduce the surface tension of the water sufficiently to allow the oil bearing clay to sink in the body of water.

The method is quite safe in that it does not involve burning or emplacing a large floating fence from a boat. Further, the method is not limited by the surface area covered by the oil slick since the clay and emulsifier, if used, can be emplaced on the slick from the air using techniques already known and employed in crop dusting and fire fighting with airplanes.

Once the clay has absorbed the oil of the oil slick, the slick is rendered substantially innocuous in that the oil bearing clay can come in contact with land, boats, other property, and wildlife without depositing substantial amounts of oil thereon. Thus, the oil bearing clay can be forced up to and even onto the shoreline, skimmed from the surface of the water, and hauled away in dump trucks without damaging the shoreline or any wildlife that may come into contact with the clay during this procedure.

If desired, the oil bearing clay can be disposed of without ever bringing same into contact with land by adding thereto at least one emulsifier in an amount which reduces the surface tension of the water sufficiently to allow the oil bearing clay to sink in the body of water. The oil bearing clay can then be left in the body of water for the bacteria and/or enzymes which naturally occur in the water to digest the oil. In this procedure, it is preferable to use a biodegradable emulsifier so that the same bacteria and enzymes which digest the oil also digest the emulsifier leaving only the clay.

Lindstrom Process

A process developed by *O.B. Lindstrom; U.S. Patent 3,749,667; July 31, 1973* permits disposing of oil spilled at sea by first burning the oil and thereafter applying an inorganic sinking agent. The sinking agent particles, less than 50 mm in size, are dispersed over the burning oil and become coated with the oil residue which is absorbed onto the particles as they sink. The sinking agent particles may be sand, gravel, chalk, gypsum, slag of heavy materials like iron ore, and the like.

Phillips Petroleum Company Process

A process developed by *W.W. Crouch and C.W. Childers; U.S. Patent 3,591,494; July 6, 1971; assigned to Phillips Petroleum Company* is one in which hydrocarbons floating on an aqueous body are removed by contacting them with a fine powder which is a mixture of a polymer compatible with hydrocarbons and a water-insoluble inorganic filler, whereby the powder absorbs the hydrocarbons and thereafter sinks to the bottom of the aqueous body.

Particularly suitable mixtures for hydrocarbon removal are butadiene-styrene copolymers and barite, polyisoprene and iron sulfide, natural rubber and portland cement, polybutadiene and barium sulfate, polybutadiene and dolomite, butadiene-styrene copolymer and lithopone, polyisoprene and kaolin clay, and natural rubber and magnetite. Specific examples of useful mixtures are butyl rubber and calcium molybdate, ethylene-propylene rubber and magnesium orthoborate, ethylene-propylene-diene rubber and iron boride, 90:10 butadiene-acrylonitrile copolymer and lead sulfate, crosslinked polystyrene and lead dioxide, and impact polystyrene and andalusite.

Various methods can be employed to effect an intimate admixture of the polymer and the water insoluble inorganic filler. For example, the solution or latex containing the polymer can be deposited in the form of small droplets or a spray on the surface of the filler while the mixture is being subjected to heat, reduced pressure, and a mixing action. The heat and reduced pressure drive off the polymer solvent or the water from the latex while the mixing action ensures homogeneity of the polymer-filler mixture and also minimizes agglomeration of polymer particles. Alternatively, heated streams of the filler and polymer latex or solution are passed through a mixing type nozzle which also subjects the mixture to an atomizing action. It is desirable that large lumps of polymer or polymer-filler mixture be eliminated or minimized. Any large lumps that do form can be reduced by attrition or grinding devices, such as ball mills.

Likewise, the polymer-filler mixtures can be contacted with the hydrocarbons to be removed by a variety of methods. For example, on small spills a single surface vessel can travel around the spill, spraying, blowing, or dusting the powdery material toward the center, until the spill has been sunk. On larger spills, several surface vessels can be employed. Aerial spraying can also be employed. Any conventional spraying or powder dispensing device can be employed with these oil-sinking agents. A mixing action is advantageous in promoting intimate contact of the powdery material and the oil. Normal wave action is generally sufficient for this purpose.

It is also within the scope of this process to add microorganisms which consume or degrade hydrocarbons to the polymer-filler powder prior to the application of the powder to the hydrocarbon spill. Examples of such microorganisms are species of the genera Bacillus, Pseudomonas, and Nocardia. The added microorganisms along with those naturally occurring accelerate the biodegradation of the sunken hydrocarbon.

GELLING/COAGULATION

Gelling agents may be applied over the surface or periphery of a floating oil slick and are intended to absorb, congeal, entrap, fix, or make the oil mass more rigid or viscous so as to facilitate subsequent physical or mechanical pick up. The gelling concept is also undergoing extensive study for stabilizing petroleum cargoes aboard a stranded or heavily-damaged tanker at sea subject to mass spillage.

Possible gel agents include molten wax or soap solutions, lanolin, liquid solutions of natural fatty acids, soaps of the alkaline metals, treated colloidal silicas, the amine-isocyanates, and the polymer systems. One manufacturer indicates a cost of $3/gal of gel agent, a use ratio of about 1:1, and the ability to mix the recovered gel mass with fuel oils serving as replacement bunker fuel. This gel agent is applied to the surface of the water by a high-pressure spray system to provide sufficient agitation and mixing of the gel-oil mass.

Preliminary research on the gelling of tanker cargoes tends to show that a 3 to 10% gel agent will be required at a materials cost of 13 to 40 cents per gallon of tanker crude oil gelled. However, total operational costs, and the ability to salvage and reuse the gelled cargoes, are not fully known at this time.

The gelling approach for treating oils on water, although promising, must provide greater attention to application and distribution, lower materials and operational costs, and suitable means of picking up the amorphous oily masses. Bunker C, heavy crude oils, and some gel agents by themselves may clog intakes, pumps and suction lines. The major difficulty is the ability to harvest the congealed mixtures since gelled oils cannot be easily collected by mechanical or manual means. Necessary improvements are needed in the gelling approach in line with a total operational cleanup system.

Agency of Industrial Science & Technology Process

A process developed by *G. Kondoh, S. Honda and Y. Murakami; U.S. Patent 3,536,616; October 27, 1970; assigned to Agency of Industrial Science & Technology, Japan* for removing oils floating on the surface of water comprises spraying a solution of a synthetic polymer dissolved in a low-boiling point organic solvent over the oil floating in thin film form on the surface of water, thereby forming a thin film or fine droplets of the solution on the surface, allowing the floating oil to adhere to the thin film or fine droplets, and collecting these by an appropriate means.

Theoretically, synthetic polymers having a wide angle of contact with water and a small angle of contact with oil should be suitable; in this sense, polyester, vinyl chloride, polystyrene, and polyethylene polymerized under high pressure are thought effective, but in view of organic solvent required to dissolve the synthetic polymer and cost of polymers, polyethylene is most effective among them.

Chevron Research Company Process

A process developed by *R.L. Ferm; U.S. Patent 3,703,464; November 21, 1972; assigned to Chevron Research Company* for treating petroleum product spills is one in which coconut husk material is spread on the spill to absorb it. When the spill is on fresh or salt water in the form of a slick, the coconut husk material coagulates the film, keeps it from sinking, and forms a mass which lends itself to easy removal from the water by mechanical pickup and the like.

International Synthetic Rubber Company Limited Process

A process developed by *K.I. Wyllie and E.W. Duck; U.S. Patent 3,265,616; August 9, 1966; assigned to The International Synthetic Rubber Company Limited, United Kingdom* involves cleaning or collecting a surface layer of oil from an area of water by applying a synthetic rubber latex to the oil, coagulating the latex plus absorbed oil in the presence of an aqueous salt solution or an aqueous acid solution to produce oil containing rubber film, crumbs or particles and then skimming said crumbs or particles from the water surface.

According to one embodiment of the process, an oil patch is sprayed with a rubber latex which preferentially contains between 5 and 25% of rubber though higher and lower contents are not excluded. On contact with the oil layer the rubber in the latex, helped to a certain extent by emulsifiers present, absorbs the oil. The emulsifiers present are those required for latex manufacture. The most satisfactory are fatty acid soaps. Rosin acid soaps are often used. Latices can also be prepared with cationic emulsifiers but are not often produced. In the presence of salt water the latex plus absorbed oil coagulates to produce lumps or crumbs. The coagulation process can be enhanced by a subsequent spraying with a diluted aqueous solution of acidic brine or an acidic solution containing no salt.

The acid solution, if used alone, should have a pH of less than 6.5. For fresh water it is essential that either the salt solution or the acidic solution should be added since soluble salts are not otherwise present in sufficient quantity to cause coagulation. The salt solution should have a concentration of at least 1 to 2% if coagulation is to be efficient.

After that treatment the oil-containing rubber may be gathered up by means of a net or a wire guaze. In cold areas a half-inch mesh is quite sufficient to entrap the oil-containing rubber but in warmer areas a finer mesh may be advantageous in order to help trap any oil that escapes the rubber. The net or guaze may be used in any convenient manner. For example, the coagulated rubber mass can be pushed using a vertical gauze or it may be dragged using a net or gauze. This process may well be easier if the coagulated rubber is allowed to harden for a period, for example one day, before removing it.

Though in principle any nonoil-resisting rubber latex could be used in this fashion a cheap, readily available latex is that of a styrene-butadiene copolymer known as styrene-butadiene rubber (SBR) prepared synthetically in an aqueous emulsion. Other suitable rubbers are polybutadiene and butyl rubber.

Phillips Petroleum Company Process

A process developed by *H.E. Alquist and A.C. Pitchford; U.S. Patent 3,785,972; Jan. 15, 1974; assigned to Phillips Petroleum Company* permits containing oil on the surface of water and removing the oil from the water surface by increasing the liquification temperature of the oil to 50° to 80°F above the temperature of the water on which it is floating by incorporating a wax into the oil to form a crust-like fused mass which will act as a boundary against extension of the oil mass and which can be easily skimmed from the water surface.

This method consists of applying to the surface of the oil (1) a wax in a mixture with (2) a volatile substance which can be inflammable, allowing the volatile substance to evaporate or igniting the inflammable substance to form a coherent mass of the wax and oil.

For large spills a containing ring of fused polymer 6 to 10 ft in radial thickness at the perimeter of the spill can be developed. This method can be used to build a ring 6" to a foot or more in vertical thickness which will contain the oil in the center until it can be pumped out.

The ultimate objective is to convert heavy oil floating on the surface of water into a substance hard enough that it can be readily skimmed or lifted from the water or to convert the perimeter of a large spill into a solid sufficiently hard to contain the larger center core of liquid oil until the liquid can be pumped out by a skimmer. A satisfactory solid is produced when the liquification temperature of the spilled oil is raised from about 50° to about 80°F above the temperature of the water upon which the oil is floating.

The method is subject to two major variables in treating oil spills. These variables are the temperature of the water on which the oil is floating and the inherent pour characteristics of the oil. For example the temperature of seawater in the commercial shipping lanes where spills most often occur can vary from 20° to 60°F. Heavy crude oils usually have pour points varying between +40°F and -60°F or even lower. When a high pour point oil is spilled on cold water very little wax is required to meet the 50° to 80°F differential in the liquification temperature of the oil and the water temperature. On the other hand, a high water temperature and spilled oil with an inherently low pour point require a large amount of wax to achieve the desired differential between liquification point of the oil and the water temperature.

Waxes suitable for use in this process include any low temperature melting wax such as the paraffinic waxes, naturally occurring ozocerite, carnauba wax, C_{12} to C_{25} alcohols and fatty acids such as stearic or palmitic. The concentrations of the waxes will vary depending on the melting point of the wax, its solubility in the particular oil with which it is used and the temperature of the sea water. Additions of wax in the order of 5 to 50% by weight of the oil are effective. Preferably a range of 10 to 15% by weight of wax added to the oil should be used.

Texaco Inc. Processes

A process developed by *E.L. Cole and H.V. Hess; U.S. Patent 3,614,873; October 26, 1971; assigned to Texaco Inc.* involves cleaning up marine oil spills by freezing the surface layer of oil, preferably with particles of Dry Ice or the like, to enable the layer to be screened off the surface as a cake.

Figure 212 shows the recovery process in a schematic way. The spill is shown more or less diagrammatically as in **12**, presumably having so recently occurred that it is, in effect, a massive and relatively thick layer of oil just commencing to spread. Vessel **10** which, in the embodiment shown, takes the general form of a military landing vessel, is provided on an upper deck portion with a gun or projector **14**, adapted to broadcast or project particulate material controlledly over a substantial distance.

The details of such gun, being known in the art, from no specific part of the process and are not disclosed herein other than to say that the gun is continuously and regularly supplied with particulate Dry Ice or frozen carbon dioxide, from any suitable source, not shown, in the form of granules of about 1/32 to 5/32 inch average particulate dimension. The gun may be supplied with a continuous source of compressed air **16** to project the particles or may use the pressure of sublimed Dry Ice.

The operator, by aiming the gun, continuously directs the particles of carbon dioxide just inside the margin of the spill as indicated at **18**, so that they first reach and cover a marginal area inside the periphery to the point represented by the dotted line **20**.

Sufficient of the particles are cast in this marginal area to form a frozen barrier and thus confine the unfrozen oil. When this has been done, steps are taken to pick up and recover the frozen oil, at the same time continuing to broadcast the freezing agent into the central, unfrozen portions of the oil spill, as well as in the frozen marginal portions to supplement and continue the freezing effect on the oil which has not yet been recovered.

FIGURE 212: RECOVERY OF OIL SPILLS BY FREEZING

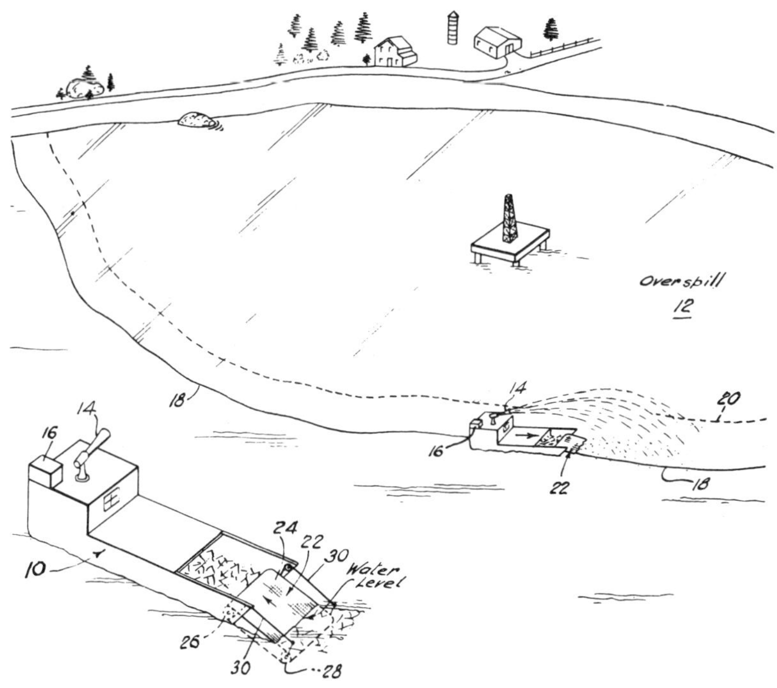

Source: E.L. Cole and H.V. Hess; U.S. Patent 3,614,873; October 26, 1971

While the pickup of the frozen or semisolid oil may be handled in various ways, as for example, by surface plows, skimmers or the like, it is preferred to do this by means of a travelling screen such as that arranged at the bow portion of the vessel or landing barge and indicated by the reference numeral **22.**

The screen comprises a continuous belt **24** of screening material, such for example as ⅛" mesh bronze screening which rides on rollers **26** and **28.** The rollers are mounted on a rigid internal frame not shown, which is pivotally attached at its rear to the bow of the landing barge. Actually the assembly may be pivoted on the axis of roller **26** which extends athwartships and is located above the water level and rotated continuously by a drive motor not shown, in such a direction that the upper run of the screening moves continuously into the vessel. The forward roller **28** is located outwardly of the bow and preferably lowered into the water by means of cables **30.** As a result, therefore, the outer end of the screen as it passes over the roller dips into the water below the frozen oil cake, which is deposited on the upper surface of the mesh as the vessel moves forward and is conveyed into the vessel as indicated.

The process as above intimated, is of specific application to relatively heavy oils such as heavy crude oils or reduced crudes or relatively heavy ends such as the heavier grades of fuel oil which have a substantial thickness and viscosity and thus tend to be more readily immobilized by freezing so that they can be removed as a solid or semisolid cake.

A process developed by *F.C. McCoy, H.V. Hess and R.L. Sung; U.S. Patent 3,732,162; May 8, 1973; assigned to Texaco Inc.* is one in which oil spills are removed from the surface of a body of water by contacting the oil with a coagulating amount of a coagulant such as asphalt and mixtures of wax or asphalt with anitcaking agents. Enough coagulant is used to form a floating, semisolid mass with the oil. The coagulant may be used in finely divided form or in a molten state.

Winkler Process

A process developed by *J. Winkler; U.S. Patent 3,567,660; March 2, 1971* permits conversion of oil spills into improved rubberized fiber-fortified asphaltic materials by coagulating oil spills with previously ground, spent automotive rubber tires which have preferably been previously premixed with powdery polystyrene or asphalt, the asphalt derived from naturally occurring asphaltite from petroleum or coal tars; harvesting by mechanical means the resulting nonliquid conglomerate and reacting and concentrating this mixture by distillation to an asphaltic consistency.

Yosemite Chemical Co. Process

A process developed by *E.R. De Lew; U.S. Patent 3,198,731; August 3, 1965; assigned to Yosemite Chemical Co.* comprises applying to an oil while it is floating on the water an agent which congeals the oil. Included among such agents are liquid metal soaps, molten wax and lanolin. These materials form a water-in-oil emulsion when applied to the oil slick in contact with the water which is said to be removable from the water by fine mesh nets, screens or perforated clam shell buckets.

MAGNETIC SEPARATION

Avco Corporation Process

A process developed by *R. Kaiser; U.S. Patent 3,635,819; January 18,1972; assigned to Avco Corporation* involves dispersing a hydrocarbon base ferrofluid containing an oil-soluble, water-insoluble surfactant and a stable colloid of magnetic solids, e.g., magnetite into the oil slick, then using a magnetic field to attract and pick up the oil spill, which is now magnetically responsive.

Pfizer Inc. Process

A process developed by *J. Warren; U.S. Patent 3,717,573; February 20, 1973; assigned to Pfizer Inc.* involves removing petroleum hydrocarbons from the surface of fresh or saline waters by dispersing magnetic iron oxide in the presence of wetting agent on the surface of oil-contaminated water, and removing oil-adsorbed metallic particles by means of magnetic attraction.

It has been found that the dispersion of magnetic metallic particles of iron oxide on the surface of oil in the presence of a suitable wetting agent results in adsorption of the oil by the magnetic particles to create a thick, viscous sludge which can then be removed by application of a suitable magnetic force. This process for oil removal contains such basic features as reasonable cost, rapid recovery rate, and, most importantly, compatibility with marine life.

Light and heavy oils (Diesel No. 2, vacuum pump A and 90 W transmission oil) were layered on the surface of fresh water and salt water in tests. When powdered iron oxide (Fe_2O_3) is dispersed on the oil surface, a slimy mass is formed, quickly with agitation, more slowly if left undisturbed, and portions of the iron oxide fall through the film to the bottom of the container causing a small loss of oil. However, when a wetting agent is added either to the iron oxide or to the surface of the oil, the iron oxide particles become rapidly covered, the polar attraction of the iron oxide for oil exceeds that of the oil-water phase, and a heavy cohesive sludge forms which remains on the surface in the case of more viscous oils, or drops to the bottom with lighter oils. The iron oxide-oil sludge is easily removed with a magnet.

The wetting agent or detergent, which may be a mannide monooleate, a nontoxic complex mixture of polyoxyethylene ethers of mixed partial oleic esters of hexitol anhydrides, or a quaternary ammonium compound, is added to the oil at an optimum ratio to that of the dispersed oil which will vary with the viscosity of the oils. Sufficient magnetic iron oxide having an average particle size of about 0.05 to 0.25 micron is added to form a monolayer on the oil-water surface. Crude magnetic iron and iron oxides with larger particle sizes may also be effectively used.

The recovered oil may be removed from iron oxide sludge by treatment with a detergent, a hydrocarbon solvent or heating. This permits reuse of the iron oxide if economically feasible. For large scale operations it is visualized that the coarse iron oxide powder and wetting agent will be blown or sprayed from a bulk ship equipped with horizontal booms. This will be followed by a towed magnetic rake or grid to pick up the magnetic oxide-oil sludge. A conveyer belt will dump this in a holding tank where the iron oxide can be reclaimed by treatment with a detergent and magnetic separator.

Turbeville Process

A process developed by *J.E. Turbeville; U.S. Patent 3,657,119; April 18, 1972* is one in which buoyant, water resistant ferromagnetic particles are distributed over the polluted area to adhere to the oil. A magnetic field generated via a magnetic net or parallel series of magnetic grids is then applied to collect the oil coated particles and, if desired, transport them to a more convenient area for disposal.

The particles used may comprise any low density expandable plastic beads such as polystyrene beads which are commercially available. One method for making such beads ferromagnetic, is to coat them after expansion with iron or like ferromagnetic material. Alternatively, the ferromagnetic material may be incorporated into the plastic before it is expanded. This latter approach enables easier storage and transport because of the smaller volume. The plastic is then expanded immediately prior to or simultaneously with their being spread onto the polluted surface of the body of water.

The creation of the ferromagnetic layer has additional advantages. It more positively maintains the pollutant in a cohesive body so that it can be more readily controlled. Also, it the pollutant is a low grade high density type fuel oil, which may have a natural tendency to sink in brackish water, the added buoyancy imparted by the particles will cause the oil to float. Indeed, since the density of oil and of water are functions of temperature, the added flotation imparted by the particles overcomes any adverse change in temperature of the surrounding environment.

Figure 213 shows a detail view of a particle as used in the present process. In this instance, the particle **14** may be made ferromagnetic by coating it with adhesive **22** and metal filings **24**. The ferromagnetic particles are formed by pretreating polystyrene beads in an unexpanded or expanded form. Alternatively, the particle is formed by inserting metal filings into the bead while in an unexpanded state. The beads are commercially available and are particularly applicable to the process due to their buoyancy, light weight and affinity for such surface pollutants as oil.

The ferromagnetic particles may be expanded simultaneous to dispersal in the polluted area. Therefore, large numbers of these ferromagnetic particles can be efficiently and rapidly transported to a sight wherein oil spillage or other surface pollutants exist.

FIGURE 213: SECTION OF SORBENT MAGNETIC PARTICLE

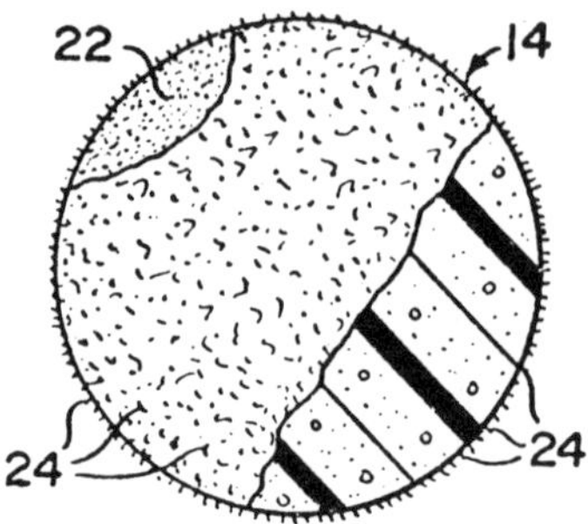

Source: J.E. Turbeville; U.S. Patent 3,657,119; April 18, 1972

COMBUSTION

Another way of disposing of oil on water is to burn it. The success of this method depends on supplying the blaze with sufficient oxygen and keeping it hot enough. One problem is that the thin layer of oil is cooled by the water, making it nearly impossible to ignite. A wicking agent, such as polyurethane foam, helps insulate the oil from the cool water below and keep the oil ignited.

There are disadvantages to burning. It results in air pollution, and it may create a fire hazard. For these reasons, burning is not usually practical in sheltered waters, although it may be desirable to burn oil on the open sea.

Burning of oil on water or land by special methods and materials seems to offer an attractive and perhaps inexpensive means of eliminating large amounts, providing of course, the many significant hazards are also recognized. Freshly spilled oils and crudes containing volatile components are relatively ignitible. If a thick layer of oil on water is present, the oil will sustain burning until the volatiles and a portion of the heavier fractions are combusted. Conversely, with fresh oil spills within a harbor or confined area, a significant fire danger exists when the level of hydrocarbon vapors is within the range of flammability, (e.g., gasoline, aviation fuels, or light crude oils).

Wood, debris or other material caught within an oil slick can serve as a wick to start or sustain an oil fire. The cooling of a layer of oil by the water body beneath will greatly deter burning. However, the wick will withdraw the oil and insulate the burning oils from the cooling action of the water, and at the same time provide a renewal and vaporization surface for combustion.

Experience by certain investigators has indicated that floating oils on the sea with thicknesses less than 3 mm (0.12 inches) will not burn. It is also reported that layers of kerosene, gas oil, lubricating oil and fuel oil on water will not burn at all without a wick. In one instance, attempts were made to ignite fresh Iranian crude oil five minutes after a spill without success. Once an oil spillage has spread, the material quickly loses its volatile components and ignition is extremely difficult. Weathered oils are consequently reported to present almost no fire hazard.

In experiments carried out at the Edison, N.J. Laboratory of E.P.A., a heavy fuel oil and a lightweight crude oil were placed atop a layer of water contained in metal tanks with 24 square feet of exposed surface area. Attempts were made to combust the oils using different burning agents. It was concluded from these experiments that the light crude, freshly applied in a floating thickness of 2.5 mm (0.1 inches), required external support with burning agents and an ignition source to burn near completion.

When the No. 6 fuel was used for testing, one of the agents used would not sustain burning at a thickness of ½ to ⅔ inch. Another agent generously applied over the surface of the No. 6 fuel caused sporadic burning and the minimum required oil thickness to sustain burning was ⅓ to ½ inch. It is important to note that a thickness of ⅓ to ½ inch is equivalent to an oil slick of 7 million gal/sq mi. It was evident after this burning that appreciable oil was still remaining. A third agent used performed well with the No. 6 fuel at a thickness of ⅒ to ¼ inch. The manufacturer producing this material, however, suggests that in order to have complete burning of all the oil, the oil layer must be completely covered with the material with no broken patches. Required amounts of this material (very low bulk density) may be as high as 1 lb for each 12 to 15 ft^2 of oil slick.

Field scale oil slick burning experiments, with and without special agents were undertaken in 1970 by both the U.S. Navy and the Edison Water Quality Laboratory. Preliminary data from these experiments indicate the following:

(1) Burning of free-floating or uncontained oil slicks is extremely difficult unless the thickness of oil is 2 mm or greater.

(2) Adequate automated seeding methods for both the powder and nodule-type burning agents are lacking. Spreading of the burning agent on the oil slick had to be accomplished by hand. This conclusion was also reached by the Navy, which conducted burning experiments in May 1970.

(3) Contained South Louisiana crude oil was successfully burned (80 to 90% reduction) without the application of burning agents and/or priming fuels. Bunker C could not be ignited under these same conditions.

(4) Bunker C was successfully burned (80 to 90% reduction) when the slick was seeded with burning agents and an appropriate priming fuel. It was discovered that South Louisiana crude oil performed better as a priming agent than did gasoline or lighter fluid.

(5) Use of magnesium type flares and gasoline torches to ignite the burning-agent-treated slick proved unsuccessful. Success was achieved, however, using a blow torch once it was learned how to manipulate the torch in such a manner that the torch gas pressure did not push aside the oil and seed material so as to expose the water surface.

Burning agents were considered when the Torrey Canyon was in the final stages of destruction off the British coast in March of 1967. As a last resort, the British Government attempted simultaneous and complete burning of the 15,000 to 20,000 tons of oil remaining in the badly broken tanker by aerial bombing, incendiaries, and catalyst-oxidizing devices. The major objective was to penetrate and lay open the decks by explosive surgery, exposing the oil in the storage compartments to large amounts of oxygen required for burning. High-explosive, 1,000 lb bombs filled with aluminum particles, thousands of gallons of aviation fuel, napalm bombs, rockets, and sodium chlorate devices served to produce a massive fire, if not a sustained fire. The British concluded that no appreciable amounts of oil escaped burning, but if any did, it was lost to the open sea.

Controlling the burning oil mass and ensuing air pollution problems would appear to preclude intentional burning except where the oil mass is distant from the coastline, offshore facilities, vessels, etc. The safety and welfare of all parties, however, remote from the burning site are of utmost importance. The possible loss of additional oil, and the loss of a drilling platform or vessel at the source of the spill must be recognized. Because of potential merits in controlled burning, further research is desired on new methods, techniques and procedures both in the laboratory and in the field, together with additional guidelines for burning.

Commercial burning agents are available for promoting combustion of an oil slick. However, in most cases it is apparent these agents are designed for a relatively thick oil layer which is sufficiently contained. These agents are intended to serve one or more functions such as providing increased surface area exposed to burning; addition of catalysts, oxidizers and low-boiling volatile components; absorbence and entrapment of the oil; or creating a wicking mechanism via surface diffusion and capillary action by the material added.

According to EPA, burning agents are available from the following sources:

(1) Eduard Michels GmbH, Essen, Germany: Kontax, the commercial name for this agent, ignites spontaneously when it comes in contact with water. In 1969, the Dutch conducted field experiments which included burning of oil on beaches, and in open waters. Results of this investigation indicated that the quantity of agent required was dependent upon wind, condition of sea and continuity and thickness of oil.

Dutch engineers also reported that a method should be developed to jettison, hurl or catapult the agent from a vessel in such a way that there is not the slightest risk of having the agent come in contact with rainwater or spray. Dropping from an airplane is worth considering provided special packaging requirements are met.

(2) Pittsburgh Corning, Pittsburgh, Pa.: Known as Seabeads, these cellulated glass nodules are available in sizes from ⅛" to ¼" in diameter. By capillary action, the nodules become coated with oil. Depending upon the type and age of the spilled oil, combustion is accomplished by using an incendiary device alone, or in combination with a primer fluid such as gasoline. After burning, Seabeads still remain, therefore, they must be collected or left to break up by abrasion. During the Arrow incident in 1970, this burning agent was used, with varying degrees of success, on small patches of spilled oil.

(3) Guardian Chemical Corporation, Long Island City, N.Y.: Pyraxon, a powder material, is a catalyst containing small amounts of oxidizing materials. Pyraxon liquid is used as the priming agent or starter fluid, with the powder being sprayed, blown or dropped onto the oil, in and around the flame.

(4) Cabot Corporation, Boston, Mass.: Cab-O-Sil ST-2-0 promotes combustion by acting as a wicking agent. Produced from fumed silica, this powdery-like material, is surface treated to render it hydrophobic. It may be applied directly to the slick or by spraying in a stream of water. Combustion is best accomplished by using an incendiary device in conjunction with a primer fluid. Reportedly, this product was used successfully for handling a 2,000 gal spill at Heard Pond, Wayland, Mass.

Atlas Copco AB Process

A process developed by *A. Molin and O. Carlsson; U.S. Patent 3,586,469; June 22, 1971; assigned to Atlas Copco AB, Sweden* involves destroying drifting oil layers on the surface of water basins by sustained combustion in a zone contiguous to the oil layer and in relative motion with respect thereto. In the method for thus combating drifting oil, a plurality of jets of combustion sustaining gas, in particular compressed air, are blown against the oil layer in the zone for sustaining combustion therein. In the combating means a hollow element is connected to a source of pressure gas, in particular compressed air, and kept afloat at the surface of the water with longitudinally spaced discharge openings on the element blowing the pressure gas against the oil layer in the zone for sustaining combustion therein.

Figure 214 shows one version of the application of the present process; using shore-based equipment for the combustion of a spill adjacent to the shore. As shown there, lines of interconnected oil booms **17** are moored by means of weights **40** to the bottom of the sea near the shore **43**. An oil patch drifting with wind or currents **10** is caught by the oil booms and guided towards an element **23** laid out between the lines of the booms and provided with discharge openings spaced along the length of the element. More particularly, the element consists of a single floating tube **24** of plastic material provided with an asbestos covering. By means of weights **41** there is furthermore immersed and moored at the bottom of the sea a perforated tube **37** for forming, when supplied with compressed air, an air bubble curtain and thereby an above-surface barrier **36** above the tube which barrier guides the oil layer **11** towards the combustion zone at the element. For purposes of increasing the combustion intensity gaseous oxygen additive may be led over to the compressed air hose **26** from an oxygen tank **39** for liquid oxygen placed ashore, whereby the oxygen content in the compressed air may be increased. In case of need fuel agents such as gasoline, spirit, coal gas and the like may be supplied to the combustion zone via special supply conduits, not shown, or as a dosage to the compressed air in the hose in case the combustion tends to become sluggish when encountering large accumulations of heavy oil in the combustion zone.

FIGURE 214: ATLAS COPCO PROCESS FOR BURNING OFF WATER SURFACES

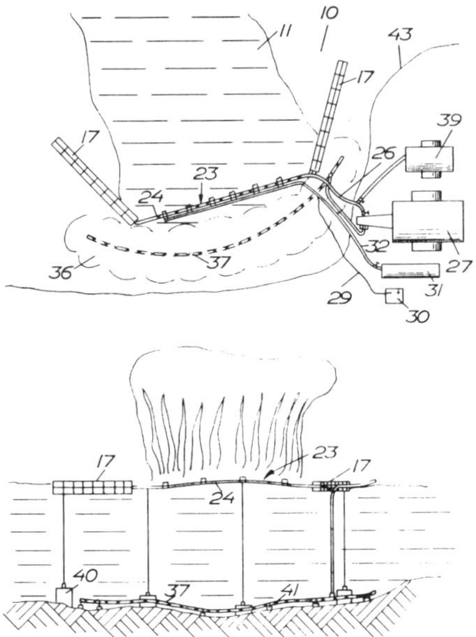

Source: A. Molin and O. Carlson; U.S. Patent 3,586,469; June 22, 1971

From the compressors **27** air under pressure is supplied to the element for sustaining, by the oxygen contained therein, the combustion in the zone above the tubes when the oil layer reaches the latter and is washed thereover by drifting with wind and water currents towards the element. In the zone above the tubes the oil layer is subjected to a plurality of fine gas jets from the discharge openings hitting the layer transversely from below and vehemently mixing the oil with gas and conditioning the zone for continuous sustained combustion. The oil zone above the tubes, thus effectively blown through and oxygenized, is ignited for example by means of torches or matches. Preferably there are arranged ignition coils on the tubes which coils are fed by an electric heating current via a cable **29** from a current source **30**.

If an increase of the intensity of combustion is desired, a fuel gas container **31** may be provided. The container supplies fuel gas such as coal gas via a conduit **32** to the fuel nozzles arranged at the tubes adjacent to the ignition coils. The element is displaced relative to the layer preferably at such rate that the layer flowing-on thereagainst will have time to burn away continuously and substantially completely in the zone above the tubes.

Cabot Corporation Process

A process developed by *P.R. Tully, W.J. Fletcher and H. Cochrane; U.S. Patent 3,556,698; January 19, 1971; assigned to Cabot Corporation* is one in which particulate solids are applied to the spill and the resulting system is thereafter fired. Such treated spills are more easily ignited and the combustion thereof is more complete than experienced with untreated spills. When certain conditions pertaining to the type and amount of treating agent applied to the spill are met even further benefits accrue to the process. The benefits reside in improved physical character of the burned residue which is more amenable to physical removal thereof from the water or land mass than the burned residuum of untreated spills.

The particulate solid materials useful as the treating agents are generally any substantially hydrocarbon and water insoluble particulate solid having an average ultimate particle diameter of less than about 250 mμ (preferably less than about 100 mμ), an apparent density of less than about 50 lb/ft^3 (preferably less than about 15 lb/ft^3) and a specific surface area (as determined by the BET-N_2 method) of greater than about 10 m^2/g (preferably greater than about 50 m^2/g). Specific examples of such materials are: carbons such as carbon black, activated carbons, chars and the like; metal and metalloid oxides produced by way of various precipitation, arc, plasma or pyrogenic processes such as silica, titania, alumina, silica-alumina, silica gel, alumina gel, iron oxide, zirconia, vanadia, chromia, magnesia, zinc oxide, copper oxides, and the like.

In addition, there exist various naturally occurring siliceous clays and minerals such as chrysotile asbestos which can be specifically treated so as to fall within the ambit of the above-recited solubility, particle diameter, density and surface area criteria. Certain of such naturally occurring materials, in particular, asbestos, are usually acicular, lamellar or fibrous in form rather than roughly spherical. However, such acicular or fibrous materials are to be considered as meeting the particle diameter criteria set forth hereinabove when the average cross-sectional dimension of the ultimate particle thereof is less than about 250 mμ.

Additionally, particularly when it is desired that a substantial amount of physical integrity be imparted to the combusted residuum resulting form the firing of the treated oil spills, it is preferred that the particulate solid utilized be possessed of a substantial degree of structure. Structure is that property of a particulate solid which signifies the extent to which primary particles thereof tend to form into a chainlike network. Accordingly, the higher the structure of a particulate solid, all other factors being equal, the greater the reinforcement capabilities thereof will normally be when dispersed into a suitable matrix. Generally, the structure of a particulate solid is roughly proportional to the ability of the solid to absorb oils, such as linseed or mineral oil or dibutyl phthalate.

The term oil absorption factor is to be considered as the minimum amount of dibutyl phthalate (in cc) required to cause the coalescence of 100 grams of a given particulate solid into single spherical structure by working incremental amounts of the oil into the solid by means of hand stirring with a spatula. Thus, those particulate solids which, in addition to meeting the limitations previously set forth, also bear oil absorption factors of greater than about 100 cc/100 g of solid are normally greatly to be preferred. Further, when such preferred solids are utilized, it is advantageous in terms of eliciting the maximum reinforcing function in the burned oil slick residuum, to treat the oil slick with at least about 2% by weight thereof of the highly structured particulate solid and even more preferably with between about 4 and about 10% by weight thereof.

In parctice the particulate solids utilized are also preferably rendered hydrophobic prior to application thereof to the oil spill. When such particulate solids are utilized, potential loss of the solid material into the water phase is normally vastly lessened. In fact, water may be effectively utilized as a carrier liquid for the conveyance of such hydrophobic particulate solids to the oil slick. In particular, metal and metalloid oxides, such as silica or titania, which have been rendered hydrophobic by treatment thereof with various organo-silicon compounds have been found to be especially preferred treating agents.

The application of the treating agents to the oil spill may be undertaken in any suitable manner. For instance, the particulate solids may be applied by hand, by air drops, or other procedures well known to the particulate material application arts. When, however, the much preferred hydrophobic particulate solids are utilized the application to water-borne spills can be made advantageously either from beneath or above the surface of the spill. A water stream can be utilized (in accordance with the Bernoulli pump principles) to aspirate the hydrophobic solid from its storage area and convey same to the spill. When this stream is played upon the upper surface of the slick the water carrier sinks therethrough leaving, however, the preponderance of the hydrophobic particulate solid in or on the slick.

Further, again referring to water-borne spills, the hydrophobic solids can be conveyed into the water phase from beneath the oil slick and thereafter released. Due to the hydrophobic/olephilic nature of the particulate solid the solid rises to the oil/water interface and is entrained substantially entirely into the slick proper. Where air or other gas is also entrained in this subsurface application method, the gas tends to rise through the slick, thus providing an additional mild and often desirable dispersing function. Additional benefits to be derived from the use of hydrophobic particulate solids and application thereof to the oil slick from beneath the water surface also reside in the substantial obviation of potential losses of the materials due to convection, breezes, wind gusts, etc. which are often encountered in the environment above the surface of the slick as well as providing improved capability of precision and uniformity of application.

Subsequent to the application of the treating agents to the oil spill, the resulting oil/solid system is ignited in any suitable manner. Such ignition may be expeditiously achieved by applying to a localized area of the treated slick a small amount of a highly flammable liquid, such as lighter fluid, kerosene, or the like and igniting this so-called initiator. The resulting flame thereafter progresses into the treated slick proper and propagates across the surface of the oil/solid system.

Castellucci-Krouskop Process

A process developed by *N.T. Castellucci and N.C. Krouskop; U.S. Patent 3,661,497; May 9, 1972* involves spreading a layer of substantially spherical ceramic nodules on the upper free surface of the layer of combustible liquid. The nodules are wetted with the combustible liquid and the combustible liquid is ignited on the upper surface of the nodules until combustion is self-sustaining. The combustible liquid on the upper surface of the nodules consumed by combustion is continually replaced with combustible liquid from the layer until substantially all of the combustible liquid in the layer is consumed. The cellular ceramic nodules have a multiplicity of separate closed cells and the outer surface of the nodules has a plurality of cup shaped recess portions.

The cellular ceramic nodules suitable for use in this process may be prepared in accordance with the process described in U.S. Patent 3,354,024 from a pulverulent glassy material and a celluating agent or from other pulverulent materials and a cellulating agent in accordance with the process described in U.S. Patent 3,441,396. A description of the process for providing a textured surface on the nodule may be found in U.S. Patent 3,493,218, and entitled "Tower Packing Element". The cellular ceramic nodules enable and enhance combustion of the combustible liquid to be removed from the body of water through interaction of the physical characteristics of the nodules, such as the surface morphology, the density the impermeability, the chemical composition, the thermal characteristics, and the like.

The nodules may have an apparent density of between about 6 and 30 lb/ft^3 and a thermal conductivity of between about 0.40 to 0.50 Btu/hr/ft^2/°F/in at 75°F. The nodules can be made in many different sizes. Nodules of a size between one-eighth and one-half inch with an apparent density of between 10 and 20 lb/ft^3 were found suitable.

In U.S. Patent 3,354,024 the nodules are made by admixing relatively fine pulverulent glass with a cellulating agent such as carbon black or the like. A binder is then added to the mixture which is then pelletized and subsequently coated with a parting agent that serves to maintain the pellets discrete during the cellulation process. The coated pellets are heated in a rotary furnace or kiln to a cellulating temperature and the pellets cellulate to form substantially spherical cellular ceramic nodules with a continuous outer skin. Although pulverulent glass is a preferred constituent of the cellular ceramic nodules, other glassy materials as described in U.S. Patent 3,441,396 may be used. The term ceramic is intended to encompass both pulverulent formulated glass and other suitable pulverulent glassy materials.

The cellular ceramic nodules thus produced have a core of individual completely closed cells of ceramic material and a continuous outer skin of ceramic material. For use with the herein described process, it is preferred that the cellular ceramic nodules produced as described above be abraded or otherwise treated to remove the relatively thin continuous outer skin and a portion of the layer of underlying closed cells to expose, over the entire surface of the nodule a portion of the layer of cells therebeneath. The cells on the abraded surface are opened to form a surface having a plurality of contiguous individual cup-like recessed portions or cell fragments. With their low thermal conductivity, the nodules function as thermal insulators during combustion thereby preventing loss of heat to the underlying water and confining and concentrating the available heat to the region of combustion in the thin film of liquid on the surface of the nodules.

Continental Oil Company Process

A process developed by *D.D. Sparlin; U.S. Patent 3,698,850; October 17, 1972; assigned to Continental Oil Company* is one in which particles of foamed water-soluble and dispersible alkali metal silicates are distributed over oil slicks to absorb the oil. The oil is then burned after which the water-soluble and dispersible particulate foamed alkali metal silicate particles solubilize and disperse.

Gulf Research & Development Company Process

A process developed by *R.J. McGuire, E. Mitchell and J.P. Pellegrini, Jr.; U.S. Patent 3,696,051; October 3, 1972; assigned to Gulf Research & Development Company* is one whereby oils floating on the surface of open bodies of water can be removed by burning them in situ in the presence of an oleophilic particulate material such as vermiculite which has been treated with a metallo cyclopentadienyl compound such as dicyclopentadienyliron.

Halliburton Company Process

A process developed by *R.F. Rensvold; U.S. Patent 3,705,782; December 12, 1972; assigned to Halliburton Company* is one whereby an oil slick is destroyed by applying thereto finely divided particles of a compound capable of generating a combustible gas, upon contact with water, allowing the particles to contact the underlying body of water so that bubbles of combustible gas rise through the oil film and admix therewith, so as to enhance the combustibility of the oil, and then igniting the oil-gas mixture to burn and destroy the film, e.g., calcium carbide to form acetylene gas.

Phillips Petroleum Company Processes

A process developed by *F.J. Shell; U.S. Patent 3,607,791; September 21, 1971; assigned to Phillips Petroleum Company* involves removing hydrocarbons from the surface of a body of water by placing a polypropylene fabric sheet over and in contact with the hydrocarbons and combusting those hydrocarbons passing onto the upper surface of the sheet.

It has been found that initial combustion of hydrocarbons on the upper surface of the polypropylene sheet does not destroy the sheet and during the combustion, hydrocarbons located beneath and in contact with the sheet pass upwardly through the sheet and are combusted on the upper surface. The polypropylene sheet thereby functions as a wick and causes the hydrocarbons to be removed and separated from the surface of the water and supports the hydrocarbons in a spaced relationship from the water surface during combustion. Ignition of the hydrocarbons can be by several methods whereby a flame is brought into contact with the hydrocarbons.

A process developed by *J.W. Marx, U.S. Patent 3,677,982; July 18, 1972 assigned to Phillips Petroleum Company* is one in which petroleum oil floating on the surface of water is removed therefrom by adsorbing the oil on a treated cellulose sponge and then burning the adsorbed oil from the sponge while it remains in contact with the water. During the combustion, the treated cellulose sponge continues to adsorb oil and deliver it to the combustion zone.

Pittsburgh Corning Corporation Processes

A process developed by *W.D. Johnston; U.S. Patent 3,661,495; May 9, 1972; assigned to Pittsburgh Corning Corporation* is a process for the substantially complete combustion of a combustible liquid including the combustion of a layer of the combustible liquid floating on a body of water. Cellular ceramic nodules having a core of separate closed cells and substantially continuous outer surfaces or skins with a coating of particulate alumina parting agent adhering thereto are treated to remove a substantial portion of the particulate alumina parting agent. The nodules are preferably washed in a water or dilute acid bath with mild agitation, to loosen and remove the particles of alumina parting agent adhering thereto without substantial abrasion or fracturing of the continuous outer skin.

The treated nodules are relatively dust free and have a smoother outer surface. A layer of the treated cellular ceramic nodules is formed on the upper surface of the combustible liquid with a substantial number of the nodules in contiguous relation with adjacent nodules in the layer. The upper exposed surfaces of the nodules are wetted with the combustible liquid to form a film or layer thereon and the wetted films on the exposed surfaces of the nodules are ignited until combustion is self-sustaining. The combustible liquid films on the exposed upper surfaces of the nodules consumed by combustion, are continually replaced with combustible liquid from the bulk of the liquid until substantially all the combustible liquid is consumed.

Another process developed by *W.D. Johnston; U.S. Patent 3,661,496; May 9, 1972; assigned to Pittsburgh Corning Corporation* is one in which cellular ceramic nodules are prepared by coating uncellulated pellets with a particulate carbonaceous parting agent and cellulating the coated pellets in a rotary furnace or kiln. The cellular ceramic nodules obtained by the above process have a relatively thin coating of the carbonaceous parting agent thereon and a relatively smooth continuous outer skin.

These particles are used in the same manner as those described just above which were prepared using an alumina parting agent.

A process developed by *R.B. Heagler; U.S. Patent 3,663,149; May 16, 1972; assigned to Pittsburgh Corning Corporation* is one in which a generally U-shaped, buoyant, self-propelled vessel floats partially submerged in a body of water and has a longitudinal channel portion with a front opening. The vessel has an open bottom portion beneath the longitudinal channel portion. As the vessel advances into a body of water, a band of water with the layer of combustible liquid floating thereon, enters the channel portion of the vessel. The rate at which the combustible liquid, as a layer, enters the channel portion of the vessel is dependent on the forward speed of the vessel and the speed is controlled so that substantially all of the layer of combustible liquid is removed by burning before the band of water passes under the rear portion of the vessel.

As the vessel advances, the band of water with the layer of combustible liquid moves through a mixing chamber within the channel portion where a monolayer of cellular ceramic nodules are positioned on the top surface of the layer of combustible liquid. The layer of combustible liquid with the nodules floating thereon, moves rearwardly with the forward advance of the vessel into a combustion chamber where the layer of combustible liquid is ignited and burned. The nodules within the combustion chamber are recycled to the mixing chamber where they are again positioned as a monolayer on the upper surface of the layer of combustible liquid. Combustion air is provided for the combustion chamber and the combustion gases may be subjected to a secondary burning in the stack to remove the combustible materials in the combustion gases and provide a substantially smoke-free waste gas. Apparatus is provided to seal the combustion chamber and mixing chamber if the burning of the combustible liquid tends to spread beyond the receiver.

The reader is referred to Figure 215 for plan and elevation drawings which give more details of the present apparatus and process. There is illustrated a vessel **10** having a generally U-shaped configuration with a central longitudinal channel portion **12**. The vessel preferably has a pair of buoyant side portions **14** and **16** with an open bottom portion below the channel portion to provide an area within the vessel that confines a layer of combustible liquid floating on the surface of the body of water. The channel portion has a front opening **18** which permits the entry of the layer of combustible liquid floating on the surface of the water into the channel portion. The channel portion may be described as having a mixing chamber **20** and a rear combustion chamber **22**. The mixing chamber is located in front of the combustion chamber so that the layer of combustible material may be prepared for burning before it enters the combustion chamber.

The vessel, particularly the side portions and the enclosure for the combustion chamber is preferably fabricated from a light material to permit the vessel to float on the body of water without displacing the latter and layer of combustible liquid within the central channel portion. The vessel may be fabricated from an admixture of cellular ceramic nodules and Portland cement in a ratio of about 5 parts by weight of cellular ceramic nodules to 1 part by weight of Portland cement. Other materials having sufficient buoyancy may also be employed.

When floating in a body of water, the vessel is partially submerged as illustrated so that a band of the layer of combustible liquid floating on the top surface of the body of water will enter the front opening and move rearwardly through the channel portion as the vessel advances in the body of water. The layer of combustible liquid moves rearwardly into the mixing chamber within the confines of the vessel and within the mixing chamber. A layer of cellular ceramic nodules **24** is positioned on the top surface of the layer. The cellular ceramic nodules are distributed onto the upper surface of the combustible liquid within the mixing chamber by means of a distributor device **26**, such as, for example, a vibrating screen.

A gate member **28** extends along the front portion of the mixing chamber and is arranged to close the channel portion and isolate the mixing chamber from the forward portion of the channel portion. The gate is arranged to prevent the combustion of the layer of combustible liquid from spreading into the forward portion of the channel and possibly to the layer of combustible liquid beyond the confines of the vessel. Propelling means **30** are provided for the vessel and are preferably arranged adjacent the rear portion of the vessel to maintain the layer of combustible liquid within the channel portion in a quiescent state as the layer moves rearwardly within the channel portion. Suitable controls **32** are provided for the propelling means and the other apparatus for recycling and distributing the cellular ceramic nodules.

The layer of combustible liquid within the mixing chamber having the layer of cellular ceramic nodules positioned thereon, moves rearwardly in the mixing chamber and through an opening **34** into the rear combustion chamber. Suitable igniting means **36** are provided within the rear combustion chamber to ignite the layer of combustible liquid with the cellular ceramic nodules positioned thereon.

The rear combustion chamber is preferably enclosed with a front vertical wall **38**, a rear vertical wall **40**, a pair of spaced vertical side walls **42** and **44**, and a top wall or roof **46**. With this arrangement, an enclosure is provided for the rear combustion chamber that has a bottom opening which permits the layer of combustible liquid floating on the surface of the body of water to be moved into the enclosure as a layer floating on the body of water and burned while remaining as a layer on the surface of the body of water.

FIGURE 215: FURNACE SHIP FOR BURNING OIL SLICKS FROM WATER

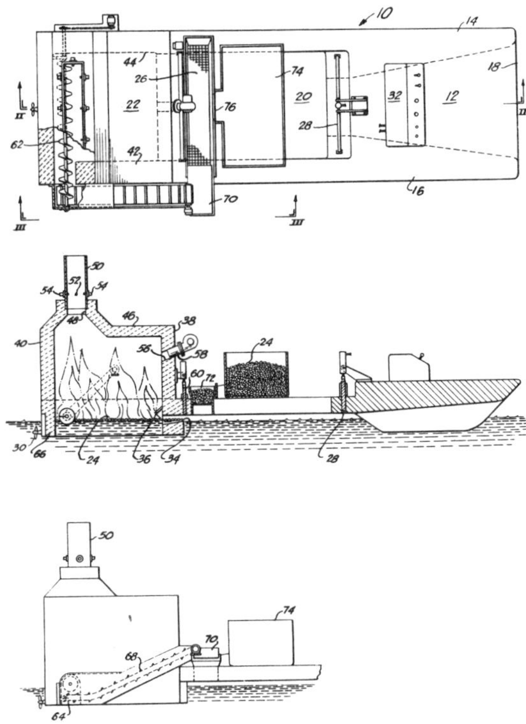

Source: R.B. Heagler; U.S. Patent 3,663,149; May 16, 1972

The roof has a stack opening **48** in which a preferably metallic stack **50** is positioned. The metallic stack has a plurality of radial openings **52** arranged to receive burner outlets **54**. Although not illustrated, the burner outlets are attached to a suitable source of combustion gas and may be utilized to ignite and burn the combustible material in the combustion gas and provide a relatively smoke-free gaseous product of combustion.

The combustion chamber front wall has an opening **56** in which a blower **58** is positioned to supply combustion air to the rear combustion chamber. It should be understood that other suitable means may be provided to supply combustion air to the combustion liquid within the combustion chamber. A gate **60** is provided to close the opening into the rear combustion chamber in the front wall to again prevent the spread of the combustion of the layer of combustible liquid beyond the confines of the vessel. It should be understood that the gates are safety devices and remain in a normally open position since the layer of combustible liquid without the layer of cellular ceramic nodules thereon, is difficult to ignite and it is seldom that the layer of combustible liquid will sustain ignition.

The layer of cellular ceramic nodules on the surface of the body of water, because of the forward motion of the vessel, tend to collect against the base of the combustion chamber rear wall. A suitable conveying means such as the screw conveyor **62** moves the layer of cellular ceramic nodules adjacent to the rear wall transversely into an enclosed receiver **64** that is positioned outside of the rear combustion chamber. It should be noted that the combustion chamber rear wall has a bottom end **66** beneath the top surface of the water so that the nodules which float on the surface of the water remain within the rear combustion chamber and only the water previously present within the confines of the rear combustion chamber is displaced beneath the combustion chamber rear wall.

The nodules conveyed transversely by the screw conveyor and disposed in the receiver in which there is positioned a rib-type conveyor **68**. The conveyor conveys the nodules upwardly onto a vibrating feeder **70** that conveys the nodules transversely into a hopper **72** positioned above the mixing chamber. The hopper has a vibrating screen base portion **26** that feeds the nodules onto the surface of the combustible liquid within the mixing chamber. A suitable storage bin **74** has an outlet **76** that supplies makeup nodules to the hopper.

It should be understood that the size of the vessel may be varied to permit the complete combustion of the layer of combustible liquid material within the channel while the vessel moves through the body of water at a relatively fast speed. The relative position of the distributor device within the channel portion may also be advanced forwardly within the channel portion to permit the layer of cellular ceramic nodules to be positioned on the layer of combustible liquid within the channel and provide sufficient residence time of the nodules on the surface of the combustible liquid for the combustible liquid to move upwardly on the nodules. Also, to ensure complete combustion of the layer of combustible liquid within the combustion chamber, the combustion chamber may also be elongated to permit a longer residence time within the chamber for burning before the nodules are removed therefrom by the screw conveyor.

In operation, the vessel floats on the body of water and is propelled into the portion of the body of water that has the layer of combustible liquid floating thereon. A band of the layer of combustible liquid enters through the front opening into the vessel channel portion and moves rearwardly therein as the vessel continues to move forwardly in the body of water.

In the mixing chamber a layer of cellular ceramic nodules is positioned on the upper surface of the layer of combustible liquid by means of the vibrating screen type distributor. The layer of combustible liquid floating on the body of water with the layer of cellular ceramic nodules floating thereon, thereafter moves into the combustion chamber through the opening in wall **38**. Within the combustion chamber, the layer of combustible liquid with the nodules thereon is ignited by means of the igniter.

It should be understood, however, that it may be necessary to only ignite the first portion of the layer of combustible liquid with the nodules thereon upon entering the combustion chamber. Combustion will continue and will continue to spread to other portions of the layter of combustible liquid with the cellular ceramic nodules thereon as the combustible liquid enters the combustion chamber.

As the layer of combustible liquid with the cellular ceramic nodules thereon moves rearwardly in the combustion chamber, because of the forward advance of the vessel, the layer of combustible liquid is removed from the body of water by burning. The nodules adjacent the rear wall of the combustion chamber are removed from the surface of the body of water by means of the screw conveyor and are deposited in the receiver. A conveyor then conveys the nodules to a transverse vibrating conveyor of chute **70** where the nodules are again deposited in the hopper for subsequent distribution on the layer of combustible liquid therebeneath.

Combustion air is supplied to the combustion chamber be means of the blower and the secondary burners are positioned in the stack to burn any of the combustible material remaining in the combustion gases. With this arrangement, relatively smoke-free gaseous products of combustion are emitted to the atmosphere.

With the above described method and apparatus, it is now possible to quickly, efficiently and substantially completely remove a layer of combustible liquid form a body of water. With the configuration illustrated, it may be necessary to make several parallel passes through the layer of combustible liquid floating on the body of water to burn substantially all of the combustible liquid in the oil spill.

Another process developed by *R.B. Heagler; U.S. Patent 3,695,810; October 3, 1970; assigned to Pittsburgh Corning Corporation* is one in which oil residues and emulsions floating on a body of water are burned by confining the layer of residue within a furnace chamber. The furnace has combustion air inlet means adjacent the upper surface of the residue and a stack with inlets for a combustible gas. The combustible gas burns the combustible material in the gasses evolved from the combustion of the liquid residue to provide a relatively smokeless combustion process. The furnace is fabricated from a refractory material having insulating properties so that a substantial portion of the heat given off by the combustion of the residue is retained within the furnace to propagate further combustion of the residue and aid in the complete combustion of the difficult to burn portions of the residue.

The furnace is preferably fabricated from a material that permits the furnace to float partially submerged in the body of water and may be easily transported from one location on the body of water to another location thereon. The furnace may be supported from suitable pilings and the residue conveyed directly into the furnace chamber. For certain types of difficult to burn residues, a layer of cellular glass nodules with a textured outer surface is positioned to float on the upper surface of the residue within the furnace chamber. Figure 216 is a diagram showing such a furnace.

The furnace **10** is preferably fabricated from a mixture of concrete and cellular glass nodules so that the furnace will float partially submerged in a body of water. The furnace has a plurality of concentric rings **12, 14, 16, 18, 20** and **22** positioned in overlying relation with each other. The rings are preferably fabricated from an admixture of nodules and Portland cement in a weight ratio of 5:1. A typical furnace having a total height of 25", a bottom opening of 26" and a top opening of 10", weighs approximately 130 lb with an average wall thickness of 4". Where desired, suitable reinforcing wires may be used on the lowermost rings. Also where desired, the top rings may be fabricated from an admixture of cellular glass nodules and asphalt as a binder.

When positioned on a body of water, the lower ring **12** and a portion of the adjacent ring **14** are submerged below the water level. For example, a furnace weighing approximately 130 lb having the above discussed dimensions will have the lower rings submerged and the remainder of the furnace will be above the water level.

The intermediate ring **16** has a plurality of air passages **26** and an opening **28** is provided for the outlet of blower **30**. Although the air passages and the opening are illustrated as extending radially through the ring, it is preferred that the air passages and the opening for the blower extend angularly to the radius to provide a circular motion for the air entering the chamber **24**. With this arrangement, during combustion within the chamber, combustion air is drawn into the chamber through the air passages and, where it is desired, the blower provides air under pressure through the opening to the furnace chamber adjacent the top surface **32** of the liquid within the chamber.

Suitable inlet openings **34** may also be provided in the top ring **22** for conduits **36** that are connected to a source of combustible gas such as methane, or the like. The openings in the top ring are arranged to provide for combustion of the combustible materials in the gasses evolved during the combustion of the liquid within the chamber to provide relatively smoke-free combustion gasses.

As indicated above, where the layer of combustible liquid is difficult to burn and where complete combustion of the combustible liquid is desired, a layer of cellular ceramic nodules may be positioned in the chamber to float on the upper surface of the combustible liquid and enhance the combustion of the combustible liquid.

The furnace is positioned on a body of water **40** to float thereon with a portion of the furnace submerged below the upper surface. The difficult to burn combustible liquid is lighter than water and floats on the upper surface. To ignite the combustible liquid, a small amount of a primer such as Varsol or other suitable low flash point igniting agent, is poured onto the surface and is ignited by any suitable means. It should be understood that the furnace could include an automatic or remotely controlled resistance type igniter or the like to ignite the liquid on the surface.

After the combustible liquid is ignited combustion continues until substantially all of the combustible liquid within the confines of the chamber floating on the surface of the water is burned. Little, if any, residue remains. Since the furnace is floating on the body of water, it may be easily moved or transported thereon by any suitable means to be positioned with a new inventory of the combustible material within the chamber.

The furnace being fabricated from Portland cement and cellular glass nodules has refractory properties in that a minimum of spalling occurs within the furnace chamber during the combution of the combustible liquid. Also, the nodule-concrete mixture provides insulating properties for the furnace so that the heat generated during the combustion of the combustible liquid is retained within the chamber and it is believed, contributes to the substantially complete combustion of the combustible liquid so that little, if any, residue remains on the surface of the body of water within the chamber.

With the furnace, it is now possible to confine the combustion to the circumferential area of the chamber and to substantially and completely burn the combustible liquid floating on the surface of the body of water with a generally smokeless flame. Although the configuration of the floating furnace is generally described as circular, formed from a plurality of rings, it should be understood that the furnace may have other configurations, as, for example, rectangular or a polysided pyramidal shape.

FIGURE 216: FLOATING FURNACE FOR BURNING OIL SLICKS FROM WATER

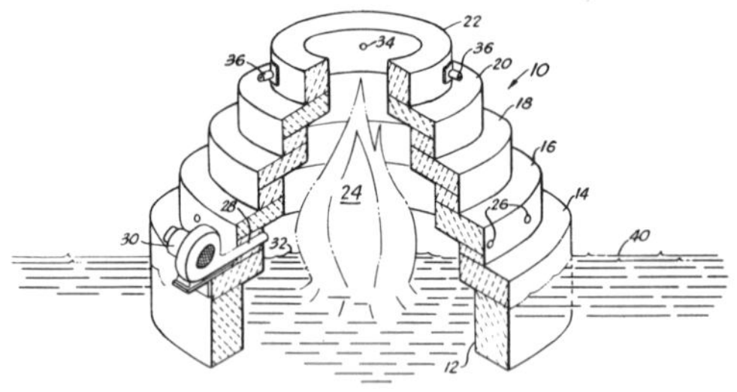

Source: R.B. Heagler; U.S. Patent 3,695,810; October 3, 1972

Stackpole Carbon Company Process

A process developed by *A.J. Shaler and W.E. Clancy; U.S. Patent 3,659,715; May 2, 1972; assigned to Stackpole Carbon Company* is one in which an elongated porous carbon member is impregnated with a combustible fluid and then floated in a generally upright position in a layer of combustible fluid on a body of water, with the lower portion of the porous member extending down in the water and with its upper portion projecting above the fluid layer. The fluid carried by the upper end of the porous member is ignited to produce a flame that is thereafter fed by combustible fluid moving up through that member by capillary action from the fluid layer, whereby to remove the fluid from the water and burn it.

Since oils wet carbon better than water does, the carbon being hydrophobic and oleophilic, the rod acts as a selective wick for the oil. The rod extends above the layer of oil and is tapered upwardly to more or less of a point. Oil will move upwardly through this tapered portion by capillary action, whereas water will not. Before the device is put in use, it is impregnated with a combustible fluid. Then, after being dropped into the water where it floats with its tapered upper portion above the oil layer on the water, the tip of the device is ignited. This produces a flame that is confined to the tip and that will be fed by oil moving up through the porous carbon as long as a layer of the oil remains on the water in contact with the device.

The outer surface of the rod, where it is in contact with the oil, should have a sufficiently large perimeter so that the rate of entry of the oil into the pores in the carbon is not so rapid as to entrain foreign particles and other debris which might otherwise clog the pores. Also, the device should be as stable as possible in agitated waters and the upper portion should project above the oil far enough to prevent the wave motion from sinking its tip from time to time below the oil layer.

By tapering the upper portion of the rod, the volume of carbon at the apex is so small that it can easily be brought to the ignition temperature of the oil. Since the thermal conductivity of the carbon-oil composite is low to begin with and the tip is distant from the cool oil layer on the water, this ignition temperature is maintained long enough, even if the device is splashed or lightly sprayed with oil or water or even rained upon, to reignite the oil flowing upwardly through the carbon in case the flame is temporarily extinguished. Thus, the tip of the device should be at a substantial height above the level of the oil layer so that it is not cooled appreciably by conduction and also so that in agitated waters, spraying, splashing and the up and down motion of the device caused by wave motion will not extinguish the flame more than momentarily.

Although porous carbon is the preferred material, other light porous materials such as foamed plastics, pumice and other ceramics, textiles and metals may be utilized. If necessary, the walls of their pores may be coated, as with silanes or other hydrophobic but not oil-repelling substances. However, none of these porous materials are as satisfactory, all things considered, as porous carbon.

One form of such a device is shown in Figure 217. The device illustrated is formed from two intersecting vertical slabs **12** and **13** of oil impregnated porous carbon that will float. The two slabs are assembled by providing one of them with a central slot extending downwardly from its upper end and providing the other slab with a central slot **14** extending upwardly from its lower end. This second slab **13** is inserted in the slot in the other one and moved downwardly to straddle the first slab. The result is a device in the form of a cross in horizontal section, as shown. To keep the device floating upright, the lower end of slab **12** carries a small weight **15**. The portion of the device extending above the oil is tapered upwardly substantially to a point. This device has a greater surface area for entry of oil per unit of length than other designs, and generally is less costly to manufacture. It also has greater stability in high winds because its cone angle is less and the ratio of its greatest width to its length is greater.

Large numbers of the components of these devices can be oil-impregnated at the factory and sealed into an efficient packing for storage on the deck of a ship. They can be readily assembled when needed by inexperienced crewmen. A single such device is capable of burning off more than 50 lb of crude oil, lubricating oil, or even of a heavy fluid fuel oil per day from the surface of an ocean, lake, or pond in heavy seas and weather, and it will continue to burn for days until the oil slick has become a film only some thousandths of an inch thick. It can easily be lighted and relighted if it is accidentally extinguished. There is no danger of the fire spreading away from the upright slab of the device, since the flame is confined to the upper portion of the latter by the cooling action of the water on its lower portions, as well as by the action of the wind. If a number of the devices is scattered over a spreading oil slick, they will tend to gather from wind drift along the downwind boundary of the slick where it is most desirable to burn off the oil and where any wind-caused water-spray is least, since upwind, the device faces a large expanse of oil-covered water. The wind does not tend to cause the fleet of devices to drift past this boundary into water that is not covered by oil, because the slick is also being wind-drifted in the same direction and also because capillary attraction holds the devices to the slick boundary.

FIGURE 217: POROUS WICK TO ASSIST IN BURNING OIL FROM WATER

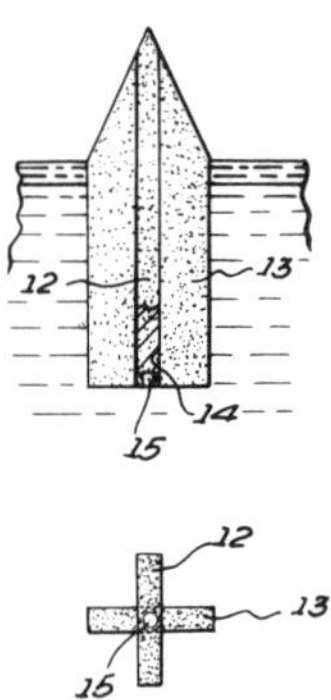

Source: A.J. Shaler and W.E. Clancy; U.S. Patent 3,659,715; May 2, 1972

WASAG Chemie AG Process

A process developed by *S. Kraemer, A. Seidl and M. Seger; U.S. Patent 3,589,844; June 29, 1971; assigned to WASAG Chemie AG, Germany* is a process for absorbing and burning away oil or other combustible liquids on water or other noncombustible liquids wherein absorbent and/or surface active noncombustible inorganic foamed particles are spread out over the combustible liquid, the combustible liquids are absorbed by the particles and the liquid absorbed by the particles is ignited.

Suitable substances of which the foamed glass particles are produced are siliceous substances such as sodium, calcium or aluminum silicate, alone or in mixtures, expanded mica or similar expanded natural products, but also oxidic materials such as Al_2O_3 and clay, and even foamed metals, especially aluminum and its alloys when they function as absorbers and remain floating on the surface of the liquid, which means that in addition to the open pores which give the material wick-like properties, it also has enough closed air cells to keep it floating.

The foamed materials which are especially suitable for absorbing and burning oil that is floating on a water surface are disclosed in U.S. Patents 3,184,371 and 3,261,894. During the production of these materials the ratio of the open pores to the closed inner cells is controlled within wide limits by the use of varying amounts of foaming agent. These foaming agents include calcium carbonate, aluminum sulfate, zinc sulfate, carbon compounds such as glycerol, sugar and others in the presence of sulfates, and also compounds containing chemically bound water. The melting point of the products thus produced are varied within wide limits, at least between 400° to 1000°C.

The melting point of the foamed glass particles is brought up to at least 400°C by addition of lead, boron etc. compounds to the aqueous silicate solution. Lead compounds suitable for this purpose are preferably lead oxides PbO, PbO_2 and Pb_3O_4, while suitable boron compounds are, for example borax and other borates. The melting points of the foamed glass are brought to a maximum of 1000° to 1100°C by the addition of alkaline earths and earth metals to the foamable silicate solution. Reference is here made to compounds of the alkaline earths and of the elements of main Group III of the Periodic Table, such as aluminum, etc., and especially in the oxide form.

By suitable adjustment of the porosities and melting points it is possible to produce products that are suitable for absorbing and burning away of oils, etc., and which after completion of the burning away process become sintered at their surfaces and in the manner acquire sufficient density to sink in the water. In a similar manner other products are obtained which will retain their closed air cells in sufficient abundance to keep them floating on the surface after the burning has ended.

The products of combustion are preferably allowed to escape into the air while some of the ash remains in the foam. After the burning, the foam remains floating on the surface if it retains sufficient porosity to keep the densities of the floating particles less than the densities of the liquids. This is accomplished by using glasses, metal oxides or expansible clay of high melting point for the production of foamed particles. The foamed material will sink if there is sufficient pore loss to increase the density above that of the noncombustible liquid. This result is produced by the use of low melting glasses as starting materials.

It is preferred that the foamed particles remain floating after the combustion because they can then be brought on land and collected without soiling the coast or acting unfavorably upon the marine fauna. If, however, economic considerations are controlling, then those foamed particles should be preferred which will sink after the burning, because the original raw materials are very cheap.

The treatment of the foamed particles to render them hydrophobic is accomplished in many ways with industrial silicone oils (produced from organosiloxanes derived from alkylhalogen silanes) which are dissolved in organic solvents, the foamed particles being sprayed with the silicone solution or immersed therein. The silicone oils are also applied as aqueous emulsions, or in the form of their precursors (e.g., mono-, di- and trichlorosilanes) which by hydrolysis with steam are coated upon the foamed particles. The water-repelling agents are also vaporized upon the foamed particles under vacuum. From this it is seen that silicone oils or their precursors are applied in many ways. The foamed particles may be also rendered water-repellent by other known methods, e.g., by coating them with oils such as heating oil, asphalt, mineral fats, waxes such as montan wax or ozocerite, but also with salts of fatty acids, e.g., calcium stearate in solution, suspension, emulsion or dispersion.

The substances used in the process have the following characteristics, they are inorganic silicious, oxidic or metallic foam. They are incombustible, buoyant, insoluble or only slightly soluble in water, insoluble in oil, absorbent for liquids that do not dissolve in water, stable in the presence of water and only slightly swellable or not swellable at all. Their affinity for combustible liquids is greater than their affinity for water.

OIL REMOVAL FROM BEACHES

When oil pollutes a sandy beach, no single form of contamination takes place; it depends on the type of oil, length of time at sea, temperature, time the oil has been on the beach, and type of sand. Some oils, sufficiently long at sea, will arrive at the beach as pebbles or streaks, and can be removed easily by a beach cleaner. Other types of oil (particularly crudes) which have been at sea for a long time are water-oil emulsions that are somewhat similar to butter, and look like chocolate mousse. These emulsions, while on the beach, are altered by environmental and biological impact; they become putty-like and finally brittle. This type of pollution can also be cleaned up by a beach cleaner or dry screening. Fresh crude (and many fuel oils) will penetrate the sand, coating sand particles and filling some of the interstitial voids in the beach.

Beach materials vary greatly in the ease with which oil can wet them. Quartz sand is difficult to wet with oil in the presence of water, while many shell materials are more readily wetted. Consequently, when a beach is contaminated by oil a complex situation may exist where several forms of contamination occur. Hence, a cleanup which does not consider the broad spectrum of contamination will not be successful.

Experience with liquid oil falls indicates that the depth of penetration and position of the contaminant is not easily ascertained from the surface. Uncontaminated sands may bury the contaminated part, and the width and depth of the oil contamination may vary markedly within short distances; thus, finding the contaminated sand can be expensive. Modern practice has been to take large swathes of the beach and, as spots of contamination remain, to either take a second cut of sand or dig out the contaminated spots by hand. This results in a large amount of sand in which relatively small sections are contaminated. Thus, any cleanup procedure must either concentrate contaminated sands or be very economical in the treatment of the contaminants.

A figure of $5.00/ton for the replacement of sand on a beach is a reasonable estimate. This price includes removal of contaminated sand, finding clean sand, transportation, and addition of the sand to the beach. Any process used in beach restoration must consider that cleaning costs greater than something like $5.00/ton are in competition with simply removing the sand, disposing of it, and replacing it with fresh sand.

In any treatment of the sand to remove the oil, care must be taken not to trade oil pollution for other types of pollution. Any burning techniques must have superior combustion and generate few harmful gases or particles; likewise, harmful chemicals must not be allowed to escape to contaminate the ground water or ocean. Froth flotation appears to be an inexpensive method of cleaning oil contaminated beach sand. It has the advantages of concentrating oil contaminated sand, stripping oil from sand, and working for a wide variety of oils and conditions. The question is often asked, "How do you clean an oil polluted beach?" The answer to this question is relative. What is the composition of the beach? Is it fine grain sand; is it coarse sand; is it pebble or shingle beach? What type of oil pollution has fouled the beach? Was it a #2 oil, #6 or a crude oil? How deep is the oil penetration? What is the temperature and the season? Where is the beach located? Does it have access roads? Can it be reached or traversed by wheeled vehicles or can it only accommodate tracked vehicles? Before we can even begin to answer a question on beach cleanup, we must know as many background facts as possible.

It has often been said that nature is the best cleaning agent in the world. This is especially true of rocky shores and stone beaches. But if anyone reports that nature removed oil from a sand beach, dig a little deeper, literally. What nature cannot remove, it covers up. Wind blown sand and the seasonal movement of beaches have a tendency to cover man's mistakes.

The most elementary method of beach cleanup is with the use of rakes, shovels and manpower. If the penetration of oil into the sand is less than 2 inches and the oil is not too fluid, the oil can be raked into windrows of approximately 1 ft and picked up with shovels to be placed into a front loader or dump truck. If the oil is not sufficiently weathered to be viscous, it can be made more workable through the application of a cold water spray from a garden hose or a fire hose. If the pollution damage to the beach is much more extensive, then mechanical equipment must be used.

Beach cleanup has been relatively effective when a major effort has been expended. But both the extensive physical and chemical processes used to remove high proportions of oil from shorelines kill the plants and animals which inhabit them. For this reason some beach cleanup operations have been limited to the use of sorbents such as straw, which are safer and less expensive, but do not produce a sterile environment.

A report on the evaluation of various types of mechanical equipment for the restoration of oil contaminated beaches has been presented by U.R.S. Research Company of San Mateo, California as Report PB 191,711, Springfield, Virginia, Nat. Tech. Information Service (Feb. 1970).

An *Operations Planning Manual* was prepared for use by FWPCA personnel involved in oil spill cleanup operations. The surface conditions and topography of a beach contaminated with oil and the manner in which the oil has been deposited onto the beach will dictate the choice of equipment to be utilized and the operating procedures to be followed. The procedures tested utilize motorized graders, motorized elevating scrapers, front end loaders, and conveyor-screening systems. A motorized grader and motorized elevating scraper working in combination provide the most rapid means of beach restoration; and in addition, their use results in the removal of the smallest amount of uncontaminated beach material.

Sometimes a high-use beach becomes stained with a light oil such as a #2 fuel. This becomes a problem because of instant and deep penetration. The only really effective method of cleansing this type of pollution is to expose as much of the contaminated beach sand to the sunlight and wind as possible. This can be done effectively by the use of a harrowing plow or beach cleaning machines, which are used to remove trash from beaches. Number 2 fuel, in contrast to #6 or Bunker C fuel oil, contains a relatively high amount of light ends. If these light ends are exposed to evaporation through wind and weather, they will dissipate rather rapidly.

To accelerate the evaporation and dissipation of #2 fuel from the beaches, a mat of straw should be laid on the beach, at least 1 inch thick. A disk-harrow should then be used to work the straw down into the sand column so the straw can absorb as much of the oil as possible. A beach cleaning machine should then be used to retrieve the oil soaked straw. The beach should then be harrowed or mechanically raked to hasten the dissipation of the remaining oil trapped in the subsurface of the sand.

SPILL CONTROL MEASURES

It is much easier to clean up a beach if the oil is stopped at or near the water line. An effective method of protecting a high-use beach from the onslaught of oil pollution is to throw up a sand berm approximately 3 ft high. This can be accomplished with the use of earth moving equipment, or if necessary, by hand labor. The artificial berm should be placed along the high water mark, to protect the dry sand above the intertidal zone. If the surf is strong or very active, or if there is an abnormally high tide, the artificial berm will be destroyed. Sufficient to say that the artificial berm will only prove effective in a mild surf and calm weather.

Should sufficient time exist between the notification of an oil spill and its reaching the shore face, one method of alleviating the damage to the beach is the extensive use of straw, either alone, or used in conjunction with the commercial sorbents. Straw, which has a natural absorptive capacity for oil, will adsorb between 4 and 10 times its own weight in oil. To be most effective, the straw should be laid along the low-water mark and as the tide and oil move up the beach, the straw should be worked into the oil, either through natural wave action or mechanical methods. If the oil is still washing ashore a series of deep pits can be gouged out of the sand at the water's edge to allow the oil to build up sufficient thickness to permit removal by vacuum tank truck.

CHEMICAL CONTROL AGENTS

EPA firmly believes that the use of dispersants, emulsifiers and other chemicals is entirely unjustified in the cleanup of oil polluted beaches. In both the Torrey Canyon and Ocean Eagle incidents, it was noted that the use of chemicals on sand caused the sand to become "quick", making it difficult to walk on and leaving a disagreeable odor in the treated sand. (However, other reports cite this same quicksand with oil alone. As the literature on this subject is very vague, more research by the government and private industry seems indicated.)

In oil penetration tests conducted at Sandy Hook, N.J., it was concluded that various types of persistent oil alone penetrated no more than 2 inches into the sand while oil mixed with chemicals (dispersants, emulsifiers, etc.) caused penetration of the mixture into the sand at least three times the depth of the untreated oil. After the application of chemicals to an oil soaked beach and subsequent hosing down of the mixture with seawater, the sand appeared cleansed of oil. However, further investigation revealed that this observation was deceptive, as the oil and chemical mixture was found between 4 and 12 inches, in irregular bands, below the surface of the sand.

A report on the use of chemical agents in cleaning beaches has been published by the Federal Water Pollution Control Administration as Report PB 189,172, Springfield, Virginia, Nat. Tech. Information Service (August 1969).

Oil-dispersing chemicals were tested for cleaning persistent type crude oil from experimentally contaminated New Jersey coastal beaches and were found to be generally ineffective. Although they completely cleaned the surface of the oiled sand, they removed little of the total oil. Instead they caused the oil to penetrate more deeply into the underlying sand, thereby compounding the pollution problem by expanding the zone of pollution, complicating any subsequent mechanical removal and, possibly, causing the oil to persist longer. Chemical treatment failed to induce quicksand or cause perceptible erosion of beach sand. A decrease in the cohesiveness of the sand was observed, but this also occurred in the presence of oil alone and could not be attributed to the presence of chemical.

COMBUSTION METHODS

Many attempts have been made to burn oil on the shore face. Through an extensive literature search and personal observation it is concluded that the burning of oil in situ on a contaminated beach is not very practicable. It has been tried with commercial burning agents, flame throwers, oxygen tiles and various mixtures of kerosene and gasoline, but all have met with no success. Small, fresh pools of neat crudes or light oil can be burned successfully if their lighter ends have not evaporated, but this is a patchy operation at best and is not recommended for cleanup of massive spills.

J. Wardly Smith, Ministry of the Environment, United Kingdom, has conducted a series of research experiments involving the burning of oil on beaches. The conclusions reached by these studies were:

The type of heavy oil normally contaminating beaches and the foreshore is very difficult to burn by ordinary means and combustion ceases when the source of heat is removed.

The addition of solid oxidants aids the combustion of part of the deposit, but a sticky, tarry residue is left with medium heat and a residue of dry, black carbon with prolonged intense heat. About 30% by weight of oxidant has to be employed to achieve this and when the deposit occurs on sand, shingle or pebbles the amount of heat necessary to raise the whole to ignition temperature is so great that, in practice, the cost of removal would probably be excessive.

The application of an oil absorbent, combustible material, such as sawdust impregnated with an oxidant, resulted in a much steadier burning rate and an increased removal of the deposit. The additive, however, burns away before the higher boiling, tarry fractions begin to burn.

Although some improvement was obtained by adding materials which function as a wick, the same limitation regarding the amount of heat to be supplied still prevailed.

A major disadvantage of burning as a method for the removal of oil deposits on beaches, especially when dealing with the large semisolid lumps, is that, when heat is applied, the oil becomes mobile and penetrates deeper into the beach, thus increasing the contaminated area if it becomes exposed at a later date.

On the basis of these small-scale experiments it was considered that, for all practical purposes, burning beach oil deposits would be less efficient and more costly than mechanical or manual removal, and that further investigation would be unprofitable.

A feasibility analysis of incinerator systems for the restoration of oil contaminated beaches has been prepared by the Envirogenics Company Division of Aerojet-General Corporation and published as EPA Report 15080-DXE-11/70.

The feasibility of employing a combustion process for restoring oil contaminated beaches was investigated. Beach access problems and the handling characteristics of shore materials limited the potential application to recreational (sand) sites. Thermodynamic arguments required that a system design be adopted in which the contaminated sand would undergo combustive processing in a confined arrangement. The design selected, from those analyzed, proved to be a three-effect combustor based on the rotary kiln principle. Provided that the sand to be cleaned is carefully enough collected to furnish a reasonable ($\geqslant$6%) oil content and is moved away from the surf and drained to an acceptable moisture level ($\leqslant$6%), basic processing costs would be highly attractive. In comparison with uncontaminated sand, the cleaned product exhibits only a slightly greyish hue.

Culpepper Process

A process developed by *G.O. Culpepper, Jr.; U.S. Patent 3,734,774; May 22, 1973* involves a system for burning a hydrocarbon coating, such as an oil coating from the surface of particulate solid material such as sand in a pressurized firing zone or firing chamber where the solids are introduced under pressure into the firing zone. An air-fuel mixture is introduced into the firing zone and burned therein. Immediately downstream of the firing chamber is a doughnut restriction which confines the combustion to the firing chamber. The ratio of air to fuel in the air-fuel mixture is always such that the fuel will be completely burned. The system will be provided with controls to shut the unit down when the temperature goes above or below a predetermined temperature and also to shut the system down when a lack of complete combustion of the fuel is indicated.

FROTH FLOTATION TREATMENT

As reported by G.D. Gumtz in Report EPA-R2-72-045 (September 1972), based on laboratory studies, a 30 tph pilot plant was built for cleaning oil contaminated beach sands. The plant utilized the principle of froth flotation. Extensive field testing considered different oils, feed concentrations, both brackish and seawater, and a range of processing conditions. Forty-one field tests were conducted at the U.S. Navy's Fleet Anti-Air Warfare Training Center at Dam Neck, Virginia. These varied from nominal runs with sand feed rates of 30 tph and oil concentrations of 0.5% to oil/water separations at high capacity.

Using the test results, a mobile unit was designed, constructed, field tested, and delivered to the Environmental Protection Agency. Data was obtained on the effects on cleaning efficiency of relevant process parameters: sand feed rate, feed steadiness, oil type, oil concentration, sand age, feed homogeneity, water rate, water type, slurry density, residence time, aeration, temperature, surfactant effects, organic solids effects, and oil deposition on wet or dry sand. The mobile unit operated successfully under a wide range of conditions. This device should prove a valuable adjunct to existing oil spill cleanup procedures.

The primary conclusion from the entire project was that froth flotation can be used to clean oil contaminated beach sand and that a mobile beach cleaner as supplied to the EPA can be used as a cleanup device during appropriate oil spill emergencies. It was further concluded that:

The processing cost was about $0.77/ton of sand. This compares favorably to the estimated costs incurred in removing and disposing of oily sand during spill incidents such as the one in San Francisco Bay during the spring of 1971 even though actual spill experiences at 100 to 200 cents/ton of oily sand do not consider replacement of the sand removed from a beach.

The Mobile Beach Sand Cleaner is indeed mobile and can be operated with minimum logistical support and personnel.

The unit does away with the problem (especially in the long run) of securing and maintaining permanent storage for oily sand.

The unit also assures that valuable sand (e.g., Hawaiian sand at about $20/ton) on recreational beaches is not depleted.

DISPOSAL OF RECOVERED SPILL MATERIAL

Following pickup, a mixture of oil and water generally remains. The oil should be separated from this mixture so it can be stored for later disposal. Oil/water separation devices are integrated into the systems of many types of skimmers. These devices range from settling tanks and gravity separators to absorbent rollers which are squeezed at different pressures to separate water and oil. Once it is separated, oil may be temporarily stored in barges or other available storage equipment, such as portable units which expand as the oil is pumped into them, until it can be moved to larger facilities for storage or disposal.

Many absorbents, such as polyurethane foam, can be pressed to remove the oil. The recovered liquid is generally 80% oil. However it is recovered, the oil will vary greatly in quality. In some cases the oil can be rerefined for future use. For the disposal of oil which cannot be rerefined, several techniques exist. Burning is one possibility, but considerations of safety and air pollution limit its use. Burial is another disposal technique, but is subject to government regulations.

Land spreading has been described as a conserving and nonpolluting method of disposing of oily wastes by G.K. Dotson, R.B. Dean, W.B. Cooke and B.A. Kenner in Report PB 213,749; Springfield, Va., Nat. Tech. Information Service (1972).

ECONOMICS OF OIL SPILL PREVENTION AND CONTROL

As indicated elsewhere in this volume, spill prevention is preferable to spill cleanup. However all preventive measures do cost money also. The costs of preventive measures which might be incorporated into ships or offshore or shore establishments must be weighed against the costs, both tangible and intangible, which arise from disastrous spills. Such cost evaluations may be expected to guide the development of both ships and facilities, with effects on their future sizes and operating characteristics. On the whole, economics and good sense commend an effort to prevent pollution rather than accept the cost of its occurring.

Cleaning up an oil-contaminated area is time-consuming, difficult and costly. To the costs of the cleanup proper must be added the other costs of the oil invasion, the destruction of fish and other wildlife, damage to property, contamination of public water supplies, and any number of other material and esthetic losses. Depending on the quantities and kinds of oil involved, these losses may extend for months or years, sometimes for decades, with correspondingly heavy costs of restoring the area to its prior condition.

With cost of accidents comes the question of liability. As suggested earlier in this volume, in the event of any oil spill, the responsible vessel should be held until responsibility for such a spill has been determined, the oil adequately removed, and the vessel operator has provided evidence of financial responsibility for such environmental damage as may have occurred.

In a study by Arthur D. Little, Inc., the costs of cleaning up offshore oil spills of various sizes are projected. Selected results of that study are summarized in Table 20. The costs of some previous OCS accidents involving oil spills are reported in Table 21. These costs include all those costs incurred directly in stopping or cleaning up the spilled oil. They do not include the costs of fines (if any), costs of lawsuits for damages, or costs borne by other parties, for example, by commercial fishermen.

The CEQ Report on outer continental shelf operations (1974) does not describe incremental costs of various applications of current technology to environmental protection. Such data would be useful and one hopes that such a study will be initiated. In some instances the costs of safe operation and environmental controls may increase the cost of extraction beyond the level at which operations are economically attractive. In such a case, resources should be developed elsewhere under circumstances where total costs, with environmental costs properly taken into account, are less. Importantly, the fact that environmental controls in such a case are costly should not be used as grounds for reducing the level of control, but rather should indicate that the development of that resource should be deferred to a time when the costs of environmental control are reduced through technological advances or the value of the resource increases.

TABLE 20: COST ESTIMATES FOR OIL SPILL CLEANUP OFFSHORE

	-- 2,380 Barrels (100,000 gal) --		- 245,000 Barrels (35,000 tons) -	
Combatant Method	**Direct Costs***	**Capital Costs****	**Direct Costs***	**Capital Costs****
Chemical Dispersion	$ 80,000	$ 51,600	$6,200,000	$ 862,500
Absorption (straw)	113,000	79,300 (579,300)***	8,625,000	1,405,000 (6,405,000)***
Sinking	64,900	56,600	4,505,000	1,385,000
Combustion	82,200	49,500	6,171,000	675,000

*Direct costs are those incurred directly for a specific spill, including material and operating costs.

**Capital costs are initial equipment and warehouse costs. Boom costs were not included. Persons conducting the study assumed spills in the open ocean which spread over areas so large as to make booms ineffective.

***Includes equipment for the collection of spent materials.

Source: Arthur D. Little, Inc., *Combating Pollution Created by Oil Spills—Volume One: Methods,* Report to the Department of Transportation, U.S. Coast Guard (June, 1969).

TABLE 21: COST OF SOME PREVIOUS OCS ACCIDENTS

Description/Location	Total Cost (in millions)
Union (Santa Barbara)	$10.6
Shell (Gulf of Mexico)	30.0
Amoco (Gulf of Mexico)	15.4
Chevron (Gulf of Mexico)	15.0

Source: *Energy Under the Oceans,* Univ. of Okla. Press (1973)

FUTURE TRENDS

As pointed out in this volume, trends for the future include vast increases in the amount of drilling on the outer continental shelf (OCS) and vast increases in the sizes of tankships. The increase from tankers of the present 100,000 to 200,000 ton size to very large crude carriers (V.L.C.C.'s of 200,000 to 400,000 ton capacity) and to ultralarge crude carriers (U.L.C.C.'s of over 400,000 tons capacity) certainly increases the spill potential in case of a serious accident to one of these large vessels. Thus both the frequency and quantities involved in drilling as well as the larger hazard potential in case of tanker accident makes preventive measures more critical. Further, the world does not really possess adequate knowledge of the effects of oil as a guide to necessary control levels.

Thus, as pointed out by the Council on Environmental Quality in 1974, research is needed on: Basic studies of population life histories for many species, including identification of survivorship, fecundity, larval life style, migrations, behavior, etc.; community response at the species level following pollution incidents or in controlled experiments; petroleum degradation processes and weathering rates as a function of temperature, light, nutrient concentration, etc.; physical-chemical relationship of oil hydrocarbons and sediment materials, including transport; the effect of sediments on degradation of low-boiling aromatic fractions; and identification of specific toxic hydrocarbons.

In addition further research is needed on: Adaptations of organisms to oil, including genetic changes; sources of oil in the ocean environment; amounts and impacts of heavy metals discharged during operations; long-term effects of oil on marine organisms; impacts of oil during sensitive stages of species development; release and fate of carcinogenic components of oil; effects of oil on commercial fisheries; effects of oil on the food web; correlation of drift bottle data and movement of oil; and data on currents in areas where great increases in outer continental shelf drilling are planned.

In the matter of containment, booms are available which can contain spills in shallow water and harbors but booms are clearly not available to contain spills in high seas under severe wave, current and wind conditions. Thus for OCS operations, prevention must surely take precedence over cures.

As regards removal of spills, more development needs to be done on combustion techniques since the use of sorbent materials really presents the only ecologically viable alternative. Neither sinking nor dispersion are wholly acceptable.

The future is thus one of severe challenge when one combines the problems posed by supertankers and greatly increased offshore drilling with containment and removal techniques which are imperfect at best.

COMPANY INDEX

The company names listed below are given exactly as they appear in the patents, despite name changes, mergers and acquisitions which have, at times, resulted in the revision of a company name.

INVENTOR INDEX

U.S. PATENT NUMBER INDEX

NOTICE